Product Development

Product Development

A Structured Approach to Consumer Product Development, Design, and Manufacture

Second Edition

Anil Mital
Mechanical Engineering and Manufacturing Engineering and Design, University of Cincinnati, Cincinnati, Ohio, USA

Anoop Desai
Mechanical Engineering, Georgia Southern University, Statesboro, Georgia, USA

Anand Subramanian
JF Associates, Vienna, Virginia, USA

Aashi Mital
History Department, University of Cincinnati, Cincinnati, Ohio, USA

ELSEVIER

AMSTERDAM • BOSTON • HEIDELBERG • LONDON • NEW YORK • OXFORD
PARIS • SAN DIEGO • SAN FRANCISCO • SINGAPORE • SYDNEY • TOKYO

Elsevier
Radarweg 29, PO Box 211, 1000 AE Amsterdam, Netherlands
The Boulevard, Langford Lane, Kidlington, Oxford OX5 1GB, UK
225 Wyman Street, Waltham, MA 02451, USA

First edition 2008
Second edition 2014

British Library Cataloguing-in-Publication Data
A catalogue record for this book is available from the British Library

Library of Congress Cataloging-in-Publication Data
A catalog record for this book is available from the Library of Congress

ISBN: 978-0-12-799945-6

For information on all Elsevier publications
visit our website at store.elsevier.com

Dedication

To our families and colleagues.

Contents

Preface to the second edition

Our intention in preparing the first edition was to provide a comprehensive coverage of the product development process—from conceiving a product to designing it and, finally, manufacturing it. The coverage also included a critical aspect of the product development process, that is, design of the facility to manufacture the final product. We felt that such coverage was lacking; most product design books on the market tend to be single dimensional, covering only one aspect of the product development process. The coverage tends to focus on manufacturing, marketing, materials, quality, or some similar aspect. The excellent reviews the first edition received from professional journals and peers, and its adoption by many universities in their curricula motivated us to undertake the preparation of the second edition.

As we noted in the first edition, much of the information focuses on the fundamentals of the product development process; it is time-tested and covers the basics of issues such as the product design process, selection of materials, and choice of the manufacturing method, and is, therefore, not subject to drastic changes from edition to edition. The reader will find that much of the basic information has remained unchanged, or has undergone only minor changes in the second edition as well. We have added and corrected material throughout the text. As new research has come to light, we have made additions accordingly. For instance, a new chapter has been added integrating design guidelines pertaining to concurrent consideration of product usability and its functionality. Basic information in Chapter 1 has also been updated. The changes in the second edition, thus, are incremental in nature.

As before, the chapters are divided into three parts. One can focus as much attention on each part as one desires. We would like to emphasize that the product development process involves a wide variety of expertise that simply cannot be provided by a single individual. It is a team process rather than a singular effort. And yet there are aspects that rely more on the creativity of an individual than the whole team—for instance, conceiving the physical form or the preliminary design. In a classroom setting, therefore, one can emphasize both individual creativity as well as team effort.

It is our hope that this revised edition will be as useful as the first edition and will continue to provide an overview of the entire product development process to individual practitioners, students, and researchers.

We are very thankful to our colleagues and reviewers who have encouraged us to undertake the preparation of the second edition by providing positive feedback. This feedback is particularly appreciated as it has come unsolicited. For this we are very grateful. We hope the revisions included in the second edition prove to be as useful as the materials in the first edition and that we have not failed our readers.

Preface

Manufacturing is essential for generating wealth and improving the standard of living. Historically, developed countries have devoted at least 20% of their gross domestic product (GDP) to manufacturing. It is unlikely that any nation would achieve the "developed" status without a significant proportion of its GDP-related activities devoted to manufacturing. Furthermore, the manufacturing activities must culminate in production of high-quality products that people need and want, globally. The emphasis on a global market is critical in today's economy, characterized by shrinking national boundaries and globalization of the marketplace. Not only should the products manufactured be wanted, these should be high-quality products that are reliable, economical, and easy to use and produce, and are brought to the market in a timely manner.

Efforts to develop, design, and manufacture a consumer product knowledge base, by and large, have been fragmented and can be categorized into two main domains. The first domain primarily comprises product developers who emphasize issues such as identifying the market, defining product features, and developing promotional strategies for the market. The second domain comprises mainly manufacturing and design engineers involved in the technical details of product design and manufacture. In this context, the emphasis to date has been on only manufacturing processes; to a very limited extent engineers have focused on issues of product assembly and maintenance.

As is evident, the development, design, and manufacture of consumer products entails not only the interests of people in both domains but also those of the consumer and the user (the two are not necessarily the same). Among their interests are attributes such as a product's usability, its functionality, and how its function can be maintained and repaired. From the design and manufacturing perspective, there are many other important considerations, such as how the product components are assembled, how the product will be disassembled during the course of routine maintenance or troubleshooting and at the end of its life, and how the material–manufacturing–cost configuration will be optimized. Such a comprehensive approach to product development, design, and manufacture is lacking at present. Also, no books are available that propagate teaching such a comprehensive product development and design approach.

This book provides a comprehensive approach to product development, design, and manufacture and attempts to fill the existing void. While this comprehensive approach has been outlined in archival research publications and taught at the University of Cincinnati at the graduate level in its College of Engineering, it is yet to become widely available to students at large. This book is intended to share our perspective on the entire product "development to manufacture" spectrum and emphasizes the "how-to" process.

Chapters 1 through 3 outline the importance of manufacturing in the global economy, what kinds of products to develop, and what is the general product design process. In other words, they discuss why manufacture, what to manufacture, and how to design what to manufacture. Then Chapters 4 through 10 discuss and describe specific methodologies dealing with the selection of material and processes, and designing products for quality, assembly and disassembly, maintenance, functionality, and usability. In Chapters 11 through 13, we cover some basics of manufacturing cost estimation, assessing (forecasting) market demand, and developing preliminary design of the facility to manufacture the developed product. While not directly related to product development and design, we consider this information critical in the overall product manufacture cycle.

While this book is intended for senior and starting level graduate students, it should prove useful to any product designer interested in cradle-to-grave design. It should be particularly useful to all design and manufacturing engineers, production engineers, and product design researchers and practitioners.

We wish to thank our numerous colleagues and many former students who have encouraged us to undertake the writing of this book, telling us time and again how much such an effort was needed. We hope we have not failed them and have met their expectations, partially if not fully.

Biographies

Anil Mital is Professor of Mechanical Engineering and Manufacturing Engineering and Design at the University of Cincinnati. He is also a Professor of Physical Medicine and Rehabilitation at the University of Cincinnati. Formerly, he was a Professor and Director of Industrial Engineering at the same institution. He holds a B.E. degree in Mechanical Engineering from Allahabad University, India, and an M.S. and Ph.D. in Industrial Engineering from Kansas State University and Texas Tech University, respectively. He is the Founding Editor-in-Chief Emeritus of the *International Journal of Industrial Ergonomics*, and the Founding Editor-in-Chief Emeritus of the *International Journal of Industrial Engineering*. He is also the former Executive Editor of the *International Journal of Human Resource Management and Development* and author/coauthor/editor of over 500 technical publications, including 24 books and over 200 journal articles. His current research interests include Design and Analysis of Human-Centered Manufacturing Systems, Application of DFX Principles to Product Design, Economic Justification, Manufacturing Planning, and Facilities Design. He is the founder of the International Society (formerly Foundation) of Occupational Ergonomics and Safety and winner of its first Distinguished Accomplishment Award (1993). He is a Fellow of the Human Factors and Ergonomics Society and recipient of its Paul M. Fitts Education Award (1996) and Jack A. Kraft Innovator Award (2012). He is also a recipient of the Liberty Mutual Insurance Company Best Paper Award (1994) and a Fellow of the Institute of Industrial Engineers and Past Director of its Ergonomics Division. In 2007, he received the Dr. David F. Baker Award from the Institute of Industrial Engineers for lifetime research activities. He has also received the Eugene L. Grant Award from the American Society of Engineering Education and the Ralph R. Teetor Award from the Society of Automotive Engineers.

Anoop Desai is an Associate Professor of Mechanical Engineering in the College of Engineering and Information Technology at Georgia Southern University. He received his Ph.D. in Industrial and Manufacturing Engineering from the University of Cincinnati in 2006. His main research interests are product life cycle management and design. His research deals extensively with design for "X" principles, focusing on green design, environmentally conscious manufacturing, and design for maintainability. He is also actively involved in research and teaching in the areas of Engineering Economy, New Product Development, CIMS, and Quality Control. He has written over 70 articles, including over 25 journal papers, and his work has been widely cited.

Anand Subramanian is a Senior Managing Engineer at JFAssociates, Inc. based in the Washington, DC, area. He received his doctoral and master's degrees in Industrial Engineering from the University of Cincinnati, OH, and bachelor's degree in Production Engineering from the University of Bombay, India. He is a Certified Six-Sigma Black Belt and also a registered Certified Professional Ergonomist with the Board of Certified Professionals on Ergonomics. Since 2003, he has been associated with JFAssociates, Inc., where his responsibilities include leading and conducting time and motion studies, experimental design, data collection, statistical data analysis, data modeling and reporting, identifying and evaluating specific needs and requirements of users, performing ergonomic evaluations of office workstations and manufacturing workstations, developing and implementing ergonomic programs, and preparing reports and presentations. His areas of specialty include time and motion studies, process improvement, ergonomic evaluations, economic analyses, facilities planning, and warehouse design. He is coauthor of a number of national and international journal publications and has made presentations at a number of prestigious Industrial Engineering and Ergonomics conferences. Currently, he is the coeditor of the *Industrial and Systems Engineering Review* (ISER), an open access journal aimed at the advancement of industrial and systems engineering theory and practice as applied to any enterprise system. He also serves as a member of the Board of Directors of the Institute of Industrial Engineers (IIE) National Capital Chapter since 2011. From 2007 to 2011, he served as the managing editor of the *International Journal of Industrial Engineering: Theory, Applications, and Practice*.

Aashi Mital currently attends the University of Cincinnati, where she continues her advanced studies in Nineteenth-Century American History and Classical Archaeology. Her research and publication interests focus on Received Memory in Post-Civil War Southern Culture, Radical Reconstruction, the Industrial Revolution, and Geopolitical Relations of the Greek Archaic Period. She is the Assistant Historian at the Cincinnati Observatory and serves as a professional consultant for the collaborative Museums and Historic Sites of Greater Cincinnati organization. All of this allows her to fulfill an ever-growing passion for outreach education, museum operations, and public relations, as well as groundbreaking research. Throughout her ongoing duties for the *International Journal of Industrial Engineering: Theory, Applications and Practice*, she strives to create parallels, by means of interdisciplinary work, in order to reexamine historical methodologies and create new approaches within the industry. The young historian finds inspiration through her interests, including astronomy, sailing, cooking, wine tasting, and the arts.

Part One

The Significance of Manufacturing

1

1.1 Globalization and the world economy

Globalization of the marketplace is synonymous with, or akin to, the free flow of goods and services, labor, and capital around the world. Aided by huge improvements in global communication and the transport industry, the barriers to free trade are being eroded, and most countries are advancing on the path to embracing market capitalism. This includes not only traditional capitalist nations such as the United States and United Kingdom, but communist giants such as China and social republics such as India. In countries such as India and Brazil, large pools of inexpensive and relatively skilled workers are putting pressure on jobs and wages in the rich countries in Europe and North America and, lately, China (a machine operator in China earns about $6405 compared to $4817 in India; Time, 2013). For consumers, the benefits of free trade are reflected in cheaper and better quality imports, giving them more for their money. This, in turn, forces the domestic producers to become increasingly competitive by raising their productivity and producing goods that can be marketed overseas.

For a long time, the West (North America and Western Europe) dominated the world economy by accounting for most of the global output of products and services. This picture has undergone a major change in the last few years; currently over half the global economic output, measured in purchasing power parity (to allow for lower prices in economically poorer countries), is accounted for by the emerging world. Even in terms of GDP (gross domestic product), the emerging world countries (also referred to as the *Third World* or *poor countries*) account for nearly one-third the total global output and more than half the growth in global output. The trend clearly indicates that economic power is shifting from the countries of the West to emerging ones in Asia (King and Henry, 2006; Oppenheimer, 2006). At the present time, developing countries consume more than half the world's energy and hold nearly 80% of the foreign exchange reserves; China leads the pack, with nearly $3.66 trillion in foreign exchange reserves (The Wall Street Journal, 2013). The exports of emerging economies in 2012 were approximately 50% of total global exports. Clearly, this growth in the emerging world countries, in turn, accelerated demand for products and services from traditionally "developed" countries. Globalization, therefore, is not a zero-sum game: China, India, Brazil, Mexico, Russia, and South Korea are not growing at the expense of Western Europe and North America. As individuals in emerging economies get richer, their need and demand for products and services continue to grow.

Product Development. DOI: http://dx.doi.org/10.1016/B978-0-12-799945-6.00001-6

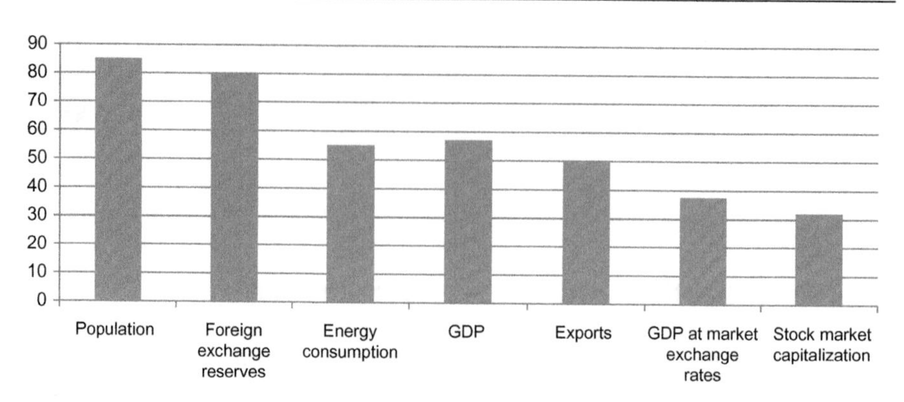

Figure 1.1 Emerging economies as a percent of the world total.
Source: Adapted from The Economist, August 4, 2011.

As the emerging economies have become integrated in the global economy, the Western countries' dominance over the global economy has weakened. Increasingly, the current boost to global economy is coming from emerging economies, and rich countries no longer dominate it. With time, industrial growth in the developing countries, as indicated by the growth in energy demand (oil), is getting stronger. Figure 1.1 shows emerging economies in comparison to the whole world using a number of measures. For instance, growth in emerging economies has accounted for nearly four-fifths of the growth in demand for oil in the past 5 years. Further, the gap between the emerging economies and developed economies (defined by membership in the Organization for Economic Cooperation and Development prior to 1994), when expressed in terms of percentage GDP increase over the prior year (growth rate), has widened (Figure 1.2). Between 2003 and 2013, the emerging economies have averaged nearly 8.5% annual growth in GDP (International Monetary Fund, 2013) compared to just over 2.5% for the developed economies. Figure 1.3, for instance, shows the trend in the US GDP growth. If such trends continue, the bulk of future global output, as much as nearly two-thirds, will come from emerging economies.

When the current and anticipated future GDP growth are put in historical perspective, the post-World War II economic growth and the growth during the Industrial Revolution appear to be extremely slow. It would be fair to say that the world has never witnessed the pace of economic growth, it has undergone in the last two decades. Owing to lower wages and reduced capital per worker, the developing economies have the potential to raise productivity and wealth much faster than the historic precedent. This is particularly true in situations where the know-how and equipment are readily available, for instance, in Brazil, Russia, and India; China has been losing the wage advantage as labor costs there are getting increasingly higher.

Associated with fast economic growth are higher living standards for the masses and greater buying power. While, on one hand, this has increased the global demand for products and services, on the other hand, it has created a fear of job and industrial

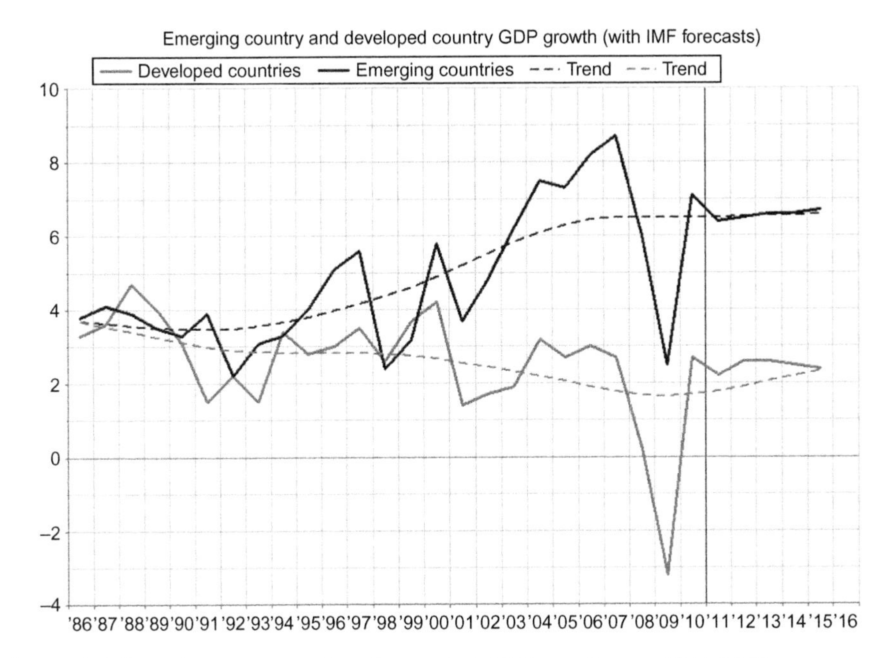

Figure 1.2 Emerging versus developed countries' GDP growth rates 1986–2015.
Source: Adapted from International Monetary Fund, World Economic Outlook Database, Hopes, Realities, Risk, 2011.

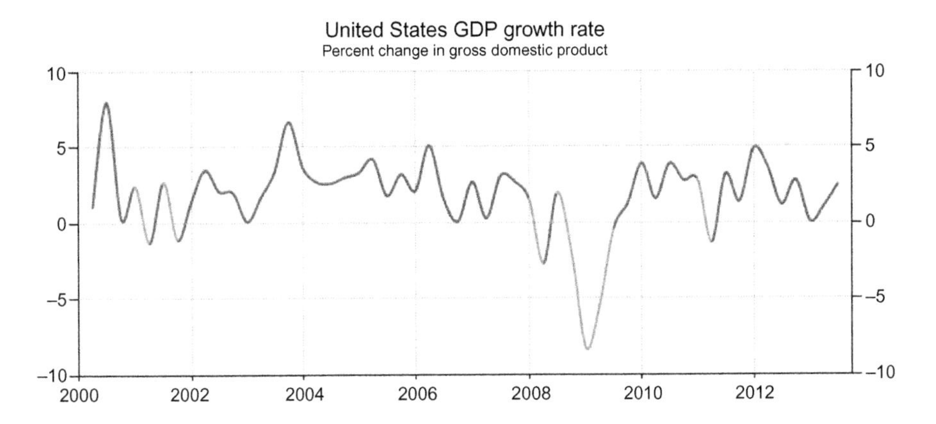

Figure 1.3 United States GDP growth in recent years.
Source: Adapted from Bureau of Economic Analysis, U.S. Department of Commerce, 2012.

output migration to less capital-intensive emerging economies. Such fears are baseless, as the increased demand in emerging economies is creating greater demand for products and services from both internal *and* external sources in the newly developing markets. The huge and expanding middle-class markets in China and India just prove the point. It is anticipated that the global marketplace will add more than a billion new consumers within the next decade. And, as these consumers mature and become richer, they will spend increasingly more on nonessentials, becoming an increasingly more important market to developed economies (Ahya et al., 2006).

While the integration of emerging economies is resulting in redistribution of income worldwide and a lowering of the bargaining power (lowering of wages and shifting of jobs to low wage countries) of workers in the West, it should be realized that emerging economies do not substitute for output in the developed economies. Instead, developing economies boost incomes in the developed world by supplying cheaper consumer goods, such as microwave ovens, televisions, and computers, through large multinationals and by motivating productivity growth in the West through competition. On the whole, growth in emerging economies will make the developed countries better off in the long run. Combined with innovation, management, productivity improvements, and development of new technologies, the developed economies can continue to create new jobs and maintain their wage structures. If wages remain stagnant or rise more slowly, this would have more to do with increasing corporate profit than competition from emerging economies. Figure 1.4 makes the point that corporate profits in the G7 countries have been increasing in the last four decades (U.S. Department of Commerce, 2012). Increased competition, however, should reduce profits and distribute benefits to consumers and workers over a period of time. An estimate by the Petersen Institute for International Economics states that globalization benefits every American family to the tune of $10,000 per year or nearly 10% of the family annual income (Bergsten, 2010). This translates into almost $1 trillion in benefits to the American economy and a tremendous boost in output.

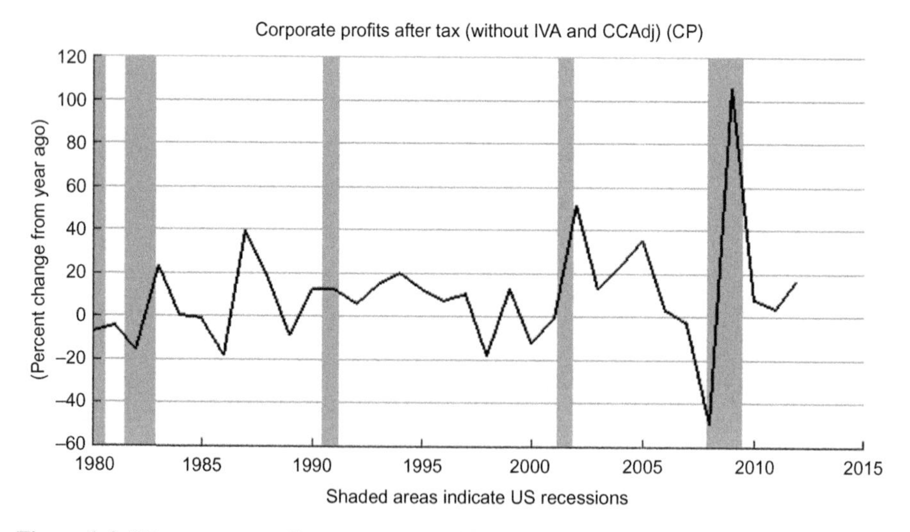

Figure 1.4 G7 corporate profits as a percent of GDP.
Source: Adapted from Bureau of Economic Analysis, U.S. Department of Commerce, 2012.

1.2 Importance of manufacturing

The synopsis of globalization and the state of the world economy presented in the previous section leads to a simple conclusion: global output will continue to rise, and at a faster pace as the consumer markets around the world get bigger and bigger. This presents both emerging and established economies with an unprecedented opportunity to boost national prosperity by efficiently producing high quality products that are needed and wanted. Any shortcoming in achieving this outcome most certainly is going to result in a loss of competitiveness in the global market. For emerging economies, the stakes are much higher, as this will jeopardize the very prospect of these economies ever achieving a "developed" status. In fact, manufacturing activities are essential for any nation for the creation of wealth, raising the standard of living of its population and, ultimately, achieving a high economic status. In fact, no nation in the world has ever achieved developed status without a manufacturing base that comprises at least 20% of GDP and provides at least 30% of the goods traded between nations (Mital et al., 1994). The importance of manufacturing in the context of globalization is evident.

For many countries, such as Japan, Switzerland, and Taiwan, that have no natural resources of consequence, manufacturing is the only means of survival. These countries must generate wealth by trading high value-added products with the rest of the world and use that wealth to meet their need for energy and staples.

The manufacturing, however, must be competitive. That is, the unit labor cost must be held down and the output must be of the high quality that consumers want. Further, the output must make it to the global market in a timely manner. It is imperative to realize that poor quality products can result in the loss of national prestige, and the stigma associated with producing low quality products is neither easy nor inexpensive to overcome. Producing innovative products of high quality also requires avoiding intellectual stagnation and loss of creativity. These are the essential ingredients of remaining competitive. Making lots of stuff that relies on core technologies from elsewhere will not lead either to continued economic prosperity or the ability to compete with the best in the world. Countries such as China, India, Brazil, and South Africa must be creative, developing and mastering new technologies, training workers and management in necessary skills, and developing product brands for global consumers. On the other hand, to benefit fully from globalization, countries such as the United States and United Kingdom must produce higher value-added goods and services while keeping their markets open and flexible.

The importance of output, and thereby manufacturing, is further demonstrated by the vigor of consumers in the developing countries. It was a commonplace belief that American consumers, by virtue of their anemic savings culture, keep the global economy humming. It was said that, if the United States catches cold, Japan, which must survive on the strength of its exports, gets pneumonia. This is no longer so. Japan is no longer dependent on the United States as the primary market for its exports, as seen in Figure 1.5. Whereas exports from the United States and United Kingdom to emerging economies have stagnated, exports from Japan have flourished. And

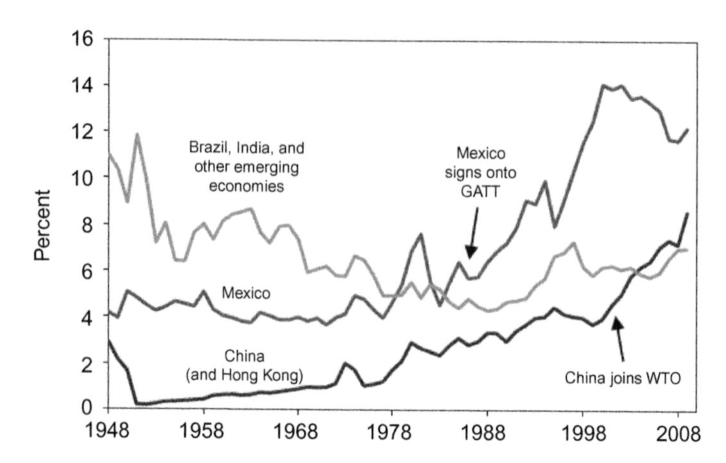

Figure 1.5 US Exports to emerging economies as a percentage of GDP.
Source: Adapted from International Monetary Fund, Direction of Trade Statistics, CEA Calculations.

while the US GDP is on the decline, the GDP of emerging economies is on the rise (Figure 1.2). What these trends indicate is that the emerging economies, primarily in Asia, currently drive the world economy. The Asian countries are not only producing more, they are consuming *more*. The world economy increasingly is dependent on the growth in domestic demand in Asian markets. In terms of purchasing power parity, Asia's consumer market now is larger than America's (The Economist, August 4, 2011). Keep in mind that the Asian markets are yet to develop fully; masses of new consumers are yet to appear on the scene. The growth in the Asian economies also means that the world is less vulnerable to a single economy, America's, and is likely to be more stable. All this makes for a very strong case for manufacturing, particularly for manufacturing high value-added products.

While one can see that manufacturing is important, owing to the need to increase output, it is critical to realize that manufacturing must be *efficient*. Approximately five-sixths of this planet's nearly 6 billion people live in areas considered emerging economies. As they get richer, they want more goods that improve their standard of living—houses, cars, home appliances, and the like. This, in turn, means a huge increase in consumption and demand for energy and raw materials. China alone has accounted for one-third of the increase in world oil consumption and nearly one-fourth of the increase in world metal consumption in the last 5–10 years. While some of this consumption is the result of shifting production operations from Japan, Europe, and North America to China, India, Latin America, and Africa, most of it is the result of growth in world output. As the domestic demand in China and other Asian countries, such as India, increases, the demand for oil, metals, and water increases further. Moreover, as the standard of living in developing countries improves, demand for energy and consumables per capita rise (currently, it is far below the per capita consumption levels in the United States). Given the tight supply of oil, water, and other raw materials and increasing levels of demand, it is logical to expect higher prices for energy, water, and metals. The rising prices may curb demand and slow economic growth, with adverse consequences. This is expected to stir demand for more efficient products, such as more fuel-efficient cars and products that can be recycled. Already, the movement to curb carbon emissions has gained momentum. China is a case in point: China is considering closing inefficient coal-burning power plants as opposed to building new ones at the rate of one per week (Komnenic, 2013). The concern for reducing greenhouse gases and conservation of resources mandates more efficient manufacturing. For instance, the rate of water consumption in China cannot be sustained at current levels. Nearly 60% of the water consumed in Chinese industry is not recycled. According to the Chinese Academy of Social Sciences, the number of rivers in China with a significant catchment area has been reduced from nearly 50,000 to just about 23,000. The increase in output, therefore, must be accompanied by methods that consume less energy during production and operation and designs that allow recycling of materials while minimizing or eliminating waste. In short, manufacturing must accomplish at least the following objectives:

1. Increase the output of high value-added products
2. Produce high quality goods and services, economically and quickly
3. Produce goods that are needed and wanted
4. Minimize the production of greenhouse gases
5. Maximize recycling, eliminate waste, and conserve raw materials
6. Minimize consumption of energy during production
7. Minimize consumption of energy during product operation
8. Reducing industrial water consumption and increase water recycling.

The world does not have resources that will last forever. The needs of industrialization must be met, however, and it is the manufacturing know-how that will help us accomplish the objectives of industrialization by meeting these goals.

1.3 What is manufacturing?

Historically, *manufacturing* has been defined narrowly as the conversion of raw materials into desirable products. The conversion process requires the application of physical and chemical processes to change the appearance and properties of the raw materials. A combination of machine tools, energy, cutting tools, and manual labor is applied to produce various components that, when put together (assembled) with the aid of manual effort, robots, or automated equipment, result in the final product. Manufacturing used to be considered an evil that must be carried out to undertake more meaningful business activities. Therefore, manufacturing was considered simply as a means to add value to the raw material by changing its geometry and properties (physical and chemical).

In the present-day context of economic survival and prosperity, it is insufficient to simply process some raw material into desired product shapes. The transformation must be accomplished quickly, easily, economically, and efficiently; and the resulting

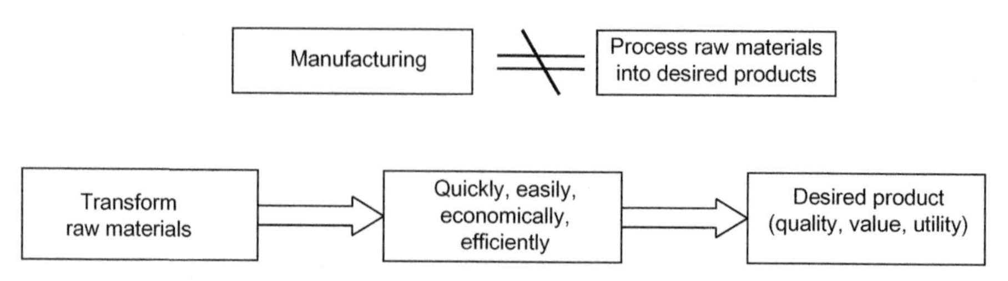

Figure 1.6 Definition of manufacturing.

product must not only be of acceptable quality but must be desired by the end user, the customer. Efficiency and economies of scale are critical for competitiveness in the global market. Further, it is important that a product make it to the market quickly, so as to capture as large a market share as possible. From this standpoint, a product should be innovative and have value and utility for the customer; a "me-too" product has a low probability of survival in today's global market. Figure 1.6 shows the essential requirements of modern manufacturing.

The terms *manufacturing* and *production*, though often used interchangeably, are not the same. While *manufacturing* generally refers to activities that convert raw materials into finished products by using various shaping techniques, *production* is a general term associated with output and can apply to the output of coal mines and oil fields as easily as to power plants and farms.

The type of products that are manufactured generally are classified into two broad categories: consumer products, such as automobiles, coffeemakers, lamps, and televisions, and producer capital goods such as drilling machines, lathes, railroad cars, and overhead cranes. Whereas consumer products are directly consumed by the public at large, producer goods are used by enterprises to produce consumer goods. Enterprises and organizations that employ capital goods to produce consumer goods are known as *manufacturing industries*. Specific activities used to convert raw materials into finished products, such as milling, grinding, turning, and welding, are known as *manufacturing processes*.

Manufacturing engineering, by definition, involves the design, planning, operation, and control of manufacturing processes and manufacturing production. A manufacturing system is an organization that comprises not only the manufacturing processes and production but also activities such as marketing, finance, human resources, and accounting for the purpose of generating output. The entire manufacturing infrastructure involves all activities associated with generating output. Figure 1.7 shows the entire manufacturing enterprise wheel.

As shown in Figure 1.7, the customer is the center of the manufacturing infrastructure. Whatever technologies and resources are utilized and whatever activities are undertaken, it is with the understanding that the customer is the center of attention. One can, therefore, restate that manufacturing is the use of the appropriate and optimal combination of design, machinery, materials, methods, labor, and energy to produce desirable products quickly, easily, economically, and efficiently. And this knowledge is essential for wealth generation, global competitiveness, and economic

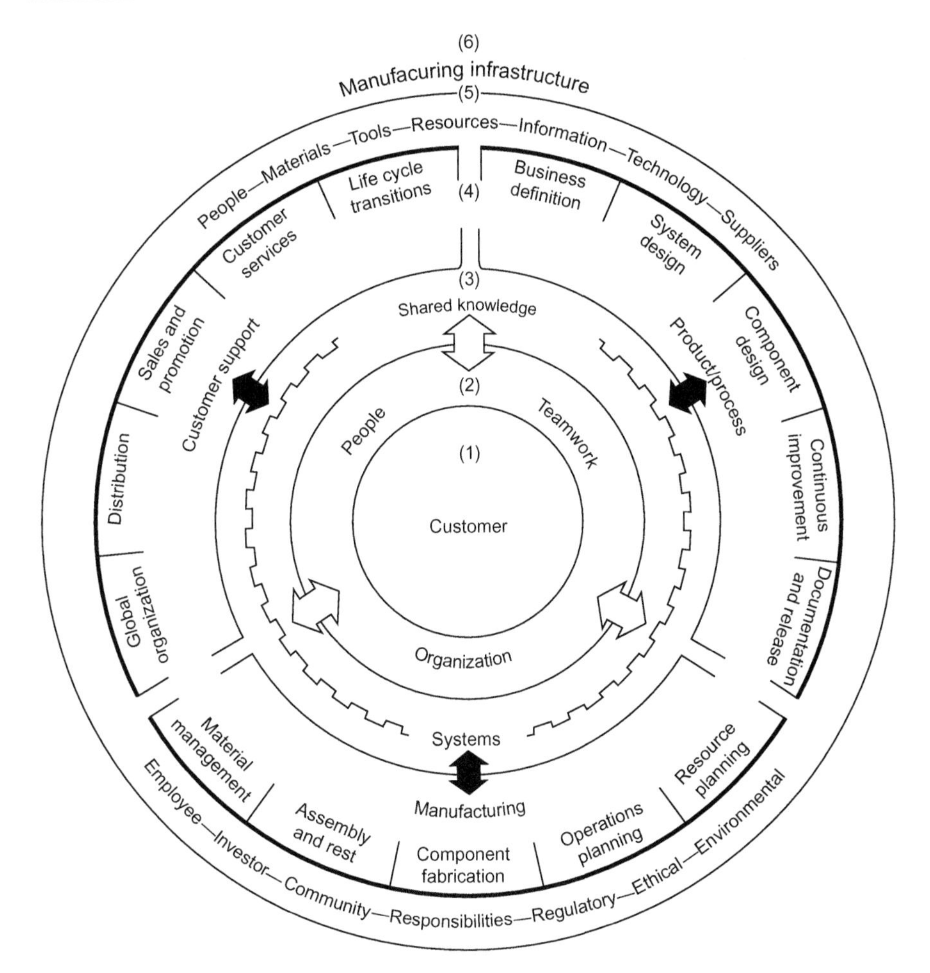

Figure 1.7 The manufacturing enterprise wheel (as outlined by the Society of Manufacturing Engineers, 1994).

survival. Just having resources is not sufficient, as is the case with many countries. Japan and Taiwan have shown the importance of manufacturing and its impact on economic growth.

1.4 Some basic concepts

In this section, we discuss some basic concepts that are important in the overall understanding of the process of product development, design, and manufacture. Specifically, we define the following terms: *capital circulation* or *the production turn*, *manufacturing capability*, *mass production*, *interchangeability*, *product life cycle*,

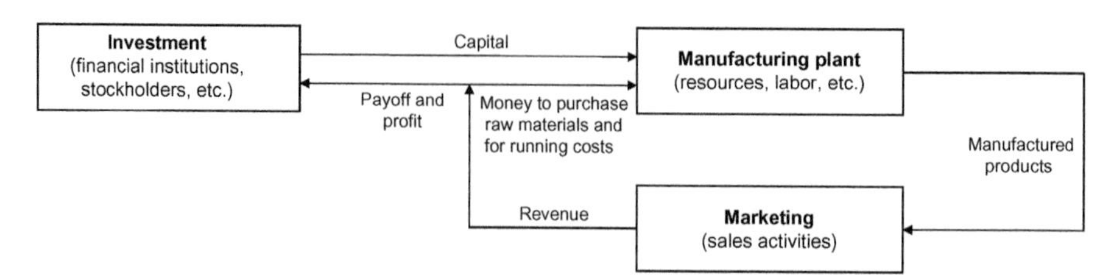

Figure 1.8 Capital circulation or the production turn.

the S curve or the technology growth cycle, *simultaneous or concurrent engineering*, *design for "X,"* and *the engineering problem-solving process*.

1.4.1 Capital circulation or the production turn

Businesses exist to make money. Manufacturing and production activities are no different. The exceptions are nonprofit activities and government activities, such as those undertaken in the defense of the country or for the welfare of the citizenship, such as bridge or highway building. Even these activities must provide benefits that are at least equal to the costs incurred.

As explained by Karl Marx, capital is utilized to acquire the means of production, which, with the assistance of labor, produce goods that are sold. The proceeds from the sale (revenue) are used to accumulate capital (profit). In the context of modern manufacturing activities, a manufacturing enterprise invests capital, by borrowing from either a bank or other source (stockholders or profits from other projects), in a manufacturing plant, produces goods by employing manufacturing activities, sells the manufactured goods with the help of a sales and marketing force, and generates revenue. Part of this revenue is returned to the lending institution (or stockholders in the form of dividends) and part is retained as profit for other ventures. Figure 1.8 shows the circulation of capital or the production turn. The cycle works most efficiently when the cost of production is minimized (profits are maximized) and goods are produced and sold quickly. The cycle obviously is less efficient when the production costs are high, or production takes longer, or products cannot be sold easily or quickly and inventory builds up.

1.4.2 Manufacturing capability

The combined limitations on the size and weight of products that can be processed, the manufacturing processes available, and the volume (quantity) that can be produced in a specified period of time are collectively referred to as the *manufacturing capability* of a manufacturing plant. Not all manufacturing plants are equipped with machine tools that can undertake processing of all kinds of materials. In other words, plants generally have only a limited number of manufacturing processes available and, therefore, can process only a limited number of materials. A plant equipped to manufacture airplanes cannot produce pharmaceutical products. Similarly, machine tools have limitations and can accommodate only products of certain shapes and sizes. The number and variety of machine tools and the size of the labor force also limit the number of units that can be produced in a specified time: per hour, per day, per month, and per year.

1.4.3 Mass production

Mass production refers to the production of large quantities of the same kind of product for a sustained or prolonged period of time. Generally speaking, the production quantity has to be in at least thousands (preferably millions) and is unaffected by daily fluctuations in sales. Television sets, computers, and automobiles are typical examples of products of mass production. Mass production is associated with a high demand rate for a product, and the manufacturing plant typically is dedicated to the production of a single type of product and its variations (e.g., production of two-door and four-door automobiles in the same plant). The machine tools involved are special purpose tools that produce only one type of part quickly and in large numbers and generally are arranged sequentially in a line and in the order in which manufacturing operations must take place (some variations, such as cellular layouts, also exist). The product flows through these machine tools until completed. The layout of machine tools is called a *product layout*.

1.4.4 Interchangeability

When the tires of a car wear out, we simply go to a tire shop and replace the old worn-out tires with new ones. We assume that, if we provide the size of the tire, not only can the replacement tire be easily obtained, it would fit the car wheel properly. This is possible as a result of the concept of interchangeability, which requires that parts must be able to replace each other and, as much as possible, be identical. Interchangeability is achieved by ensuring that each part is produced within a specified tolerance so that replacement can be undertaken without the need of performing any fitting adjustments. In other words, the production of a part is standardized by minimizing variation in size between parts; the variation must be acceptable, as defined by an acceptable level of tolerance. Since interchangeability has dire economic consequences, many countries have established national standards to promote interchangeability. In many instances, the standards are international in nature and adopted by most countries.

1.4.5 Product life cycle

The time period between conceiving a product and the point at which manufacturing it no longer is profitable is defined as the *product life cycle*. As shown in Figure 1.9, the sales volume for a new product rises after its introduction. Once the customers recognize and accept the product, sales increase rapidly (growth). This is followed by a maturity period, when sales increase further. Eventually, competitive products appear on the market and sales decline. As the market saturates and the product no

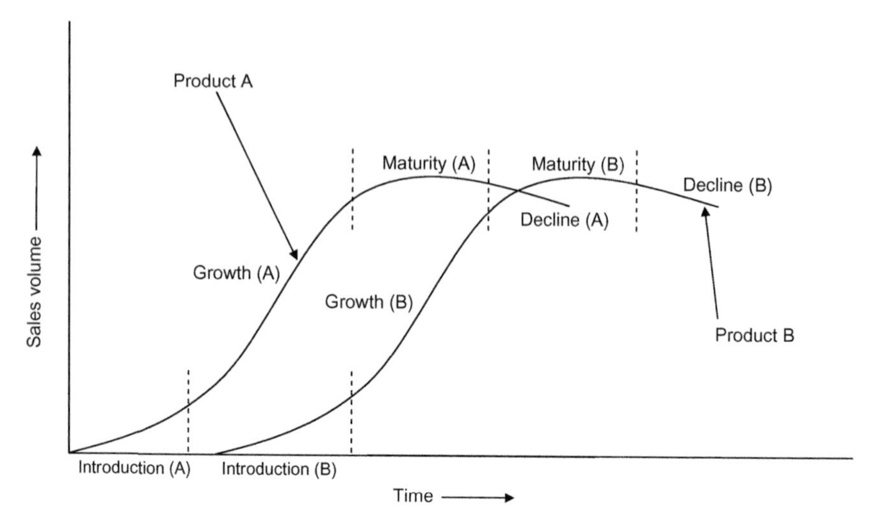

Figure 1.9 The product life cycle.
Source: Adapted from Kotler, 1988.

longer is fresh, sales and profits decline further and it no longer is profitable to produce the product. During this period, profits should be maintained by making minor modifications to the product and relaunching it as new and improved. Businesses (designers and manufacturers) must understand this cycle to maximize profits. Efforts should focus on extending the maturity period as much as possible. Also, for businesses to grow, they must launch new products in such a way that a new product approaches sales maturity just when the ones launched earlier are in decline.

1.4.6 The S curve of the technology growth cycle

The growth of technology is an evolutionary process, following an *S* curve. It has three phases: a slow growth phase, followed by a rapid growth phase, and finally a leveling off phase. The progression of these three stages looks like letter *S* stretched to the right (Figure 1.10). Once the third stage is reached and growth is exhausted, a paradigm shift occurs and new technology evolves. Initially, it takes a lot of effort (time) to understand and master the technology, but as knowledge and experience accrue, progress becomes rapid. Eventually, however, technology is fully exploited and a state of exhaustion is reached; little is gained in performance (new product development), even with considerable effort. It is critical for any research and development (R&D) program to recognize this moment of paradigm shift and come up with new technology on which newer products can be based. A successful R&D program is able to negotiate this technological paradigm shift successfully by introducing new technologies just when the older ones are becoming exhausted. This is a dire necessity for the continual growth of the enterprise. Companies that are unable to provide continuity

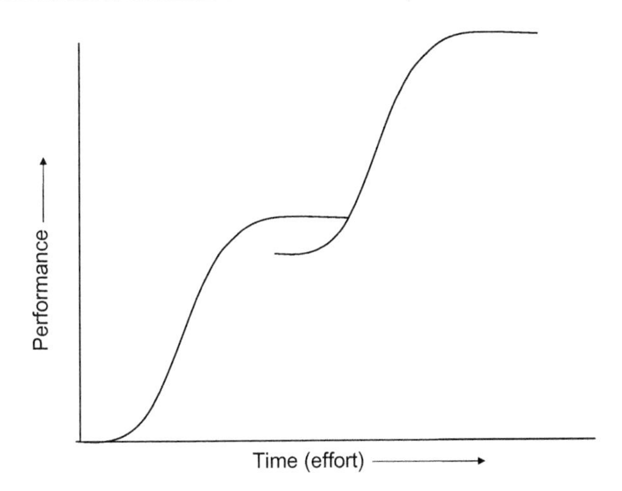

Figure 1.10 The *S* curve of technology evolution.

from one *S* curve to another lose market to their competitors. Cincinnati Milacron is a good example. At one time, the company owned nearly three-fourths of the world robot market; it has been out of the robot manufacturing market now for many years as it failed to realize that market needs shifted from general-purpose robots to special purpose robots. Companies that provide continuity are able to make the transition from one *S* curve to another successfully and thrive (e.g., Boeing, IBM, Motorola, and Microsoft).

1.4.7 Simultaneous or concurrent engineering

A product that functions in a limited, unexpected, or unsatisfactory manner does not enjoy consumer confidence. With this in mind, over the years, the following two primary criteria have dominated the thinking of product designers: functionality and performance. However, in today's competitive marketplace, consideration of functionality and performance alone in product design and manufacture is insufficient. A designer must deal with realistic market constraints, such as costs, timing, and the current state of technology; the availability of technology (material, process, etc.) frequently is dictated by factors such as production volume and production rate. For instance, production processes such as die casting, which are suitable for large volumes, are totally unsuitable for small volumes; methods such as those involving metal removal (machining from a solid) may have to be used for smaller production volumes.

While many of the factors mentioned previously have conflicting requirements (e.g., cost containment needs may dictate the selection of a cheaper material), it has been contended that manufacturing is the most significant factor in product design and must dominate trade-offs when conflicting requirements of various other factors are considered. It is also contended that a product designed for manufacturability is most likely to be dependable, perform satisfactorily, and succeed commercially (Corbett et al., 1991). This is so because the design for manufacturability (DFM) philosophy requires designers to aim at designing products users want and that can be produced economically, easily, and quickly, and can function reliably. Further strengthening this statement are the needs to optimize production, buyers' expectations of product variety, concerns for the environment (green design), and compliance with product liability laws.

It is now widely recognized that the design and manufacturing functions must be closely associated if these goals are to be met. This close, and now inseparable, association is referred to by many names in the published literature: *design for manufacturability* (manufacture), *design for excellence*, *concurrent engineering*, or *simultaneous engineering*. The term *integration engineering* is popular in some circles. Under the simultaneous or concurrent engineering, the design of a product is based on concurrent integration of the following major activities (Chang et al., 1991):

1. Design conceptualization and design axioms
2. Identification of product functions
3. Product modeling and CAD (graphical and analytical representation of the product)
4. Material selection (material properties and associated manufacturing processes)
5. Design for efficient manufacturing (minimizing positional requirements and considering assembly)
6. Specification of dimensions and tolerances (selection of machinery).

1.4.8 Design for "X"

It is our belief that terms such as DFM, *concurrent engineering*, and *simultaneous engineering*, as defined in the published literature, even though considerably more detailed than the conventional product design process (where only the form, function, material, and process are considered) still are not detailed enough to yield the maximum benefits of the overall philosophy. As shown in Figure 1.7, competitive manufacturing requires clearly understanding the needs of customers, which way the market is heading, how to design products that fulfill the needs of customers, how to utilize materials and processes so that high quality products can be manufactured quickly and economically, and how to design and fabricate products that are safe, usable, and easy to inspect and maintain. In addition to how a product should be built, product designers must ponder the question, How should it function? The product designers must also be sensitive to the fact that the product design process takes into consideration the issue of mass production. Equally important is the issue of market demand. Specifically, the goal of a product design team should be to design a product that meets the users' needs and, over the life of the product, can be sold economically. The DFM concept must also include careful and systematic study of all these issues and should mandate concurrent integration of all relevant information (a strategic, or systematic, approach to product design). It is much more than manufacturing processes. It is an effective integration of user and market needs, materials, processes, assembly and disassembly methods, consideration of maintenance needs, and economic and social needs.

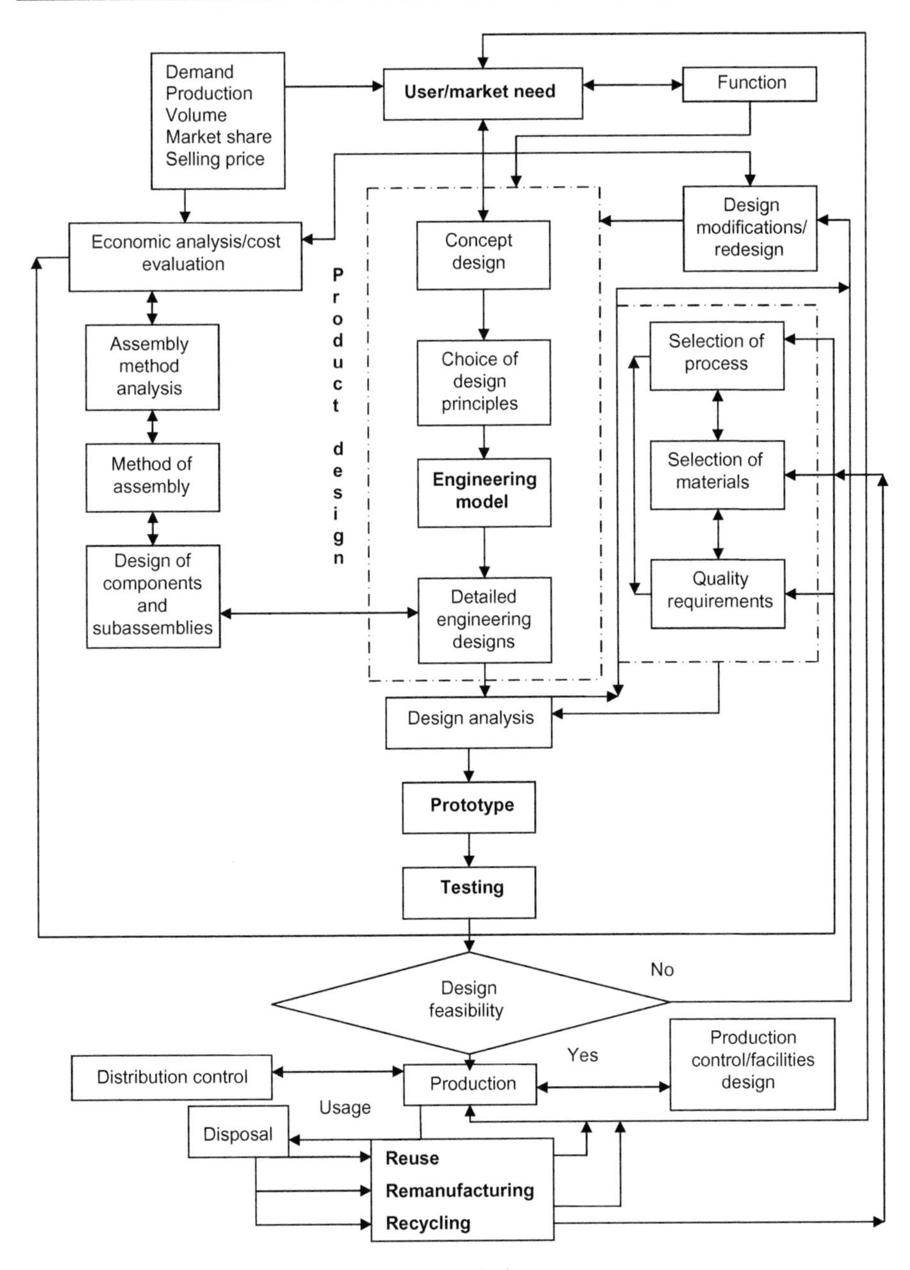

Figure 1.11 An integrated approach to product design.

Figure 1.11 shows a considerably more detailed and integrated approach to product design and reflects more accurately the DFM philosophy. We call it DFX or design for "X." The major activities included are

1. User and market needs and function
2. Concept designs and choice of design principles
3. Identification of materials and processes
4. Design and process analysis and design modification
5. Quality requirements
6. Analysis of assembly and disassembly methods
7. Engineering models and detailed engineering designs
8. Economic analysis and production cost estimation
9. Development of a prototype
10. Engineering testing and redesign
11. Design feasibility
12. Production
13. Production and distribution control.

Figure 1.11 indicates that the product design process (DFX) is very interactive, with feedback required during its various stages. The arrows in the flowchart not only show the feedback, they indicate the direction of the design progression.

1.4.9 The engineering problem-solving process

The basic engineering problem-solving process, outlined by Krick (1969), has five steps:

1. Formulate the problem
2. Analyze the problem
3. Search for alternative solutions
4. Decide among the alternative solutions
5. Specify the solution.

Unless a problem is recognized and clearly defined, it is not possible to solve it, for we must know what we are trying to solve. Engineers as problem solvers must determine if the problem is worth solving. That is, determine the consequences of ignoring the problem—minor to major expense. Next, the problem must be analyzed in detail by gathering as much information as possible, both quantitative and qualitative. This helps in developing a clear understanding of the problem.

Once the problem has been clearly understood, one must seek alternative solutions. Engineers often are satisfied with a single solution; they must seek alternatives to determine the economic attractiveness of various solutions. The final solution must not only solve the original problem, it must be affordable (economically attractive). The goal is to solve the problem in the least expensive manner.

The final step is to specify the solution by properly documenting the steps of the solution. This perhaps is the most important step in the entire process. A poorly documented solution is ineffective and the problem, for all practical purposes, will persist.

1.5 Summary

Manufacturing is critical for the economic well-being of nations. A country rich in resources but without the manufacturing know-how is unlikely to prosper, while countries that are resource poor but have this knowledge will grow rich. Globalization is leading the surge for output, and only the countries that have the knowledge to apply manufacturing technologies efficiently will remain competitive.

In this chapter, we provided a synopsis of the world economy and the impact of globalization. We discussed why it is important to pay attention to manufacturing. We also discussed the broad meaning of manufacturing; it is much more than simply converting some raw materials into finished products by means of processes. Finally, we defined and discussed some of the basic terms that are important in the overall understanding of the product design, development, and manufacture process.

References

Ahya, C., Xie, A., Roach, S., Sheth, M., Yam, D., 2006. India and China: New Tigers of Asia, Part II. Morgan Stanley Research, Bombay, India.

Bergsten, C.F., 2010. World trade and the American economy. Keynote Address, Petersen Institute for International Economics, Los Angeles. May 3.

Chang, T.C., Wysk, R.A., Wang, H.P., 1991. Computer-Aided Manufacturing. Prentice-Hall, Englewood Cliffs, NJ.

Corbett, J., Dooner, M., Meleka, J., Pym, C., 1991. Design for Manufacture: Strategies, Principles, and Techniques. Addison-Wesley, New York, NY.

International Monetary Fund, 2013. World Economic Outlook Database. International Monetary Fund, Washington, DC.

King, S., Henry, J., 2006. The New World Order. HSBC Global Econ. Q1.

Komnenic, A., 2013. China to shut down four coal-fired power plants. Mining News (Mining. Com). October 7.

Kotler, P., 1988. Marketing Management: Analysis, Planning, Implementation, and Control. Prentice-Hall, Englewood Cliffs, NJ.

Krick, E.V., 1969. An Introduction to Engineering and Engineering Design, 2nd ed. John Wiley & Sons, Inc., New York, NY.

Mital, A., Motorwala, A., Kulkarni, M., Sinclair, M.A., Siemieniuch, C., 1994. Allocation of functions to humans and machines. Part II: the scientific basis for the guide. Int. J. Ind. Ergon. 14 (1/2), 33–49.

Oppenheimer, P., 2006. The Globology Revolution. Goldman Sachs, London.

Society of Manufacturing Engineers, 1994. Process Reengineering and the New Manufacturing Enterprise Wheel. SME. Dearborn, Michigan, USA, January 01.

The Economist, 2011. August 4.

The Wall Street Journal, 2013. China's foreign exchange reserves jump again. October 15.

TIME, 2013. China Makes Everything. Why Can't It Create Anything? November 18.

U.S. Department of Commerce, 2012. Bureau of Economic Analysis.

Developing Successful Products

2

2.1 Introduction

Successful companies in the business world constantly operate in a state of innovation in terms of products they manufacture, frequently introducing new products or modifying and improving existing products as needed and desired by the customers. The overall process of conceptualizing a product and designing, producing, and selling it is known by a generalized and comprehensive process called *product development*. In this chapter, we discuss the initial steps of the product development process; Chapters 4–11 are devoted to the designing and manufacturing aspects of product development, while in Chapters 12–14 we discuss components of the overall industrial process associated with the product development process. The marketing and sales aspects of the product development process, while important, are considered beyond the scope of this book and are mentioned in this chapter only in passing.

The key to new product development is the information that indicates what people want, what features of the product are considered absolutely essential, what price they are willing to pay for it, what features are desirable but can be sacrificed for a lower price, current and potential competitors, and likely changes in the market size. Knowing what the market needs is essential in order to develop innovative new products; this knowledge is what leads to developing a successful business strategy. Any product development strategy that is not based on market needs will lead to failure.

Before a successful product can be developed, someone has to come up with, or develop, an idea for conceptualizing it. There cannot be just one idea; several promising ideas need to be developed and analyzed before the detailed plans for a new business activity can be generated. Figure 2.1 shows the progression of actions in the development of a new business activity.

In developing the overall business strategy, a company has to develop and manage its entire product portfolio. Such a portfolio includes not only new-to-market products but also modifications of the existing line of products as well as products that are in the maturity part of sales (Figure 1.9). Concurrently, the company has to ensure that research and development of new technological platforms continues so that the transition from one S curve to another can take place smoothly.

In the following sections, we discuss the attributes of a successful product development process, what successful new products have in common, what steps are necessary to develop a successful portfolio of products, how to identify customer and market needs, and how to develop plans for a new product development.

Product Development. DOI: http://dx.doi.org/10.1016/B978-0-12-799945-6.00002-8

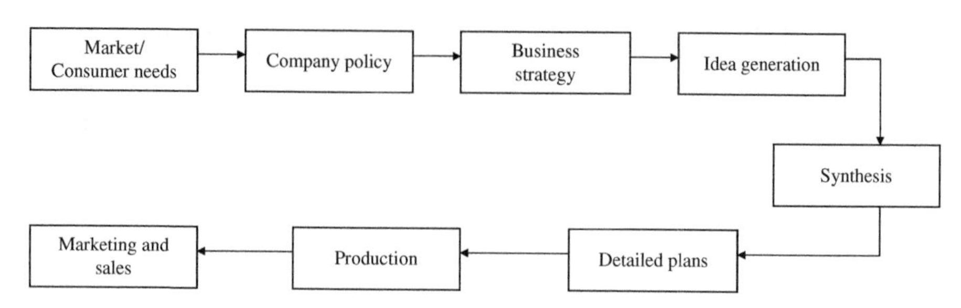

Figure 2.1 Progression of actions in a new business activity.

2.2 Attributes of successful product development

Products that sell well and make a healthy profit (measured by the minimum attractive rate of return a business establishes for itself) reflect a successful product development process. However, as shown in Figure 1.9, sales grow slowly; consequently, it takes time to assess profitability. As a result, we must rely on the definition of manufacturing (Figure 1.6) to establish the attributes of a successful product development process.

According to Figure 1.6, a business should develop the high quality products the market desires quickly, economically, easily, and efficiently. This definition leads to the following attributes that define a successful product development process:

1. **Cost**: Both the cost of producing the product and the total cost of developing it.
2. **Quality**: The quality of the product.
3. **Product development time**: From assessing market needs to product sale.
4. **Development of know-how**: The ability to repeat the process for future products.

The product cost determines its selling price and, to a large extent, its market attractiveness. This is not to say that price is the sole determinant of what the buyers find attractive about a product; cheaper but inferior quality products tend to fall by the wayside. The price does determine profitability, however, and it is in this context that product cost is important. Product cost is a function of both fixed costs, such as tooling and capital equipment costs, and variable costs, such as material and labor costs. How much money the business spends on developing the product, from concept to prototype, also determines the profitability. A product does not become profitable until the development costs are fully recovered. Figure 2.2 shows the relationship between the cumulative cash flow and product life cycle.

Unless a product satisfies customers' needs and is considered dependable, it will not succeed in the marketplace. The quality of the product, therefore, is the ultimate determinant of the price customers are willing to pay for it. The share of the market a product gains is reflected by its quality. For instance, there is a growing movement to seek products that make less noise. As many as half of all consumers may be willing to pay more for less noisy products (The Economist, 2013). In this case, reduced

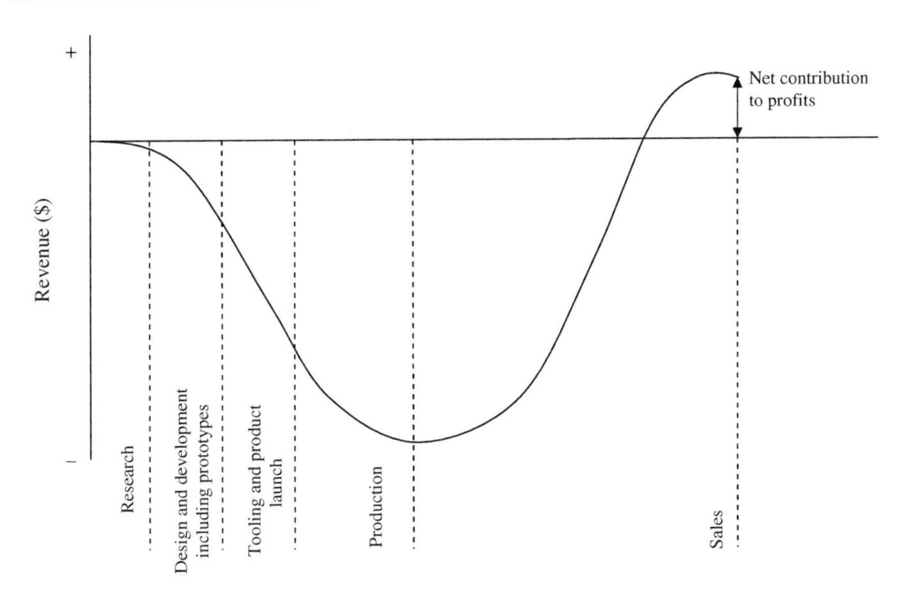

Figure 2.2 Relationship between cumulative cash flow and product life cycle.
Source: Modified from Corfield, 1979.

noise emitted by a product is not only associated with a better quality product, it is in reality cheaper to operate as it consumes less energy. According to Boeing, its 787 Dreamliner is not only the most fuel-efficient airplane; it is the quietest aircraft in that size class.

How quickly a product makes it to the market also determines the overall economic returns on the investment and can be used as a measure of the success of the development effort. Quickness to market, however, cannot come at the cost of the quality.

In contrast to "one-time wonders," development of successful innovative products one after the other reflects the know-how a business has acquired over a period of time. Such companies are able to perform efficiently, effectively, and economically. This, in turn, is reflected in reduced development time, lower development costs, and products that capture significant market share and become profitable.

2.3 Key factors to developing successful new products

To succeed, a business must develop and market new products. However, not all new products that are developed succeed in the marketplace. What separates successes from failures? Cooper (1993, 1996) and Montoya-Weiss and Calantone (1994) identified several factors from a large number of studies that make new product winners. We discuss each of these factors briefly.

2.3.1 Uniqueness

Products that succeed in the marketplace are unique and clearly superior to reactive "me-too" products, which lack any distinguishing characteristics and provide only marginal benefits. The winners

- Provide excellent value for the money spent not only to buy the product but to operate it as well
- Have excellent quality in comparison to their competition as perceived by customers
- Meet customers' needs more fully than competing products, have unique features, and avoid problems associated with similar other products
- Have highly visible and perceived useful benefits and features.

The product development process, therefore, must aim at developing products that are superior in value, distinct in features, and provide clear and unique benefits to the user. Top successful brands tend to excel in this regard.

2.3.2 Customer focus and market orientation

Focus on customer wants is critical to the development of successful products. Such focus improves success rates and profitability (developing economies must focus on the international consumer in order to become globally competitive, economically; this requires product and technology innovation). To achieve a strong market orientation during the product development, businesses must

- Develop a thorough understanding of the nature of the market. As markets differ from region to region, a one-size-fits-all philosophy is very likely to fail.
- Understand the competition, which can be local, regional, or global. Maintain the market orientation.
- Devote resources to activities that determine customers' wants. Marketing activities are critical in this regard.
- Develop a relationship between product attributes and user needs.
- Seek customer input throughout the product design, development, prototyping, testing, and marketing (e.g., Boeing in the development of its 777 and 787 aircraft).

The purpose of a strong market orientation is to leave nothing to chance by seeking customer inputs and incorporating them in product design.

2.3.3 Doing the homework

Work preceding actual product design is critical in determining if a product will be successful. This includes the decision to proceed with the project, a quick study of the market for the product, technical assessment of the capabilities and requirements, detailed market research, and the in-depth financial analysis (developing the pro forma: capital needed, sources, potential sales, etc.). According to Cooper (1996), only about 7% of developmental money and 16% of development effort are devoted to these "critical" activities. This lack of attention to predevelopment work significantly increases chances of product failure; company's name, reputation, and size of the sales force do not necessarily help if the "homework" is skipped.

2.3.4 Sharp and early product definition

Predevelopment work, or homework, leads to a sharp and early product definition and is essential for reducing the time to market. A product definition includes:

- An outline of the concept and the benefits to be provided
- A list of product attributes and features, ranked in the order "essential" to "desirable," and how these compare with competitors' products
- A description of potential users and attributes of the market (size, demographics, etc.)
- An outline of the business strategy (how the product will be placed vis-à-vis competitors).

In the absence of a clear product definition, the chances of failure increase by a factor of 3 (Cooper, 1996). A sharp product definition forces attention to predevelopment work and sets clear goals for product development. It also forces all parties involved in the development to commit themselves to the project.

2.3.5 Execution of activities

Product development teams that succeed consistently do a better job across the activities identified under homework and market orientation. These teams do not skip market studies and do undertake trial sales (using test markets to see how the products will fare). There is no rush to market to capture that illusive share which maximizes profits. Some exceptions to this practice exist, however; Sony, for instance, primarily believes in "creating" a market rather than identifying one. In general, the completeness, consistency, and quality of predevelopment work are crucial to reducing development time and achieving profitability. The quality of execution is not limited to predevelopment work, however; it has to be an integral part of all development activities, from concept development to delivery to market.

2.3.6 Organizational structure and climate

For product development teams to succeed, they must be multifunctional *and* empowered. This means that

- Teams comprise members from all basic functions: research and development, engineering design, production, quality, sales and marketing, and so forth.
- Each member of the team represents the team and his or her "function," not the department and its "territory."
- Teams devote most of their time to project planning and product development.
- Team members share excellent communication and are in constant contact with each other.
- The entire team is accountable for the entire project.
- The team is led by a strong and motivating leader.
- Company management strongly supports the project, the team, and the team leader.

Although these points seem obvious, many businesses do not get the message. It is important that the operating climate be supportive, recognizing effort and rewarding success. The urge to punish failure or discourage risk taking should be avoided. Further, top management should trust the team and the team leader and avoid micromanaging the project; once it has appointed a strong team leader and picked qualified team members, it should provide proper encouragement and make the necessary resources available. The corporate structure should encourage employees to provide ideas for new products, new technological platforms, and new ventures.

2.3.7 Project selection decisions

Many companies are involved in too many projects at one time, scattering valuable resources among many candidate projects. However, not all projects are likely to materialize. Product selection helps narrow down the choice of projects so that resources may be directed to those projects most likely to succeed and become profitable. This requires making tough "go" and "no-go" decisions, where projects that have only marginal value are "killed" so that those with merit may get the necessary resources and focus so good products are developed. Superiority of the product in comparison to competitors' products, product attributes that meet consumers' needs, and market attractiveness are some of the factors that need to be considered in making selection decisions.

2.3.8 Telling the world you have a good product

Having a good product is not enough; it must be promoted properly in the marketplace. New products must be launched at appropriate forums and adequate resources must be allocated to market them. The launch and marketing efforts must be supported by a professional staff that can troubleshoot and service the product promptly if needed (ideally, if the product is designed properly and has high quality built in, this would not be an issue). It would be foolhardy to assume that a good product will sell itself by word of mouth. The launching of the iPhone and iPad by Apple is a case in point. The launch had wide publicity and was covered by major media worldwide.

2.3.9 Role of top management

As stated under Section 2.3.6, the primary role of top management is to support the product development team and provide it with the necessary resources. Management must realize that lack of time, money, and human resources are the main causes of failure. Top management must also clearly articulate the strategy for the business as it pertains to the development of the new product. It must define the goals for the new product. These goals typically include types of product, percentage of market share to be captured, profits from the new product, technologies on which to focus, direction of research and development, and long-term goals. It is also worth emphasizing, again that micromanagement can have a very negative effect, as can pushing the development team in the direction of a favorite project.

2.3.10 Speed without compromising quality

The quickness with which a product makes it to market is an important determinant of profitability. The advantage of speed is lost, however, if it means compromising product quality or the quality of executing essential activities. Since time to market is important, it can be reduced without sacrificing quality by

- Performing many activities concurrently; the lines of communication among the team members and management must be kept open. It is critical not to have just one form of communication (e.g., only written).
- Mapping out the entire project development on a time scale, ensuring that all activities are given adequate time and the precedence of activities is not violated (some activities cannot be performed unless other activities are completed first, e.g., estimating product cost without first completing details of materials and processes).
- This activity–time map is sacrosanct and not violated.

It must be realized that violation of the timeline represents a lack of discipline, needed resources, or both. Regardless, this means delaying the delivery of the product. Also, there is no assurance that the timeline will be violated only once. What stops it from being violated again and again once the process starts? It is better to redirect more resources to the project than to violate the timeline; the timeline must be considered sacred.

2.3.11 Availability of a systematic new product process

Cooper (1993) outlined a stage gate process adopted by many companies. This process formalizes the new product development process, from concept development to launch, by dividing it into logical steps (stages) with strict go and kill decision criteria (gates). These criteria are established by the project team and generally are listed in terms of deliverables for each stage. Each stage can include several concurrent activities, but each activity must meet certain criteria to proceed to the next stage. These criteria, or gates, serve as the quality control checkpoints (ensuring the quality of execution of activities) and cannot be violated or deferred to the next stage. According to Cooper, this sort of strict action to enforce the product development plan results in many advantages, including improved teamwork, early detection of failure, higher success rate, better launch of the product, and a shorter time to market.

2.3.12 Market attractiveness

The market for launching the new product should be attractive; however, this is easier said than done. Nevertheless, some market attributes can help identify an attractive market. Among the desirable market attributes are:

- The market is large and the product is essential for customers (it is important to realize that consumer needs may vary from region to region).
- The market is growing rapidly or has the potential for rapid growth.
- The market economic climate is pro-product (positive).
- The market demand for the new product is not cyclic (seasonal) or unstable.
- The customers are receptive to adopting the new product (typically younger customers) and can easily adopt the product in their lifestyle (e.g., smart phones and electronic books).
- The customers are more eager to try new products and less concerned about their price.
- The customers have sufficient disposable income (the size of the middle class is frequently discussed in this regard).

While it is desirable to have an attractive market, the success of a new product is less sensitive to external environment than to what the development team does to understand the market and customer needs and incorporate them into the product design. These positive actions are more responsible for success.

2.3.13 Experience and core competencies

It is very unlikely that a business will succeed right away in a totally new area of expertise. Some experience in the basic technologies needed, management capabilities, knowledge of product category, market needs for the nature of the product, resources necessary for developing newer products in the area is necessary for success. This requirement may be termed *synergy* or *familiarity with the business*. In general, the stronger the fit between the requirements of the new product development and core competencies (expertise) of the business, the greater are the chances of success. Specifically, the fit must be in terms of

- Technical expertise, both in terms of production capabilities and future research and development
- Management capabilities, particularly the ability to handle complex projects in different business climates
- Marketing, selling, and customer service resources
- Market and customer needs in different regions.

In general, it is easier for a business to succeed if it is on familiar territory. However, this does not mean that a business does not or should not venture into different fields. If such a need arises, it should be pursued cautiously. One way to acquire core competencies in a new area is to acquire existing businesses in that area. This kind of action has become quite commonplace in the world today. Numerous examples of businesses acquired other businesses in the same and totally different fields. Sony's venture in the entertainment business, TATA business house, involving ventures ranging from automobiles to management consulting to running hospitality business, and GE's involvement in manufacturing activities ranging from aircraft engines to medical devices are some prominent examples of companies acquiring core competencies by buying other businesses.

2.3.14 Miscellaneous factors

Some of the factors listed in this section have unexpected effects on the success of a new product. Among these factors are order of entry, innovativeness, and the nature of benefits.

Order of entry has mixed results as far as success in the marketplace is concerned. While a very innovative new product may have some initial success, in general it is better to introduce a high quality product rather than be the first on the market. A poor quality product may capture some market initially, but the bad experience associated with it can have a lasting effect on customers, which may not be possible to overcome, ever. Obviously, it is highly desirable that an innovative, high quality product be the first to market—such as, for instance, the Chrysler minivan. This example also represents a situation where the competition simply followed with me-too type products instead of offering innovative variations, such as minivans with two sliding doors; it was left up to Chrysler to come up with that variation. There is a myth that highly innovative products are risky. In fact, if a new product offers an innovative solution to customers' needs, there is no reason for it to fail. Some caution here is necessary: products that have less innovation or too much innovation are less likely to succeed than products with a moderate degree of innovation; customers are more likely to accept such products, treating products with little innovation as me-too products and those with a very high degree of innovation as fancy gadgets.

A business simply cannot introduce a product and, on the basis of price advantage alone, expect to succeed. Unless a new product provides good value for the price, it is bound to fail, as price alone is an inadequate benefit for success.

In summary, successful consumer products have the following attributes:

1. They offer entirely new benefits that existing products do not.
2. They offer a new secondary benefit in addition to the new primary benefit.
3. They are comparable to what the competition offers.
4. They eliminate an important negative in existing market products.
5. They offer a higher quality features than available in the market.
6. They harness contemporary societal trends.
7. They offer a price advantage in comparison to the competition.

2.4 Strategy for new product development

The primary objective in establishing a strategy and a business plan for developing a new product is to ensure that all concerned parties "buy into" the effort and a consensus is reached on the fundamental inputs to the plan. As mentioned earlier, these inputs include information regarding the market, sources of capital, business pro forma, information about the nature of the product, and information about the market. However, before a development plan can be put together, certain activities must be performed in order to develop an overall new product development strategy. These activities include:

1. Determining the company's growth expectations from the new products
2. Gathering information of interest regarding capabilities, market, and the customers
3. Determining what opportunities exist
4. Developing a list of what new product options exist
5. Setting criteria for inclusion of new product(s) in the company's portfolio of products
6. Creating the product portfolio (new, modified, and existing)
7. Managing the product portfolio to maximize profitability.

2.4.1 Determining the company's growth expectations from new products

A company's mission typically provides some insight into its business objectives. The business objectives and the overall business plan delineate the role the company expects new products to play in its growth. The company 3M, for instance, expects products developed in the last 5 years to contribute 25% to its profits. The role of new products can be a similarly worded target, indicating what new products are expected to contribute to the overall business goals. Setting such a target is important in deciding what resources to direct to new product development. It also helps in reviewing a company's technical and financial capabilities, what product concepts are within company's ability to develop and are attractive to its customers, what are the risks and how these risks can be spread by diversifying product portfolio, and how well the company's short-term and long-term goals are being met.

2.4.2 Gathering strategic information

While the company may already have information regarding its customer base, market needs, business and technical capabilities, and the competition, it helps to periodically update this information. New market research, information on emerging competition, development of new markets, updating internal documents on customer needs, and so on must be carried out from time to time. Most useful is the compilation of all this information in a meaningful form, for instance, comparing the company's technical capabilities and product and sales profiles with those of its competition (benchmarking). This building of the corporate knowledge base is not a one-time effort but a dynamic process that allows the company to constantly update its strategic and business plans.

2.4.3 Determining existing opportunities

As mentioned already, the challenge is to present the information gathered in a meaningful form so that strategic and business plans may be revised and new opportunities identified. The presentation of information should be such that different product options and opportunities are easily identified. Two tools are helpful in this process: a matrix scoring model and a map of the opportunities.

The matrix scoring model is useful in situations where a number of options are available and the best one must be chosen. An example of this kind of analysis is comparing different sites for locating a facility using a number of selection criteria. In choosing a potential product concept for development from among several possibilities, the concepts can be compared using a variety of criteria with weights assigned to each criterion. The scores for each criterion of each concept are added and the totals compared to make the final selection. Table 2.1 shows how the matrix scoring model works.

The scoring scale used in Table 2.1 ranges from 1 (poor) to 10 (excellent) and is somewhat arbitrary; a 5-point, 7-point, or other scale with fewer or more gradations can be used. A larger scale with more gradations increases the sensitivity of the evaluation process; shorter scales with fewer gradations reflect lower evaluation sensitivity. The weights chosen for different criteria indicate the relative importance of the various criteria. One can use a 10-point total, a 100-point total, or different total points for weight as long as the distribution among the criteria is relative. The method also

Table 2.1 Matrix scoring model

		Product concept scores		
Criteria	**Weight (w)**	**A**	**B**	**C**
Financial	3	$3 \times 3 = 9$	$2 \times 3 = 6$	$2 \times 3 = 6$
Customer needs	4	$8 \times 4 = 32$	$5 \times 4 = 20$	$7 \times 4 = 28$
Production ease	2	$4 \times 2 = 8$	$3 \times 2 = 6$	$5 \times 2 = 10$
Core competency	2	$3 \times 2 = 6$	$4 \times 2 = 8$	$8 \times 2 = 16$
Total score		55	40	60

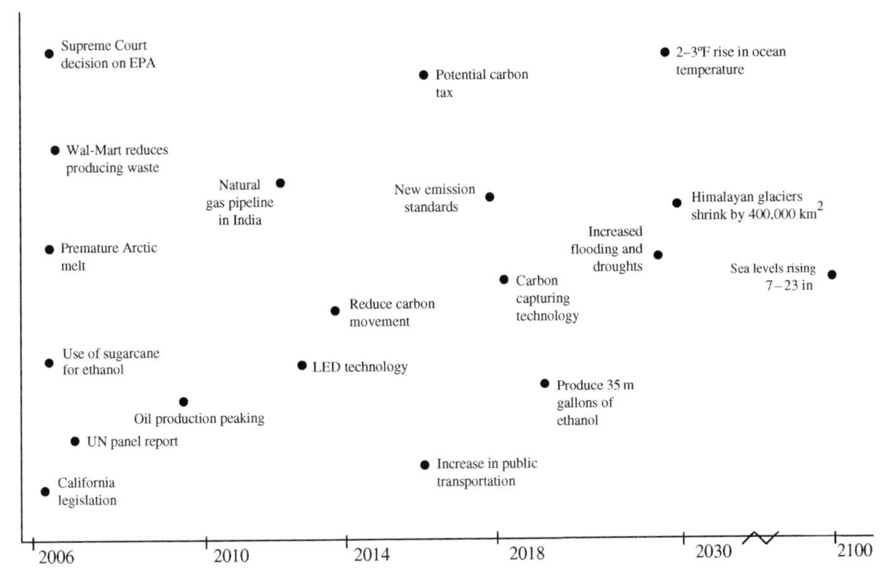

Figure 2.3 Mapping of events and trends in global warming to point out opportunities for energy efficient and alternative energy products (times are approximate; not all events are shown; some events are anticipated).

allows using as many criteria as one chooses as long as it is realized that more criteria reduce the relative importance of each, as the weight then gets distributed over a larger number.

The second method for identifying opportunities for new products requires developing a map of all events and trends and linking them on a time horizon. This method was developed by Motorola (Willyard and McClees, 1987); it helps a company identify new product opportunities. Figure 2.3, for instance, shows opportunities for developing energy efficient products by linking events and trends associated with global warming.

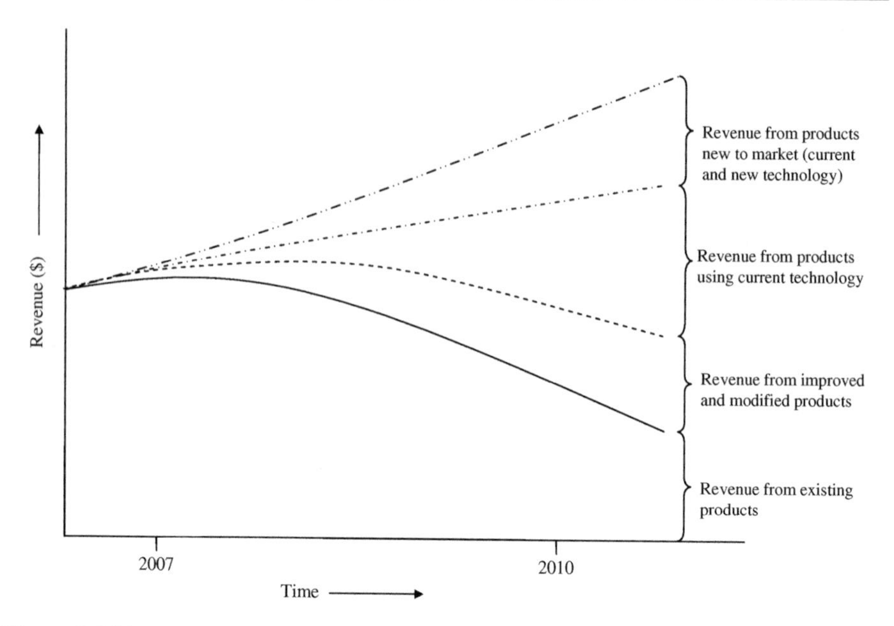

Figure 2.4 Forecast revenue to be generated from sales of new products.

The event map can be used to make decisions regarding which products a company should pursue to achieve forecast revenues (generated from existing products, modified and improved products, products from existing technological platforms, and totally new products). This is known as *gap analysis*; gaps in forecast revenue are to be filled by sales from new products (Figure 2.4).

2.4.4 Developing a list of new product options

After mapping trends and events and reviewing sales goals for new products, the company is ready to develop a list of new product ideas and options. These have to be consistent and compatible with core competencies and the strategic information gathered. The list should be as complete as possible so that all available product options may be considered before selection is finalized. The possible options should be listed in an easy-to-compare format (Table 2.1 shows one option). Important information includes, but is not limited to, information on the financials (investment, cost of production, etc.), risks, available technologies, production capabilities, status of the concept, uniqueness of the features, production goals, expected profitability, product lifespan, potential for derivatives, development team expertise, and synergy with existing products and programs.

2.4.5 Setting criteria for product inclusion in the portfolio

The company expects that including a new product in its portfolio of products will increase its sales revenue and profitability. Typically, all businesses expect a minimum return on investment (ROI). If a product option fails to meet the economic criteria established by the company, it should not be considered any further.

In addition to economic criteria, the company should also look at how well the new product option conforms to its short-term and long-term goals. Can the company develop other products from the technology developed for this product? Is the product concept so new that it exposes the company to unacceptable risks? Are the investment requirements disproportionate? Would the option open new markets to the company? Answers to these kinds of questions help the company develop portfolio criteria. For the final selection, both economic and portfolio criteria must be established and considered.

2.4.6 Creating the product portfolio

As shown in Figure 2.4, the company's product portfolio includes existing, modified, and new products. New products that meet the selection criteria fulfill the new product target, address customer and market needs, promote the company's mission, and meet its business objectives should be included in the portfolio. The final decision should be taken by appropriate people from management and the project development team and should be based on the best information available. A consensus among the participants is necessary so that everyone buys into the process.

2.4.7 Managing the portfolio

Managing the portfolio typically includes assembling the right product development teams, making resources available, ensuring that research and development efforts are focused on developing technological platforms from which new products can be developed, developing appropriate marketing and sales strategies, and the like.

2.4.8 Developing new product plans

Once the preceding activities (also known as *strategic development activities*) are completed, it is time to develop plans for a new product. The first step in this process is the development of a statement of customer needs. This, in turn, requires understanding the customer. In this regard, customer connection and the early definition of customer value are the two most important practices that separate high-performance companies from low-performance ones (Deck, 1994).

2.4.8.1 Understanding consumers and their needs

The key question here is: Will the new product excite consumers into spending money? To answer this question, we have to understand the customers. We have to know at least the following:

- What are their critical needs and how well are these being met at present?
- Who are the consumers, what products do they use, how do they use products, and under what conditions are the products used?

- How are current market products received and perceived with respect to meeting their needs?
- How do consumers know that a product is working, what signals covey that a product is working?
- Are products being used for physical reasons (e.g., hammering a nail), emotional reasons (e.g., feel-good products), or both?

Market research techniques help in understanding consumer needs. These techniques can be broadly classified into two categories: qualitative techniques and quantitative techniques.

Under qualitative techniques, the following techniques generally are included: focus group interviews, one-on-one or in-depth interviews, and in-home visits. The focus group approach is used most frequently. Typically, a group comprising 6–10 individuals of similar background and demographics, led by a moderator, discusses a topic. The one-on-one, or in-depth, interview focuses on one individual at a time in order to learn his or her habits, motivations, needs, and so forth. In-home visits allow for a closer interaction with consumers in their own environment. However, to be effective, households have to be representative of the general population or the population that is to be the focus of the new product development.

Qualitative techniques, while providing information that can be quite enlightening and creative in nature, are limited by small group size or numbers. In such cases, it is better to use quantitative techniques. Among the widely used quantitative techniques are brand image research, segmentation research, and conjoint analysis. Brand image research, using a scale of "important" to "not at all important," attempts to determine which attributes are most important to customers, their opinion of all key brands regarding these attributes, and which attributes tend to best predict overall brand opinion. In segmentation research, consumers are grouped into segments and their response patterns examined. Conjoint analysis is based on the concept that a brand, product, or concept can be considered as a bundle of attributes that make some contribution to overall customer acceptability. In contrast to qualitative techniques, quantitative techniques provide objective and reliable information leading to an understanding of the consumer.

To test the concept idea among consumers, it is important to know how many consumers are likely to buy the product in order to try it. This number of consumers who will try the product is determined as follows:

$$\text{Consumer trial} = \text{Interested universe} \times \text{Consumer awareness} \times \text{Retail distribution}$$

The interested universe is determined by using the purchase interest scale (a weighted 5-point scale from "definitely would buy" to "definitely would not buy"). This universe is adjusted downward by multiplying by factors for the expected consumer awareness and retail distribution the company expects to achieve.

It is also important to know how many consumers who try a new product will buy it. Consumer behavior research indicates that consumer buying behavior is both regular and predictable (Uncles et al., 1994). Their loyalty, however, is greater in the

Table 2.2 Laundry detergent purchase trends in the United States (Information Resources, Inc., 1985)

Brand	Market share (%)	Repeat purchase (%)
Tide	25	71
Wisk	10	62
Bold	8	58
Era	6	55
Cheer	5	55
A&H	5	57
All	5	53
Ajax	2	50
Dash	1	54

Table 2.3 Customer needs statement for rechargeable electric toothbrush

Small, compact, good fit in hand, nonslip grip
Attractive modern styling
Easy to charge; charge should last at least 7 days
Solid base; to double as charger
Both 110 and 220V operation
Brushing head to have rotational and reciprocating movement
Interchangeable and variable size cleaning heads
Price to be <$100
Attractive colors and packaging
Brand name

case of big brand name customers than small brand name customers, as shown by laundry detergent purchase data gathered by Information Resources, Inc., shown in Table 2.2.

The market research techniques described here should lead to a clear and concise statement of customer needs from a customer perspective. Table 2.3 shows an example of customer needs statement for an electric toothbrush. Another example of a customer needs statement is shown in Table 2.4.

2.4.8.2 Understanding the market

To properly evaluate the potential for new product success, it is necessary to understand the market. This market understanding should clarify how the product will benefit both the consumer and the company and should focus on the following factors:

1. Market fit with the overall mission of the company
2. Synergy between the market and the company
3. Attractiveness of the market.

Table 2.4 Customer needs statement for blender

Ability to puree
Ability to mix ingredients evenly
Ability to crush ice
Sturdy base to prevent tipping
Variable-speed motor (three to five speeds) and quiet operation
Attractive styling and availability in assorted colors
Easy to use
Easy to disassemble and clean
Dishwasher safe components
Detachable, nonrusting blades
Weight <2.5 pounds
Motor not to overheat quickly
Pulse grind operation
Easy operating controls
Wide mouth jar for easy loading
Pour spout on jar for easy pouring
Easy to read English and metric graduations on jar
40–60 ounce capacity
Spill-proof lid
Clear jar for easy visibility
Cord storage area
110/220V operation
Price under $35

While the first factor is obvious, synergy between the market and the company is determined by answering a series of questions, such as

Is the market new to the company?
Will the company have to learn a new business?
How well do its management, talent, and skills apply to this market?
Can the company use its technological skills in this market? Its production facilities?
Will the company have to learn new technologies?
Have the technologies been acquired by acquisition? How well does the acquired management fit with the market?
Are significant capital investments required to enter the market?
Can the existing marketing and sales forces and strategies be used? Or must new ones be developed?

A poor fit between the core competencies of the company (knowledge, experience, and capability bases) and unfamiliarity with the market is a recipe for disaster. Davis (1996) provides a couple of examples: Fruit of the Loom, a well-known garment manufacturer, introduced the Loom laundry detergent in 1977 and discontinued it in 1981, and Procter & Gamble acquired the Orange Crush business but had to withdraw from it as it never understood the role of bottlers in the market. Both examples indicate that it is important for survival that the relationship between the old and new business be properly understood.

Attractiveness of the market is also determined by answering a series of questions:

Is the market a large one? How large? Geographical and population sizes?
What are the past market trends? Growing? Stagnant? Shrinking?
What competitors are already in the market? What are their strengths and weaknesses?
Does one competitor dominate the market? Which one?
What are the cost and sales profiles of the leading competitor? Marketing and sales strategies? Technologies? Patents? Reaction to new competition?
How do consumers perceive the competition? How well are their needs being met?
What market share is held by the competition?
What are the trends in consumer needs? What are the features of the new product?
How are the pricing and features likely to attract the consumers?
Is the timing of new product introduction good?

Responses to these questions can be weighted to determine the market attractiveness and market share one expects to gain, initially and over a period of time.

2.4.8.3 Product attributes and specifications

Once the assessment of customer needs and market conditions is complete, it is time to develop product specifications. The elements that should be included in product specifications include the following (Rosenthal, 1992):

- **Performance**: Primary operating characteristics of the product
- **Features**: Characteristics of the product
- **Reliability**: Mean time between failures
- **Durability**: Product life estimate
- **Serviceability**: Ease of repair, part replacement, maintenance
- **Esthetics**: Look, feel, sound
- **Packaging**: Packaging requirements, labeling, handling
- **Perceived quality**: Subjective reputation of the product
- **Cost**: Manufacturing, servicing.

The details pertaining to all elements are tabulated, resulting in product specifications. Table 2.5 shows product specifications for an electric toothbrush.

2.4.8.4 Schedules, resources, financials, and documentation

Developing an agreeable schedule is the next step in the product development planning effort. While the schedule and achievable milestones should be realistic, attention should be paid to competition and profitability realities. Tools such as the program evaluation and review technique (PERT), critical path method (CPM), and Gantt charts should be used to determine a realistic working schedule. The schedule also should be kept in mind in allocating and phasing resources (types, quantity, timing, etc.).

The next logical step is the development of financial data, also known as *product pro forma* in some circles. The kinds of financial details needed are

- Developmental cost: investment, hours, people, and so forth
- Cost of developing prototypes

Table 2.5 Product specification for rechargeable electric toothbrush

Performance
Effective cleaning effects; 2-min cleaning cycle time Long-lasting battery life (rechargeable); at least 30 min
Features
Capability to reach different areas with ease Small, lightweight design; no more than 1.3 ounce Ergonomic grip, comfortable; 1.5-in. circumference Exceeds the American Dental Association requirements for storage and replacement Timer Waterproof assembly
Reliability
Effective cleaning each use Cleaning head to last 12 weeks
Durability: All components to last 2 years under normal usage
Serviceability
Easy to replace head design Easy to replace floss design
Esthetics: New ergonomic styling concept
Packaging
Small and compact box packaging Lightweight package design and packing material; no more than 1 lb Attractive labeling and graphics RFID (radio frequency identifier) for added security and easy inventory tracking
Cost
Manufactured cost <$80 Service and warranty cost <$15

- Capital costs, tooling costs, setup costs, training costs, and the like
- Direct labor, materials, and overhead costs
- Packaging, distribution, marketing, and sales costs
- Product manufacturing cost estimate, selling price determination
- Production volume, revenue, profits, and so on
- ROI, profitability, time frame for recovering investment, and the like.

Putting all this information in a concise form, preferably on a single sheet of paper, is a product plan. Tables 2.6 and 2.7 show examples of product plans.

Table 2.6 **Product plan for rechargeable electric toothbrush**

<table>
<tr><td colspan="2">Customer needs
Small, compact, attractive styling
Easy to use
Easy to clean
Price <$100</td></tr>
<tr><td colspan="2">Key product attributes</td></tr>
<tr><td colspan="2">Rotating and reciprocating head design
Overall weight <1.3 ounce
Overall length no more than 8 in.
Polymer-based ergonomic contoured handle
Soft bristles, pressure sensor limiting force 1.5 newtons
Dual head: front, brushing; rear, tongue cleaning
Rechargeable battery with 1-h life cycle
LCD screen for charge display, brushing cycle time, sanitization status
Quiet (<60 dBA)</td></tr>
<tr><td colspan="2">Product financials</td></tr>
<tr><td colspan="2">Development costs, $1,781,000
Tooling and capital, $3,500,000
Manufacturing cost, $70.70
Distribution and administration costs, 15%
Margin for profit, 20%</td></tr>
<tr><td colspan="2">Market and competition</td></tr>
<tr><td colspan="2">Gain >25% of market share
Penetrate all leading retail chains</td></tr>
<tr><td colspan="2">Development schedule</td></tr>
<tr><td>Phase</td><td>Completion</td></tr>
<tr><td>1. Customer needs
2. Product concept
3. Product design
4. Prototype development
5. Manufacturing
6. Product release</td><td>February
May
August
September
October
December</td></tr>
<tr><td colspan="2">Resource requirements</td></tr>
<tr><td>Weeks</td><td>Phase</td></tr>
<tr><td>6
24
12
16
10</td><td>Marketing and product management
Design engineering
Computer-aided design
Manufacturing engineering
Quality and test engineering</td></tr>
<tr><td colspan="2">Key interfaces</td></tr>
<tr><td colspan="2">Marketing
Research and development</td></tr>
</table>

Table 2.7 **Product plan for blender**

Customer needs

Ability to puree, crush ice, mix, grate
Fast, quiet, and safe operation
Easy to use, disassemble, clean
Price <$35
Heavy-duty motor
Clear, wide mouth jar with spout

Key product attributes

Mixes ingredients evenly
Overall weight <2 pounds
Overall height no more than 15 in.
Quiet (75 dBA) and fast operation
Five-year life
Available in various colors
Automatic shutoff if motor reaches 150°F
Dishwasher safe, nonrust blade

Product financials

Development costs, $575,000
Tooling and capital, $4,070,000
Manufacturing cost, $21
Service cost, $10
Distribution and administration costs, 15%
Rate of return, 10.4%

Market and competition

Gain 5% of market share
Advertise on TV
Market to young
Use wholesale outlets

Development schedule

Phase	*Completion*
1. Customer needs	January
2. Product concept	February
3. Product design	March
4. Prototype development	May
5. Manufacturing	August
6. Product release	December

Resource requirements

Hours	*Phase*
500	Marketing
5000	Design engineering
4000	Computer-aided design
8000	Manufacturing engineering
2000	Quality and test engineering

Key interfaces

Marketing in India and China
Research and development

2.5 Summary

In this chapter, we attempted to lay out a step-by-step procedure for developing new-to-market products. The critical steps in this procedure are developing a consumer- and market-friendly business strategy, identifying consumer needs, and recognizing market conditions. The outcome of the process is a concise, preferably a single page, product development plan. It should be recognized that this process requires the participation of a number of individuals with a wide range of expertise. It would be foolhardy to think that one individual, single-handedly, could accomplish this task. No matter how capable an individual, the key lies in recognizing that people must buy into the final outcome and the outcome must have synergy with the company's mission and business objective.

It is also worth mentioning at this point that information technology (IT) now is the dominant technological item with which the younger generation defines itself. How this technology is handled in the development of new products and how it is customized, to a considerable extent determines the success of a company in developing products people like. For instance, cellular telephones today are fashion items. What sort a person has defines that person. People nowadays have a tendency to replace their cell phones long before they wear out. In that regard, cell phones are like cars. These products not only define people, they bring them together and, for the younger generation, serve as symbols of independence, lifestyle, and mobility. Cars and cell phones are examples that suggest how the development of products may proceed in the future. Here, one should consider at least two phenomena: the product features are as much a function of social factors as technological factors and there is no convergence in product design.

The first phenomenon dictates that, as technology develops and becomes more affordable, features once limited to high end items, such as touch screen and cruise control in cars, become widely available. Color screens now are common features in cell phones and cameras. The second phenomenon, convergence in product design, suggests that there is no point in looking for an ideal product. Should all cars converge in the direction of a single ideal car—and look identical? How about telephones, cameras, televisions, lamps, furniture? People are different with different needs, tastes, and preferences. As long as individuals differ, there will be a need for diverging products. The question is: How do we manufacture divergent products that address consumer needs and make these products profitable?

References

Cooper, R.G., 1993. Winning at New Products: Accelerating the Process from Idea to Launch. Addison-Wesley, Reading, MA.

Cooper, R.G., 1996. New products: what separates the winners from the losers The PDMA Handbook of New Product Development. John Wiley & Sons, Inc., New York, NY.

Corfield, A.P., National Economic Development Office, 1979. Product Design (The Corfield Report). National Economic Development Office, London.

Davis, R.E., 1996. Market analysis and segmentation issues for new consumer products The PDMA Handbook of New Product Development. John Wiley & Sons, Inc., New York, NY.

Deck, M., 1994. Why the best companies keep winning the new product race. R&D, 4LS–5LS.
Montoya-Weiss, M.M., Calantone, R., 1994. Determinants of new product performance: a review and meta-analysis. J. Prod. Innov. Manage. 11 (5), 397–417.
Rosenthal, S., 1992. Effective Product Design and Development. Business One Irvin, Homewood, IL.
The Economist 7 September 2013. The Sound of Silence.
Uncles, M.D., Hammond, K.A., Ehrenberg, A.S.C., Davis, R.E.A., 1994. A replication study of two-brand loyalty measures. Eur. J. Oper. Res. 76, 375–384.
Willyard, C.W., McClees, C.W., 1987. Motorola's technology roadmap process. Harvard Bus. Rev. 13–19.

The Structure of the Product Design Process

3.1 What is design?

Design is the act of formalizing an idea or concept into tangible information. It is distinct from *making* or *building*. Taking the concept for an artifact to the point just before the process of converting it into a physical, or embodied, form begins may be described as the process of designing it. According to Caldecote (1989), design is the process of converting an idea into information from which a product can be made.

From an engineering perspective, the application of scientific concepts, mathematics, and creativity to envision a structure, a machine, a system, or an artifact that performs a prespecified function is the definition of *design*. *Design* is used pervasively, its meaning being somewhat different for an engineer than for an industrial designer. While an engineer is more concerned with the arrangement of parts, the mechanics of the arranged parts, and their functionality when put together, an industrial designer is more concerned with the appearance of an artifact. Since, in designing consumer products, both form and function are important, both disciplines (engineering and industrial design) are crucial in the development of the final information from which a product can be made. The degree to which a product design depends on engineering or industrial design is determined by the product itself. A product that relies mostly on esthetics, such as textile products, greeting cards, and furniture, is within the design spectrum of an industrial designer, while products that are function dominant, such as automobile engines, building foundations, and gear trains, are within the domain of engineers. Consumer products depend on both engineers and industrial designers for success—on engineers for function and on industrial designers for esthetics. The degree to which each discipline dominates the design varies from product to product. Figure 3.1 shows the design spectrum for both disciplines.

In general, the relative cost of products determines the extent to which a discipline contributes to the overall design. The design of a fighter plane is going to depend more on the principles of aerodynamics than on requirements of appearance or what pleases the eye. The design of fabric, on the other hand, is dominated by needs of appearance. Most consumer products, however, have significant contributions of both industrial and engineering design content. Figure 3.2 shows the relative cost shares of engineering and industrial designs in a typical product design. Generally, engineering design costs tend to be 10- to 100-fold more than industrial design costs, due to the very nature of engineering design: investment in functionality, reliability, and the like.

The outcome of the design process is the information that can be used to build it. The format of this information has changed over the years. It used to be informal

Product Development. DOI: http://dx.doi.org/10.1016/B978-0-12-799945-6.00003-X

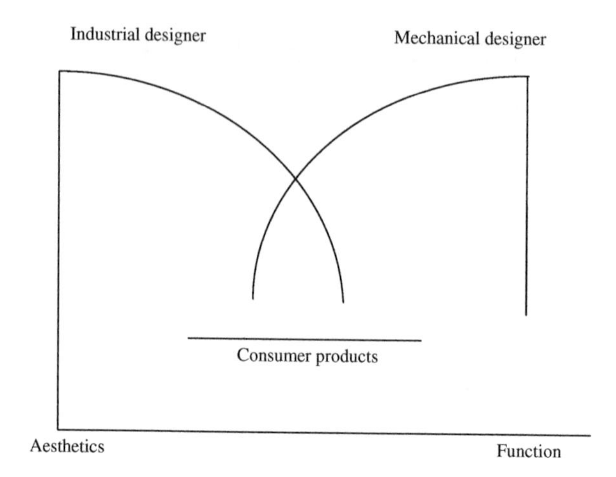

Figure 3.1 The design spectrum for most consumer products.
Source: Adapted from Caldecote (1989).

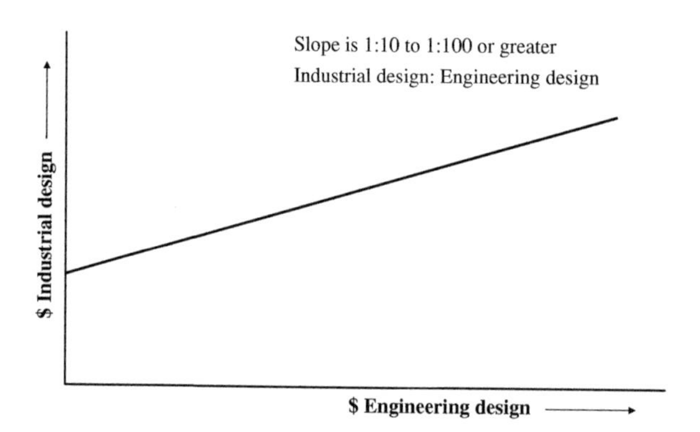

Figure 3.2 Relative costs of engineering and industrial designs.
Source: Adapted from Caldecote (1989).

drawings, leading to formal blueprints, to design drawings, to drawings in electronic format (CAD) that can be stored on electronic media. Some of the information formats, such as CAD drawings, can be fed directly to machines to produce the object (e.g., CAM–CAD interfaces).

3.2 The changing design process

In a primitive society, people designed things without being conscious of the effort. Stone Age tools, doors to mud dwellings, and protection for the feet are examples of products that were developed with no formal drawing or awareness of the design

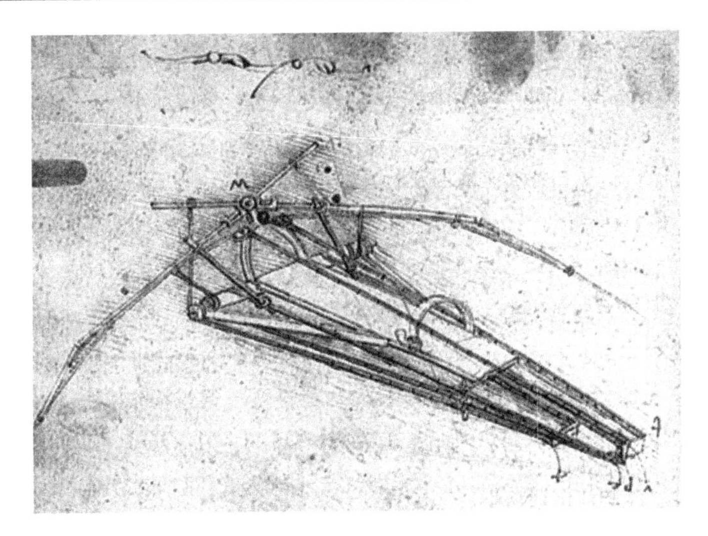

Figure 3.3 Leonardo da Vinci's flying machine.

process. Things were created without anyone designing them; the existence of technology was not a necessity.

The modern design process emerged with the growth of the industrial society. While design by drawing or sketching has existed in some form or the other for more than 5000 years, it became more formalized with time. Figure 3.3 shows the sketch of a flying machine by Leonardo da Vinci. With the growth of the industrial society, this process has become more sophisticated. Figure 3.4, for example, shows a typical engineering drawing from the twentieth century. In the last few decades, humans have been aided by computers; and computer-aided drawings, such as the one shown in Figure 3.5, have become commonplace. These days, the design drawings need not even be on paper; storage of drawings on electronic media has become routine.

It has been argued, for instance by Jones (1970), that design by drawing provided a designer much greater flexibility to manipulate the design compared to the preindustrial society craftsmen. This flexibility allows manipulation of the design without incurring the high cost of building the product and then changing it. The design-by-drawing process, however, does not guarantee success. According to Alexander (1969), this self-conscious process is limited, creates misfits and failures, and cannot replace centuries of development and adaptation in the preindustrial society that led to necessary inventions. Further, the extent of innovation required is beyond the average design-by-drawing designer and, therefore, a new design process, something totally different from design by drawing, is needed. Jones (1970) recommends that we shift our emphasis from product design to system design to avoid the design failures that created large, unsolved problems such as pollution and traffic congestion. Jones recommends a hierarchy of design levels, from components to products to systems to the community.

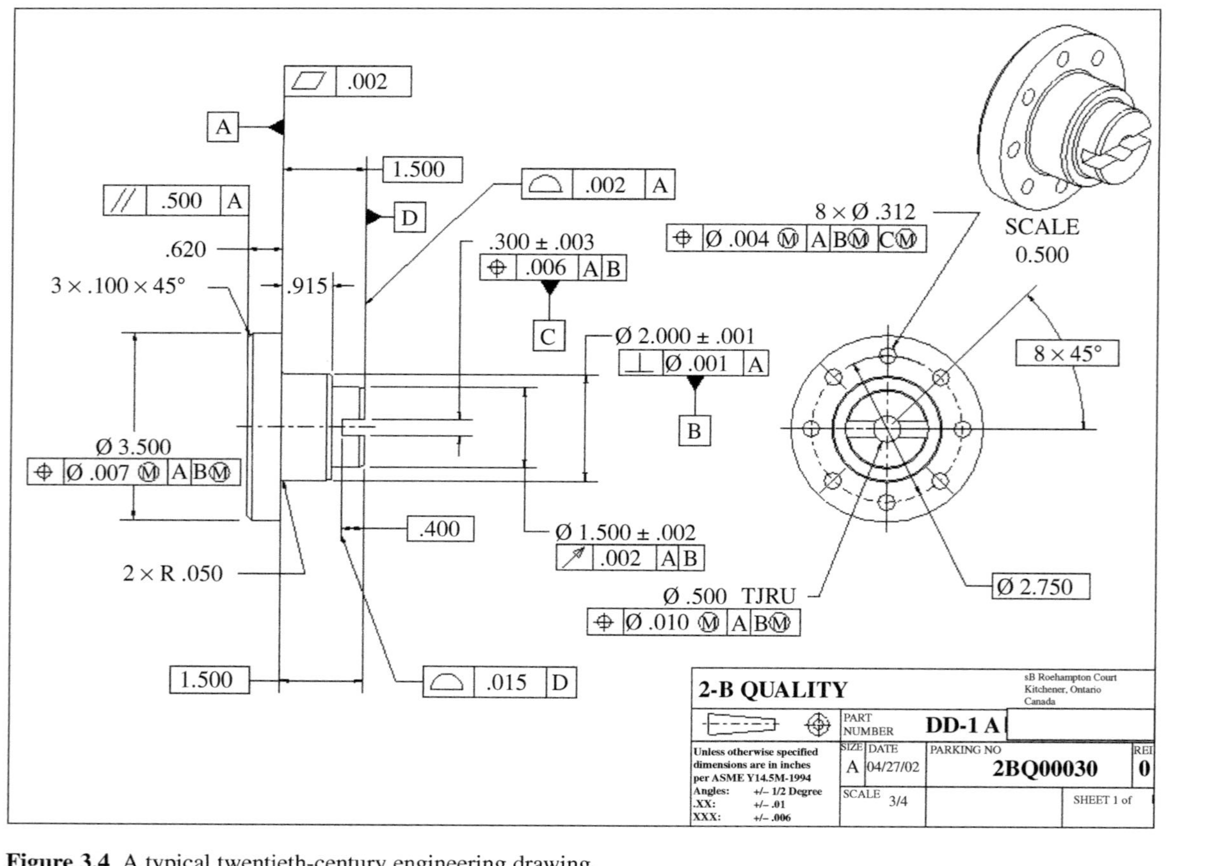

Figure 3.4 A typical twentieth-century engineering drawing.

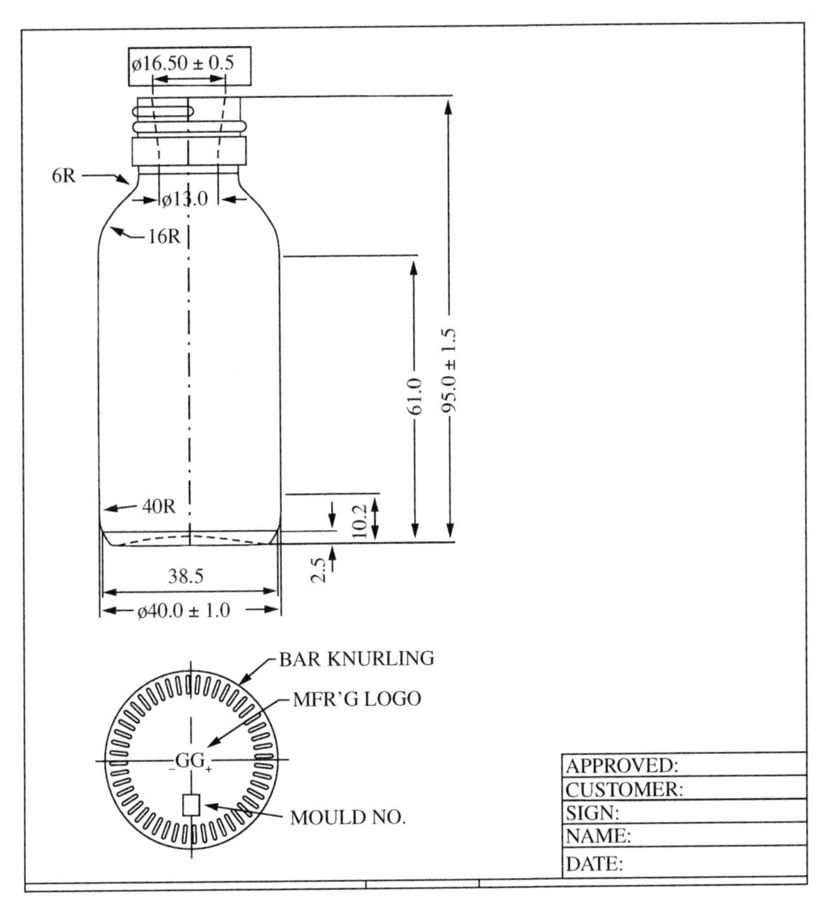

Figure 3.5 A contemporary computer-generated design drawing.

The views of Jones are foreshadowed by Schon (1969; quoted by Cross, 1989) who argued that we are experiencing a postindustrial emergence as indicated by the following elemental shifts:

- From component to system to network
- From product to process
- From static organizations and technologies to flexible ones
- From stable institutions to temporary systems.

In the context of design process changes, this means design becomes a central corporate function open to all in the corporation, instead of just the designer. This externalization of the design process allows all stakeholders (e.g., users and businesses) to see what is going on and to contribute to it in a way that is beyond the capabilities, knowledge, and experience of the designer. In a way, this shift is as significant as the shift from craftwork to design by drawing. For the corporation, such a shift means that it need not commit itself to a single product line or technology; its commitment could be to a major human function and the technologies and organizational relationships necessary to carry it out. Both Schon and Jones argued that we are in the midst of this changing design process, and increasingly the designers have to focus on designing the systems and subsystems (e.g., transportation systems instead of automobiles) instead of just products, which become obsolete. Corporations such as General Electric, Siemens, and Boeing, to some degree, reflect this changing design process philosophy.

3.3 Design paradigms

Design paradigms are models or quintessential examples of designed solutions to problems. The term *design paradigm* is used within engineering design to indicate an archetypal solution. Thus, a Swiss Army knife is a design paradigm illustrating the concept of a single object that changes configuration to address a number of problems. Design paradigms can be used either to describe a design solution or as an approach to design problem solving. In this section, we briefly discuss the following important concepts (based on Petroski, 1994) through some unfortunate but classic examples:

- The need for a model
- The need for redundancy
- The scale effect
- Avoiding starting problem analysis in the middle
- Avoiding confirming a false hypothesis
- Avoiding tunnel vision.

3.3.1 The need for a model

In sixth-century Crete, large stone columns needed to be moved across long distances. These columns were too heavy for the axles of four-wheel carts, putting too much pressure on roads. A six- or eight-wheel cart presented problems in distributing the load over the axles. Sledges were an alternative, but wider wheels were considered better, widest being the best. Wider stone columns were thought to provide a workable solution. Chersiphron, an architect, cut out the center of the column at each end, fitted lead/iron pivots into the cut out center, and put a wooden frame around the column for pulling it.

The scheme worked well for round columns but not for rectangular ones. Chersiphron's son, Metagenes, used the entire column as an axle by building wide wheels around it at each end. Since this concept worked well too, Paconius, an almost contemporary Roman engineer, bid on the job of moving the pedestal for a statue of Apollo. The pedestal was a stone block, 12 × 8 × 6 ft.—about 50 tons of stone. The job was unusual, and Paconius had his own scheme for handling the stone block. He built a great 15-foot-diameter horizontal wooden spool around the pedestal and wrapped it with rope. The end of the rope came over the top of the spool and was attached to several yokes of oxen. As the oxen pulled the rope, the spool was to roll forward, playing out the rope. It seemed to make sense. As the rope uncoiled, it did indeed cause the wheels to turn, but it could not draw them in a line straight along the road. Hence, it was necessary to draw the machine back again. Thus, by this drawing to and fro, Paconius got into such financial embarrassment that he went bankrupt. Had Paconius built a model, he could have tested his concept and avoided the embarrassment as well as the financial ruin.

3.3.2 The need for redundancy

In 1961, the first of many "tower blocks" began to be built to accommodate the housing needs of the thousands of local people of West Ham in east London, an area badly damaged during World War II. The design chosen was the Larsen-Nielsen method of using precast reinforced blocks "slotted" into place on site, then bolted and cemented together. This was seen as a safe, quick way to provide new homes while minimizing on-site construction. To avoid shoddy construction and expensive work delays, all walls, floors, and stairways were precast.

On the morning of May 16, 1968, a freak gas explosion caused the collapse of one corner of a 23-story block of the Ronan Point apartments in Clever Road, Newham, in east London. The construction of Ronan Point began on July 25, 1966, and the building was handed over to Newham Council on March 11, 1968; it cost approximately £500,000 to build. It was 80 × 60 ft. in area, 210 ft. high, and consisted of 44 2-bedroom apartments and 66 1-bedroom apartments, five apartments per floor. However, at 5:45 a.m. Thursday morning, an explosion occurred in apartment 90, a southeast corner apartment on the 18th floor of the new building, blowing out sections of the outer wall. The modern design apparently proved to have a major fault (insufficient support) which allowed a domino-style collapse of wall and floor sections from the top of the building to the ground (Figure 3.6). Officially, the design faults were investigated and took the blame for much of the disaster. It was determined that there was no redundancy; there was a need to effectively join all components of the structure to take up the load during such an eventuality as a gas explosion.

3.3.3 The scale effect

In the late 1800s, a railway bridge across Scotland's Firth of Tay swayed and collapsed in the wind. Seventy-five passengers and crew on a passing night train died in the crash. It was the worst bridge disaster in history. So, when engineers proposed bridging the even wider Firth of Forth, the Scottish public demanded a structure that looked like it could never fall down. Chief engineers Sir John Fowler and Benjamin Baker came up with the perfect structural solution: a cantilevered bridge, with a span of 8276 ft. The Firth of Forth Bridge (Figure 3.7) is made of a pair of cantilevered arms "sticking out" from two main towers. The beams are supported by diagonal steel tubes projecting from the top and bottom of the towers. These well-secured spans actually support the central span. This design makes the Firth of Forth Bridge one of the strongest—and most expensive—bridges ever built. The bridge was opened in 1890.

The Firth of Forth Bridge, which was thought to have been overdesigned, was the basic model for the Quebec Bridge. This bridge across the St. Lawrence River was

Figure 3.6 The collapse of the Ronan Point apartment complex in the United Kingdom.

Figure 3.7 The Firth of Forth Bridge was completed in 1890 in the United Kingdom.

Figure 3.8 The Quebec Bridge under construction in 1907.

the brainchild of the Quebec Bridge Company. In 1903, the Quebec Bridge Company gave the job of designing of the bridge to the Phoenix Bridge Company. The company also contracted a renowned bridge builder, Theodore Cooper from New York, to oversee the engineering design and construction.

The peculiarities of the site made the design of the bridge most difficult. Because the St. Lawrence was a shipping lane, the 2800-foot bridge was required to have a 1800-foot single span to allow the oceangoing vessels to pass. Further, the bridge was to be multifunctional and 67 ft. wide to accommodate two railway tracks, two streetcar tracks, and two roadways. The key to the cantilevered bridge design was the weight of the center span.

In late 1903, P.L. Szlapaka of the Phoenix Bridge Company laid out the initial drawings for the bridge. His design, which had the background of the overdesigned Firth of Forth Bridge, was approved with very few changes. The estimated weight of the span was calculated based on these initial drawings. In 1905, the working drawings were completed and the first steel girder was bolted into place. These working drawings took over 7 months for final approval. In the meantime, the work had begun. It was not until Cooper received the drawing that he noticed that the estimated weight of the span was off, on the low side, by almost 8 million pounds. Cooper had two choices: condemn the design and start over or take a risk that there would be no problem. Telling himself that the 8 million pounds was within engineering tolerances, Cooper let the work continue (Figure 3.8). After all, he wanted to be known as the designer of the greatest bridge in the world. Also factored into the decision was that the Prince of Wales (later to be King George V) was scheduled to open the bridge in 1908 and any delay in construction would upset the planning.

Figure 3.9 The collapsed Quebec Bridge.

On June 15, 1907, an inspecting engineer noted that two girders of the anchor were misaligned by a quarter of an inch. Cooper called this a "not serious" problem. In the inspection report in August 1907, it was noted that the girders had moved out of alignment a bit more and "appeared bent." Although this condition was a bit more disconcerting, the work continued. On August 27, 1907, the warning bells finally went off when the inspection team noted that, over the weekend, the girders had shifted a "couple of inches" and were more obviously bent. At 5:32 p.m. on August 29, the girders trembled with a grinding noise and gave way. The bridge structure plunged over 150 ft., taking with it the lives of 75 workers (Figure 3.9).

The members of the Royal Commission of Inquiry investigating the collapse wrote in their 1908 report, "A grave error was made in assuming the dead load for the calculations at too low a value... This error was of sufficient magnitude to have required the condemnation of the bridge, even if the details of the lower chords had been of sufficient strength." The lower chord members had a rectangular section of 5 ft., 7.5 in., not much smaller than the overdesigned Firth of Forth Bridge chord's circular intersection. The load at the center span was much greater than anticipated, and even though the lower chord section was comparable to the Firth of Forth Bridge chord, it was not sufficient due to the size of the middle span. Clearly, the scale of the center span should have been considered and the chord section enlarged. Extrapolation from other bridge data simply did not work in this case. The bridge finally was completed with the help from the British engineer who had worked on the Firth of Forth Bridge.

The significance of the scale factor also is reflected when we wonder why we do not have any giants. To be able to stand up, the bones of a 40-foot giant must support a load that would require bones of much bigger size or of a different material; larger bones would also interfere with limb movements.

3.3.4 Avoiding starting problem analysis in the middle

Liberty ships were cargo ships built in the United States during World War II. Eighteen American shipyards built 2751 Liberties between 1941 and 1945. They were British in conception but adapted by the United States, cheap and quick to build. The ships were used primarily for carrying troops.

The initial design was modified by the U.S. Maritime Commission to conform to American construction practices and to make it even quicker and cheaper to build. The new design replaced much riveting, which accounted for one-third of the labor costs, with welding. No attention was paid to how welding would affect the structure.

Early Liberty ships suffered hull and deck cracks. Almost 1500 instances of significant brittle fractures were recorded. Nineteen ships broke in half without warning. Investigations focused on the shipyards, which often used inexperienced workers and new welding techniques to produce large numbers of ships quickly. A researcher from Cambridge University demonstrated that the fractures were not initiated by welding but instead by the grade of steel used, which suffered from embrittlement. It was discovered that the ships in the North Atlantic were exposed to low temperatures that changed the mechanism of cracking from ductile to brittle, causing the hull to fracture easily. The predominantly welded (as opposed to riveted) hull construction then allowed cracks to run large distances unimpeded. One common type of crack nucleated at the square corner of a hatch that coincided with a welded seam, with both the corner and the weld acting as stress concentrators. Had the effect of welding on structural stiffness been considered, the problem could have been avoided.

Similar was the case with NASA's shuttle booster joint. The design was based on Titan III joint and included two O-rings instead of one. This was thought to be safer. The outcome indicated otherwise. After the 1986 space shuttle disaster (*Challenger*), the booster rocket was redesigned with three O-rings (Figure 3.10).

Both examples indicate that it is unwise to solve a problem by beginning in the middle; it is important not to lose sight of the goal and start the process at the very beginning.

3.3.5 Avoiding confirming a false hypothesis

It is easy to validate a hypothesis by means of several examples. However, only one counterexample is necessary to invalidate it. It is inevitable in a data-driven statistical study that some false hypotheses will be accepted as true. In fact, standard statistical practice guarantees that at least 5% of false hypotheses are accepted as true (the probability of type I error being 5%). Therefore, out of the 800 false hypotheses 40 will be accepted as true, that is, statistically significant. It is also inevitable in a statistical study that some true hypotheses will not be accepted as such. It is hard to say what the probability is of not finding evidence for a true hypothesis, because it

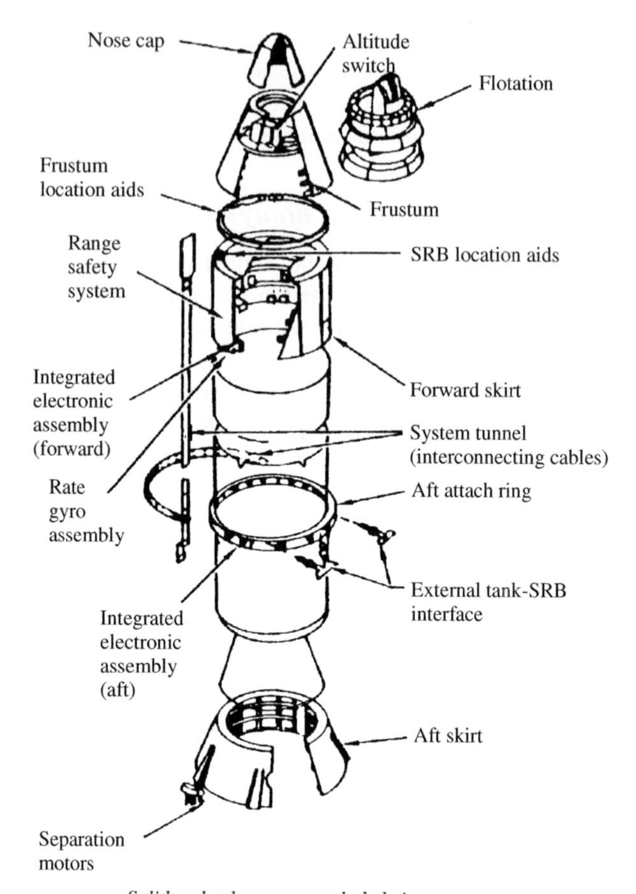

Figure 3.10 The redesigned solid rocket booster for the space shuttle (NASA).

depends on a variety of factors, such as the sample size. Our goal, therefore, should be to collect as much data as possible to reduce the probability of accepting a false hypothesis or to come up with a proper explanation, such as why a failure occurs. It is not uncommon in the field of medicine to find a clinical study whose conclusions are proven to be false when a larger study is undertaken later. In engineering design, similar occurrences are known to happen.

The jet transportation age began on May 5, 1952, when the de Havilland Comet 1 (Figure 3.11) began scheduled flights from London to Johannesburg. The Comet had a cruising speed of 490 mph at 35,000 ft. and a range of 1750 miles with a payload of 44 passengers. The cabin was pressurized equivalent to 8000 ft. at an altitude of 40,000 ft. This gave a pressure differential of 8.25 psi (56 kPa) across the fuselage, twice the value previously used. De Havilland conducted "many tests" to ensure the structural integrity of the cabin. However, three accidents occurred in which Comet aircraft disintegrated in flight, and all Comet 1 aircraft were subsequently withdrawn from service:

Figure 3.11 The unfortunate de Havilland Comet airplane of BOAC.

- G-ALYV after leaving Calcutta (now Kolkata) in May 1953. Violent storms were thought to be involved and some wreckage was recovered. No firm conclusions were drawn as to the cause.
- G-ALYP over Elba in January 1954 after 1286 cabin pressurization cycles. Little wreckage was recovered and no major problems were found in fleet inspection. Fire was assumed to be the most likely cause and modifications were made to improve fire prevention and control. The aircraft was returned to service.
- G-ALYY, flying as SA 201, after leaving Rome in April 1954. About 70% of the aircraft was recovered and reconstructed at Farnborough. The engines were recovered more or less intact, showing that engine disintegration was not the cause of the accident, and neither was any evidence of fire found.

The Comet G-ALYU, which had experienced 3539 flying hours and 1221 cabin pressurization cycles, was subjected to full-scale flight simulation testing at Farnborough. The fuselage was hydraulically pressurized in cycles, while the wings were flexed with jacks to simulate the flight loads. Water was used for this pressurization because calculations had indicated that the energy release under cabin rupture with air as the pressurization medium was equivalent to the explosion of a 500-pound-force bomb in the cabin. The cabin also was supported in water to avoid extraneous weight effects. After the equivalent of a total of 3057 (1836 simulated cycles) flight cycles, a 2-mm crack near the escape hatch grew to failure. This was repaired, and after 546 flight cycles, a 4.5-m section of the cabin wall ruptured due to fatigue cracking. It was concluded that explosive cabin failure had caused the loss of the three Comet aircraft. Developing a detectable crack 6 mm long consumed some 95% of the cyclic life.

The Royal Navy was charged with getting the relevant fuselage piece of G-ALYP from the sea (using simulation trials, based on the way the aircraft was now thought to break up in flight) to establish the likely position of this part of the aircraft on the seabed. This showed unmistakable signs of fatigue. The fatigue crack was associated with the stress concentrations of the rather square rear ADF window cutout (stress of 315 MPa at the edge of the window) and with a bolt hole around the window (although the stress at the bolt position was only 70 MPa).

The chief designer at de Havilland had wanted to glue the windows in position, but the tooling for the square shape was too difficult to make. A lower stress concentration shape would have been easier to manufacture.

The manufacturer had performed fatigue tests of the forward cabin area at about 10 psi (with cracking occurring at 18,000 cycles), but these were carried out after static tests of to up to 16.5 psi (twice operating pressure) had previously been applied. Cracks also were known to be present after manufacture, and the remedy was to drill 1.6-mm holes at the crack tip to "arrest" them (such an arrested crack was present near the rear ADF window, which had not propagated until the final failure). In the end, the following causes were identified:

1. New technology introducing new load cases (high-altitude flight for turbojet engines requiring cabin pressurization).
2. Mismatch between service loads and fatigue test procedure.
3. Possible contribution from out-of-plane bending loads (biaxial stresses) resulting from the following design failures:
   - Improperly understood failure mode assessment procedures necessitated by implementation of new technology.
   - Poor configuration due to wing root engine placement (very few other aircraft have had engines in this position), affecting uprating potential, fire hazard, and structural integrity in the event of engine disintegration.

3.3.6 Avoiding tunnel vision

The 1940 Tacoma Narrows Bridge, a very modern suspension bridge with the most advanced design, collapsed in a relatively light wind. The final investigation revealed that the designers did not pay attention to the light weight of the road and supporting system in their concern for vertical and horizontal "flexibility."

The state of Washington, the insurance companies, and the US government appointed boards of experts to investigate the collapse of the Narrows Bridge. The Federal Works Administration appointed a three member panel of top ranking engineers. In March 1941, the panel of engineers announced its findings. "Random action of turbulent wind" in general, said the report, caused the bridge to fail. This ambiguous explanation was the beginning of attempts to understand the complex phenomenon of wind-induced motion in suspension bridges. Three key points stood out:

1. The principal cause of failure of the 1940 Tacoma Narrows Bridge was its "excessive flexibility."
2. The solid plate girder and deck acted as an airfoil, creating "drag" and "lift."

3. Aerodynamic forces were little understood, and engineers needed to test suspension bridge designs using models in a wind tunnel.

"The fundamental weakness" of the Tacoma Narrows Bridge, said a summary article in *Engineering News Record*, was its "great flexibility, vertically and in torsion." Several factors contributed to the excessive flexibility: the deck was too light. The deck was too shallow, at 8 ft. (a 1:350 ratio with the center span). The side spans were too long, compared with the length of the center span. The cables were anchored at too great a distance from the side spans. The width of the deck was extremely narrow compared with its center span length, an unprecedented ratio of 1–72.

The pivotal event in the collapse of the bridge was the change from vertical waves to the destructive twisting, torsional motion. This event was associated with the slippage of the cable band on the north cable at midspan. When the band slipped, the north cable became separated into two segments of unequal length. The imbalance translated quickly to the thin, flexible plate girders, which twisted easily. Once the unbalanced motion began, progressive failure followed. Wind tunnel tests also concluded that the bridge's lightness, combined with an accumulation of wind pressure on the 8-foot solid plate girder and deck, caused the bridge to fail.

3.4 The requirements for design

Designing is the application of technical and scientific principles to arrange components of a device. When the device is adapted and embodied to achieve a specific result, it must satisfy the six requirements as outlined by Pye (1989). These requirements are as follows:

1. It must correctly embody the essential principle of arrangement.
2. The components of the device must be geometrically related to each other and to the objects, in whatever particular ways suit these particular objects and this particular object.
3. The components must be strong enough to transmit and resist forces as the intended results require.
4. Access must be provided.

   The requirement for ease and economy is:
5. The cost of the result must be acceptable.

   The requirement of appearance is:
6. The appearance of the device must be acceptable.

Design is the process of satisfying these requirements.

As Pye states, these design requirements are conflicting in nature, and therefore, the final design, no matter what, cannot be perfect. It is the designer's responsibility to determine, in consultation with the client, the degree and location of "failures" (compromises)—for instance, the conflict between the requirements for economy and durability, between speed and safety, between usability and functionality. Since the final design is the result of many compromises, design may be considered a problem-solving activity; the requirement of appearance makes it an art. Therefore, design is both a problem-solving activity and an art.

3.5 The design process

In this section, we discuss the steps of the actual design process. However, before we discuss these steps it would be prudent to discuss the problem that confronts the designers.

3.5.1 Problem confronting the designers

A product has certain properties that make it useful to people. The properties can be physical, such as size, weight, or strength, or chemical, such as composition, heat tolerance, or rust resistance. Some of the properties are intrinsic, some are extrinsic, and some are the result of the physical form of the product (geometrical form). Table 3.1 shows the various intrinsic, extrinsic, and design properties of a product. As a result

Table 3.1 Properties of a Product

Design	Internal	External	System
Structure	Strength	Operational properties	Space requirement
Form	Manufacturing properties	Ergonomic properties	Durability, life
Tolerance	Corrosion resistance	Aesthetic properties	Weight/mass
Surface	Durability	Distribution properties	Maintenance
Manufacturing methods		Delivery and planning properties	Operation
Materials		Law conformance properties	Surface quality
Dimensions		Manufacturing properties	Color
		Economic properties	Appearance
		Liquidation properties	Storage space
		Function	Transportability, packing
		Functionally determined properties	Delivery deadline
			Laws, regulations, standards, codes of practice
			Quality
			Operational costs
			Price
			Wastes
			Recycling
			Function
			Reliability

(Adapted from Hubka and Eder, 1988)

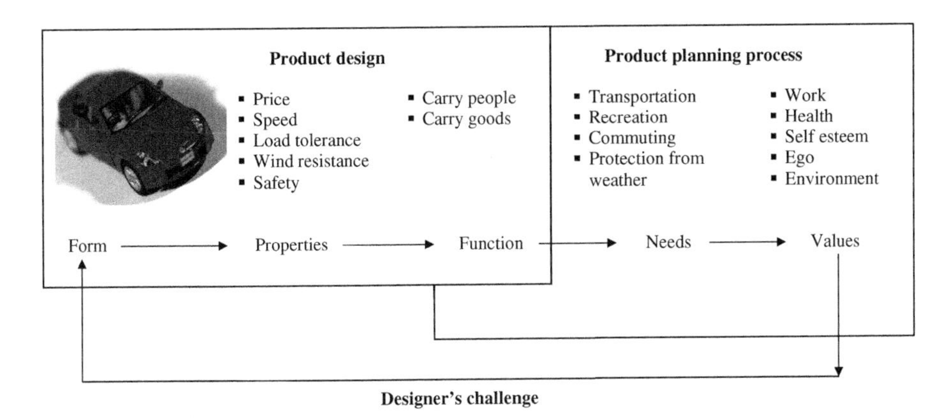

Figure 3.12 Link between a product's form, its properties, its function, and human needs and values.

of these properties, the environment in which it operates, and the geometrical form it has, a product can perform certain functions. The fulfillment of these functions satisfies human wants and needs and helps the product achieve one or several values. The achievement of these values is what makes a product useful to people. Figure 3.12 shows this progression.

Knowing the form of a product, it is possible to derive its properties, the functions it can perform, the human needs it can satisfy, and the values it will achieve. The process of design, however, does not require predicting properties and functions from the form, for the form is unknown. Rather, it is to achieve an embodied form that, by virtue of its intrinsic and extrinsic properties, performs certain functions that satisfy human needs. In other words, the challenge for a designer is to move from right to left in Figure 3.12. Transition from function to form, to a considerable degree, depends on the ability, imagination, and creativity of the designer. This, then, is the problem that confronts designers: to embody properties in a geometrical form such that the embodied form, when used as intended in the specified environment, can perform the intended functions.

While a specific product can perform only certain functions, it is possible to come up with a number of forms that perform the same set of functions. Conceiving these forms and choosing the final one is the challenge for designers. While design methods can help, creativity and imagination are crucial in the transition from function to form.

3.5.2 Steps of the engineering design process

The basic engineering design process is not unlike the engineering problem-solving process, described briefly in Chapter 1, as designing is not much different from solving any problem. In fact, "designing" is a special form of problem solving, as the cycle of activities that must be undertaken to come up with a design is similar. Hall (1968) outlined these basic activities as

- **Problem definition**: Studying needs and environment
- **Value system design**: Stating objectives and criteria
- **Systems synthesis**: Generating alternatives
- **Systems analysis**: Analyzing alternatives
- **Selecting the best system**: Evaluating alternatives against selected criteria
- **Planning for action**: Specifying the selection.

These activities are reflected in many models of the basic design process. Figure 3.13 shows one of the earliest such models, developed by French (1971). A variation of the model, somewhat more detailed and shown in Figure 3.14, is presented by Pahl and Beitz (1984). The activities in Figure 3.13 are replaced by phases in Figure 3.14. The German professional engineers' body, Verein Deutscher Ingenieure (VDI) produced a general design process guideline, VDI 2221, shown in Figure 3.15 (1987). This is a somewhat detailed variation of the engineering design process model shown in Figure 3.14. The VDI 2221 model outlines a procedure that first emphasizes analyzing and understanding the problem in detail, then breaking the problem into subproblems, finding solutions to these subproblems, and finally

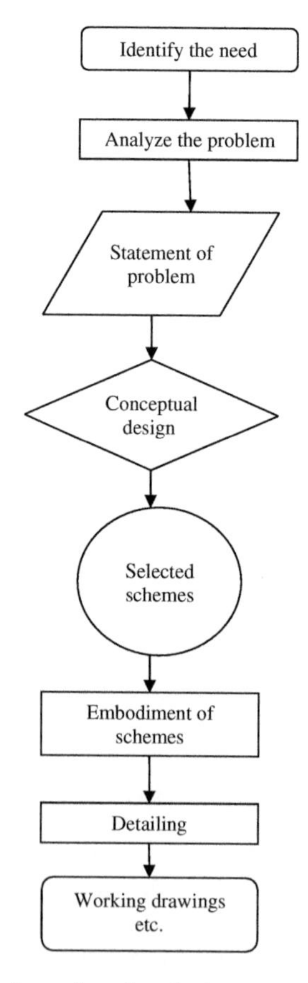

Figure 3.13 French's (1971) basic engineering design process.

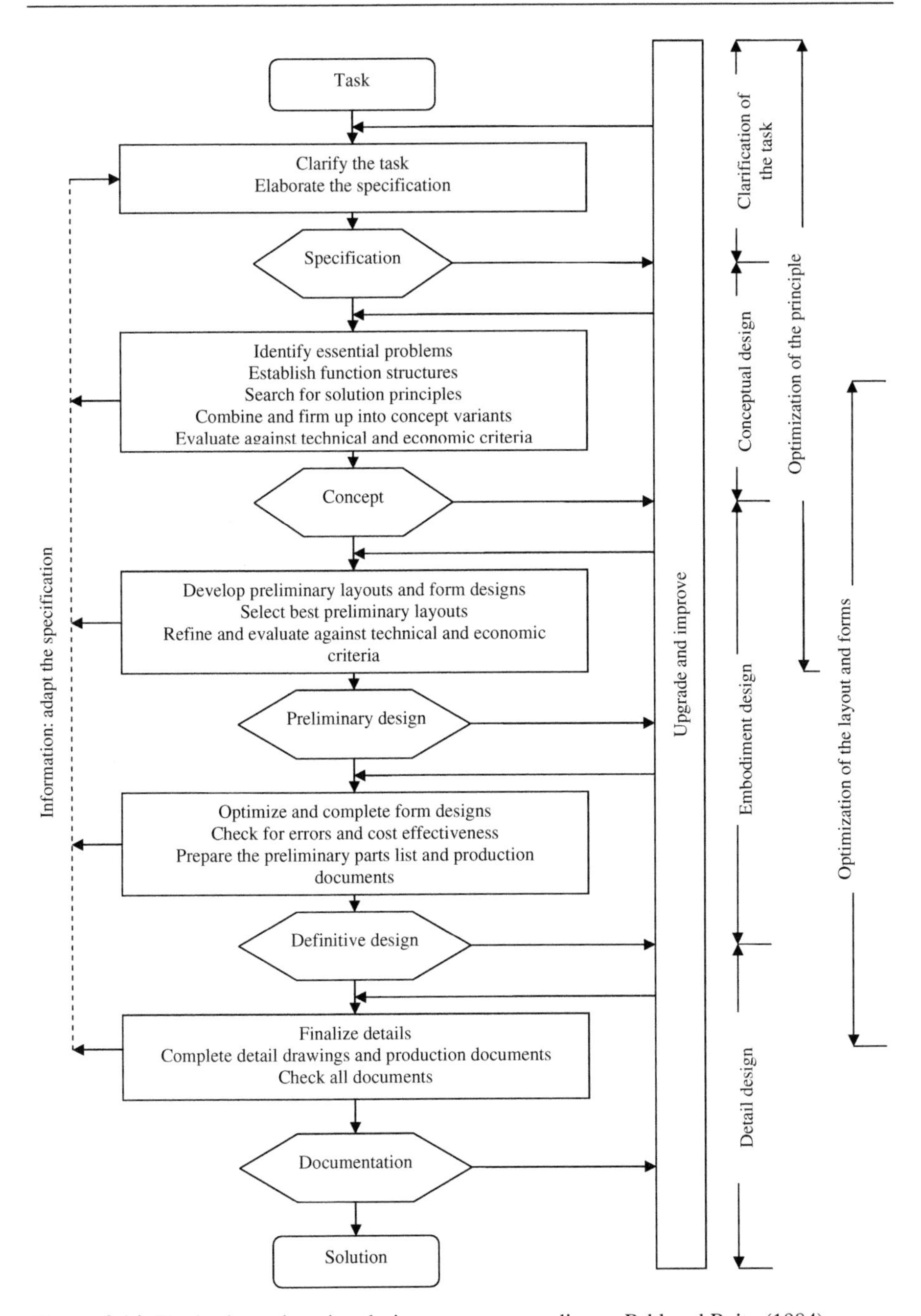

Figure 3.14 The basic engineering design process according to Pahl and Beitz (1984).

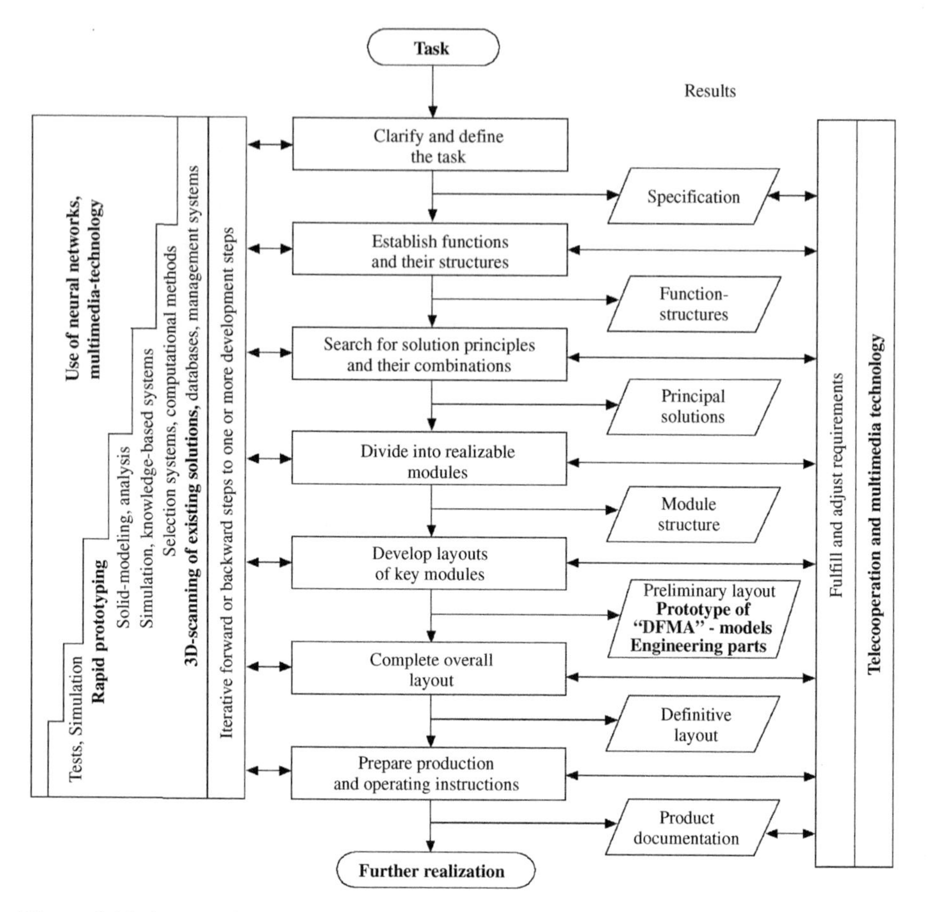

Figure 3.15 A general approach to engineering design (VDI 2221, 1987).

combining these solutions into an overall solution. All phases of this model require coming up with a variety of solutions (divergence in the design phase). From this variety of solutions, the best option is picked (convergence in the design phase). This concept is shown in Figure 3.16.

The principles used for the solutions to subfunctions generally are basic scientific and engineering principles. Pahl and Beitz (1984) provided examples of solution principles to subfunctions, as shown in Figure 3.17.

3.5.3 Defining the problem and setting objectives

A problem is the result of an unfulfilled need. Unless the need is clearly defined, the problem cannot be formulated; unless a problem is properly formulated, there is either no solution or no guarantee that the proposed solution will solve the problem. It is typical, however, that the need, when expressed initially, is vague. The problem, therefore, is ill defined.

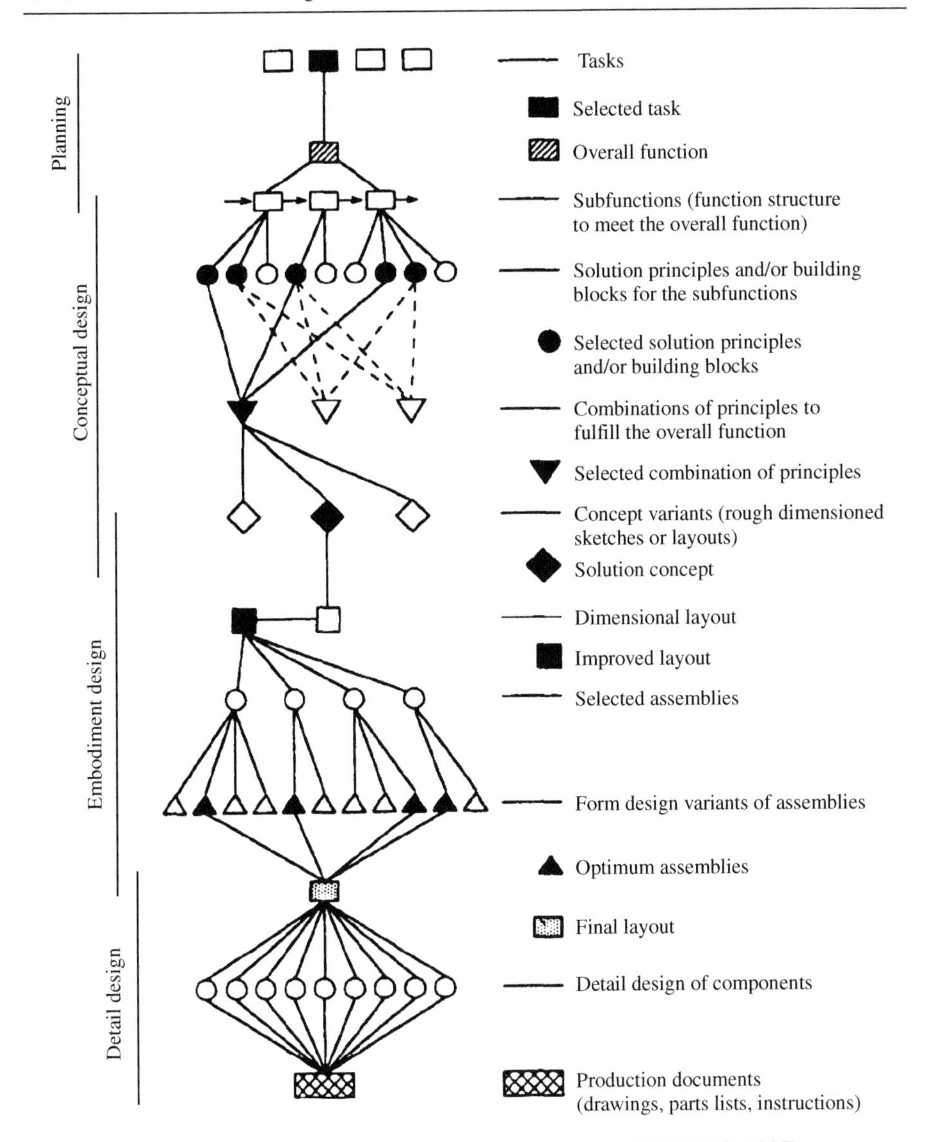

Figure 3.16 Divergence and convergence in the design process (VDI 2222, 1987).

Since the start of the problem, and therefore the design solution, generally is vague, the first step is to state the general objective and gradually clarify it. It is possible that, in the process, the initial objective may change or be altered significantly. Such a change or alteration reflects a better understanding of the problem and, eventually, a more fitting design solution.

As the objective changes, from a broad goal to specific goals, the means to accomplish the result may change as well. With each stage of the change, the objective should be reiterated in clear and precise language. As the objective becomes more

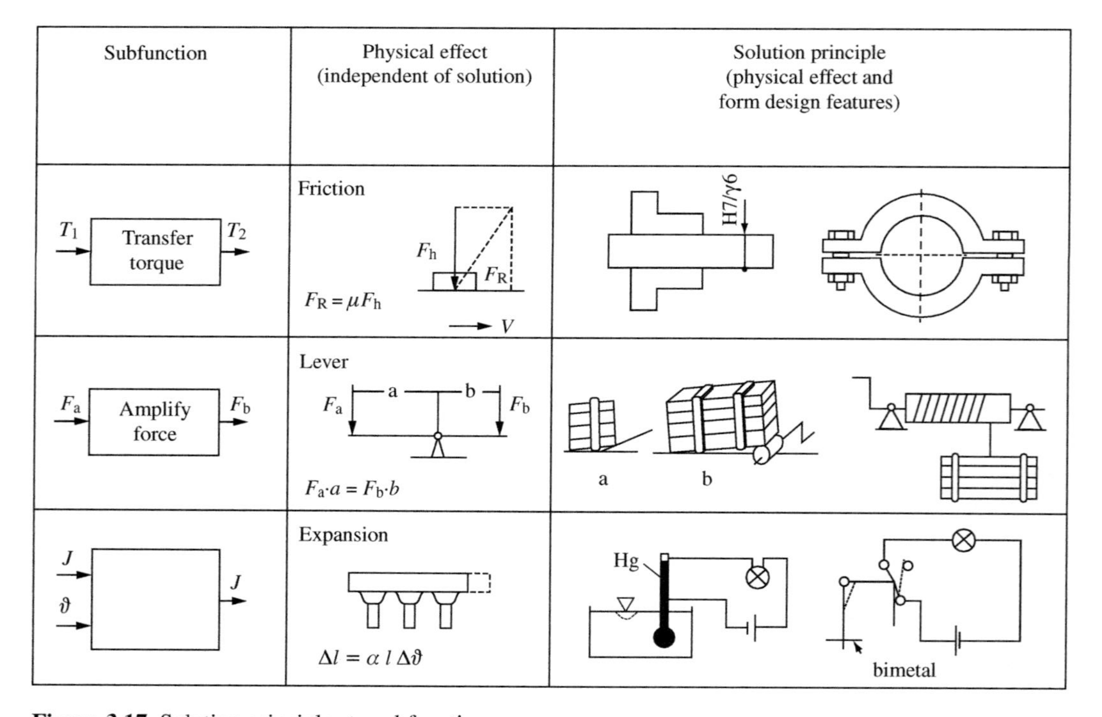

Figure 3.17 Solution principles to subfunctions.
Source: Examples from Pahl and Beitz (1984).

specific or is broken down into subobjectives, the criteria for evaluating design solutions emerge. These are classified as design specifications.

The objective tree method is a convenient format for developing and clarifying objectives. Beginning with a general statement or broad objective, the intent is to prepare a sequence of objectives and subobjectives. Some of these could be scaling (e.g., price has to be as low as possible) and some nonscaling (e.g., the price must be under $100); some could be in the form of requirements (e.g., the operating temperature must not exceed 210°F); and some could be standards (requirements imposed by an external authority, such as the local, state, or federal government or industry agreements). The objectives are arranged in a tree format showing hierarchical relationships and interconnections.

Figure 3.18 shows an objectives tree for a comfortable, safe, and attractive child car seat. The figure represents a top-down approach, starting from a higher level objective and progressing to lower level objectives. There is no clear end point in the top-down approach, but achieving a lower level objective is a means to achieving the higher level objective. When the lower level objectives are used as a means to achieving higher level objectives (an end) and the tree is redrawn from a bottom-up approach, the result is a means–end chain or cause–effect chain. This method helps clarify objectives. Figures 3.19 and 3.20 show top-down and bottom-up trees for the safety of automobile travel (Keeney, 1992).

A checklist approach may be used to develop a comprehensive list of objectives. Pugh (1990) provided a list of 24 factors that may be used in a checklist format to come up with a comprehensive list of subobjectives. These factors were summarized by Roozenburg and Eekels (1995):

1. **Performance**: Which function(s) does the product have to fulfill? By what parameters will the functional characteristics be assessed? Speed? Power? Strength? Accuracy? Capacity? Noise?
2. **Environment**: To which environmental influences is the product subjected during manufacturing, storing, transportation, and use? Temperature? Vibration? Humidity? Which effects of the product on the environment should be avoided?
3. **Life in service**: How intensively will the product be used? How long does it have to last?
4. **Maintenance**: Is maintenance necessary and available? Which parts have to be accessible?
5. **Target product cost**: How much may the product cost, considering the price of similar products?
6. **Transportation**: What are the requirements of transport during production, and to location of use?
7. **Packaging**: Is packaging required? Against which influences should the packaging protect the products?
8. **Quantity**: What is the size of the run? Is it batch or continuous production?
9. **Manufacturing facilities**: Should the product be designed for existing facilities? Are investments in new production facilities possible? Will the production or a part of it be contracted out?
10. **Size and weight**: Do production, transport, or use put limits as to the maximum dimensions? Weight?
11. **Esthetics, appearance, and finish**: What are the preferences of the consumers? Should the product fit in with a product line or house style?

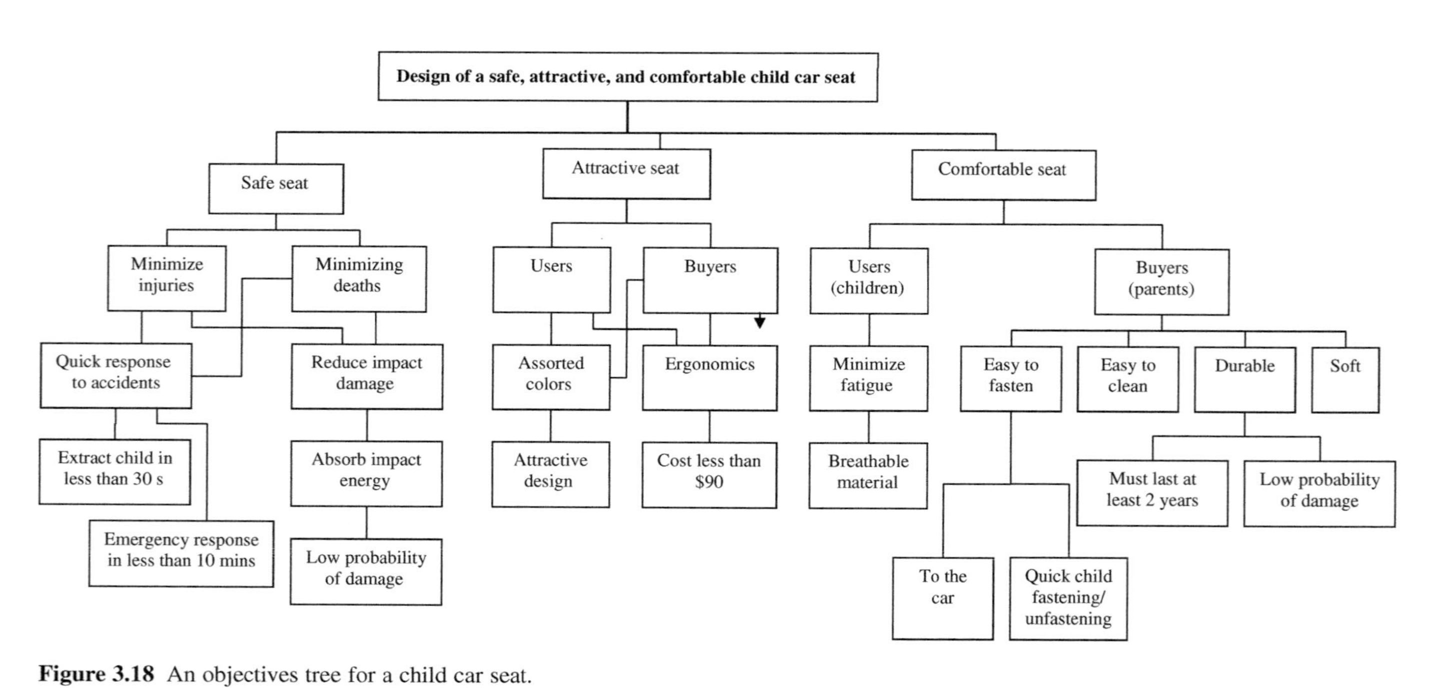

Figure 3.18 An objectives tree for a child car seat.

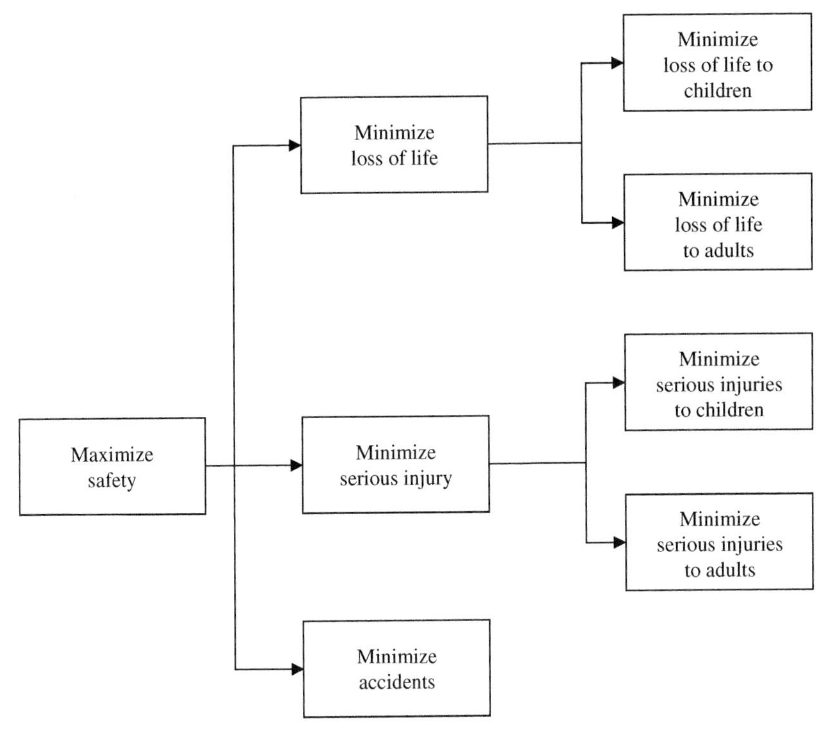

Figure 3.19 An objectives tree for safe automobile travel, top-down approach.
Source: Adapted from Keeney (1992).

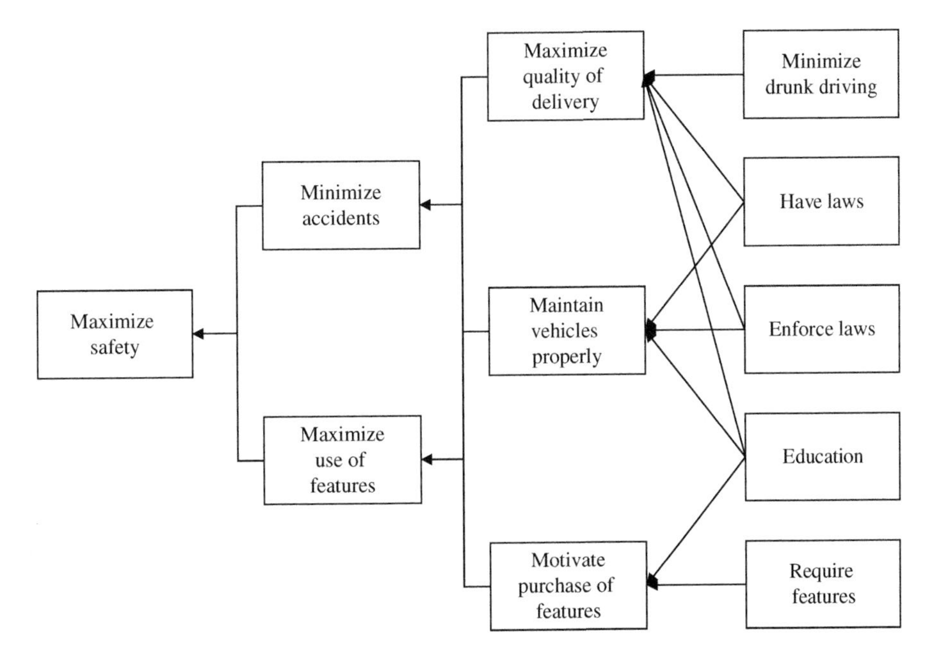

Figure 3.20 A means–end tree for safe automobile travel, bottom-up approach.
Source: Adapted from Keeney (1992).

12. **Materials**: Are special materials necessary? Are certain materials not to be used?
13. **Product life span**: How long is the product expected to be produced and marketable?
14. **Standards**: Which standards apply to the product and its production? Should standardization within the company be taken into account?
15. **Ergonomics**: Which requirements (perception, use, handling, etc.) does the product have to meet?
16. **Quality and reliability**: How large may mean times between failure and mean times to repair be? Which failure modes and resulting effects on functioning should not occur?
17. **Shelf life and storage**: During production, distribution, and use, are there periods of time in which product is stored? Does it require specific measures?
18. **Testing**: To which functional and quality tests is the product submitted within and outside the company?
19. **Safety**: Should any special facilities be provided for the safety of the users? Nonusers?
20. **Product policy**: Does the current and future product range impose requirements on the product?
21. **Social and political implications**: What is the public opinion with regard to the product?
22. **Product liability**: For which unintended consequences of production, operation, and use can the manufacturer be held responsible?
23. **Installation and operation**: Which requirements are set by final assembly and installation outside the factory and by learning to use and operate the product?
24. **Reuse, recycling, and disposal**: Is it possible to prolong the material cycle by reuse of materials? Parts? Can the materials and parts be separated for waste disposal?

The list of objectives should be analyzed to remove similar objectives and objectives that are not biased. Objectives should be further stated in terms of performance and their hierarchical relationships should be verified. Requirements and standards should be checked for acceptable values and a means–end relationship should be drawn.

3.5.4 Establishing functions, setting requirements, and developing specifications

Once the objectives list has been reexamined, consolidated, and edited, a function analysis must be performed to determine what the product should achieve (not how). For this purpose, the product can be treated as a black box that converts specified inputs into outputs. The overall function should be as broad as possible, and this is what the product design accomplishes: conversion of inputs into outputs. The designer should ask questions about the inputs and outputs (e.g., What are the inputs? Where do they come from? What are the outputs for? Stages of conversion? Sequence of conversion?) to develop a complete list of inputs and outputs. These inputs and outputs can be flows of materials, energy, and information. For instance, a coffee bean grinder takes beans (matter), energy (electrical), and information (signal) as inputs and grinds beans. The ground beans (matter), heat (energy), and information (signal) are the outputs. In this case, the inputs and outputs are specified but not how the beans are ground.

Next, the function in the black box is replaced with a block diagram showing all subfunctions and their links with each other and with inputs and outputs. There can be no loose inputs and outputs or subfunctions. Eventually, the designer must look for proper components for accomplishing each subfunction. The component can be a

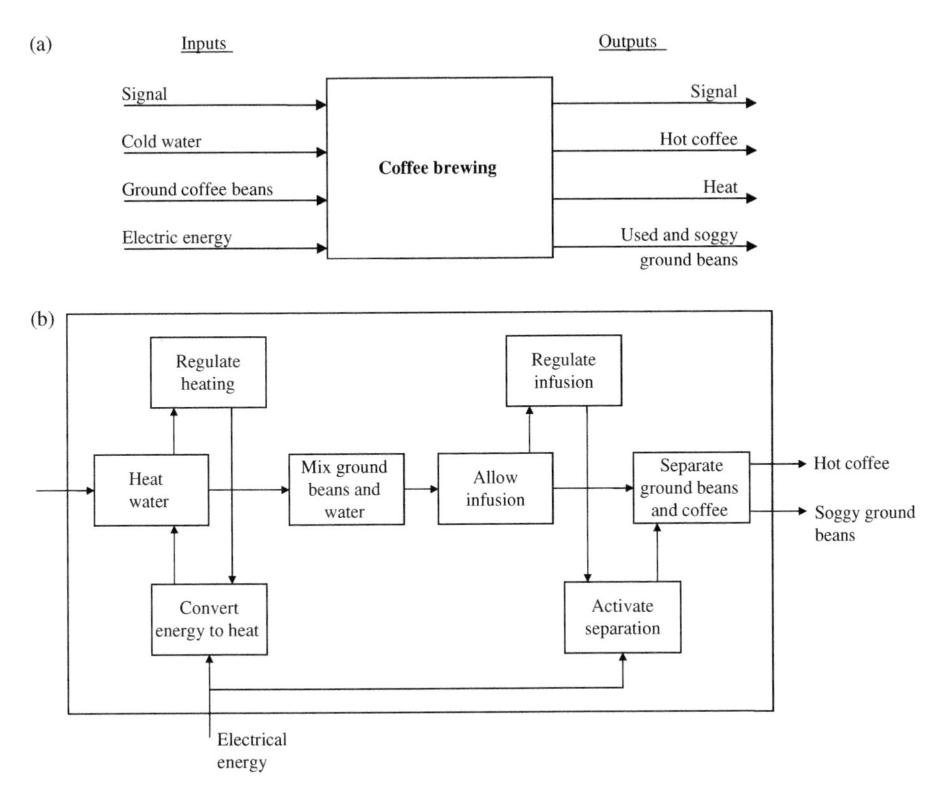

Figure 3.21 (A) A black box model for coffee brewing; (B) function analysis for an automatic coffeemaker.
Source: Adapted from Hubka et al. (1988).

person or a mechanical or an electrical device. Figure 3.21, modified from Hubka et al. (1988), shows the black box model and function analysis for an automatic coffeemaker.

The function analysis specifies what a design must achieve. However, no limits are set. Setting precise limits on the functions is called *setting performance specifications*. This means that we must set limits on the range of acceptable solutions. Performance specifications should be neither too narrow (otherwise, normally acceptable solutions are eliminated) nor too broad (which may lead to inappropriate solutions). The design objectives and functions are the source of performance specifications. The list of performance attributes contains all conditions the design must meet. These include requirements (functional as well as nonfunctional) and standards. Additionally, if one wishes, performance specifications for desired attributes may be included as well. For each attribute, the designer must determine what the product must achieve or do as far as performance is considered. It is most desirable to express this quantitatively, if at all possible. For instance, the toothbrush cleaning cycle must be 120 s long or the product handle diameter should be 1.5 in.; vague or qualitative specifications, such as "approximately 10 pounds" or "as low as possible" should be avoided. In general, performance specifications should be independent of any solution. If a range would satisfy

Table 3.2 Design specifications for a small city car

Characteristics	Required	Desired
1. General Characteristics		
Car for city	X	
Number of seats: 2(4)	X	
Number of wheels: 4	X	
Space utilization: Maximum		X
Range: >100 km		X
Economical	X	
2. Working conditions: City area	X	
3. Dimensions		
2.5 m length	X	
1.5 m width		X
1.6 m height		X
4. Weight		
Maximum net: 400 kg		X
Loading capacity: 200–300 kg	X	
Gross weight: 600–700 kg		X
5. Luggage capacity		
Minimum volume: 150 dm^3	X	
With dropped back: 359 dm^3		X
6. Speed		
Maximum: 70 km/h	X	
7. Motor type		
8. Safety: As high as possible: active and passive	X	
9. Pollution		
Meet standards	X	
Zero		X
10. Form and esthetics		
Pleasing	X	
Convertible		X
11. Production: 500 units/year		X
12. Price: $2000 to $2500		X

Source: Adapted from Pighini et al. (1983).

performance specifications, it should be specified instead of a fixed number. Table 3.2 provides design specifications for a small city car developed by Pighini et al. (1983).

One way to generate product specifications is to use the method based on the principles of quality function deployment (QFD). The method requires using the "house of quality" interaction matrix with product attributes as rows and technical parameters (also known as *engineering characteristics*), such as weight, volume, material, and force required, as columns. The technical parameters have to be related to the basic engineering characteristics. For instance, torque is determined by the gear ratio of the transmission train and the motor power. These, in turn, are affected by variables such as voltage and resistance. It is these basic variables that have to be related to product attributes to determine the specifications. Each attribute is assigned a weight, or relative importance, factor. Next, the relationship between the product attribute and technical parameters is determined using notations for strong positive, medium positive, neutral, medium negative, and strong negative, and a value is assigned to each of these relationships. Multiplying the weight and the relationship score produces a value for the strength of the product attribute–technical parameter relationship. The sum for each column (parameter) indicates the priority of that technical parameter in the product design. At the bottom of the matrix, the measurement units of technical parameters are listed. A target value for each parameter is set on the basis of its importance and how the values of the attribute for competitors' products would compare against the target.

3.5.5 Developing provisional designs

The methods for developing design solutions to problems are called *design methods*. Any procedures, techniques, aids, or tools for designing are a part of the design methods. While some methods are informal, many are formal and rely on decision theory and techniques used in management science. Jones (1981) outlined as many as 35 design methods. In general, the design methods are classified under the following categories: association methods, creative confrontation methods, and analytic methods.

3.5.5.1 Brainstorming

The most widely known creative association method is brainstorming. This method, invented by Osborn (1963), is very effective in generating a large number of ideas. Even though most of the ideas originally generated are rejected, some ideas will be worth pursuing. The method requires a small group of people (four to eight) of diverse backgrounds to participate in the solution-generating activity. According to Osborn, the following four rules must be followed:

1. There should be no criticism of the ideas generated. The participants should not think about the utility or worth of the idea and certainly should not criticize them. Such a criticism stifles the idea generation process.
2. Any idea is welcome; no matter how outlandish. Such divergent thinking is productive. The participants must feel secure in proposing any wild ideas they think relevant.
3. Adding to other ideas to improve on them is welcome.
4. The goal is to generate as many ideas as possible.

The procedure requires that the problem statement be provided to the selected group of participants in advance of the brainstorming session. Background information, some examples of solutions, and information regarding the participation rules also should be provided. A preliminary meeting explaining the rules and purpose may be held.

During the actual session, participants provide their ideas through the moderator. Nonparticipants also can provide their ideas in writing. Once the list of ideas is complete, participants discard silly ideas and focus on those that seem relevant. The relevant ideas also may be elaborated by others later.

Some variations of the original brainstorming method are also practiced, for instance, the brainwriting-pool method. In this variant, five to eight people write their ideas on a sheet of paper in silence, once the problem has been explained. The sheet of ideas is placed in a pool. Participants who already put their sheet in the pool pull a sheet from the pool and add ideas to it. After about 30 min, the ideas on the sheets are evaluated as in the original method.

Another variation of the original method is the 6–3–5 method, in which six participants write three ideas on a sheet. The sheets are passed five times among the participants, during which ideas are added. As many as 108 ideas can be generated in this manner.

3.5.5.2 Analogies and chance

Many innovative ideas can be attributed to chance. For instance, vulcanized rubber was invented by Charles Goodyear when he accidentally added sulfur to rubber when working in his wife's kitchen. Similarly, John Boyd Dunlop invented the pneumatic tire after observing the behavior of the garden hose when watering his garden.

Cyanoacrylate, also known as superglue or Krazy Glue, developed by Harry Coover at Eastman Kodak during World War II, is another example of a chance discovery. Coover was searching for a way to make plastic gunsight lenses when he discovered cyanoacrylate, which did not solve the problem since it stuck to all apparatuses used to handle it. It was patented in 1956 and developed into Eastman 910 adhesive in 1958. The new glue was demonstrated in 1959 on the television show *I've Got a Secret*, when the host Garry Moore was lifted into the air by two steel plates held together with a drop of Eastman 910. Cyanoacrylates are now a family of adhesives based on similar chemistry.

The Post-it note was invented in 1968 by Dr. Spencer Silver, a 3M scientist who stumbled on a glue that was not sticky enough. In 1974, a colleague, Arthur Fry, who sang in a church choir, was frustrated that the bookmarks kept falling out of his hymnal, so he applied some of Silver's glue to his markers. 3M launched the product in 1977 but it failed, as consumers had not tried the product. A year later, 3M swamped Boise, Idaho, with samples. Of people who tried them, 90% said that they would buy the product. By 1980, the product was sold nationwide and a year later it was launched in Canada and Europe.

The examples cited here demonstrate the role of analogy and chance in solving problems. Bright people often can come up with a design solution when two situations are accidentally confronted. Seemingly, the two situations have nothing to do with each other or have some indirect relationship. Gordon (1961), and later Prince (1970), formalized the mechanics of the process which now is known as *synectics*.

Synectics is different from brainstorming: instead of generating ideas, the group tries to work collectively toward a particular solution. The session is much longer and much more demanding. The group is pushed to use particular types of analogies by making the strange familiar and the familiar strange. The types of analogies encouraged are

- **Direct analogies**: A biological solution to a similar problem is sought. As an example, Velcro was designed using an analogy with plant burrs.
- **Personal analogies**: The team members put themselves in the problem being solved; for example, How would I feel if I were a transmission?
- **Symbolic analogies**: The team members use metaphors and similes, such as the "jaw" of a clamp.
- **Fantasy analogies**: The team members dream of an ideal solution to the problem; for example, a child's fantasy, such as a door opening by itself when the owner reaches it.

The synectics method involves the following steps:

1. State the problem (e.g., how to fly and stay in place)
2. Use a direct analogy (e.g., hummingbirds stay flying in one place)
3. Analyze the analogy (e.g., flap wings rapidly, 15–80 times per second; how to achieve this?)
4. Force a fit (e.g., use multiple horizontal rotors)
5. Generate ideas
6. Develop the ideas.

In addition to synectics, other methods in this category include random stimulus, intermediate impossible, and concept challenge. In the random stimulus method, a word, object, or image, chosen at random, is linked to the original problem. In the intermediate impossible method, an ideal solution to the problem is thought up, and from it, gradually, a practical solution is developed. The concept challenge method undermines the problem statement in order to envision a new solution.

3.5.5.3 Analytic methods

Analytic methods are also called *systematic methods*. The method described in this section is a combination of function analysis (discussed in Section 3.5.4) and the morphological chart method. A morphological chart is a summary of subsolutions to subfunctions.

As mentioned in Section 3.5.4, the main function in the black box connecting inputs to outputs is replaced with several subfunctions, showing interrelationships among them as well as the inputs and outputs (Figure 3.21B); subfunctions can be added step by step. Next, the function structure is elaborated by changing, splitting, or combining subfunctions until the best function structure results. In the third step, the subfunctions are replaced with general function symbols. Figure 3.22 shows the symbols recommended by Pahl and Beitz (1984). Roth (1970) also provides general function symbols, shown in Figure 3.23.

Figure 3.24, from Pahl and Beitz (1984), shows how the general function (harvesting potatoes) in the black box is replaced with subfunctions and these subfunctions with the general function symbols shown in Figure 3.22. Figure 3.25, from Roth (1970),

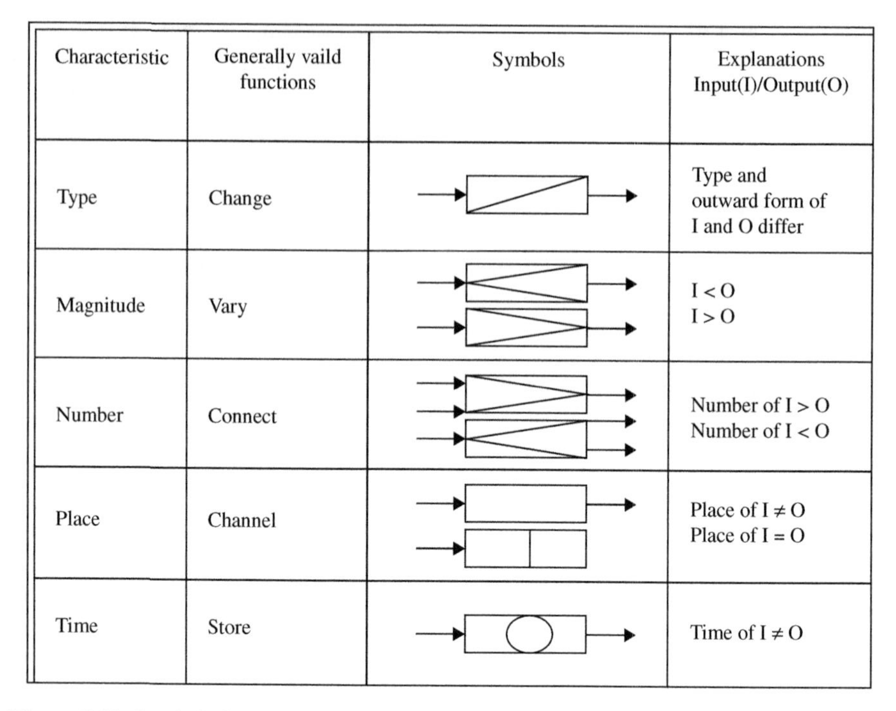

Characteristic	Generally vaild functions	Symbols	Explanations Input(I)/Output(O)
Type	Change		Type and outward form of I and O differ
Magnitude	Vary		I < O I > O
Number	Connect		Number of I > O Number of I < O
Place	Channel		Place of I ≠ O Place of I = O
Time	Store		Time of I ≠ O

Figure 3.22 Symbols for general functions.
Source: Adapted from Pahl and Beitz (1984).

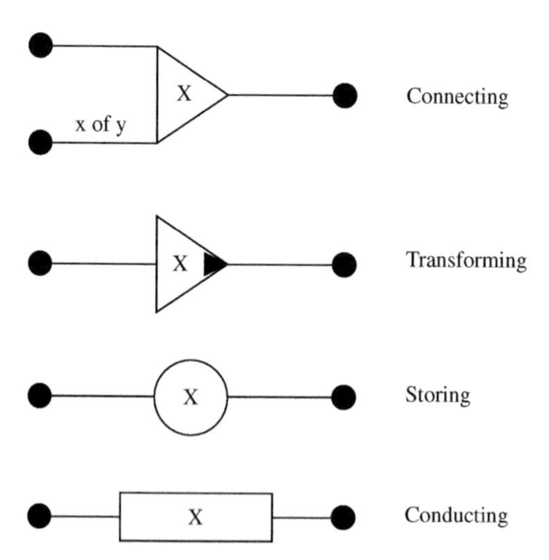

Figure 3.23 Symbols for general functions.
Source: Adapted from Roth (1970).

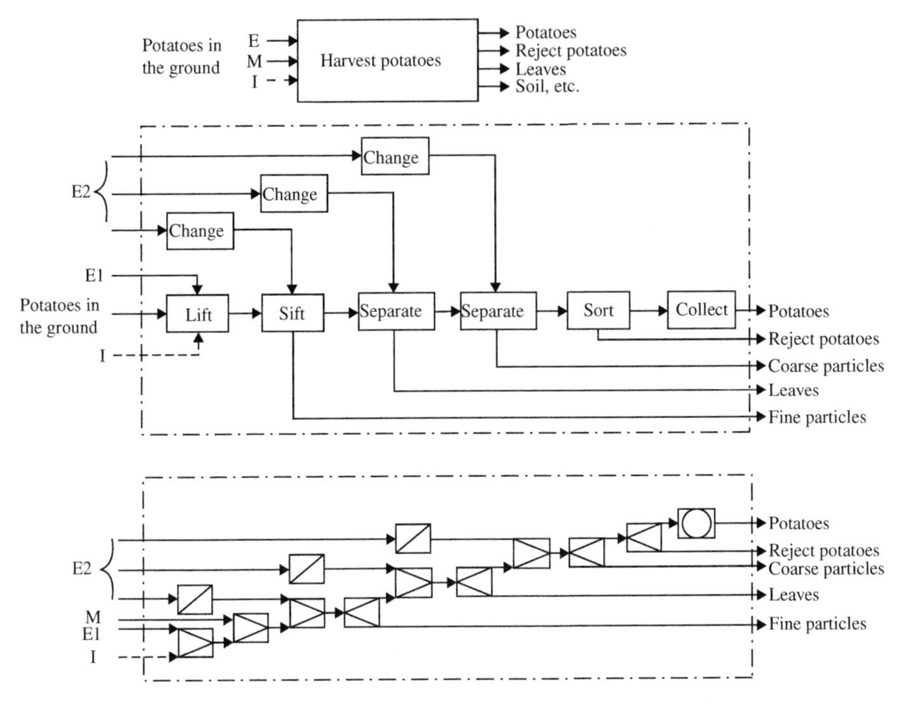

Figure 3.24 Function analysis and structure for a potato harvesting machine.
Source: Adapted from Pahl and Beitz (1984).

using his symbols provided in Figure 3.23, shows how subfunctions can be added step by step.

Once the function analysis is complete, the task at hand is to develop a matrix of subfunctions as rows and possible solutions as columns. The best means to achieve each subfunction is identified. The selected solution box is highlighted. The combination of chosen solutions to subfunctions should yield the design solution to the problem. Figure 3.26 shows the morphological chart for potato harvesting from Pahl and Beitz (1984).

3.5.6 Evaluation and decision making

Once a number of alternative designs have been developed, the designer must choose the best one. Consideration of the objectives is essential in making the final choice. Since there usually are multiple objectives, each with a different value, it is necessary to weight the objectives so that all designs may be compared objectively and across a wide range of objectives. Generally, a matrix of objectives with associated weights and the various evaluation criteria, such as cost, performance, and comfort, is prepared. Each design is evaluated against these criteria using a scale (a 5-point, 7-point, 10-point, or other). The scale graduations may be descriptive, such as excellent

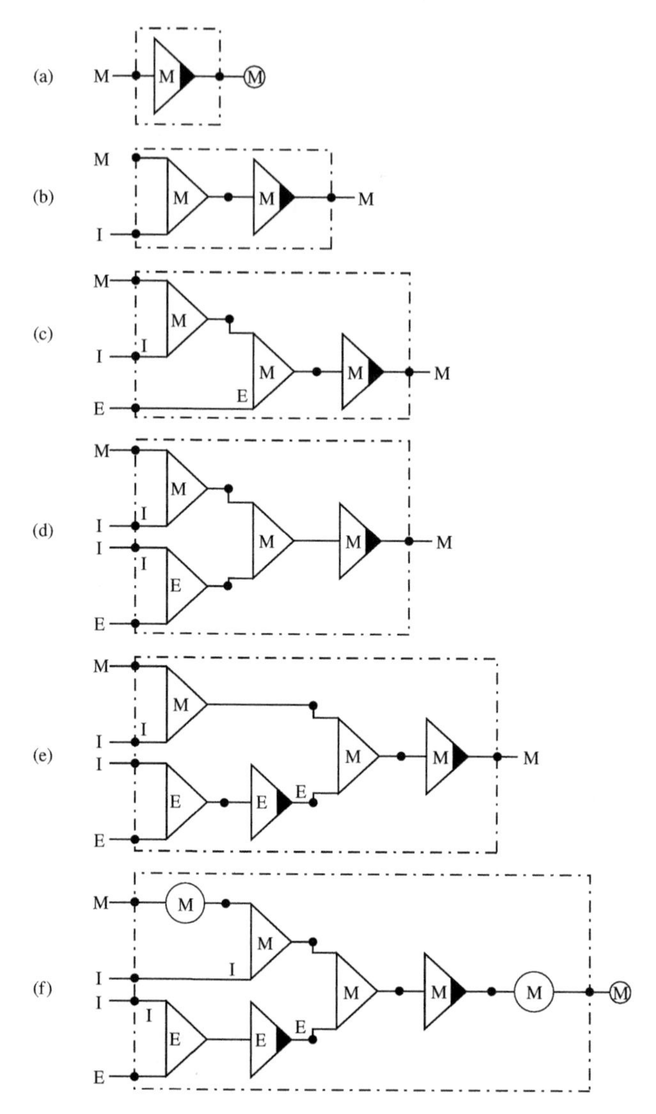

Figure 3.25 Step-by-step addition of subfunctions for an electric coffee mill.
Source: Adapted from Roth, 1970.

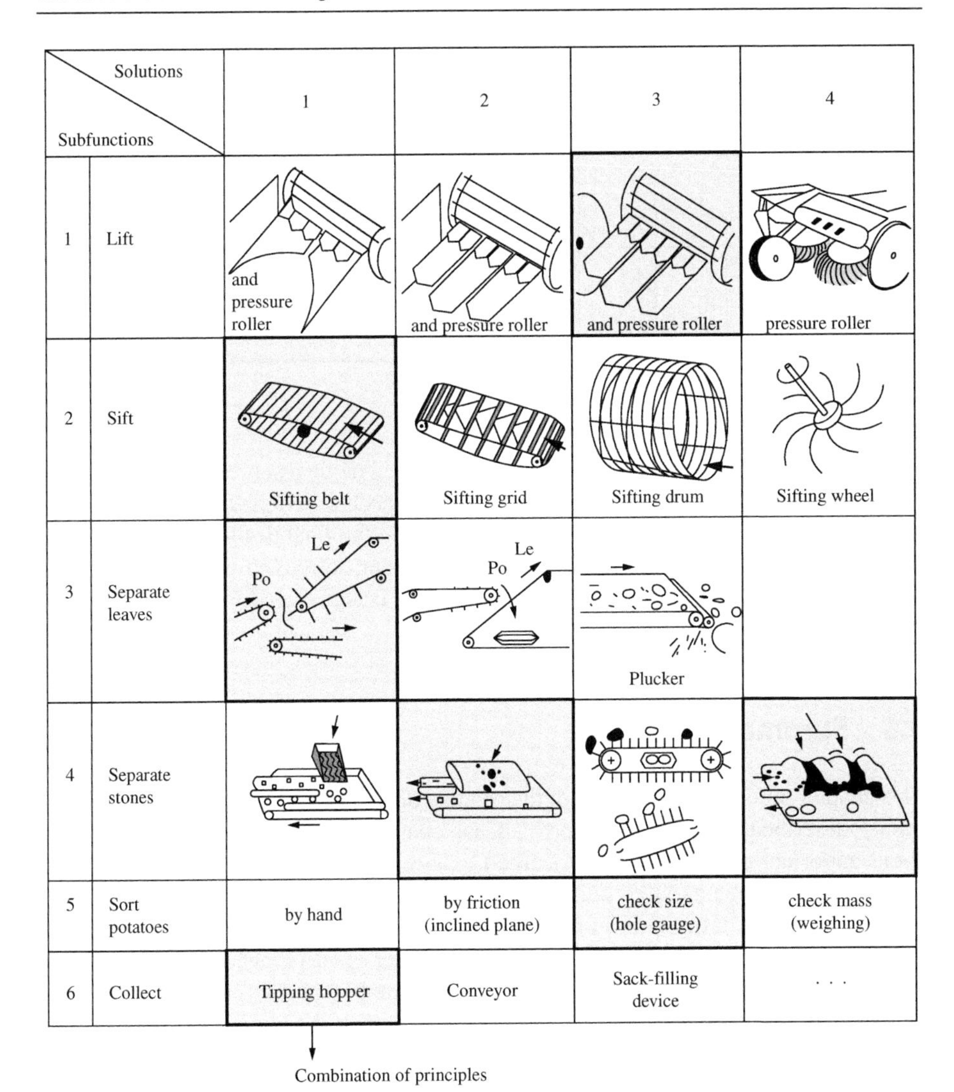

Solutions
Subfunctions
1
2
3
4
1
Lift
and pressure roller
and pressure roller
and pressure roller
pressure roller
2
Sift
Sifting belt
Sifting grid
Sifting drum
Sifting wheel
3
Separate leaves
Le
Po
Le
Po
Plucker
4
Separate stones
5
Sort potatoes
by hand
by friction (inclined plane)
check size (hole gauge)
check mass (weighing)
6
Collect
Tipping hopper
Conveyor
Sack-filling device
. . .
Combination of principles

Figure 3.26 Morphological chart for a potato harvesting machine.
Source: Adapted from Pahl and Beitz (1984).

Table 3.3 **A simplified design decision matrix**

Design	Design criteria									Total
	Cost			Comfort			Reliability			
	W	*S*	*T*	*W*	*S*	*T*	*W*	*S*	*T*	
1	0.5	9	4.5	0.3	7	2.1	0.2	8	1.6	8.2
2	0.5	8	4.0	0.3	8	2.4	0.2	8	1.6	8.0
3	0.5	9	4.5	0.3	9	2.7	0.2	9	1.8	9.0

W, weight; *S*, score; *T*, $W \times S$.

solution, poor comfort, and very economical design. The final score for each design is the sum of the scores for each design weighted for each criterion. The design scoring the highest is the winner. This, however, may not be the final design, as the designer may incorporate features that rated higher from other designs to modify the chosen design. Table 3.3 shows an example of automobile designs rated for various criteria. In this case, Design 3 is selected.

3.6 Summary

This chapter outlined the structure of the product design process. It defined *design* and briefly described how the design process has changed over time. We also discussed some important design paradigms learned as a result of design failures. The basics of design requirements were outlined, and finally, we provided a systematic, step-by-step analytical design procedure. We realize that we have condensed a significant amount of information in this chapter, and the reader is encouraged to consult books devoted completely to the process of design. A number of references are provided next. Finally, we would like to the leave the reader with the impression that the end of this chapter is not the end of the product design process. Once the basic design is developed, it must be refined from the standpoint of quality, materials, processes, assembly and disassembly, maintenance, functionality, and usability. This refinement is considered in detail in Chapters 4–11.

References

Alexander, C., 1969. Notes on the Synthesis of Form. Harvard University Press, Cambridge, MA.

Caldecote, V., 1989. Investment in new product development. In: Roy, R., Wield, D. (Eds.), Product Design and Technical Innovation Open University Press, Milton Keynes, UK.

Engineering News-Record, January 21, 1982. A bridge to remember. McGraw-Hill, Inc. Section: Editorials, p. 190.

French, M.J., 1971. Engineering Design: The Conceptual Stage. Heinemann, London.

Gordon, W.J.J., 1961. Synectics: The Development of Creative Capacity. Harper and Row, New York, NY.

Hall, A.D., 1968. A Methodology for Systems Engineering, sixth ed. Van Nostrand, Princeton, NJ.

Hubka, V., Eder, W.E., 1988. Theory of Technical Systems: A Total Concept Theory for Engineering Design. Springer, Berlin.

Hubka, V., Andreasen, M.M., Eder., W.E., 1988. Practical Studies in Systematic Designs. Butterworth, London.

Jones, J.C., 1970. Design Methods: Seeds of Human Futures. John Wiley, New York, NY.

Jones, J.C., 1981. Design Methods. John Wiley, Chichester, UK.

Keeney, R.L., 1992. Value-Focused Thinking: A Path to Creative Decision-Making. Harvard University Press, Cambridge, MA.

Osborn, A.F., 1963. Applied Imagination: Principles and Procedures of Creative Problem Solving, third ed. Scribner, New York, NY.

Pahl, G., Beitz, W., 1984. Engineering Design. Design Council, London.

Petroski, H., 1994. Design Paradigms: Case Histories of Error and Judgement in Engineering. Cambridge University Press, New York, NY.

Pighini, U., Francesco, G., Zhang, Y.D., Vicente, S.A., Rivalta., A., 1983. The determination of optimal dimensions for a city car using methodical design with prior technology analysis. Design Studies 4 (4), 233–243.

Prince, G.M., 1970. The Practice of Creativity: A Manual for Dynamic Group Problem Solving. Harper and Row, New York, NY.

Pugh, S., 1990. Total Design Integrated Methods for Successful Product Engineering. Addison Wesley, Wokingham, UK.

Pye, D., 1989. The nature of design. In: Roy, R., Wield, D. (Eds.), Product Design and Technical Innovation Open University Press, Milton Keynes, UK.

Roozenburg, N.F.M., Eekels, J., 1995. Product Design: Fundamentals and Methods. John Wiley, New York, NY.

Roth, K., 1970. Systematik der Machine und ihre Elemente. Feinwerktechnik 11, 279–286.

Schon, D.A., 1969. Design in the light of the year 2000. Student Technologist. Quoted by Cross, N., 1989. The changing design process. In: Roy, R., Wield, D. (Eds.), Product Design and Technical Innovation. Open University Press, Milton Keynes, UK.

VDI Standards 2221 and 2222, 1987. Systematic Approach to the Design of Technical Systems and Products. Verein Deutscher Ingenieure, Dusseldorf, Germany.

[illegible] Design: The Conceptual Stage. Heinemann [illegible]
[illegible] the Development of Creative Capacity [illegible]

[illegible] Engineering [illegible]

[illegible] Technical Systems [illegible]
[illegible]

[illegible] Practical Studies [illegible]
[illegible]

[illegible] Models of Human Purpose. John Wiley, New York, NY.
[illegible] Design Methods. John Wiley, [illegible]

[illegible] Harvard
[illegible]

[illegible]

Part Two

Design Review: Designing to Ensure Quality

4

4.1 Introduction

Creativity in product design and process selection is the critical component in ensuring quality in the product and its associated processes. In previous chapters, we presented details of the design process, strategies, and tools. Now we focus on the need to integrate quality into the product design and the design review (D–R) process.

Product quality constitutes a crucial component in the product design process, directly affecting consumer loyalty and company profitability. Historically, manufacturing enterprises relied on the reactive approach of inspecting the quality of a product to ensure it conforms to design specifications. While this approach has its advantages, its principal limitation lies in the manufacturers' implicit resignation to the fact that quality needs to be inspected in since it cannot be built into the product design at the design stage. However, there has been a gradual, yet definite, transition from a reactive to a proactive strategy to managing quality by incorporating design techniques that do away with the largely unproductive inspection process. Several leading manufacturing enterprises have been successful in entirely eliminating the need to inspect by adopting a proactive approach to product design.

Quality is defined as a physical or nonphysical characteristic that constitutes the basic nature of a thing or is one of its distinguishing features. This broad definition of *quality* can be extended to the engineering and industrial realm by defining *quality* as a characteristic or group of characteristics that distinguish one article from another or the goods of one manufacturer from those of its competitor (Radford, 1992). The two different aspects of this characteristic can be classified into objective and subjective categories. The objective nature of quality is independent of the existence of mankind, whereas the subjective nature of this characteristic is expressed in terms of the general manner in which humans perceive it. Figure 4.1 shows the nine dimensions of quality and their meanings. These dimensions are relatively independent of each other; hence, a product can be excellent in one dimension, average in another, and poor in still another dimension. Therefore, quality products can be determined by using a few of the dimensions of quality.

When the expression *quality* is used, we usually think in terms of an excellent product that fulfills or exceeds our expectations. The International Standards Organization (ISO) extends this definition to include the service industry as well. The ISO defines *quality* as "the totality of features and characteristics of a product or service that bear on its ability to satisfy stated or implied needs" (ISO Standard 9000).

Product Development. DOI: http://dx.doi.org/10.1016/B978-0-12-799945-6.00004-1

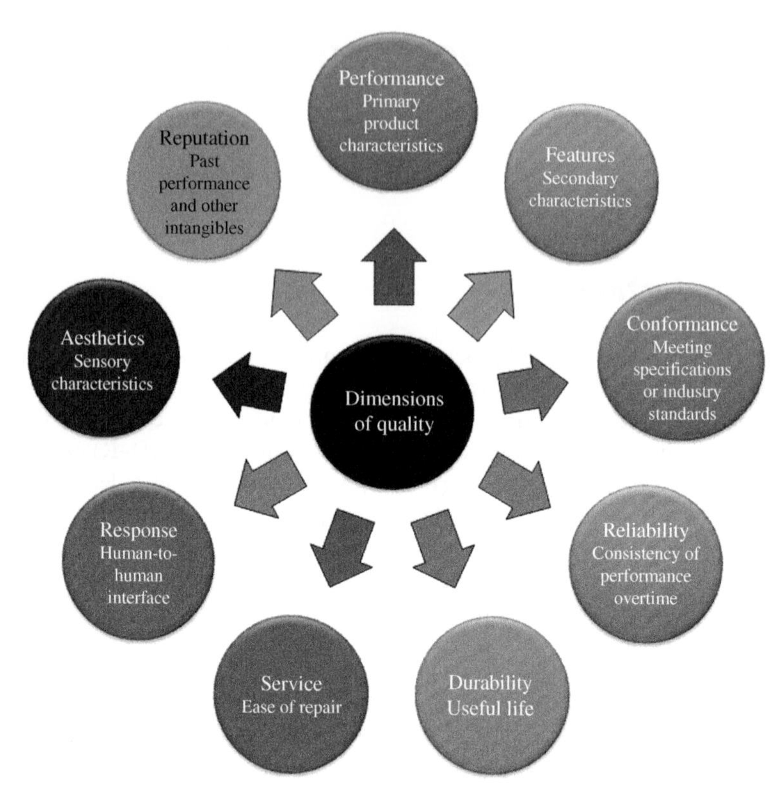

Figure 4.1 Quality dimensions.
Source: Adapted from Garvin (1988).

4.1.1 Why quality control?

The American Society for Quality Control (ASQC, 1991) commissioned a Gallup poll of over 3000 consumers in the United States, Japan, and Germany in the 1990s. The survey pointed to significant commonality in customer thinking regarding quality. Performance, price, and reputation were three of the most valuable attributes customers desired when selecting one company's product over another. The importance of customer satisfaction and loyalty to a company's profitability and ability to compete in a global marketplace can hardly be overemphasized. In formulating manufacturing strategies, quality is the most important factor in determining market success (Hill, 1989). "The quality of our product is excellent, but the price is too high" does not make a successful product marketing strategy. There must be a balance between quality loss and product price. The price represents the loss to the customer at the time of purchase, and poor quality represents an additional loss during the use of the product. A goal of quality control is to minimize such losses to the customer.

Given the obviously undeniable importance of incorporating quality into product or service design, it is necessary to draw a distinction between quality assurance and quality control. Quality assurance consists of all planned and systematic actions

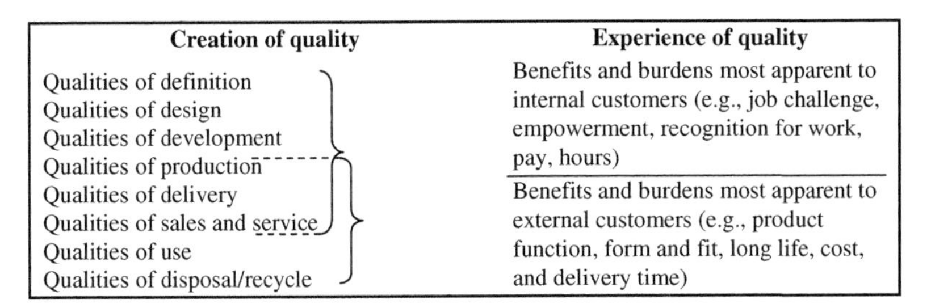

Creation of quality
Qualities of definition
Qualities of design
Qualities of development
Qualities of production
Qualities of delivery
Qualities of sales and service
Qualities of use
Qualities of disposal/recycle
Experience of quality
Benefits and burdens most apparent to internal customers (e.g., job challenge, empowerment, recognition for work, pay, hours)
Benefits and burdens most apparent to external customers (e.g., product function, form and fit, long life, cost, and delivery time)

Figure 4.2 Creation and experience of quality.
Source: Adapted from Kolarik (1999).

that are necessary to provide adequate confidence that a product or service will satisfy given requirements for quality (ISO Standard 9004). Quality control, on the other hand, is a system whereby quality is assured economically (Japan Industrial Standards, JISZ Standard 8101). Generally speaking, quality is created through processes that the manufacturer develops and maintains (Kolarik, 1999). Each of these processes plays an integral role in satisfying the eight fundamental requirements for product or service conception, development, delivery, and disposal. Figure 4.2 depicts the creation of quality as just described by the sequential achievement of eight fundamental processes. Bear in mind that this sequence of activities has to be approached systematically if customers are to have a positive quality experience.

4.1.2 Reactive versus proactive quality control

Quality control strategies can be classified into two distinct categories: reactive and proactive. The majority of quality control strategies are aimed at detecting and correcting problems that already exist. In other words, the designer of a product, process, or service incorporates a system of checks and measures to isolate and catch defects as and when they occur. By their very nature, reactive quality control strategies are better suited to identify problems and resolve them and, as such, are clearly defensive in nature. Reactive strategies try to limit losses by incorporating the largely wasteful inspection process. The reactive strategy emphasizes traditional loss accounting and data-intensive statistical inferences to justify action. This course of action is comforting for a decision maker, since action can be readily justified based on historical data (Kolarik, 1999). Traditionally, most of the classical topics of quality assurance and statistical quality control currently being taught in universities are largely reactive in nature (Banks, 1989; Duncan, 1986; Grant and Leavenworth, 1988; Montgomery, 1991).

The proactive approach to quality control is based on the cause and effect relationship and, as such, does away almost completely with historical data and related statistical analyses. This approach is based on the premise that, if a manufacturer has a reasonably good understanding of the expectation of quality of its customers (relative to its competitors), then processes aimed at creating high quality can be structured in accordance with those requirements. This concept tends to reduce risk of business failure since all operations are geared toward customer satisfaction.

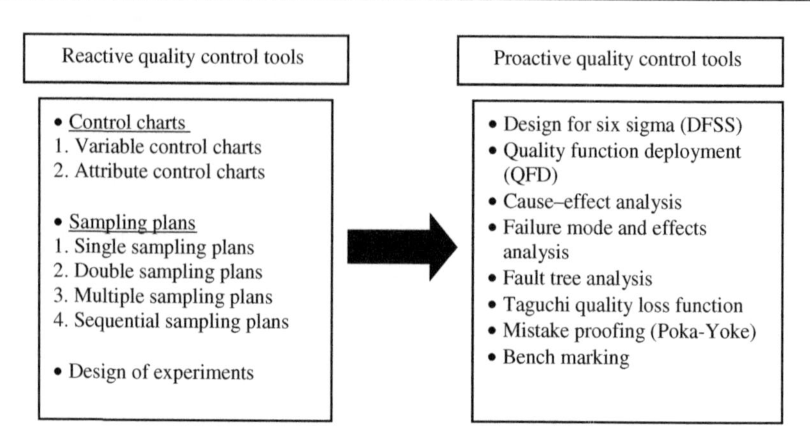

Figure 4.3 The transition from reactive to proactive quality assurance techniques.

The advantages of adopting a proactive approach to quality control, as described here, is twofold: accelerated product and process development cycles, and avoidance, not management, of losses. This leads to lower equipment downtime.

It is obvious that both advantages result in a keener competitive edge in an increasingly competitive global marketplace. Quality systems and quality transformation-related strategies have proven their importance by improving productivity, especially when appropriate leadership, training, and managerial structures are provided (Kolarik, 1999). Product and process planning strategies have myriad forms. For instance, quality function deployment (QFD) is a product planning tool used most commonly during the design stages of a new product. QFD has been observed to render extremely positive results in generating customer satisfaction and enhancing market share as well as profits (Akao, 1990; Juran, 1988). Some examples of the emergence of proactive strategies include robust design and early off-line experimental programs that support system, parameter, and tolerance design (Kolarik, 1999; Taguchi, 1986). Mistake proofing, commonly referred to as *poka-yoke* in industry, and source inspection are examples of yet more proactive strategies that stress the prevention and elimination of quality problems (Shingo, 1986).

From the preceding discussion, it is clear that proactive quality control strategies, when used strategically, enable the designer to design and build high quality into the product. This does away with time, effort, and resources wasted at the end of the production process. The proactive approach to quality control emphasizes the "do it right the first time" approach to manufacturing, quality, and ultimately customer satisfaction. Figure 4.3 is a graphical depiction of this transition.

4.2 Procedures for incorporating high quality in design stages

The reactive approach to quality control is commonly associated with suboptimal use of resources, a high amount of waste, and shrinking market share. This not only adds to the

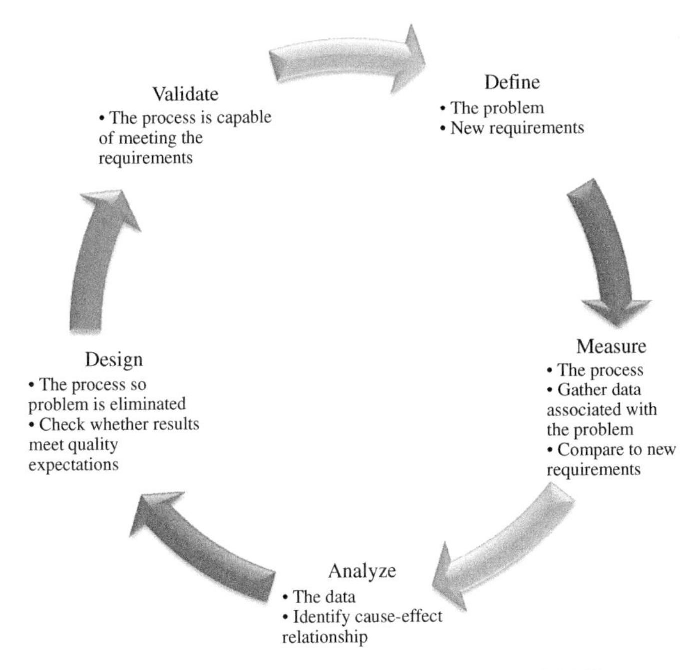

Figure 4.4 The DFSS approach to product and process design, DMADV.

product price but leads to an increase in the lead time from manufacturing to marketing the product, resulting in untimely release to the market and a loss of market share. To ensure high quality, it needs to be built into the product beginning at the design stage itself instead of inspecting for it at a later stage. Here we discuss some of the frequently used proactive quality control techniques used to solve problems in industry.

4.2.1 Design for six sigma

The term *six sigma* has several meanings. At the most encompassing level, a corporation can define it as a philosophy, a way of thinking. Technically, six sigma is a data driven approach to reducing the defects produced due to a variation in a product or process.

Design for six sigma (DFSS) can be described as a comprehensive approach to product development. It links business and consumer needs to critical product attributes to product functions to detailed designs to tests and verification. This is achieved by integrating a full suite of analytical methods from six sigma, including QFD, goal deployment, failure mode and effects analysis (FMEA), simulation, quality loss functions, and prototyping to create more reproducible, customer-driven designs faster.

Similar to six sigma, DFSS uses a structured set of steps, DMADV (define, measure, analyze, design, and validate; see Figure 4.4), to ensure repeatability and continuous improvement. DFSS focuses on translating customer requirements qualitatively and quantitatively to product specifications, then ensuring that proposed designs robustly meet customer-defined scorecards. The DMADV process is to fundamentally redesign a process. It also may be used to design a new process or product when requirements change. The goal of DFSS is to design a new process or product to replace the unsatisfactory existing process or product.

DFSS can be used to maximize the performance of products as well as services. DFSS can be utilized to enhance the speed and quality of the design processes within an organization. Organizations use DFSS when they have to design or redesign a process, product, service, or transaction. DFSS gets to the source of product and service imperfections by "designing in" performance from the earliest stages of research and development. It teaches people a methodical approach to involving the right people, asking the right questions, and using the right tools from the very beginning of any design project. DFSS can be applied to any industry and any product or process design methodology. It can be used to create new products and services, streamline software design and systems integration, or improve existing product performance. By implementing DFSS, companies can avoid costly redesign projects by ensuring that products and processes are designed, built, and launched with greater reliability and a higher performance-to-cost ratio.

4.2.2 Mistake proofing (Poka-Yoke)

Mistake proofing a product's design and its manufacturing process is an important element of design for X. Mistake proofing also is a key element in improving product quality and reliability and an element of the DFSS concept.

The Japanese concept of *poka-yoke* (mistake proofing) seeks to find and correct problems as close to the source as possible. This is because finding and correcting defects caused by errors costs more and more as a product or item flows through a process. Over time, more emphasis has been placed on the design of the product to avoid mistakes in production. Often the benefits of mistake proofing not only help with production of the product but also contribute to correcting user operation and maintenance as well as servicing the product.

The concept of mistake proofing involves finding controls or features in the product or process to prevent or mitigate the occurrence of errors and requires simple, inexpensive inspection (error detection) at the end of each operation to discover and correct defects at the source.

There are six mistake proofing principles or methods. These are listed in order of preference or precedence in the manner in which mistakes are addressed (Belliveau et al., 2002):

1. Elimination seeks to eliminate the possibility of error by redesigning the product or process so that the task or part is no longer necessary; for instance, product simplification or part consolidation that avoids a part defect or assembly error in the first place.
2. Replacement substitutes a more reliable process to improve consistency; for example, the use of robotics or automation that prevents a manual assembly error, or automatic dispensers or applicators to ensure the correct amount of a material, such as an adhesive, is applied.
3. Prevention engineers the product or process so that it is impossible to make a mistake.

4. Facilitation employs techniques and combines steps to make work easier to perform; for example, visual controls including color coding, marking, or labeling parts to facilitate correct assembly; exaggerated asymmetry to facilitate correct orientation of parts; a staging tray that provides a visual control that all parts were assembled, locating features on parts.
5. Detection involves identifying an error before further processing occurs so that the problem can be corrected quickly; for example, sensors in the production process to identify when parts are assembled incorrectly.
6. Mitigation seeks to minimize the effects of errors; for example, simple rework procedures when an error is discovered and extra design margin or redundancy in products to compensate for the effects of errors.

Ideally, mistake proofing should be considered during the development of a new product to maximize opportunities to mistake proof through design of the product and the process (elimination, replacement, prevention, and facilitation), because over 70% of a product's life cycle costs can be attributed to its design stage. Once the product is designed and the process selected, mistake proofing opportunities are more limited (prevention, facilitation, detection, and mitigation).

4.2.3 Quality function deployment

QFD was originally developed by Yoji Akao and Shigeru Mizuno in the early 1960s. They extended the original "house of quality" (HOQ) approach by deploying "hows" resulting from the top-level HOQ into lower tier matrices addressing aspects of product development, such as cost, technology, and reliability. The basic QFD methodology involves four phases that occur over the course of the product development process. During each phase, one or more matrices are constructed to plan and communicate critical product and process planning and design information. The QFD methodology flow is depicted in Figure 4.5.

Once customer needs have been identified, preparation of the product planning matrix or "HOQ" can begin. The product planning matrix is prepared as follows:

1. State the customer needs or requirements on the left side of the matrix. These are organized by category, based on the affinity diagrams. It needs to be ensured that the customer needs or requirements reflect the desired market segment(s). If the number of needs or requirements exceeds 20–30 items, the matrix is decomposed into smaller modules or subsystems to reduce the number of requirements. For each need or requirement, state the customer's priorities using a 1–5 rating. Ranking techniques and paired comparisons are used to develop priorities.
2. Evaluate prior-generation products against competitive products. Surveys, customer meetings, or focus groups and clinics are used to obtain feedback. Competitors' customers are included to get a balanced perspective. Identify price points and market segments for products under evaluation. Warranty, service, reliability, and customer complaint problems are taken into account to identify areas of improvement. Based on this, a product strategy is developed. Consider the current strengths and weaknesses relative to the competition. Identify opportunities for breakthroughs to exceed competitors' capabilities, areas for improvement to equal competitors' capabilities, and areas where no improvement will be made. This strategy is important to focus development efforts where they will have the greatest payoff.

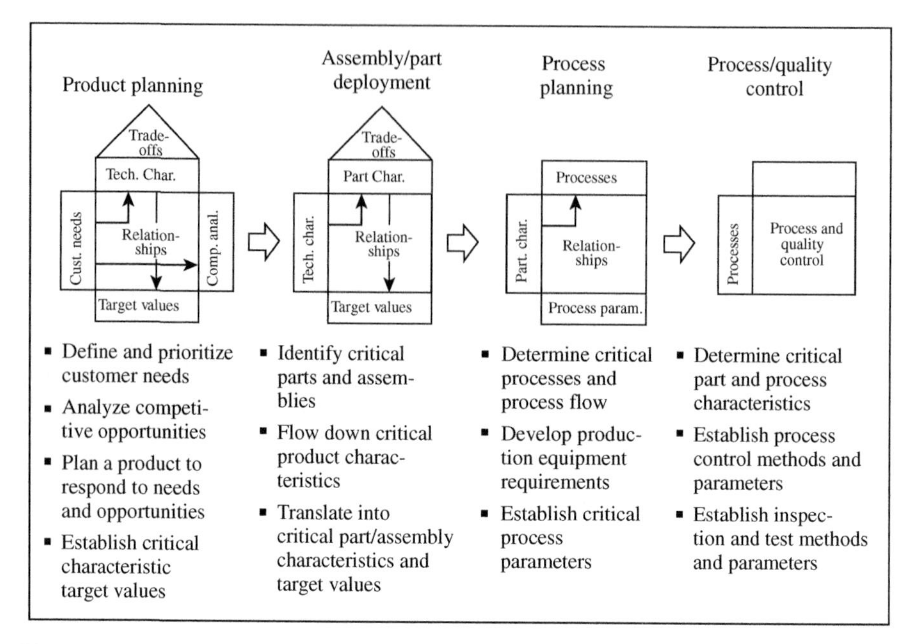

Figure 4.5 The four-phase QFD approach to building quality.
Source: Adapted from Kolarik (1999).

3. Establish product requirements or technical characteristics to respond to customer requirements and organize them into related categories. It is important to note that characteristics should be meaningful, measurable, and global.
4. Develop relationships between customer requirements and product requirements or technical characteristics. This involves using symbols for strong, medium, and weak relationships.
5. Develop a technical evaluation of prior-generation products and competitive products. Get access to competitive products to perform product or technical benchmarking. Perform this evaluation based on the defined product requirements or technical characteristics. Obtain other relevant data, such as warranty or service repair occurrences and costs, and consider this data in the technical evaluation.
6. Develop preliminary target values for product requirements or technical characteristics.
7. Determine potential positive and negative interactions between product requirements or technical characteristics, using symbols for strong or medium, positive or negative relationships. Too many positive interactions suggest potential redundancy in the "critical few" product requirements or technical characteristics. Focus on negative interactions; consider product concepts or technology to overcome these potential trade-offs or consider the trade-offs in establishing target values.
8. Next, calculate importance ratings. This involves assigning a weighting factor to relationship symbols and multiplying the customer importance rating by the weighting factor in each box of the matrix. Finally, all resulting products in each column are added.
9. Develop a difficulty rating (1–5 point scale, where 5 means very difficult and risky) for each product requirement or technical characteristic. Avoid too many difficult/high-risk items, as this will likely delay development and exceed budgets. Assess whether the difficult items can be accomplished within the project budget and schedule.

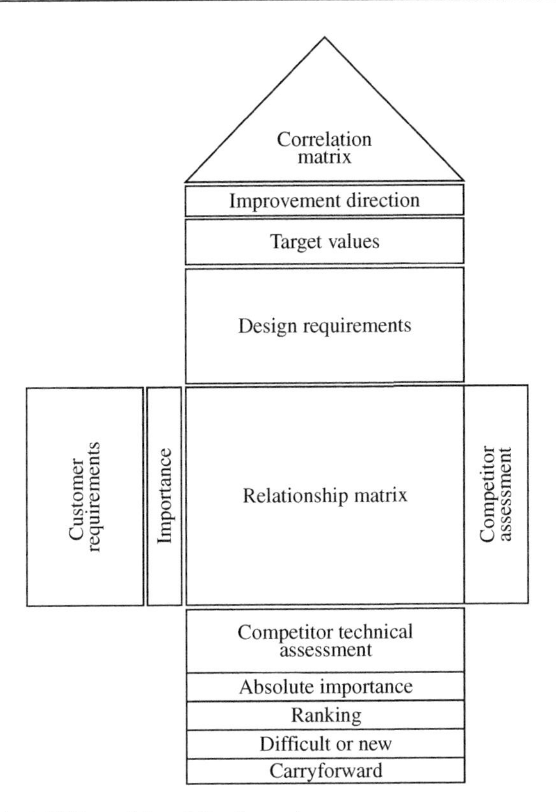

Figure 4.6 A typical QFD model and its elements.

10. Analyze the matrix and finalize the product development strategy and product plans. Determine required actions and areas on which to focus. Finalize target values. Are target values properly set to reflect appropriate trade-offs? Do target values need to be adjusted, considering the difficulty rating? Are they realistic with respect to the price points, available technology, and the difficulty rating? Are they reasonable with respect to the importance ratings? Determine items for further QFD deployment. To maintain focus on the critical few, less significant items may be ignored with the subsequent QFD matrices. Maintain the product planning matrix as customer requirements or conditions change.

The "HOQ" matrix often is called the *phase 1 matrix*. Figure 4.6 shows a typical HOQ and its elements. In the QFD process, a phase 2 matrix translates finished product specifications into attributes of design (architecture, features, materials, geometry, subassemblies, and component parts) and their appropriate specifications. Sometimes, a phase 3 matrix is used in translating attributes of design specifications into manufacturing process specifications (temperature, pressure, viscosity, rpm, etc.). Some key elements that determine the successful implementation of QFD follow:

- Management must make it clear that QFD is a high priority.
- Set clear priorities for QFD activities. Specifically, management needs to allocate resources for and insist on execution of market research and technical competitive assessment.

- Make QFD training available, preferably "just in time" to use QFD.
- Insist that decisions be based on customer requirements.
- Understand the terms used in QFD.
- Insist on cross-functional commitment and participation.
- Become leaders of QFD rather than managers.

One guideline for successful development of QFD matrices is to keep the amount of information in each matrix at a manageable level. With a more complex product, if 50 potential needs or requirements were identified and these were translated into an equal or even greater number of product requirements or technical characteristics, there would be more than 2500 potential relationships to plan and manage. Generally speaking, an individual matrix should not address more than 20 or 30 items on each dimension of the matrix. Therefore, a larger, more complex product should have its customers' needs further classified into lower hierarchical levels.

To conclude, a product plan is developed based on initial market research or requirements definition. If necessary, feasibility studies or research and development are undertaken to determine the feasibility of the product concept. Product requirements or technical characteristics are defined through the matrix, a business justification is prepared and approved, and product design then commences.

4.2.4 Design review

The purpose of a design review is to provide a systematic and thorough product-process analysis, a formal record of that analysis, and feedback to the design team for product and process improvement. According to the Japanese Industrial Standards JIS Z 8115–1981, design review is the judgment and improvement of an item at the design phase, reviewing the design in terms of function, reliability, and other characteristics, with cost and delivery as constraints and with the participation of specialists in design, inspection, and implementation. The D-R process allows for an independent critique of a product and its related process at appropriate time intervals in the product life cycle. It is an important tool to identify product-process bottlenecks using the proactive quality control strategy.

Formal design reviews (FDRs) should be performed from an independent perspective. A key requirement for a reviewer in the FDR is that he or she be an expert in the field and hold no vested interest in nor be directly responsible for the product or process under review. This constraint helps the process elicit unbiased comments, concerns, and recommendations. Development of an effective design review requires a team of functional experts (the reviewers) and a product or process life cycle plan or program (which is the subject of review).

Figure 4.7 shows the steps in a typical D-R process. In the conceptual D-R stage, designers ensure that the initial design direction maps to the business goals and user needs and review the design for alignment with broader initiatives and possible integration with other product designs. In the standards check point stage, the designs are reviewed to meet appropriate standards for consistency, accessibility, usability, and the like. The user interface (UI) design review is used to review specific interactions and provide guidance to designers on problematic issues. Finally, the creative design

Figure 4.7 D-R process and product development phases.

review is used to ensure that the visual design maps to the creative direction of the project.

Design reviews vary in both formality and structure. Reviews typically are planned into the design process, so that often we see a number of reviews conducted in sequence. Each design review focuses on one or more aspects of the design or plan as it proceeds across the product life cycle, from needs assessment to disposal considerations. These individual reports are fed back to the design and planning team for action and resolution. Each stage is recorded sequentially, and actions taken are formally entered into the record. Based on these reports, formal documentation is developed and becomes a permanent part of the quality record.

Typically, there are two forms of design review: internal and external. The internal D-R process looks into the manufacturability and design part of the product life cycle. It addresses the feasibility of the design with respect to function, form, and fit; the producibility of the design with respect to the production and process capabilities, sales and service capabilities, cost–volume–profit estimates, and the economic feasibility of the product. The external D-R process addresses the customer, target market, and their true quality characteristic demands. This review provides input on whether the product will provide customer satisfaction and feedback on what can be done to improve the design accordingly.

Some problems commonly associated with the implementation of the D-R process are:

- Unevenly matched skills and knowledge among the D-R team
- Lack of communication between product developers and the related departments
- No time to make D-R-based changes
- Lack of D-R experience

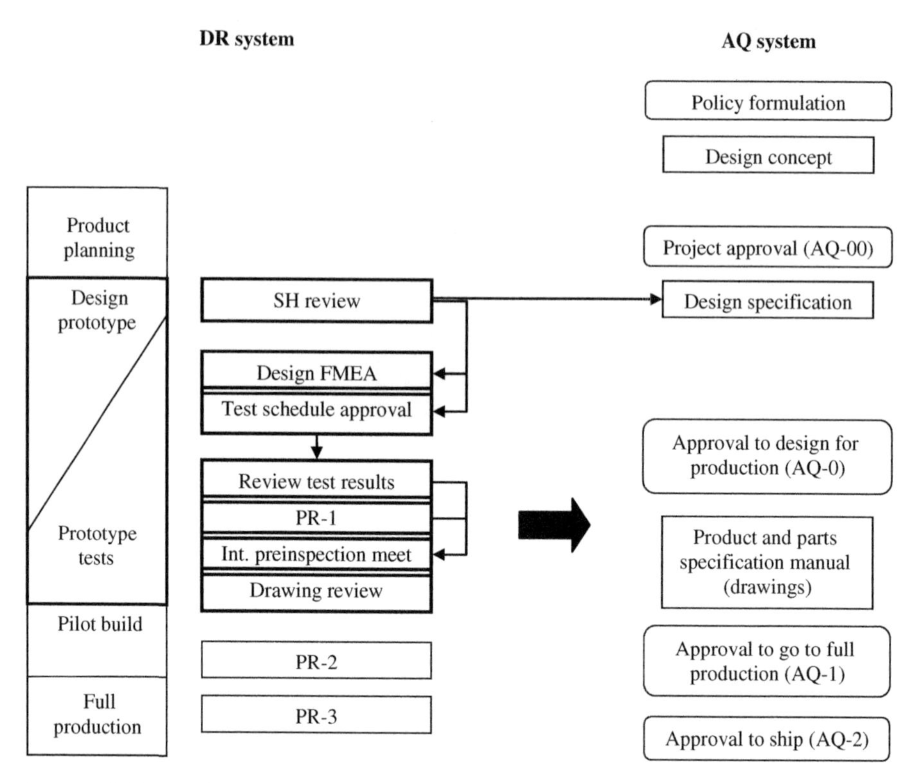

Figure 4.8 The D-R sequence.
Source: Adapted from Ichida (1996).

- Each department considers design review a separate stage and not included in the initial design process.

The D-R process, when utilized correctly, serves as a birth-to-death record of the product, processes, and the organization's diligence in serving its target market. A review of the records of a product shows the major concerns or warnings, their treatment and resolution records, and a record of how timely they were resolved.

Figure 4.8 shows the D-R sequence. The key elements of the D-R system are the soft–hard (SH) review and the FMEA processes.

4.2.4.1 SH review

The SH review addresses the need to design a product for safety, that is, in terms of the real-world conditions. The soft reviews look into the careless misuse of products by users, beyond normal wear and tear. The hard reviews look into the loss of function due to malfunctions or deterioration of each component over the estimated service life. The records generated using the SH review are an important data resource to help designers identify the various methods by which a particular product has been used throughout its life and make changes to design accordingly. Moreover, even though safety standards often exist, they do not adequately address every product use situation. SH reviews help fill in the gaps. A more important advantage of SH reviews is in determining the need for backup systems. Secondary safety backup systems might be developed to prevent sudden breakdown of components.

4.2.4.2 Failure mode and effects analysis

FMEA examines potential failures in products or processes. FMEA helps select remedial actions that reduce cumulative impacts of life cycle consequences (risks) from a systems failure (fault). FMEA can be explained as a group of activities intended to recognize and evaluate the potential failure of a product or a process and its effects, identify actions that could eliminate or reduce the chance of the potential failure, and document the process.

This method illustrates connections among multiple contributing causes and cumulative (life cycle) consequences. It is used in many formal quality systems such as QS-9000 or ISO/TS 16949 (Chrysler Corporation, Ford Motor Company, and General Motors, 1995).

The basic method is to describe the parts of a system and list the consequences if each part fails. In most formal systems, the consequences then are evaluated by three criteria and associated risk indices: severity (S), likelihood of occurrence (O) or probability of occurrence (P), and inability of controls to detect it (D).

Each index ranges from 1 (lowest risk) to 10 (highest risk). The overall risk of each failure is the risk priority number (RPN), which is the product of the severity, occurrence, and detection rankings: RPN = $S \times O \times D$.

A basic FMEA consists of a set of nine columns:

1. Function, equipment, or process identification
2. Function, equipment, or process purpose
3. Interfaces
4. Failure mode
5. Failure mechanism
6. Failure detection
7. Failure compensation
8. Failure effects
9. Preventive measures.

FMEA is a fundamental tool, useful in improving reliability, maintainability, safety, and survivability of products and processes. It encourages systematic evaluation of a product or process, recognition of hazards and potential failures, effects of the failures, the countermeasures to eliminate the failures or create secondary safety systems, and their documentation.

4.2.4.3 Experimental design

The purpose of experimental design is to provide a systematic plan of investigation and analysis, based on established statistical principles, so that the interpretation of the observations can be defended as to technical relevance. The objective is to determine those variables in a process or product that form critical parameters and their target values. By using formal experimental techniques, the effect of many variables can be studied at one time. Changes to the process or product are introduced in a random fashion or by carefully planned, highly structured experiments. Planned experiments consist of six basic steps:

1. Establish the purpose
2. Identify the variables
3. Design the experiment
4. Execute the experiment
5. Analyze the results
6. Interpret and communicate the analysis.

There are three approaches to designing experiments: classical, Taguchi, and Shainin. Most common among these is the classical approach, based on the work of Sir Ronald Fischer in agriculture during the 1930s. In the classical approach, relevant variables are clearly identified and defined. An appropriate randomization and replication structure is developed within the context of the selected model. Three major analyses are used to generate the results: descriptive (which includes summary statistics and graphs), inferential (which includes formal hypotheses), and predictive (which includes the models used to predict future responses).

When designing experiments, statisticians generally begin with a process model of the black box variety, with several discrete or continuous input factors that can be controlled. Controlling variables implies varying them at will. This is coupled with one or more measured outputs or responses. The outputs or responses are assumed to be continuous. Experimental data are used to derive an empirical or approximation model linking the outputs and inputs. These empirical models generally contain what are known as *first-* and *second-order terms*. Often, the experiment has to account for a number of uncontrolled factors, which may be discrete, such as different machines or operators, or continuous, such as ambient temperature or humidity.

Taguchi's method, commonly known as *robust design*, improves the performance of a product by minimizing the effect of the causes of variation, without eliminating the causes. According to Taguchi et al. (1989), quality is the loss a product causes to society after being shipped, other than any losses caused by its intrinsic functions. Taguchi's definition of *quality* and concept of loss minimization lead to robust design. Robust design consists of a conceptual framework of three design levels: system design, parameter design, and tolerance design. Within these design levels, quantitative methods based on loss functions are developed.

System design denotes the development of a basic prototype design that performs the desired and required functions of the product with minimum deviation from target performance values. It focuses on the relevant product or process technologies and approaches. It includes selection of materials, parts, components, and assembly systems.

Parameter design is a secondary design level, within or below the system design level. The main focus here is to ascertain the optimal levels for the parameters of each element in the system, to minimize the functional deviations of the product. The point is to meet the performance target with the least expensive materials and processes and to produce a robust product, one that is on target and insensitive to variations. It is the process of optimizing the functional design with respect to both cost and performance. Taguchi's method concentrates on designed experiments and specialized signal-to-noise ratio (SNR) measures.

Tolerance design is the process of determining the tolerance of each parameter by trading off quality loss and cost. The parameter design sets up the targets for the process parameters. The tolerance design step is a logical extension of parameter design to a point of complete specification or requirement. Obviously, this step determines the most economic tolerances, those that minimize product cost for a given tolerance deviation from target values.

4.3 Case studies

4.3.1 Design review case study

The design team of Z-Air Systems is facing a predicament. The newly launched space heater system is running into rough weather due to product failures and returns from customers. Market research indicates that about 80% of the problems are related to product performance and quality. Analysis of the problems indicates that most issues are due to design defects. Management formed the opinion that going back to the drawing board can solve the problem. The design team is under pressure to produce results, and it resorts to a comprehensive D-R program. The following details the D-R program adopted by Z-Air Systems and its results.

Figure 4.9 shows the sequence of the D-R process adopted by the design team. The entire process was initiated due to rejects and returns of the space heater systems by customers. The sequence of the process includes SH review, FMEA review, analysis of test results, performing a pilot run, reviewing design and quality, then finalizing the product run.

All the information conveyed by the marketing department regarding the complaints and reasons for rejection or return are documented and fed into the company's information system. These data are utilized in developing the SH tables. The SH tables not only allow the designers to check the product for safety but also provide inputs regarding consumer complaints, quality issues, and the like. The soft analysis provides an in-depth view of the possibilities for careless misuse of the space heater beyond its normal wear and tear. The hard analysis provides the various potentials for function loss due to malfunction or deterioration over each component's estimated service life. These results are applied to improve products' design for safety aspects.

The team observes no overlaps in the S and H factors in the review. This means no possibility of any user injury from the use of the wrong part at the wrong time. Tables 4.1 and 4.2 show a sample of the team's SH analysis table.

The next stage, after performing the SH review, is to examine the performance of the new design. The FMEA method uses the RPN index, which is used to set priorities

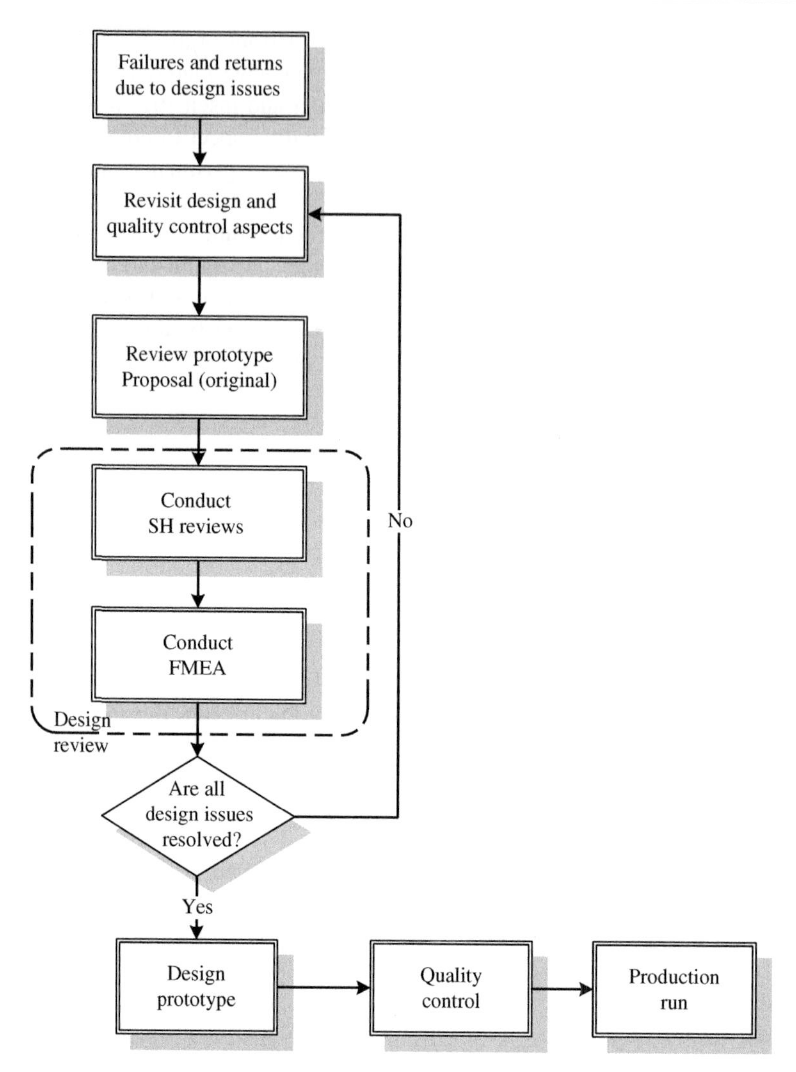

Figure 4.9 D-R sequence adopted by Z-Air Systems.

on the design and safety issues. Figure 4.10 shows a sample design FMEA for the space heater.

The implementation of the D-R method results in a sharp drop in the defect rate. Market data indicate that the new design, after the review sequence, results in increased sales and revenues. Further, the rejection and return rate is reduced by over 77% compared to the older model. This measure of the sales and market performance of the product is an indication of the success of the D-R process.

Observing the tremendous improvements in design and performance of the product, the management at Z-Air System implemented a plant-wide requirement to

Table 4.1 S factor analysis (careless misuse checklist)

Category	Item	Description	Pass/fail	Reason for rejection
Performance related	Automatic shutoff	Tip over switch activated	O	N/A
	Overheating	Motor thermal fuse	O	N/A
	Continuous operation (switch left on)			N/A
User related	Instruction not followed			N/A
	Left too close to wall (or other object)	Motor thermal fuse	O	N/A
	Operated in damp conditions	Motor thermal fuse	O	N/A

Table 4.2 H factor analysis (product safety function)

Component	Hazardous condition	Automatic response	Comment
Motor thermal fuse (improper operation)	Coil starts overheating	Motor shuts off	Replace motor
Power switch (left on)	Improper use causes overheating	Motor shuts off	Automatic shutoff function
Timer	Set at high temperature	Motor shuts off	Automatic shutoff function

Part No: 3C1749	Part Description: Space Heater Fan	FAILURE MODE EFFECT ANALYSIS					Design Lead: / Analysis Period:		Michael J. Fischer / Jan 2006 - March 2006
Component	**Function**	**Failure Mode**	**Effect of Failure**	**Causes of Failure**	**# of Occurrences**	**Severity Rating**	**Detection Rating**	**RPN**	**Preventive Actions taken**
Frame	Provide support, hold motor assembly	Corrosion and scale formation	Motor falls from frame	Metallic coating insufficient	4	5	2	40	Change metal coating process to ________
				Vapor condensation due to inadequate through flow of air	6	4	2	48	Confirm with condensation tests
				Improper brazing process	3	6	1	18	Use different joining process
		Frame breakage	Bends and breaks	Improper packaging	5	3	1	15	Improve packaging and cushion for products
				Material of frame not able to withstand vibrations due to inadequate hardness	8	8	3	192	Change material of the frame to ________

Figure 4.10 Sample design FMEA sheet.

apply the D-R process to all products. This change in management policy has yet to be quantitatively documented at the plant, but if past results are any indication, the company is headed toward a brighter future and could achieve a level of quality that would render its products superior to its competitors.

4.3.2 Six sigma case study

Classic Plastics Inc. is a leading bottling plant that supplies plastic bottles to all leading pharmaceutical companies. Because of both quality and productivity concerns, the company needs to analyze its operations and improve its output in meeting the increased market demands. The current manufacturing plant has 35 bottling machines. Cycle time was identified as the key element. The entire process is shown in Figure 4.11.

The engineering team decided to analyze the process, using the DFSS strategy to improve it. After outlining the schedule for the project, the team held brainstorming sessions to determine the factors affecting the process. On close analysis, it arrived at the tree diagram shown in Figures 4.12 and 4.13 (one- and two-level tree diagrams).

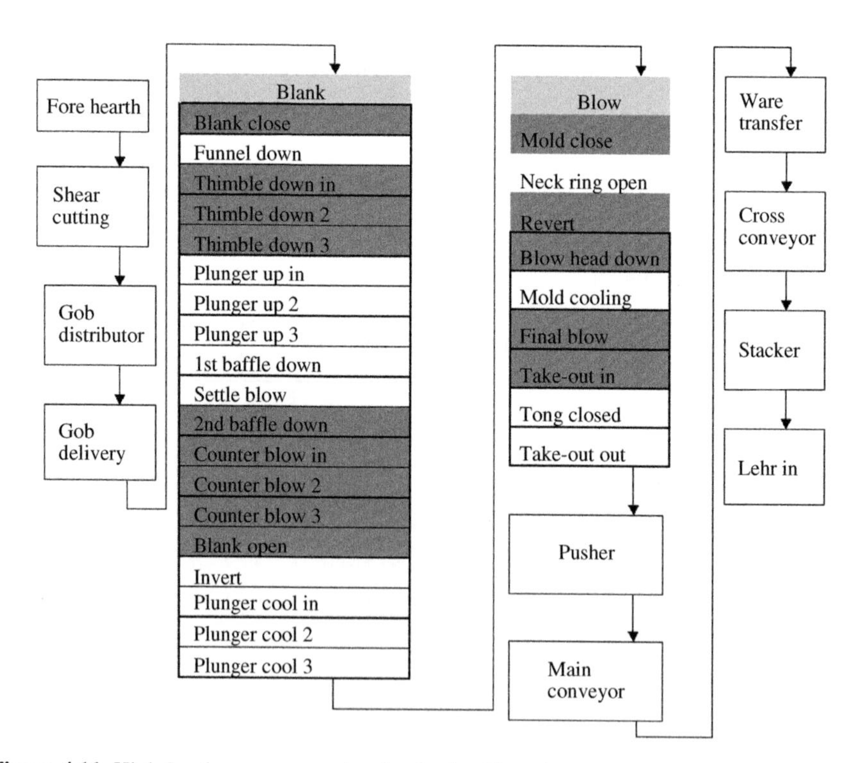

Figure 4.11 High-level process mapping for the bottling plant.

After carefully planning and conducting the experimentation, the following actions were taken:

- Cycle time was changed from 15 to 15.5 ms
- Gob length increased by 1.5 mm
- All section differentials increased by 2°
- Cooling blower pressure increased from 700 to 740 mm WC
- Pusher retract time decreased by 3°
- Number of bottles per stack row decreased from 27 to 25
- Stacker forward length increased from 320 to 330 mm

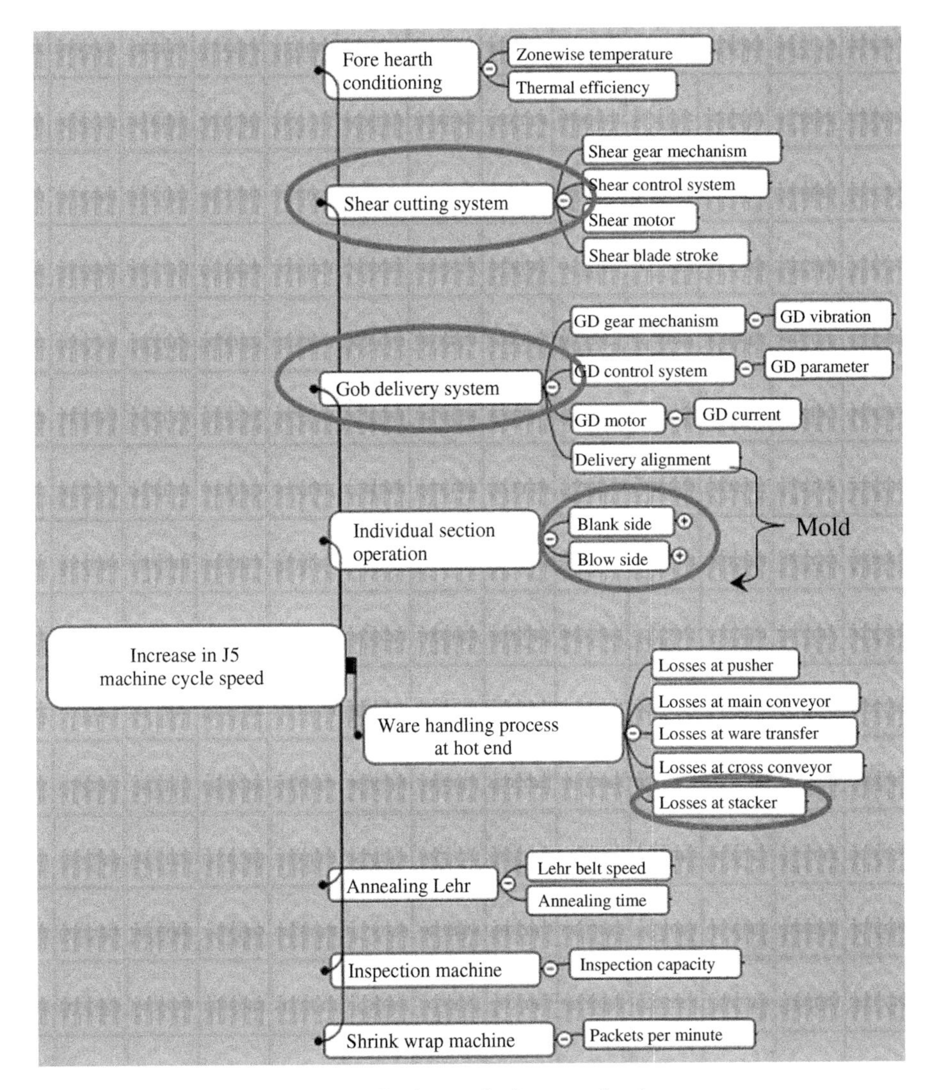

Figure 4.12 Tree diagram: factors affecting cycle time, one level.

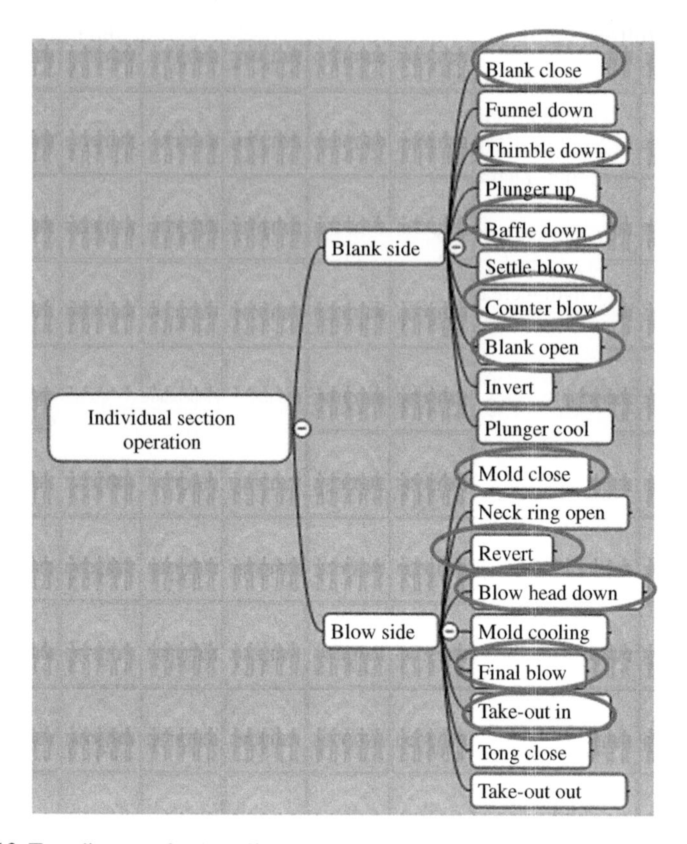

Figure 4.13 Tree diagram: factors affecting cycle time, two level.

- Stacker forward speed increased from 50% to 60%
- Lehr belt speed increased from 420 to 450 mm/min.

The results of the changed cycle time are shown in Table 4.3.

The engineers brainstorming and identifying potential problems in the process reported the following:

1. Bottle falling problem can be reduced by increasing the bearing surface
2. Proposed action: Bar knurling instead of crescent knurling
3. Hot blank can lead to blank seam and loading problems
4. Proposed action 1: Cooling fins in blank to be provided
5. Hot neck ring can lead to neck ring bulging
6. Proposed action 1: Neck ring fins to be provided
7. Proposed action 2: Number of balancing holes to be increased.

The changes, when implemented, resulted in an annual savings of $370,000. Further, they resulted in a horizontal plant-wide deployment on all 35 lines. The secondary changes in all the factors also resulted in improved esthetics throughput improvements.

Description	Actual			Proposed (15.5 cycles)			Achieved—G12 A 70 ml		
	On	Off	On time in ms (15)	On	Off	On time in ms (15.5)	On	Off	On time in ms (15.5)
Blank side									
Blank Close	5	205	2222.2	4	202	2129.0	4	210	2215.0
Plunger Up In	32	92	666.7	32	92	645.2	30	92	666.7
Plunger Up 2	32	92	666.7	32	92	645.2	30	92	666.7
Plunger Up 3	32	92	666.7	32	92	645.2	30	92	666.7
Funnel Down	4	94	1000.0	3	95	989.2	6	94	946.2
1st Baffle Down	42	84	466.7	42	85	462.4	42	84	451.6
Settle Blow	42	84	466.7	42	85	462.4	48	84	387.1
Plunger Down In	100	100	0.0	100	100	0.0	100	100	0.0
Plunger Down 2	100	100	0.0	100	100	0.0	100	100	0.0
Plunger Down 3	100	100	0.0	100	100	0.0	100	100	0.0
2nd Baffle Down	118	202	933.3	118	202	903.2	118	210	989.2
Counter Blow In	147	200	588.9	145	198	569.9	148	207	634.4
Counter Blow 2	147	200	588.9	145	198	569.9	148	207	634.4
Counter Blow 3	147	200	588.9	145	198	569.9	148	207	634.4
Thimble Down In	206	30	2044.4	204	31	2010.7	210	28	1914.0
Thimble Down 2	206	30	2044.4	204	31	2010.7	210	28	1914.0
Thimble Down 3	206	30	2044.4	204	31	2010.7	210	28	1914.0
Blank Open	205	345	1555.6	200	347	1580.6	210	345	1451.6
Invert	256	312	622.2	256	312	602.1	259	312	569.9
Plunger Cool 2	347	0	144.4	347	0	3731.2	320	10	537.6
Plunger Cool 3	347	0	144.4	347	0	3731.2	320	10	537.6
Blow side									
Mould Close	280	191	3011.1	280	188	2881.7	280	188	2881.7
Neck Ring Open	315	328	144.4	315	328	139.8	318	335	182.8
Revert	323	215	2800.0	323	215	2709.7	328	220	2709.7
Blow Head Down	335	202	2522.2	335	198	2397.8	335	198	2397.8
Final Blow	60	188	1422.2	60	186	1354.8	40	186	1569.9
Take Out In	228	260	355.6	228	260	344.1	228	260	344.1
Take Out Out	260	135	2611.1	260	135	2526.9	260	140	2580.6
Tong Closed	254	110	2400.0	254	110	2322.6	256	130	2516.1
Mould Cooling	15	160	1611.1	15	187	1849.5	0	186	2000.0

4.3.3 QFD case study

PC Solutions is a minority-owned small-scale manufacturing company that builds customized desktops and laptops and additionally provides maintenance support to its business clients. The company intends to expand its market share by increasing laptop sales over those of its competitors. The company's manufacturing engineers along with the engineering consulting team initiated the project and decided to use the QFD method to determine what improvements needed to be undertaken to improve its current laptop.

Figure 4.14 shows the process/steps that the company followed to arrive at the QFD HOQ. Figure 4.15 shows a block diagram for the HOQ.

Step 1. Voice of the customer

The company conducted an exhaustive customer analysis using its current client base and also expanded the survey to include its target market. The following tools were utilized to provide input for the QFD:

- Customer preference/focus group survey
- Satisfaction/customer service survey
- Competition product analysis
- Internal customer/design team survey.

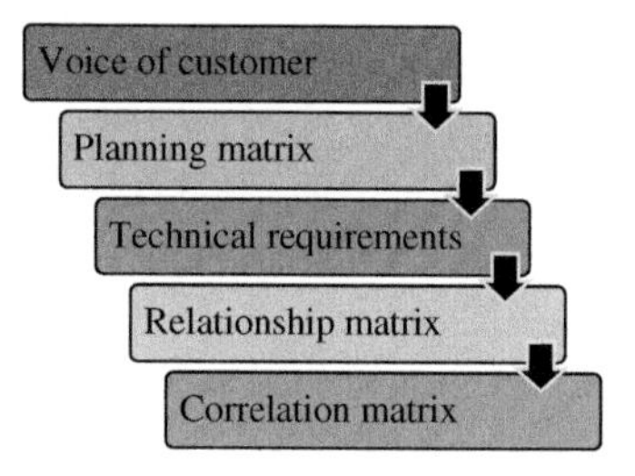

Figure 4.14 Steps in constructing QFD HOQ.

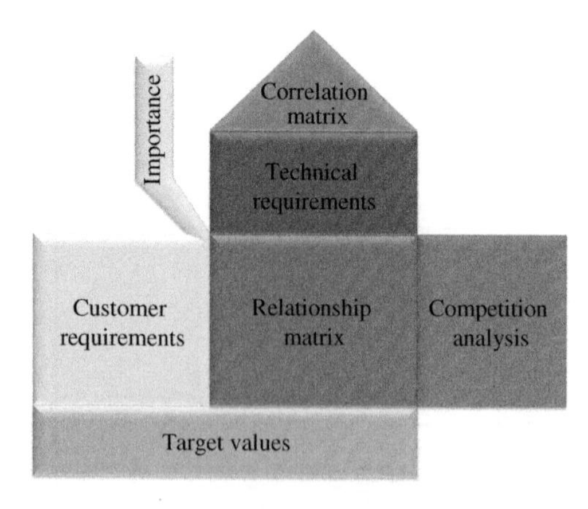

Figure 4.15 Block diagram for a QFD HOQ.

A comprehensive and detailed *ad hoc* analysis of the various surveys resulted in the following customer requirements of key value:

- Processor speed
  - Processor manufacturer (Intel/AMD/IBM/VIA)
  - Frequency (GHz)
  - Internal memory capacity (RAM Gb).
- Laptop bulk
  - Size (in.)
  - Thickness (in.)
  - Material weight (lbs).
- HDD capacity/storage space (Gb).

Step 2. Planning matrix

- In this step, we document the customer requirements and the importance ratings as perceived by the customer.
- Customers evaluated the importance of each product requirement using a scale—where 1 is low and 5 is high—to indicate the relative importance of each feature that they specify.
- These importance ratings were obtained during the customer focus group survey.
- Customer importance ratings are also paired with competitive comparisons that are reported.

Step 3. Technical requirements

- Specified technical product requirements include those characteristics such as competitive performance, process control data, field failure data, performance limits due to physics, and information about standard levels of performance or product requirements.
- The relative importance of each design requirement is given by the sum of each column's design feature-to-customer requirement indicators multiplied by its importance weight.
- Potential safety hazards and environmental effects are also flagged.
- In order to meet a customer need—whether it is spoken by an external or internal customer—the team must initiate a product requirement which will become a feature of the design.
- Design features should be grouped according to functional concept in order to identify sets of functions that will become a subassembly or module with related functionality.
- Design features imply functionality in the final product. Using the FAST (function analysis system technique) methodology, a team can determine the specific functional requirements of each feature as well as logical relationships among design features.
- Defines "how much"—or the magnitude of the design features. Specific rows will include the following information:
  - Target values of design features
  - Competitive comparison of values
  - Trends in performance improvement
  - Physical limits of performance
  - Field service return information
  - Special technical requirements
  - Applicable government standards
  - Applicable industry standards
  - Applicable environmental standards
  - Applicable safety standards.

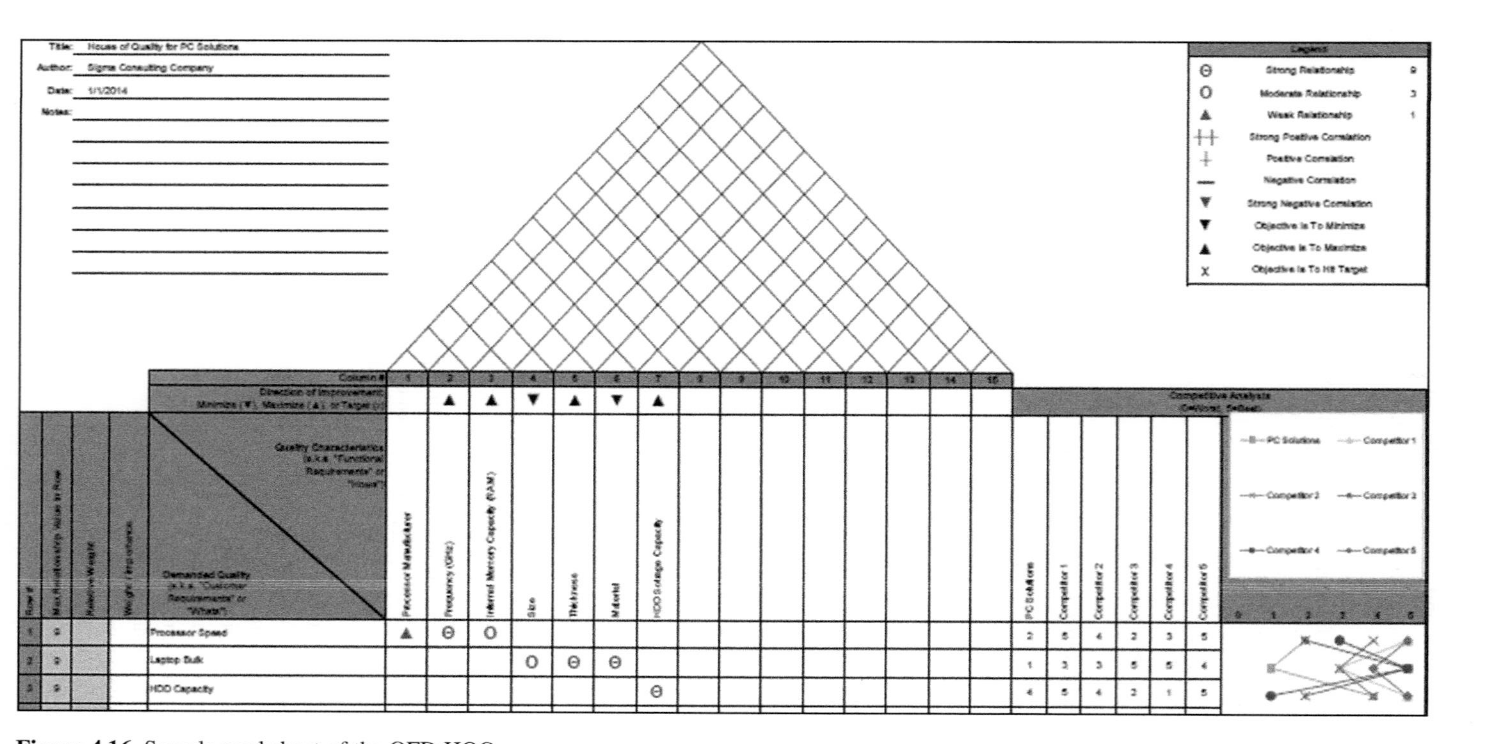

Figure 4.16 Sample worksheet of the QFD HOQ.

Step 4. Relationship matrix

- In many products, functions of some design features may interact to some degree with other features. This interaction may cause problems in the design and implementation of the product or offer opportunities for more efficient designs by integrating these features. It is always helpful to identify what features have positive or negative interrelationships.
- An arbitrary scale is used to rate the degree of relationship between the specified features:
  +9 = strong positive
  +3 = weak positive
  0 = no apparent relationship
  −3 = weak negative
  −9 = strong negative.
- A way to develop suspicions for hypotheses.
- A one point rating difference is perceivable by customers. This means that whenever a one point difference is achieved, then that feature may be used to differentiate a product. This type of feature may become a "sales point" if the value perceived by customers is significant.

Step 5. Correlation matrix

- At the intersection of each row (customer requirements) and each column (design features), the cell is used to indicate the strength of the relationship between these factors.
- The strength of the relationship is indicated by a forced weighting scale where strong = 9, moderate = 3, and weak = 1. A solid circle is often used to indicate a strong relationship, an open circle usually represents the moderate relationship, and an open triangle indicates weak relationships.
- If more than 50% of the cells have some relationships, then the level of detail in customer requirements or in the design features is too great.
- Policy decisions management must make:
  - Invest in product or process technology?
  - Proprietary or open architecture for technology?
  - Develop or acquire technology?
  - Purchase or license technology?
  - Barter intellectual property for technology?

Figure 4.16 shows a sample HOQ developed for the PC Solutions Company.

References

Akao, Y. (Ed.), 1990. Quality Function Deployment Productivity Press, Cambridge, MA.

American Society for Quality Control. ASQC/Gallup Survey, November 1991. In: Q—Official Newsletter of the American Society for Quality Control 6, no. 9.

Banks, J., 1989. Principles of Quality Control. John Wiley & Sons, Inc., New York, NY.

Belliveau, P., Griffin, A., Somermeyer, S., 2002. The PDMA ToolBook 1 for New Product Development. John Wiley & Sons, Inc., New York, NY.

Chrysler Corporation, Ford Motor Company and General Motors, 1995. Quality System Requirements QS-9000, second ed. AIAG (810) 358–3003.

Duncan, A.J., 1986. Quality Control and Industrial Statistics, fifth ed. Irwin, Homewood, IL.

Garvin, D.A., 1988. Managing Quality: The Strategic and Competitive Edge. Free Press, New York, NY.

Grant, E.L., Leavenworth., E.S., 1988. Statistical Quality Control, sixth ed. McGraw-Hill, New York, NY.

Hill, T., 1989. Manufacturing Strategy. Irwin, Homewood, IL.

Juran, J.M., 1988. Juran on Planning for Quality. The Free Press, New York, NY.

Ichida, T., Voigt, E.C. (Eds.), 1996. Product design review: a methodology for error-free product development. Productivity Press.

Kolarik, W.J., 1999. Creating Quality: Process Design for Results. McGraw-Hill, New York, NY.

Montgomery, D.C., 1991. Introduction to Statistical Quality Control, second ed. John Wiley & Sons, Inc., New York, NY.

Radford, G.S., 1992. The Control of Quality in Manufacturing. Ronald Press, New York, NY.

Shingo, S., 1986. Zero Quality Control: Source Control and the Pokayoke System. Productivity Press, Cambridge, MA.

Taguchi, G., 1986. Introduction to Quality Engineering: Designing Quality into Products and Processes. Kraus International, UNIPUB (Asian Productivity Organization), White Plains, NY.

Taguchi, G., Elsayed, E.A., Hsiang., T., 1989. Quality Engineering in Production Systems. McGraw-Hill, New York, NY.

Consideration and Selection of Materials

5.1 Importance of material selection in product manufacture

After the conception of a product idea, the questions that the research and development (R&D) personnel must ask is: What would be the best material for the product? More often this is closely followed by the question, Is the material selected easily manufacturable? In other words, what would be the best material and process combination for developing a product that not only performs the indispensable functions but is also economical to manufacture? A design criterion for the product based only on either material or process has all the ingredients of a recipe for disaster. The choice of material is a major determinant for the successful functioning and the feasible, low-cost manufacture of any product.

Materials are at the core of all technological advances. Mastering the development, synthesis, and processing of materials opens opportunities that were scarcely dreamed of a few short decades ago. The truth of this statement is evident when one considers the spectacular progress that has been made in such diverse fields as energy, telecommunication, multimedia, computers, construction, and transportation.

It is widely accepted that the final cost of a manufactured product is determined largely at the design stage. Designers tend to conceive parts in terms of processes and materials with which they are familiar and, as a consequence, may not consider process and material combinations that could prove more economical. Sometimes the designers tend to focus only on the cost aspect of materials and manufacturing and select a combination of materials and processes that lead to products of substandard quality and reduced operating life. In the long run, this leads not only to reduced brand loyalty for the product but, in many cases, to huge financial losses as a result of litigation and product liability lawsuits. The already difficult task of satisfying engineering and commercial requirements imposed on the design of a product becomes even more difficult with the addition of legislated environmental requirements. A vital cog in this product design wheel is the materials engineer. The optimal selection of material used to construct or make the product should lead to optimum properties and the least overall cost of materials, ease of fabrication or manufacturability of the component or structure, and environmentally friendly materials.

Figure 5.1 shows the various stages of the design process with their associated activities. The material selection process consists of the property, process, and environmental profiles considered concurrently at each phase of design. What happens if the material selection is not considered during each stage of the design decision

Product Development. DOI: http://dx.doi.org/10.1016/B978-0-12-799945-6.00005-3

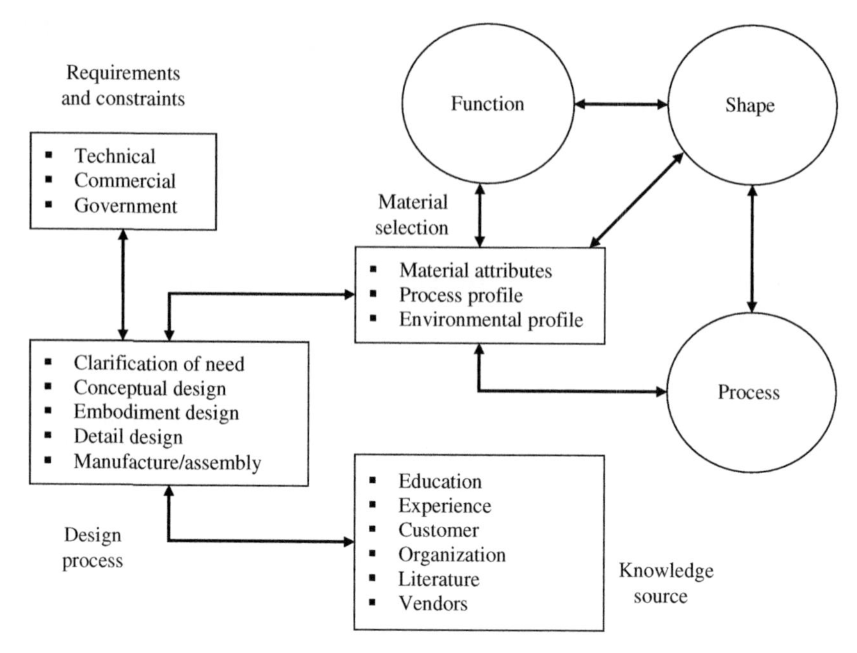

Figure 5.1 The product design phase and material selection.
Source: Adapted from Mangonon (1999).

process? The designer would be unaware of any problems around the availability of the final material, the costs associated with the manufacturing processes, or the processability of the product to be manufactured. Consider a designer who needs to design a product but has no idea of the material from which to make it. Suppose the designer designs the product considering it to be metallic, but management decides to make it of ceramics at a later stage. The processing of a ceramic product is entirely different from that of a metallic product. Ceramic and metallic products vary in structure, strength properties, manufacturability, and so on. Therefore, it is critical that decisions regarding materials to be used for manufacturing a product be made in a timely fashion (Mangonon, 1999).

The selection of an appropriate material and its conversion into a useful product with the desired shape and properties is a complex process. The first step in the material selection process is the definition of the needs of the product. Figure 5.2 shows the factors affecting the material selection process:

1. **Physical factors**: The factors in this group are the size, shape, and weight of the material needed and the space available for the component. Shape considerations greatly influence selection of the method of manufacture. Some typical questions considered by a materials designer are
   - What is the relative size of the component?
   - How complex is its shape? Does it need to be one piece or can it be made by assembling various smaller pieces?

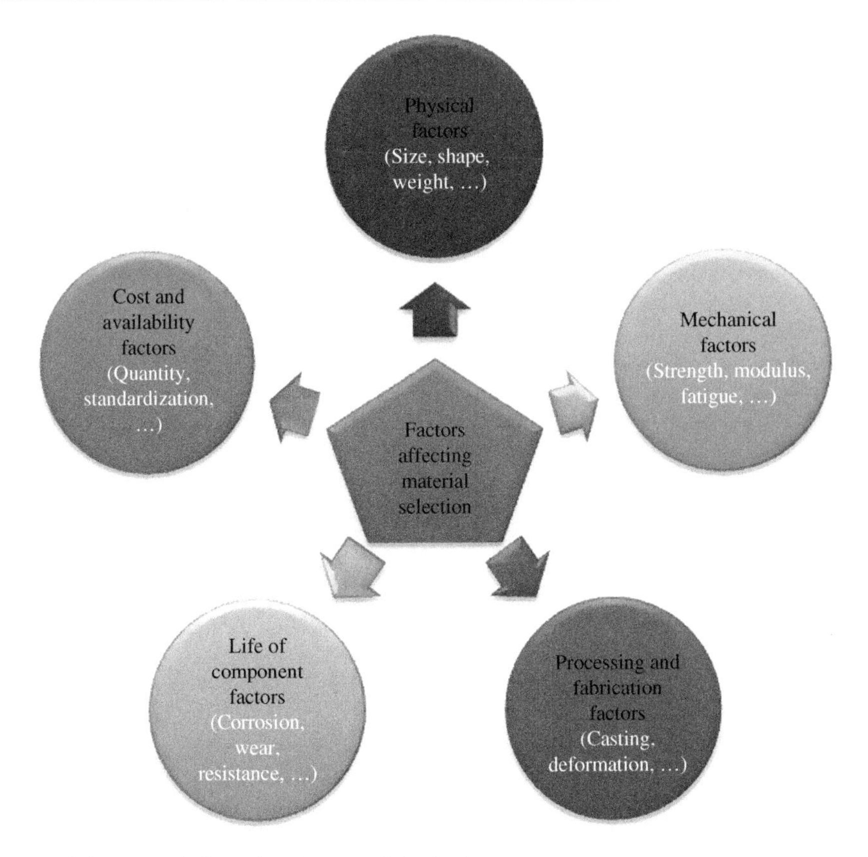

Figure 5.2 Factors influencing the material selection process.

- How many dimensions need to be specified, and what are the tolerances on these dimensions?
- What are the surface characteristic requirements for the product?

All the factors in this category interrelate to the processing of the material. For example, shape and size might constrain the heat treatment of the material. The shape of the product determines whether casting could be used. Material consideration, to a large extent, also is determined by the space available for the component.

2. **Mechanical factors**: The ability to withstand stress and strain is determined by these factors. Strength, ductility, modulus, fatigue strength, and creep are some mechanical properties that influence what material needs to be used. The mechanical properties also are affected by the environment to which the materials are exposed. Some typical questions that designers consider while narrowing down the material to be used are
   - What are the static strength needs of the product?
   - What is the most common type of loading to which the product would be subjected during its use (tensile, compressive, bending, cyclic)?
   - Is the loading static or dynamic? Would the product be subjected to impact loading?
   - Does the product require wear resistance?
   - What temperature range must the mechanical properties possess?

3. **Processing and fabrication factors**: The ability to form or shape a material falls under the processing and fabrication factors. Casting and deformation processing are commonly used. Typical questions that arise out of consideration of these factors are
   - Has the design addressed the requirements that facilitate ease of manufacture? Machinability? Weldability? Formability? Hardenability? Castability?
   - How many components are to be made? What must be the production rate?
   - What are the maximum and minimum cross-sectional dimensions?
   - What is the desired level of quality for the finished product?

   Small objects more commonly are investment casted, while intricate shapes are produced as castings. Powder metallurgy, or a sintering process, is commonly used for brittle materials like ceramics.
4. **Life of component factors**: These factors relate to the life of the materials as they perform the intended function. The properties in this group are external surface properties such as oxidation, corrosion, and wear resistance, and some internal properties such as fatigue and creep. The performance of materials based on these properties is the hardest to predict during the design stages.
5. **Cost and availability**: With reduced lead times from design to market, there is a tendency to jump to the first material that fits the selection profile. It is important to note that additional effort determining the correct material helps optimize manufacturing costs. Also, standardization of parts and materials is related to the cost of the final product. Special processing requirements or rare materials with limited availability increase the final cost and affect the timely manufacture of the product.

5.2 Economics of material selection

After developing a comprehensive list of requisite properties in a material, categorize these properties according to their level of criticality. Some property requirements may be absolute, while others may be relative. The absolute ones cannot be compromised and should be used as a filter to eliminate the materials that cannot be used.

It is apparent that no one material would emerge as the obvious choice. Here, the knowledge of a materials engineer and the handbook-type data need to be utilized. Also, the cost factor of materials needs to be closely analyzed here. Cost is not a service requirement, but it plays an important part in the selection process, both the material cost and the cost of fabricating the selected material. The final decision involves a compromise between the cost, producibility, and service performance.

Current market and economic trends force companies to produce low-cost, high-quality products to maintain their competitiveness at the highest possible level. There is no doubt that reducing the cost of a product is more effective at the design stage than at the manufacturing stage. Therefore, if the product manufacturing cost can be estimated during the early design stage, designers can modify the design to achieve proper performance as well as reasonable cost at this stage, and designers are encouraged to design to cost.

While selecting an individual operation or an entire process for producing a part or product, engineers are faced with the dilemma of selecting and analyzing a multitude of alternative methods including, but not limited to, the cost of variables such as materials, direct labor, indirect labor, tooling, utilities, invested capital, etc. These variables share a very complex relationship, and selection of one factor invariably has an effect on the determination of others.

5.2.1 Cost of materials

The unit cost of materials is a critical factor when the methods being compared involve the use of different amounts or different forms of several materials. Selecting the optimum combination of material and process cannot be performed at one certain stage of product development but should evolve gradually over the different stages. Some generic steps in material selection process are:

1. Analysis of the performance requirements
2. Development of alternative solutions to the problem
3. Evaluation of the different solutions
4. Decision on the optimum solution.

The systematic and early selection of materials and processes for manufacturing a part or a whole product is an integral part of DFM (design for manufacturing). Unfortunately, most designers tend to choose materials they are most familiar (comfortable) with. This results in exclusion of more economical material–process combinations and chances of improvements in DFM are lost.

5.2.2 Cost of direct labor

Direct labor unit costs essentially are determined by three factors: the process being used to manufacture the part or the product, the design of the part or product, and the productivity of the worker performing the operations. The general rule of thumb is that the more advanced the technology used to manufacture the part/product, the more complex is the product design with closer tolerances and advanced tooling requirements, and hence the higher is the cost of direct labor.

There is a very strong relation between the cost of direct labor and the level of automation and the number of steps in the manufacturing process. Typical of low labor content processes are metal stamping and drawing, die casting, injection molding, single-spindle and multispindle automatic machining, numerical and computer-controlled drilling, and special purpose machining, processing, and packaging in which secondary work can be limited to one or two operations. Semiautomatic and automatic machines of these types also offer opportunities for multiple machine assignments to operators and for performing secondary operations internal to the power machine time. Both can reduce unit direct labor costs significantly. Processes such as conventional machining, investment casting, and mechanical assembly including adjustment and calibration tend to contain high direct labor content.

5.2.3 Cost of indirect labor

Indirect labor is defined as work or tasks performed by personnel who do not produce products. Indirect labor costs are costs that cannot be specifically linked to the physical construction of specific products, but are necessary for producing those products. Setup employees, inspectors, material handling personnel, tool crib attendants who sharpen tools and maintain dies, janitors or housekeeping personnel, utility workers, shipping/receiving personnel, clerical workers, forklift drivers, and maintenance workers are some examples of indirect labor. Indirect labor can also apply to the salary workforce in the office, whether clerical or executive.

There exists a pervasive belief that you cannot measure indirect labor or jobs. The usual explanation is that these types of jobs are nonrepetitive and are therefore impossible to measure. Other rationales are that indirect operations may involve groups of people, the unit of output appears difficult to define, the job may entail numerous sub-operations, the work cycle is long, and the operation constantly changes geographic locations.

Advantages of an indirect labor evaluation can include operating improvements and better worker performance, and labor loads can be budgeted. The efficiency of indirect labor areas can be determined, and accurate planning and scheduling will facilitate getting the job done on schedule.

The need for indirect labor in certain processes renders an economical process more expensive. For example, the advantages of high impact forgings may be offset partially by the extra indirect labor required to maintain the forging dies and presses in proper working condition. Setup becomes an important consideration at lower levels of production. For example, it may be more economical to use a method with less setup time even though the direct labor cost per unit is increased. Single-minute exchanges of dies (SMED) are a very important step toward reducing indirect labor (setup time). However, there are additional costs to maintain the dies on a regular basis.

5.2.4 Cost of tooling

Tooling costs are increasingly an important area of focus for many discrete manufacturers today as they look for new opportunities to cut product costs without sacrificing product quality. Special fixtures, jigs, dies, molds, patterns, gauges, and test equipment can be a major cost factor when new parts and new products or major changes in existing parts and products are put into production. With high production volume, a substantial investment in tools normally can be readily justified by the reduction in direct labor unit cost, since the total tooling cost amortized over many units of product results in a low tooling cost per unit. For low-volume production applications, even moderate tooling costs can contribute to relatively high unit tooling costs.

Manufacturers also need the ability to generate highly detailed cost estimates on components. Some capabilities to look for include a detailed tooling bill of materials (BoM) with information on:

- Physical characteristics of the tool (e.g., part size, mold size, material weight, actions, lifters, number of drops, etc.)
- Materials and purchased items used in the tool (e.g., core and cavity plates, ejector box, actions and inserts, stop pins, Electron Discharge Machining (EDM) carbon, etc.)
- Labor and machine times (design, machining, assembly, finishing, tryout, labor hours by process, Co-ordinate Measuring Machine (CMM) inspection, etc.)

- Automated tooling estimates each time component is used; this provides nontooling experts with quick access to precise estimates in real time
- Refinement tools for final adjustments by tooling experts
- Ability to amortize tooling or account for separately
- Setup and calibration to specific company, equipment, rates, manufacturing rules, and operations for generating a specific plant's actual costs.

5.2.5 *Capital invested*

The best way to determine whether a manufacturing company has a moat is to measure its return on invested capital (ROIC). The upshot is it gives the clearest picture of exactly how efficiently the company is using its capital, and whether or not its competitive positioning allows it to generate solid returns from that capital. Of course, it is easier and less risky for a company to start conceptualizing a new product that utilizes an extension of existing facilities. In addition, the capital investment in a new product can be minimized if the product can be made by using available capacity of manufacturing processes currently utilized. Thus the availability of plant, machines, equipment, and support facilities should be taken into consideration as well as the capital investment required for other alternatives. If sufficient productive capacity is available, no investment may be required for capital items in undertaking the production of a new part or product with existing processes. Also, if an entire supply chain network with retailer, vendors, and distributors is readily available, it reduces the amount of investment required.

5.3 Material selection procedures

5.3.1 *Grouping materials in families*

Figure 5.3 illustrates how the kingdom of materials can be subdivided into families, classes, subclasses, and members. Each member is characterized by a set of attributes—its properties. As an example, the materials kingdom contains the family "metals," which in turn contains the class "steels," the subclass "T300 stainless steel," and finally particular member properties. It, and every other member of the materials kingdom, is characterized by a set of attributes, which include its mechanical, thermal, electrical, and chemical properties; its processing characteristics; its cost and availability; and the environmental consequences of its use. We call this its *property profile*. Selection involves seeking the best match between the property profile of materials in the kingdom and the requirements of the design.

5.3.2 *Grouping materials based on process compatibility*

Based on the examination of material families and their properties, a tentative pool of materials is made. If one material is clearly outstanding and fits all requirements, then it may be selected, but in reality this usually is not the case. Further filtering is

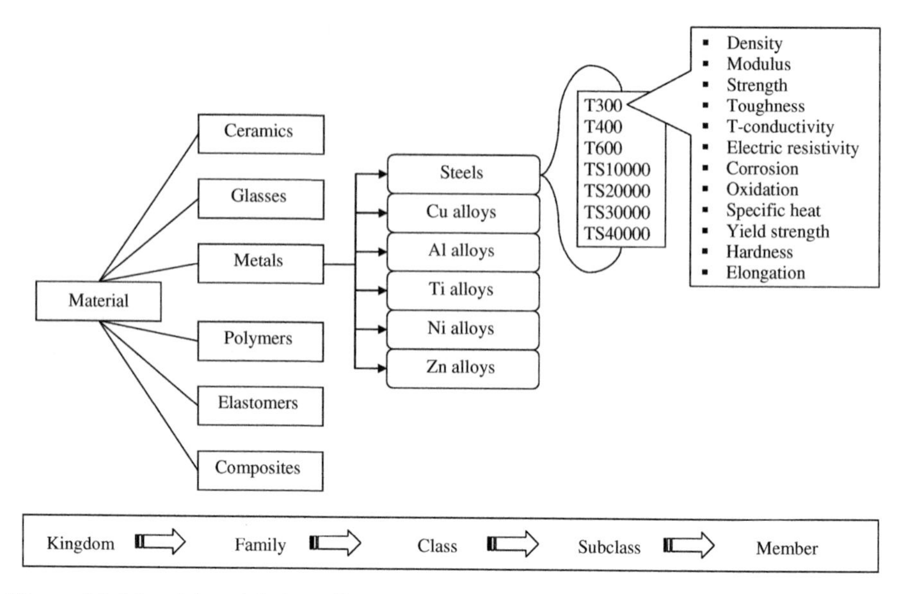

Figure 5.3 Materials and their attributes.
Source: Adapted from Ashby (2005).

required based on the fabrication process and suitability of each prescreened material to each process. The shape, geometry, surface finish, detailed specifications, and the like, to a large extent, determine which processes can and cannot be used to manufacture the product. Selecting a material based on processing requirements is a complex task because of the very large number of processing methods and sequence possibilities. The task is made even more complicated with evolving process and material combinations. Nonetheless, a decision has to be made on processing to optimize the cost and performance of the material(s) selected.

Often screening, ranking, and cost optimization processes are used to arrive at the best combination of materials. Screening and ranking eliminate candidates that cannot do the job because one or more of their attributes lies outside the limits imposed by the design. Then the manufacturing cost for a standard simple component is estimated. This standard cost is modified by a series of multipliers, each of which allows for an aspect of the component being designed. These aspects include a combination of size, shape, material, and so on. Processes then are ranked by the modified cost calculated. This allows designers to make a final choice. The final choice is made with local factors taken into account. Local factors are the existing in-house expertise or equipment, the availability of local suppliers, and so forth. A systematic procedure cannot help here: the decision must be based instead on local knowledge (DeGarmo et al., 1984).

Consider the component shown in Figure 5.4, which needs to be manufactured with the constraints shown in Table 5.1. Using the elimination technique, the best

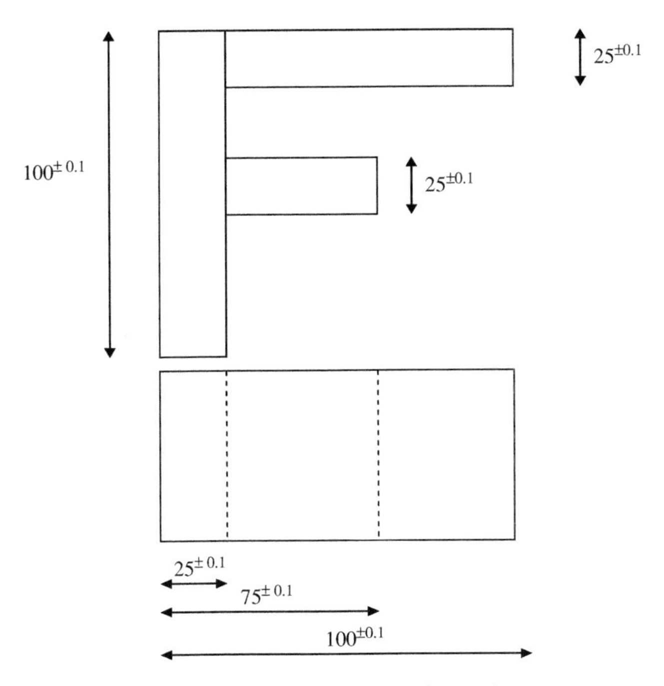

Figure 5.4 Basic drawing of the component to be manufactured.

Table 5.1 Required attributes (shape and material) for the component to be manufactured

Attributes	Condition
Shape	Required
Depression	Required
Uniform wall	Required
Uniform cross section	Required
No draft	Not required
Axis of rotation	Not required
Regular cross section	Not required
Captured cavity	Not required
Enclosed cavity	Maximum temperature 500°C
Material	Excellent corrosion resistance to weak acids and alkalis

combination of material and process can be determined. This process is shown in Figures 5.5–5.7. The desired material–process combinations can be summarized from the shape and material attributes and process relationship for the component under consideration. These are shown in Table 5.2.

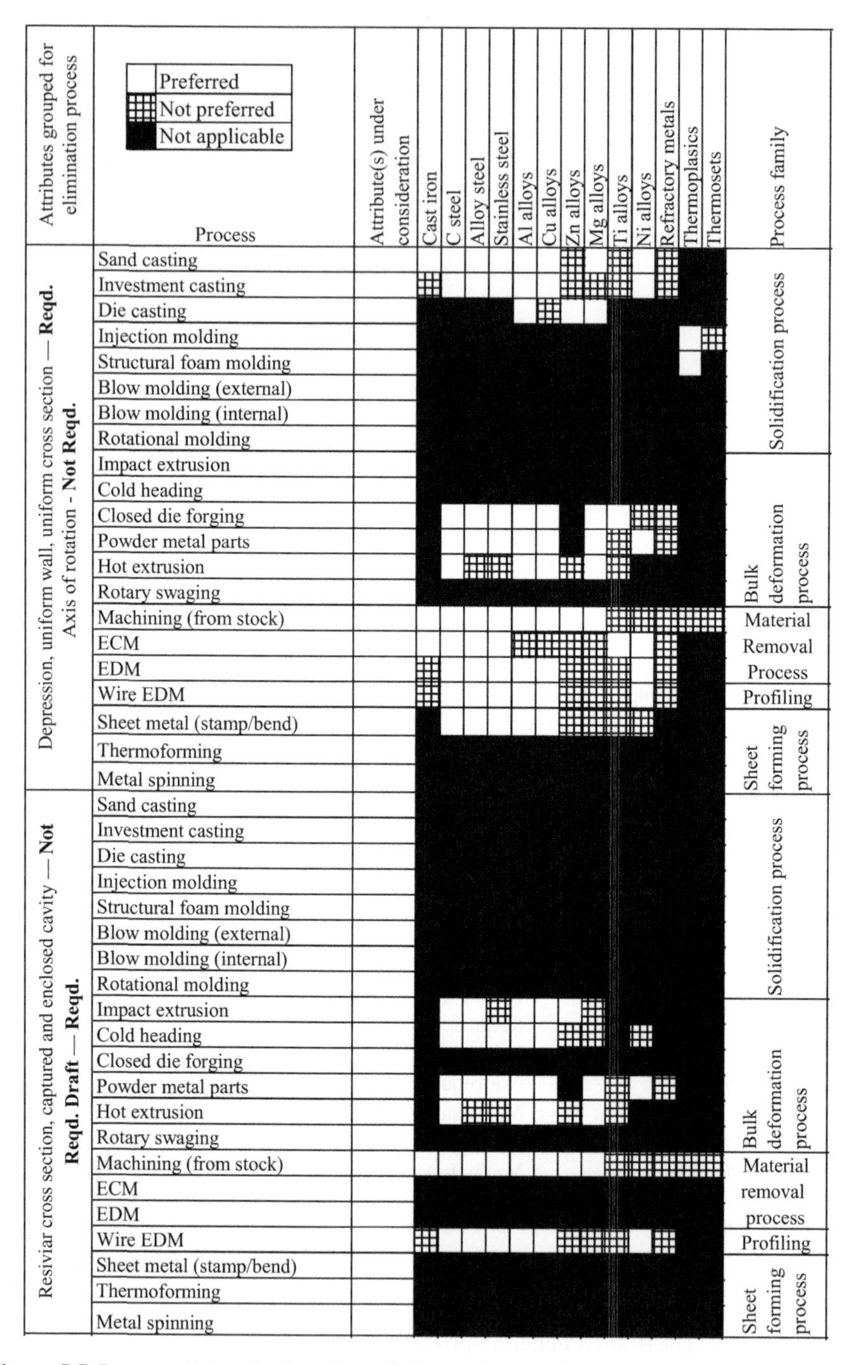

Figure 5.5 Process elimination based on all the required attributes.
Source: Adapted from Boothroyd et al. (1994).

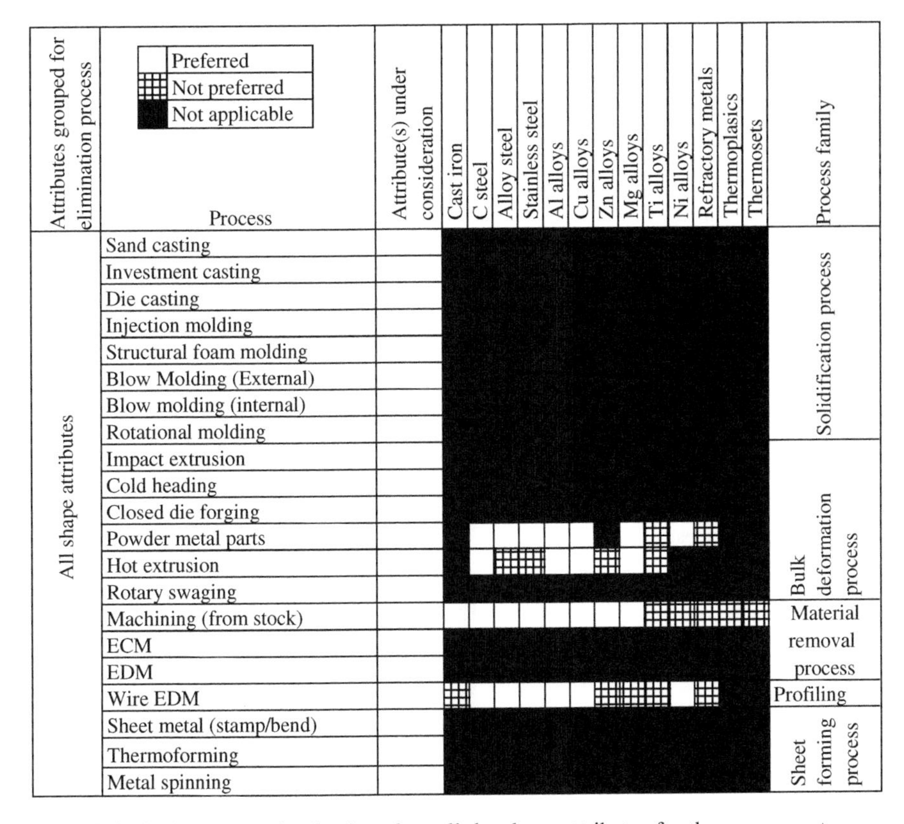

Figure 5.6 Final process selection based on all the shape attributes for the component. *Source*: Adapted from Boothroyd et al. (1994).

5.3.3 *Super materials and material substitution*

As an alternative procedure of selecting materials for a product, super materials are devised. A super material has the best attainable properties of all materials in that category. As the product attributes and process considerations are brainstormed, trade-offs in the properties of the super material are made and the choices of material suitable for that process are narrowed down. The goal of material substitution may be chosen from a combination of one or more of the following:

- To either cease or reduce the use of hazardous raw materials, such as heavy metallic pigments and dyestuffs or chlorine solvents
- Advances in technology
- Government laws, regulations, or statutes requiring use of environmentally friendly materials in production processes, to save energy, reduce waste reduction, and so forth.

Due to increasing pressure to produce high-quality products quickly, designers tend to substitute materials without substantially altering the design. This might result in improved levels of quality and cost, but the performance of the product may

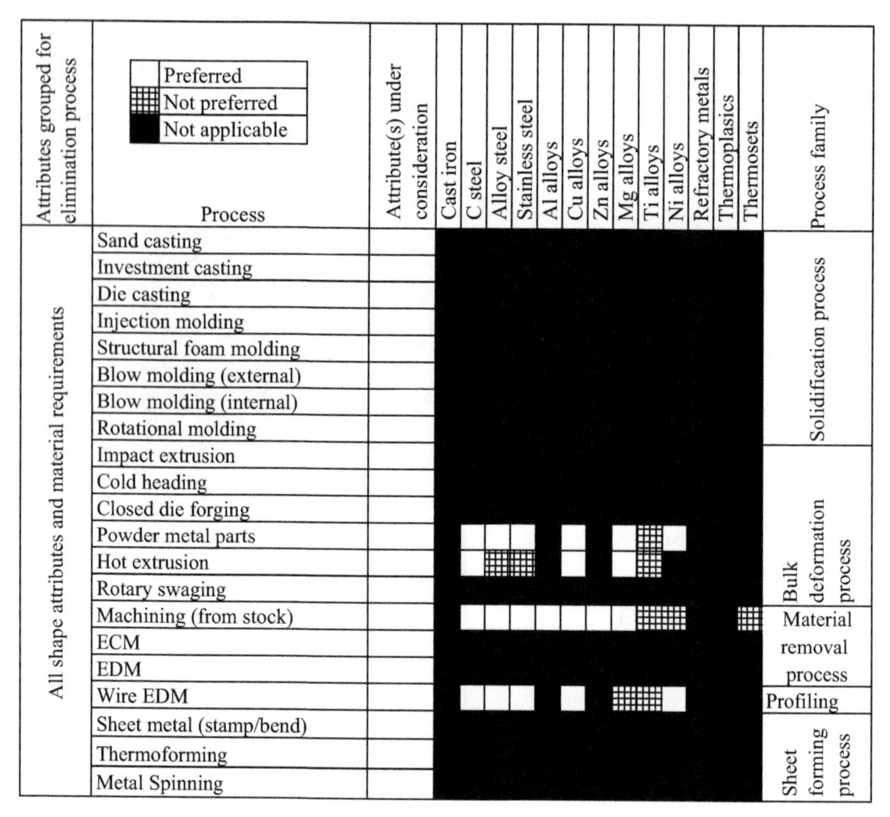

Figure 5.7 Final process selection based on all the shape attributes and material requirements of the component.
Source: Adapted from Boothroyd et al. (1994).

Table 5.2 Summary of the desired material–process combination based on the required attributes of the component under consideration

Process	Desired material	Less-desired materials
Powder metal parts	C-steel, alloy steel, stainless steel, Cu alloys, Mg alloys, Ni alloys	Ti alloys
Hot extrusion	C-steel, Cu alloys, Mg alloys	Alloy steel, stainless steel, Ti alloys
Machining from stock	C-steel, alloy steel, stainless steel, Al alloys, Zn alloys, Cu alloys	Ti alloys, Ni alloys, thermoset
Wire EDM	C-steel, alloy steel, stainless steel, Cu alloys, Ni alloys	Mg alloys, Ni alloys

be hampered. It is necessary that, when substitution of materials is considered, the designers approach it as a fresh material selection problem and perform the entire due process of revisiting all attributes and also reestablish all material–process–attribute relations.

5.3.4 Computer-aided material selection

To be of real design value, the selection of material–process combinations and their ranking should be based on information generally available early in the concept design stage of a new product, for example:

- Product life volume
- Permissible tooling expenditure levels
- Possible part shape categories and complexity levels
- Service requirements or environment
- Appearance factors
- Accuracy factors.

Due to the vast number of process–material combinations, designers often never arrive at a single right combination but are presented with a number of permutations and combinations. This problem could be solved to a great extent by use of computer-aided materials and process selection systems (CAMPS). CAMPS is a commercially available relational database system. In the selector (Figure 5.8), inputs made under the headings of "Part Shape," "Size," and "Production Parameters" are used to search a comprehensive process database to identify processing possibilities. However, it is recognized that process selection completely independent of material performance requirements would not be satisfactory. For this reason, required performance parameters can also be specified by making selections under the general categories of "Mechanical Properties," "Thermal Properties," "Electrical Properties," and "Physical Properties." As many selections as required can be made, and at each stage a list of

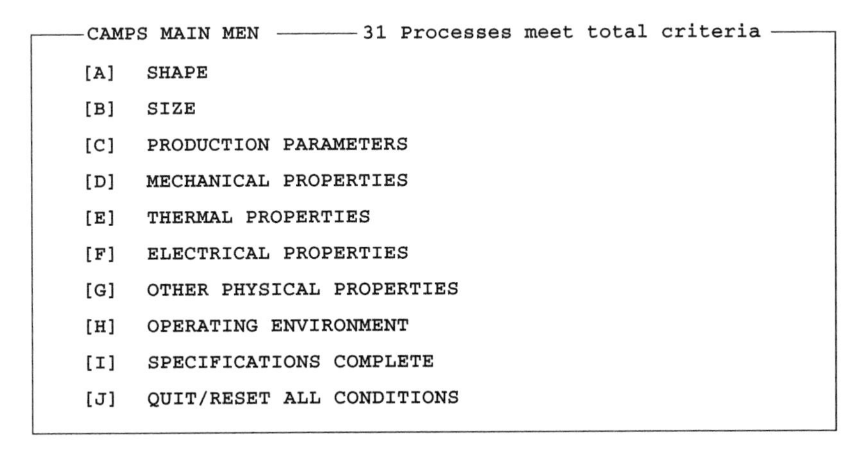

CAMPS MAIN MEN
31 Processes meet total criteria
[A] SHAPE
[B] SIZE
[C] PRODUCTION PARAMETERS
[D] MECHANICAL PROPERTIES
[E] THERMAL PROPERTIES
[F] ELECTRICAL PROPERTIES
[G] OTHER PHYSICAL PROPERTIES
[H] OPERATING ENVIRONMENT
[I] SPECIFICATIONS COMPLETE
[J] QUIT/RESET ALL CONDITIONS

Figure 5.8 Screenshot of the main selector input screen for CAMPS.

candidate processes is presented to the system user. Processes may be eliminated directly because of shape or size selections or when performance selections eliminate all the materials associated with a particular process.

During an initial search phase, when a rapid response to changes in input is essential, it would be inappropriate to search extensive material databases to identify precise metal alloys, polymer specifications, powder mixes, and the like. This would lead to unacceptably slow search procedures and provide information largely irrelevant to early process and material decision making. For example, listing all the thermoplastic resins that satisfy the specified performance requirements clearly would be premature in early discussions of the relative merits of alternative processes, their required tooling investments, and the likely size and shape capabilities. A more efficient procedure is being adopted in the CAMPS system, where, for each process, a type of super material specification, which comprises the best attainable properties of all of the materials in the corresponding category, is provided. The super material specifications are maintained automatically by the program (Boothroyd et al., 1991).

5.4 Design recommendations

5.4.1 Minimize material costs

- Use commercially available mill forms to minimize in-factory operations.
- Use standard stock shapes, gauges, and grades or formulations rather than special ones whenever possible.
- Consider the use of prefinished material as a means of saving costs for surface finishing operations on the completed components.
- Select materials as much as possible for processability; for example, use free-machining grades for machined parts and easily formable grades for stamping.
- Design parts for maximum utilization of material. Make ends square or nestable with other pieces from the same stock.
- Avoid designs with inherently high scrap rates.

Material should be selected based not only on the operating environment but also the temperature to which the product is exposed during the manufacturing process. Table 5.3 shows the maximum temperatures for various metals and nonmetals, and Figure 5.9 breaks down engineering materials by family.

5.4.2 Ferrous metals, hot-rolled steel

In choosing hot-rolled steel versus cold-finished material and in choosing the grade, the choice should be based on the concept of "minimum cost per unit of strength." Often, grades with higher carbon content or low alloy content provide lower cost parts than low-carbon grade parts, as lighter sections can be used.

When bending hot-finished steel members, the bending line should be at right angles to the grain direction from the rolling operations. Also, provide a generous bend radius. Both actions will avoid material fractures at the bend.

Table 5.3 Melting point and maximum service temperatures of selected materials

Material	°C	Material	°C
Carbon	3700	Alloy steels	1430–1510
Tungsten	3400	Stainless steel	1370–1450
Tantalum	2900	Wrought iron	1350–1450
Magnesia	2800	Cast iron, gray	1350–1400
Molybdenum	2620	Copper	1083
Vanadium	1900	Gold	1063
Chromium	1840	Aluminum bronze	855–1060
Platinum	1773	Lead	327
Titanium	1690	Tin	231
Carbon steels	1480–1520	Indian rubber	125
Beryllia	2400	Polysulfone	150–175
Silicon carbide	2310	Nylon	80–150
Alumina	1950	Polycarbonate	95–135
Mullite	1760	Polypropylene	90–125
Cubic boron nitride	1600	Polyethylene	80–120
Porcelain enamel	370–820	Felt, rayon viscose	107
Silicones	260–320	Polyurethane	90–105
Polyesters	120–310	Acetal	85–105
Glass, soda lime	290	Polystyrene	65–105
Epoxy	95–290	Cellulosic	50–105
Glass, borosilicate	260	ABS	60–100
Fluoroplastics	50–260	Acrylic	52–95
Phenolic	90–260	Natural rubber	82
Melamine	100–200	Vinyl	55–80

Source: Adapted from Bralla (1998).

When machining a hot-rolled material, it is necessary to remove sufficient stock to avoid surface irregularities such as scales, seams, deviations from straightness or flatness, and decarburization. The design recommendations (1.5–3 mm) are liberal. For moderate and high levels of production, it is worthwhile to test the actual condition of the steel being used.

5.4.3 Ferrous metals, cold-finished steel

- Use the simplest cross-sectional shape possible consistent with the function of the part. Avoid holes, grooves, and the like. Avoid undercuts, as they are more expensive (Figure 5.10).
- Use standard rather than special shapes.
- Avoid sharp corners, as they are more difficult to manufacture and may create assembly problems (Figure 5.11).
- Grooves deeper than 1.5 times the width of the part are not feasible unless the bottom radii are generous (Figure 5.12).

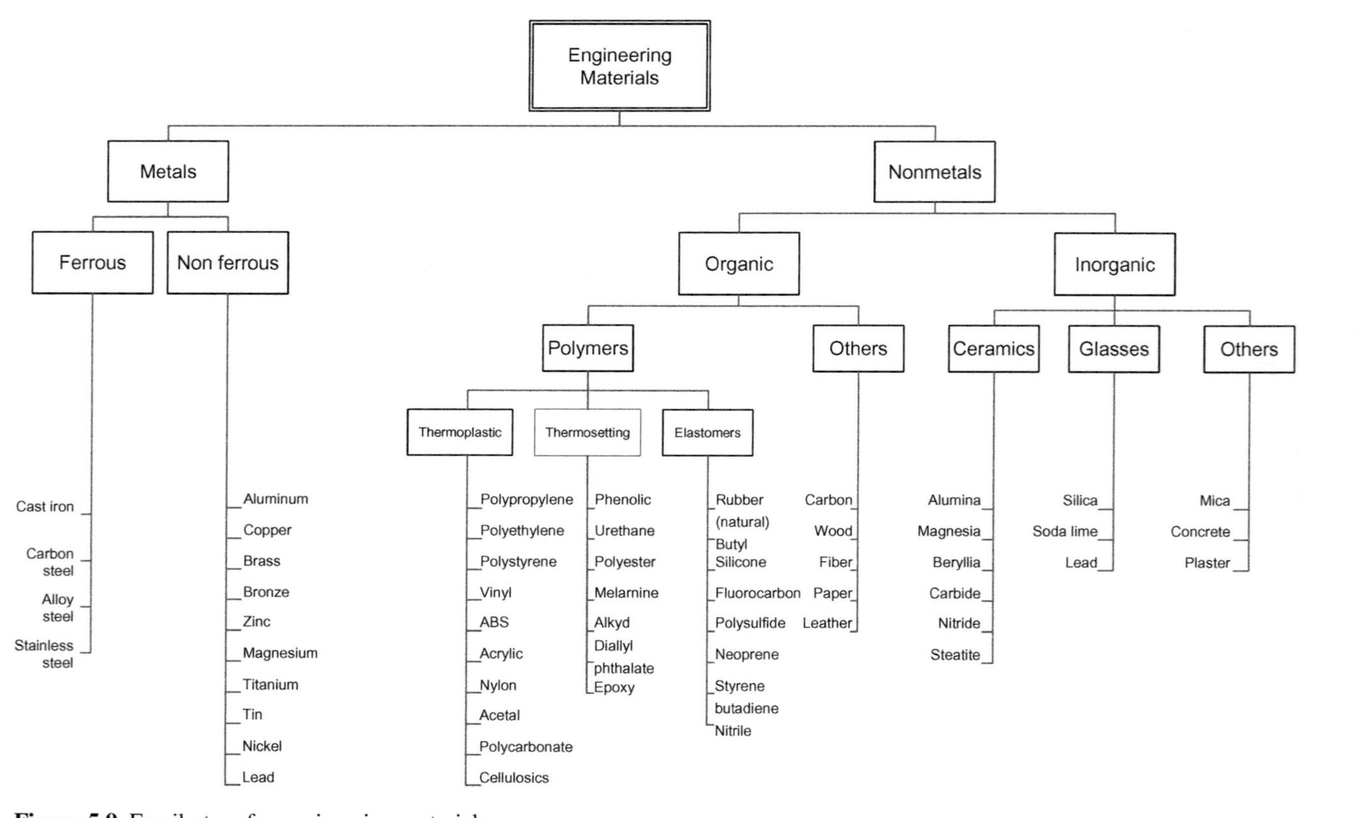

Figure 5.9 Family tree for engineering materials.
Source: Adapted from Bralla (1998).

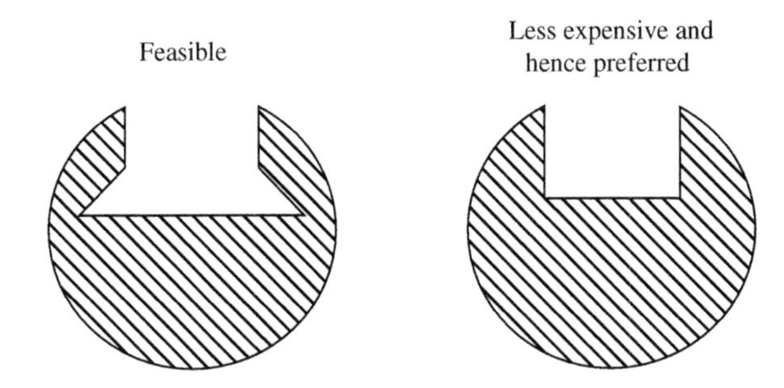

Figure 5.10 Design recommendations for ferrous metals, cold-finished steel: use the simplest cross sections.

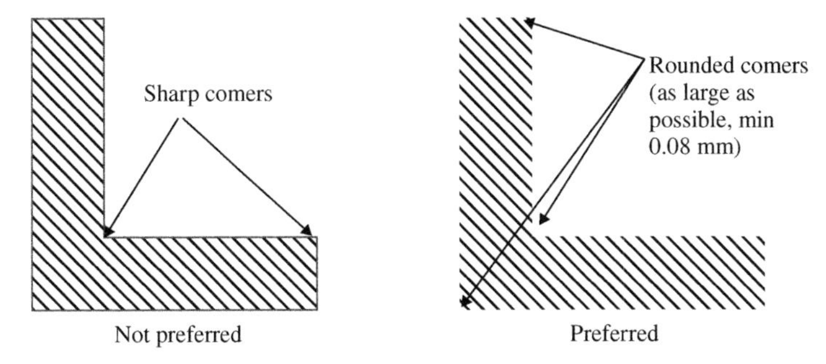

Figure 5.11 Design recommendations for ferrous metals, cold-finished steel: avoid sharp corners.

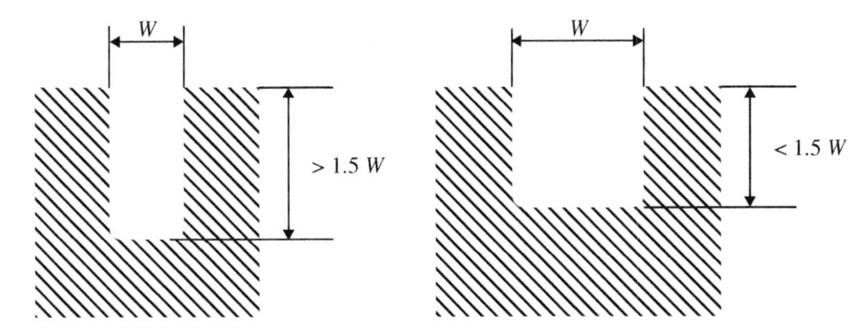

Figure 5.12 Design recommendations for ferrous metals, cold-finished steel: groove width-to-depth specifications.

- Avoid abrupt changes in the section thickness as they introduce local concentration; for example, cutting a credit card by bending it.
- Specify the most easily formed materials to minimize cost and maximize precision.
- Welded tubular sections are more economical than seamless types.

5.4.4 Ferrous metals, stainless steel (Franson, 1998)

- Use the least expensive stainless steel and product form suitable for the application.
- Use rolled finishes.
- Use the thinnest gauge required.
- Use a thinner gauge with textured pattern.
- Use a still thinner gauge with continuous backing (to avoid fracture).
- Use standard roll-formed sections whenever possible.
- Use simple sections for economy of forming.
- Use concealed welds whenever possible to eliminate refinishing.
- Use grades that are especially suited to the manufacturing process such as free-machining grades.

5.4.5 Nonferrous metals (Skillingberg, 1998)

5.4.5.1 Aluminum

- Use largest bend radii possible when forming, to avoid tearing.
- When attaching to other metal parts, the facing surface should be insulated to avoid galvanic corrosion (use zinc chromate or zinc phosphate).
- Use alkali-resistant paint on aluminum parts when joining them with wood, concrete, or masonry (this is not needed for aluminum parts embedded in concrete).

5.4.5.2 Copper and brass (Kundig, 1998)

- Use something else, as these are expensive.
- Avoid machining; use extrusion and press forming to avoid loss of material.
- Use stock sizes requiring minimum processing.
- Use the correct alloy; easily formable alloys are not easily machinable.

5.4.5.3 Titanium

- When bending titanium sheets, generous bend radii should be provided.
- Cross-section thickness should be 16 mm or more.
- Provide generous draft angles, at least 5–7°.
- Rib widths should be 10 mm or more, and the rib height should not exceed four times the rib width.
- The fillet radius of the ribs should at least be 25% of the rib height.

5.4.5.4 Magnesium

- Sharp corners, notches, and other stress raisers should be avoided.
- Strong clamping points should be provided to avoid distortion (particularly when the part needs to be clamped).
- Sidewalls should be at least half the height of the walls up to 0.25 in. and at least 13% the height of the walls 2 in. high.
- Ribs also should be fairly thick, with a top radius of at least half the thickness.
- Inside corners should be provided with a generous radius.

- Very thin-walled, large cross sections should be avoided. The length of the sections should not exceed 20 times the thickness.
- For cold press forming, the bend radii should be generous.

5.4.5.5 Zinc and its alloys

- Bends in regular commercial rolled zinc should be at right angles to the grain or rolling direction.
- The bend radius should be at least equal to the material thickness.
- For forging zinc, use a combination of zinc and magnesium, with up to 25% magnesium.

5.4.6 Nonmetals (Harper, 1998)

5.4.6.1 Thermosets and thermoplastics

- Shrinkage on cooling and curing of thermoset plastics must be taken into consideration when designing parts. Table 5.4 shows the minimum and maximum shrinkage rates during molding for various thermoset plastics and thermoplastic parts.
- Internal undercuts in a part are impossible to mold and should be avoided. External undercuts can be molded but must be avoided unless absolutely essential.
- All corners should have a radius or fillet except at set-in sections of the mold or at the parting line.
- Spacing between holes and next to sidewalls should be as large as possible. The minimum values for the holes are shown in Table 5.5.
- Molded ribs may be incorporated to increase strength or decrease warpage of thermoset parts. The width of the base of the rib should be less than the thickness of the wall to which it is attached.
- Taper or draft should be provided both inside and outside the thermoset parts. Inside surfaces should be provided with greater draft because molded parts tend to shrink toward the mold surface rather than away from it. Table 5.6 lists minimum drafts for common materials.
- Internal and external threaded holes are expensive, as they increase the cost of the part being manufactured and the mold required for manufacture.
- Thermoplastics ejector pins must be placed on the underside of the part.

Table 5.4 Minimum and maximum shrinkage rates during molding (on cooling)

Material	Percent	Material	Percent
Phenolic	0.1–0.9	Acrylic	0.3–0.8
Urea	0.6–1.4	Acrylonitrile butadiene styrene	0.3–0.8
Melamine	0.8–1.2	Nylon	0.3–1.5
Diallyl phthalate	0.3–0.7	Polycarbonate	0.5–0.7
Alkyd	0.5–1.0	Polyethylene	1.5–5.0
Polyester	0–0.7	Polypropylene	1.0–2.5
Epoxy	0.1–1.0	Polystyrene	0.2–0.6
Silicone	0–0.5	Polyvinyl chloride, rigid	0.1–0.5
Acetal	2.0–2.5	Polyvinyl chloride, flexible	1.0–5.0

Source: Adapted from Bainbridge (1998).

Table 5.5 Minimum recommended hole spacing in thermoset parts (Bralla, 1998)

Diameter of hole (mm)	Minimum distance to sidewalls (mm)	Minimum distance between holes (mm)
1.5	1.5	1.5
3.0	2.4	2.4
4.8	3.0	3.0
6.3	3.0	4.0
9.5	4.0	4.8
12.7	4.8	5.6

Table 5.6 Recommended minimum draft for some common materials

Material	Draft (°)
Polyethylene	¼
Polystyrene	½
Nylon	0–¼
Acetal	0–¼
Acrylic	¼

Source: Adapted from Bralla (1998).

5.4.6.2 Rubber

Table 5.7 lists advantages and disadvantages of some rubbers.

- Holes in rubber parts are the easiest to form and the most economical to produce during molding. Drilling holes in cured rubber by conventional means is difficult due to the flexible nature of rubber parts.
- Holes should be shallow and as wide as possible consistent with the functional needs. Avoid through holes of small size; if necessary, through holes should be at least 0.8 mm in diameter and 16 mm in depth.
- Hole-to-hole and hole-to-edge spacing should be at least one hole diameter to prevent tearing the rubber.
- Undercuts should be avoided as they increase both difficulties during demolding and production costs. If they are absolutely necessary, then they should be machined on either low- or medium-hardness rubber.
- For screw threads on rubber, it is not feasible to separate fasteners from the molded rubber; they should be placed so as to keep the rubber thickness as uniform as possible to avoid stress concentration.
- Angle inserts molded into rubber should be given generous radii at the bend to avoid cutting the rubber.

Table 5.7 **Advantages and disadvantages of some types of rubber**

Rubber	Advantages	Disadvantages	Typical applications
Natural rubber (NR)	Building tack, resilience, and flex resistance	Reversion at high molding temperature	Tires, engine mounts
Styrene butadiene rubber (SBR)	Abrasion resistance	Poor ozone resistance	Tires, general molded goods
Ethylene propylene diene monomer (EPDM)	Good ozone resistance	Poor hot tear resistance	Door and window seals, wire insulations
Nitrile butadiene rubber (NBR) or nitrile	Good solvent resistance	Poor building tack	O-rings and hose
Thermoplastic rubber	Short injection molding cycle	Poor creep characteristics	Shoe soles, wire insulation
Polyurethane	Short molding cycle and low molding pressure	Adhesion to mold	Cushioning, rolls, exterior automotive parts
Isobutylene isoprene rubber (IIR) or butyl	Low air penetration in finished parts	Voids caused by air trapped during molding	Inner tubes body mounts for automobiles
Chloroprene rubber (CR) or neoprene	Moderate solvent resistance	Sticking during processing and premature cross-linking (scorch) with some types	Hose tubes and covers, V-belts

Source: Adapted from Sommer (1998).

- The need for draft in molded rubber parts varies with both the part design and the nature of rubber. For parts having hardness below 90 Shore A, no draft is needed. Other, softer rubber must be provided draft of ¼° to 1° perpendicular to the parting line.
- Providing radii and fillets to corners is highly recommended as they reduce the stress concentrations in the parts and the mold. Fillet radius of at least 0.8 mm should be provided.
- Shrinkage of rubber products from the mold cavity temperature to room temperature varies from 0.6% to 4%, depending on the type of rubber and its filler content.

5.4.6.3 *Ceramics and glass*

Table 5.8 lists process properties of ceramics and glass.

- Ceramic part edges and corners should have generous radii or chamfers to prevent chipping and stress concentration points. Outside and inside radii should be at least 1.5 and 2.4 mm, respectively.
- Due to sagging or distortion during firing, large unsupported overhanging sections must be avoided.

Table 5.8 **Process properties for ceramics and glass, by manufacturing method**

Material and method	Technical ceramics, mostly machined	Technical ceramics, mostly pressed	Pressed glass	Blown glass	Flat glass	White ware	Refractories
Normal economic production quantities	Short to medium run	Medium to long run	Long run	Long run	Long run (without thickness change)	Medium to long run	Medium to long run
Investment required							
• Equipment	Moderate	High	Medium to high	Very high	Very high	High	High
• Tooling	Low	High	Medium to high	Very high	Very high	High	High
• Lead time to tool up for new product	1 month	3–6 months	3 months	3 months		3 months	3–6 months
• Typical output rate	Varies greatly, typically 100 pieces/shift	15,000/shift	Up to 40,000 pieces/day	150,000 containers/day to 1,000,000 lightbulbs/day	200 tons/day	6–10 pieces/day/mold	40,000 bricks/day
• Normal life of tooling	Cutter life very short compared to metal machining	Moderately long	Long	Long run	1–2 months	Plaster molds limited to 200–1000 parts (not reclaimable)	Moderately long

Source: Adapted from Mohr (1998).

- Pressed parts must be designed with uniform wall thickness. Differential wall thickness leads to nonuniform shrinkage, causing stress, distortion, or cracking. Sections should not exceed 25 mm in thickness.
- When hollow pieces are cast against a male mold, a draft angle of at least 5° must be provided to facilitate removal of the green body.
- Undercuts should be avoided in ceramic components.
- Cavities, grooves, and blind holes in pressed parts should not be deeper than half the part thickness and preferably only one-third the thickness.
- Extruded parts must be symmetrical with uniform wall thickness.
- Holes in pressed parts should be large and as widely spaced as possible. Thin walls between holes, depressions, or outside edges should be avoided. These walls should be at least as thick as the basic walls of the part.
- Ribs and fins should be well rounded, wide, and well spaced and have normal draft.
- Material removal rates are slow and the operations expensive. Hence, grinding after firing of ceramic parts is preferred and provides high accuracy.
- Holes, cavities, and deep slots can cause molding problems and should be included only when absolutely necessary. Holes are not normally punched through in the pressing operation but machined from a thin web or hollow boss.
- Walls must be of uniform thickness.
- Parts must be gently curved rather than sharp edged.
- Lettering or other irregular surface features may be incorporated as long as they are aligned in the direction of, and not perpendicular to, the mold opening.
- Ribs and flanges can be incorporated in some items such as electrical insulators. They normally are not practicable for general-purpose design and manufacture.
- Threads for bottle caps or similar connecting devices may be incorporated in blown glass parts as they are with blow-molded plastics.

References

Ashby, M., 2005. Materials Selection in Mechanical Design, third ed. Elsevier Butterworth–Heinemann, Oxford.

Bainbridge, R.W., 1998. Thermosetting-plastic parts. In: Bralla, J.G. (Ed.), Design for Manufacturability Handbook, second ed. McGraw-Hill, New York, NY.

Boothroyd, G., Dewhurst, P., Knight., W.A., 1991. Research program on material selection and processes for component parts. Int. J. Adv. Manuf. Technol. 6, 98–111.

Boothroyd, G., Dewhurst, P., Knight., W.A., 1994. Product Design for Manufacture and Assembly. Marcel Dekker, New York, NY.

Design for Manufacturability HandbookBralla, J.G. (Ed.), 1998., second ed. McGraw-Hill, New York, NY.

DeGarmo, E.P., Black, J.T., Kohser, R.A., 1984. Materials and Processes in Manufacturing, sixth ed. Macmillan Publishing Company, New York, NY.

Franson, I.A., 1998. Stainless steel. In: Bralla, J.G. (Ed.), Design for Manufacturability Handbook, second ed. McGraw-Hill, New York, NY.

Harper, C.A., 1998. Non metallic materials. In: Bralla, J.G. (Ed.), Design for Manufacturability Handbook, second ed. McGraw-Hill, New York, NY.

Kundig, K.J.A., 1998. Copper and brass. In: Bralla, J.G. (Ed.), Design for Manufacturability Handbook, second ed. McGraw-Hill, New York, NY.

Mangonon, P.L., 1999. The Principles of Material Selection for Engineering Design. Prentice Hall, Upper Saddle River, NJ.

Mohr, J.G., 1998. Ceramic and glass parts. In: Bralla, J.G. (Ed.), Design for Manufacturability Handbook, second ed. McGraw-Hill, New York, NY.

Skillingberg, M.H., 1998. Non ferrous metals. In: Bralla, J.G. (Ed.), Design for Manufacturability Handbook, second ed. McGraw-Hill, New York, NY.

Sommer, J.G., 1998. Rubber parts. In: Bralla., J.G. (Ed.), Design for Manufacturability Handbook, second ed. McGraw-Hill, New York, NY.

Selection of Manufacturing Processes and Design Considerations

6.1 Introduction

The manufacturing process is the science and technology by which a material is converted into its final shape with the necessary structure and properties for its intended use. Formation of the desired shape is a major portion of processing. The product processing could be a simple, one-step operation or a combination of various processes, depending on the processability of the material used and the specifications for the finished part, which includes surface finish, dimensional tolerances, and so forth. The method of selecting the appropriate process is closely tied to the selection of material.

What leads to a successful manufacturing process? The performance of any manufacturing process depends on

Rate: Material flow through the system
Cost: Material, labor, tooling, equipment
Time: Lead time to procure materials, processing time, setup time
Quality: Deviation from the target.

All these factors result from decisions made in selecting the process–material–part combination. As designers and engineers developing a new product, at this juncture, we already have the basic part drawing and a selection of various material–process combinations feasible for the part. The next stage is arriving at the material-manufacturing process combination that is technically and economically feasible. Figure 6.1 shows the taxonomy of manufacturing processes. The processes are arranged by similarity of function.

Manufacturing processes can be broadly classified into three categories. Based on the desired outcome, they are primary, secondary, or tertiary processes. To discuss all the process and their parameters in detail is beyond the scope of this book. We shall look into the key processes, their classification, and their specific design guidelines.

6.1.1 *Primary processes*

The primary process generates the main shape of the final product. The primary process is selected to produce as many required shape attributes of the part as possible. Such processes appear at the top of the sequence of operation for a part and include processes such as casting, forging, molding, rolling, and extrusion.

Product Development. DOI: http://dx.doi.org/10.1016/B978-0-12-799945-6.00006-5

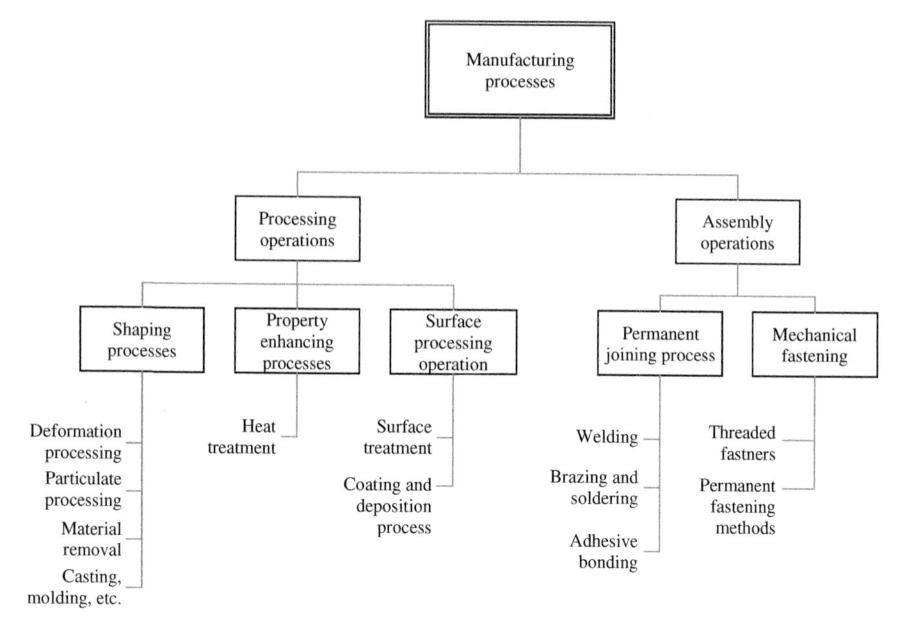

Figure 6.1 Taxonomy of manufacturing processes.
Source: Adapted from Groover (1996).

1. **Casting**: Casting is the fastest way to attain simple or complex shapes for the part from its raw material. The casting process basically is accomplished by pouring a liquid material into a mold cavity of the shape of the desired part and allowing it to cool. The different types of casting methods (for both metals and nonmetals) are shown in Figure 6.2.
2. **Forging**: Forging is a deformation process in which the work is compressed between two dies using either impact or gradual pressure to form the part. The different types of forging processes are shown in Figure 6.3.
3. **Extrusion**: Extrusion is a compression forming process in which the worked metal is forced to flow through a die opening to produce the desired cross-sectional shape. Extrusion usually is followed by a secondary process, cold drawing, which tends to refine the molecular structure of the material and permits sharper corners and thinner walls in the extruded section. The different extrusion processes can be classified as shown in Figure 6.4.

6.1.2 Secondary processes

Secondary processes, in addition to generating the primary shape, form and refine features of the part. These processes may appear at the start or later in a sequence of processes. These include all the material removal processes and processes such as machining, grinding, and broaching.

Machining is the process of removing material from a workpiece in the form of chips. The term *metal cutting* is used when the material is metallic. Most machining has a very low setup cost compared to the forming, molding, and casting processes. However, machining is much more expensive for high volumes. Machining is necessary where tight tolerances on dimensions and finishes are required.

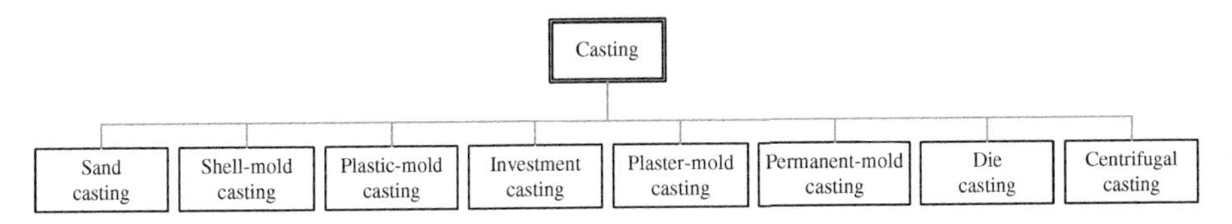

Figure 6.2 Types of casting processes.

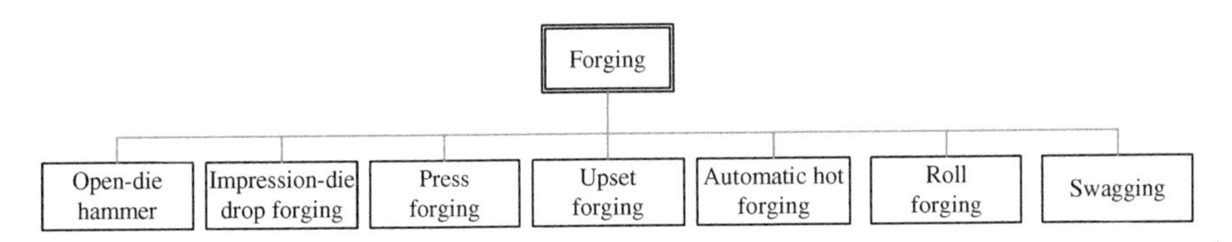

Figure 6.3 Types of forging processes.

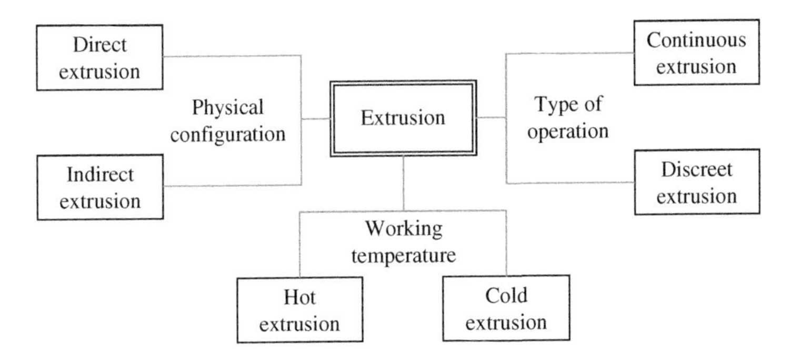

Figure 6.4 Types of extrusion processes.

The different machining processes are shown in Figure 6.5. They are commonly divided into the following categories:

Cutting generally involves single-point or multipoint cutting tools, each with a clearly defined geometry.
Nontraditional machining processes utilize electrical, chemical, and optimal sources of energy.
Abrasive machining processes are categorized under surface treatment and, hence, are discussed as tertiary processes.

Tables 6.1 and 6.2 provide summary lists of traditional and nontraditional machining processes.

6.1.3 Tertiary processes

Tertiary processes do not affect the geometry or shape of the component and always appear after one or more primary and secondary processes. This category consists of finishing processes, such as surface treatments and heat treatments. Selection of a tertiary process is simplified because many tertiary processes affect only a single

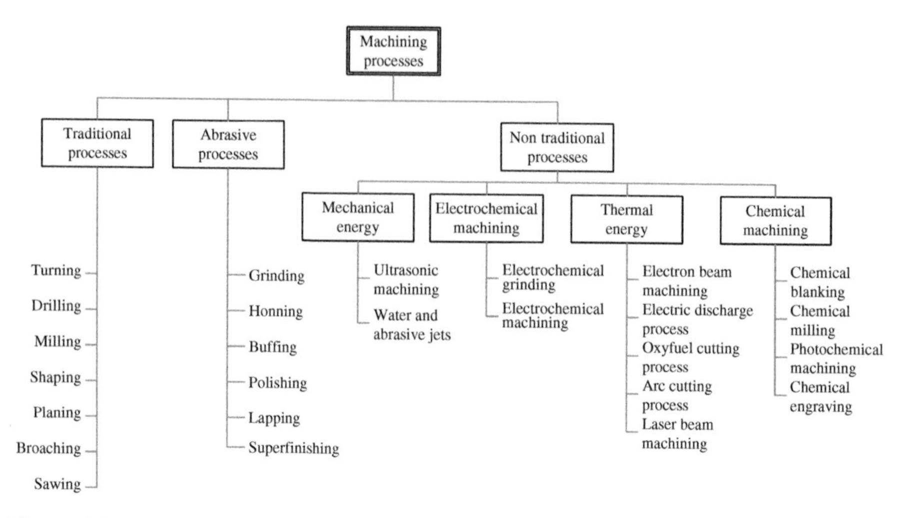

Figure 6.5 Classification of the various machining processes.

attribute of the part. Table 6.3 shows a summary of some typical abrasive machining (both traditional and nontraditional) processes.

6.2 Design guidelines

6.2.1 *Design guidelines for casting (Zuppann, 1998; DeGarmo et al., 1984)*

Shrinkages can cause induced stresses and distortion in cast components. The amount of shrinkage varies with the type of metal used for casting but can be predicted and compensated for by making patterns slightly oversized. Table 6.4 lists normal shrinkage allowances for metals used in sand casting.

Although casting is a process that can be used to produce complex part geometries, simplifying the part design improves its castability. Avoiding unnecessary complexities simplifies the mold making, reduces the need for cores, and improves the strength of the casting.

Sharp corners and angles should be avoided, since they are a source of stress concentration and may cause hot tearing and cracks in casting. Generous fillets should be provided on inside corners (Figure 6.6) and sharp edges should be blended.

Section thickness should be uniform to avoid shrinkage cavities. Thicker sections create hot spots in the casting, because greater volume requires more time for solidification and cooling. These are likely locations of shrinkage cavities. Table 6.5 provides reasonable guidelines for minimum and desirable section thickness for different material-casting process combinations. Interior walls must, however, be 20% thinner than the outside members, because they cool more slowly than the external walls. Figure 6.7 shows a part that depicts this recommendation.

Table 6.1 **Summary list of various traditional machining processes**

Process	Most suitable materials	Typical applications	Material removal rate	Typical tolerances (mm)	Typical surface roughness
Turning	All ferrous and nonferrous materials considered machinable	Rollers, pistons, pins, shafts, valves, tubings, and pipe fittings	With mild steel, up to about 21 cm^3/hp min	±0.025	125 avg.
Drilling	Any unhardened material; carbides needed for some case-hardened parts	Holes for pins, shafts, fasteners, screw threads, clearance, and venting	With mild steel, up to about 300 cm^3/min	±0.15, −0.025	63–250
Milling	Any material with good machinability rating	Flat surfaces, slots, and contours in all kinds of mechanical devices	With mild steel, up to 6000 cm^3/min at 300 hp	±0.05	63–250
Planing	Low to medium: Carbon steels or nonferrous materials best	Primarily for flat surfaces such as machinery bases and slides but also for contoured surfaces	With mild steel, up to about 10 cm^3/hp min	±0.13	63–125
Shaping	Low to medium: Carbon steels or nonferrous materials best; no hardened parts	Primarily for flat surfaces such as machinery bases and slides but also for contoured surfaces	With mild steel, up to about 10 cm^3/hp min	±0.13	63–250
Broaching	Any material with good machinability rating	Square, rectangular, or irregular holes, slots, and flat surfaces	Max. of large surface broaches about 1300 cm^3/min	±0.025	32–125

Source: Adapted from Bralla (1998); DeGarmo et al. (1984).

Table 6.2 **Summary list of various nontraditional machining processes**

Process	Most suitable materials	Typical applications	Material removal rate	Typical tolerances (mm)
Chemical machining	All common ferrous and nonferrous metals	Blank thin sheets: wide, shallow cuts	0.0025–0.13 mm (depth of material removed/min)	±0.1
Ultrasonic machining	Hard, brittle, nonconductive materials	Irregular holes and cavities in thin sections	30–4000 cm^3/h	±0.025
Abrasive jet machining (AJM)	Hard, fragile, and heat-sensitive materials	Trimming, slotting, etching, drilling, etc.	1 cm^3/h	±0.13
Abrasive water jet machining	Hard metals and nonmetals	Cutting reinforced plastics, honeycombed materials, metal sheets thicker than 13 mm	1.5–2 m/min	±0.25
Electron beam machining (EBM)	Any material	Fine cuts in thin workpieces	0.05–0.12 cm^3/h	±10% allowed on hole and slot dimensions
Laser beam machining (LBM)	Any material	Blanking parts from sheet material; machining thin parts and small holes	2.5 m/min in mild steel with oxygen assist	±0.13
Electric discharge machining (EDM)	Hardened metals	Molds	49 cm^3/h	±0.05
Wire EDM	Hardened metals	Blanking dies	130–140 cm^2/h in 5 cm thick materials	±0.05
Electrochemical machining	Difficult-to-machine metals	For making complex shapes and deep holes	Max. 1000 cm^3/h	±0.05

Source: Adapted from Bralla (1998).

Table 6.3 Summary list of various abrasive machining processes

Process	Most suitable materials	Typical applications	Material removal rate	Typical tolerances (mm)
Center-type and centerless grinding	Nearly any metallic material plus many nonmetallic	Dies, molds, gauge blocks, machine surfaces	With mild steel, up to about 164 cm^3/min at 100 hp and high-speed grinding	+0 to −0.013
Surface grinding	Nearly any metallic material plus many nonmetallic	Dies, molds, gauge blocks, machine surfaces	With mild steel, up to about 164 cm^3/min at 100 hp and high-speed grinding	+0 to −0.1
Electrochemical honing (ECH)	Hardened metals	Finishing internal cylindrical surfaces	3–5 times faster than conventional honing	+0.006–0.0125
Electrical discharge grinding	Hard materials like carbide	Form tools	0.16–2.5 cm^3/h	±0.005
Electrochemical grinding	Hardened metals and carbides	Sharpening carbide cutting tools	100 cm^3/h	±0.025

Source: Adapted from Bralla (1998).

Table 6.4 Shrinkage allowances for metals commonly cast in sand molds

Metal	Percent
Gray cast iron	0.83–1.3
White cast iron	2.1
Ductile cast iron	0.83–1.0
Malleable cast iron	0.78–1.0
Aluminum alloys	1.3
Magnesium alloys	1.3
Yellow brass	1.3–1.6
Gunmetal bronze	1.0–1.6
Phosphor bronze	1.0–1.6
Aluminum bronze	2.1
Manganese bronze	2.1
Open-hearth steel	1.6
Electric steel	2.1
High manganese steel	2.6

Source: Adapted from Zuppann (1998).

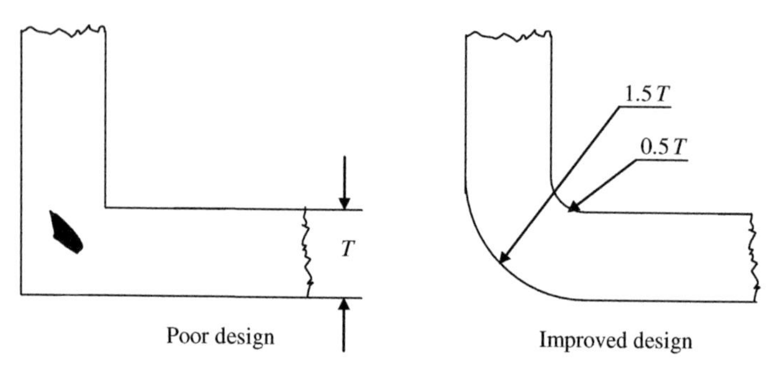

Figure 6.6 Provide generous radii to sharp corners to avoid uneven cooling and molded-in stress.

Part sections that project into the mold should have draft, depending on the type of the casting process. Draft allowances for sand cast components are about 1° or 2–3° for permanent mold processes. Table 6.6 recommends draft angles for the outside surfaces of sand-molded castings.

Tolerances achievable in many casting processes are insufficient to meet functional requirements in many applications. Almost all sand castings must be machined to some extent in order for the part to be made functional. Typical machining allowances for sand castings range between ⅙ and ¼ in.

It is desirable to minimize the use of dry sand cores, which can be achieved by changing the location of the parting plane.

Table 6.5 **Recommended minimum and desirable section thickness**

Material	Minimum (mm)	Desirable (mm)	Casting process
Steel	4.76	6.35	Sand
Gray iron	3.18	4.76	Sand
Malleable iron	3.18	4.76	Sand
Aluminum	3.18	4.76	Sand
Magnesium	4.76	6.35	Sand
Zinc alloys	0.51	0.76	Die
Aluminum alloys	1.27	1.52	Die
Magnesium alloys	1.27	1.52	Die

Source: Adapted from DeGarmo et al. (1984).

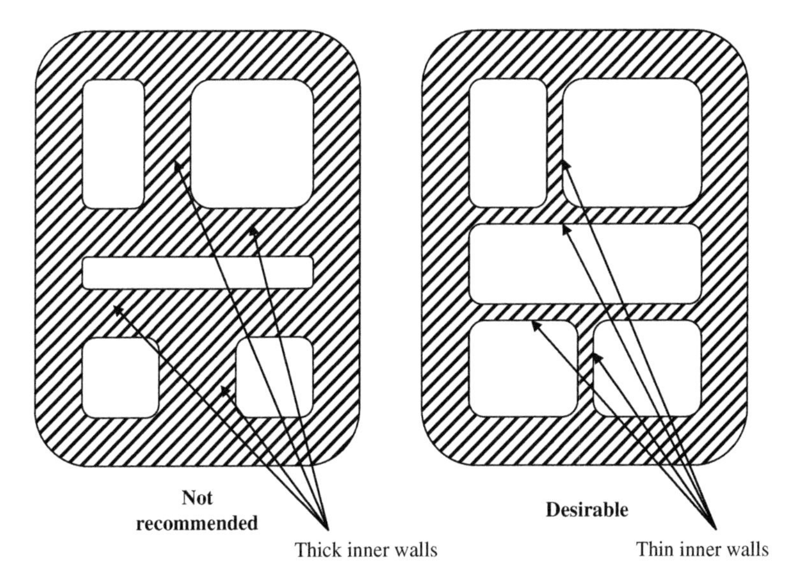

Figure 6.7 Interior walls should be 20% thinner than exterior walls, since they cool more slowly.

6.2.2 Design guidelines for forging (Heilman and Guichelaar, 1998)

The parting line should be in a plane perpendicular to the axis of the die motion. If it is not possible to have the parting line on one plane, it is desirable to preserve symmetry to prevent high side-thrust forces on the die and the press. No portion of the parting line should incline more than 75° from the principal parting plane, and much shallower angles are desirable. Undercuts cannot be incorporated into forged components since the forging must come out of the die after it is made.

Typical draft angles are 3° on aluminum and magnesium parts and 5–7° on steel parts. Draft angles on precision forgings are close to zero. Table 6.6 provides typical

Table 6.6 Recommended draft angles for outside surfaces of sand-molded castings

Ramming method	Pattern material					
	Wood		Aluminum		Ferrous	
	Pattern quality level					
	Normal	High	Normal	High	Normal	High
Hand	5°	3°	4°	3°	–	–
Squeezer	3°	2°	3°	2°	–	–
Automatic	–	–	2°	1°	1½°	½°
Shell molding	–	–	–	–	1	¼
Cold cure	3°	3°	2°	1°	–	–

Source: Adapted from Zuppann (1998).

draft angle ranges for finished forgings in the various alloy families. Low-draft or no-draft angles in products made out of aluminum and brass are possible.

Webs and ribs are difficult in metal flow as they become thinner. It is easiest when the web is relatively thick and uniform in thickness. Hence, forging components with deep ribs and high bosses is difficult, particularly so when these features do not taper.

Small corner and fillet radii tend to limit metal flow and increase stress on the die surfaces during forging. Table 6.7 shows typical minimum radii for forgings. A general rule for radii is "the deeper the impression, the larger the radius should be; both at the fillet around which the metal must flow and at the corner that must fill with metal."

Design features that promote easy forging add to the metal that must be machined away. Ample draft angles, large radii, and generous tolerances can have this effect. The machining allowance should allow for the worst-case buildup of draft, radii, and all tolerances. Machining allowances are added to external dimensions and subtracted from internal dimensions.

6.2.3 Design guidelines for extrusion (Bralla, 1998)

The major limitation and specific design recommendation for a part to be extruded is that the cross section must be same for the length of the part being extruded.

Avoid sharp corners. Provide generous radii for both internal and external corners of extruded cross sections. The minimum radii recommended for extruded sections are listed in Table 6.8. If sharp internal corners are necessary, the included angle should be as large as possible and always more than 90°.

Section walls should be balanced as much as the design function permits. Extreme changes in section thickness should be avoided, particularly in case of the less extrudable materials like steel. With steels and other less extrudable materials, holes in nonsymmetrical shapes should be avoided.

Table 6.7 **Minimum radii for forgings**

Depth of rib or boss (mm)	Minimum radius	
	Corner (mm)	Fillet (mm)
13	1.6	5
25	3	6.3
50	5	10
100	6.3	10
200	16	25
400	22	50

Table 6.8 **Minimum recommended radii for extruded sections**

Material	Minimum radius	
	Corner (mm)	Fillet (mm)
Al, Mg, and Cu alloys		
As extruded	0.75	0.75
After cold drawing	0.4	0.4
Ferrous metals, Ti and Ni alloys		
As extruded	1.5	3
After cold drawing	0.75	1.5

Source: Adapted from Bralla (1998).

In case of steel extrusions, the depth of an indentation should be no greater than its width at its narrowest point. Further, for the cross-sectional length of any thin-walled segment, the ratio of length to thickness of any segment should not exceed 14:1. For magnesium, ratios of 20:1 are recommended.

Symmetrical cross sections are preferred to nonsymmetrical designs to avoid unbalanced stresses and warpage. Figure 6.8 shows good and bad practices for cross sections of parts to be extruded.

6.2.4 Design guidelines for metal stamping (Stein and Strasse, 1998)

Ensure maximum stock utilization. Shapes that can be nested close together are preferred because this reduces costs by reducing the scrap rate. Figures 6.9 and 6.10 show improved material utilization.

The diameter of pierced holes should be no less than the stock thickness. In the case of alloy steels, the diameter of hole should at least be twice the thickness of the stock.

Spacing between holes should be a minimum of twice the stock thickness. The minimum distance between the lowest edge of the hole and the other surface should be one and a half times the stock thickness.

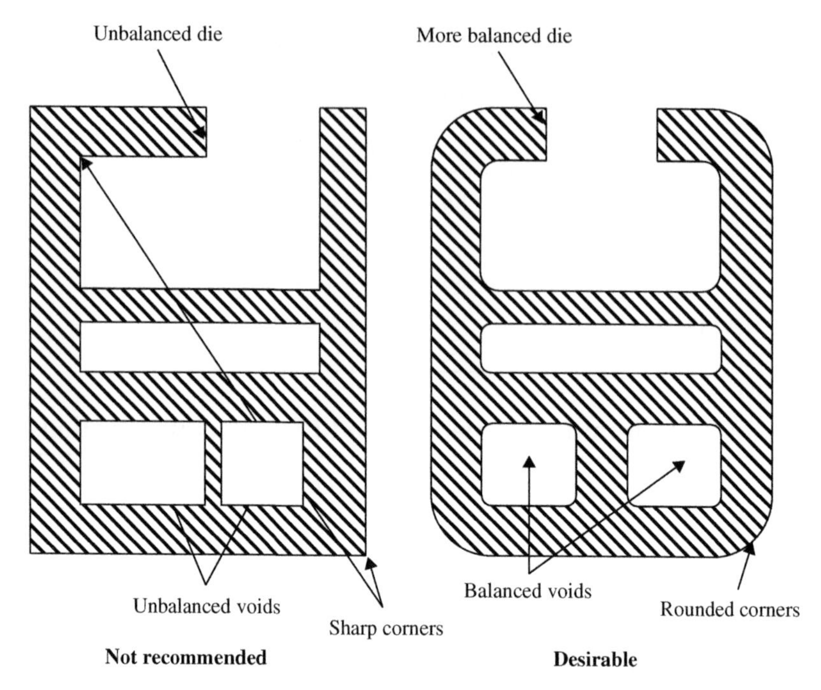

Figure 6.8 Desirable and undesirable practices in the design of cross sections to be extruded.

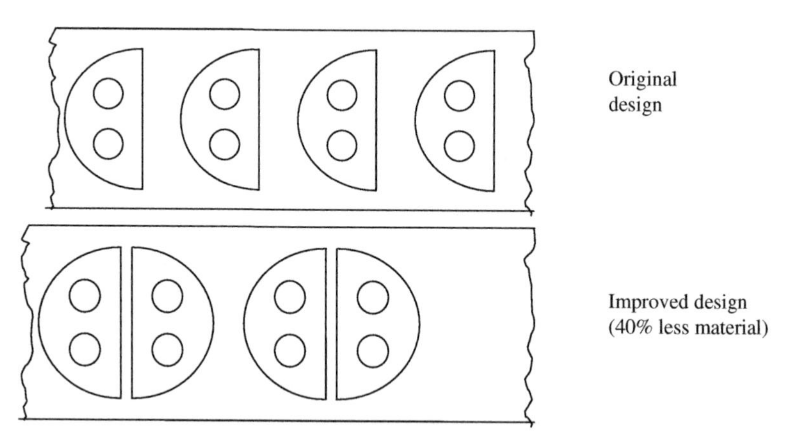

Figure 6.9 Example of a part that was redesigned to provide better nesting of blanks and thus improved material utilization.

Sharp corners, both internal and external, should be avoided. A general rule is to allow a minimum corner radius of one-half the stock thickness and never <0.8 mm.

Designers should take into consideration the grain direction, as this determines the strength of the component.

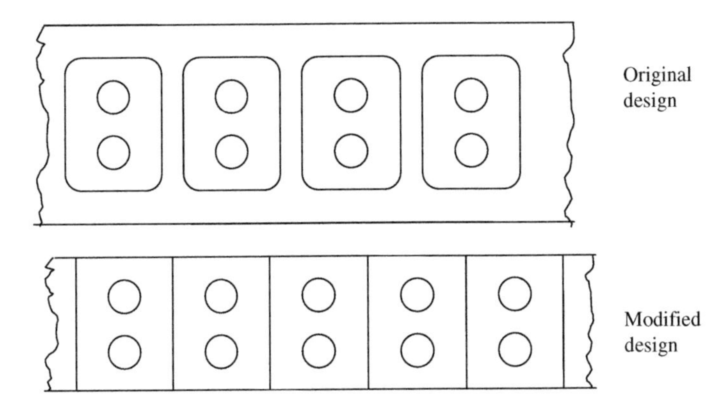

Figure 6.10 Use of strip stock of the width of the part with a better utilization of material.

Long narrow projections should be avoided because they are subject to distortion and require thin, fragile punches. Long sections should not be narrower than one and a half times the stock thickness.

6.2.5 Design guidelines for powdered metal processing (Swan and Powell, 1998)

Draft is not desirable and, in production, usually produces problems. Lack of draft is an advantage because die walls can be absolutely parallel to each other, enabling component faces to be parallel and of close tolerance. An exception is the sidewalls of recesses formed by a punch entering the top side of a part. In these cases, a draft of 2° or more is advisable.

The minimum recommended wall thickness is 1.5 mm (Figure 6.11). The minimum distance between the sidewalls and a hole or between two holes also is 1.5 mm. The normal maximum ratio of wall thickness to length is 18:1.

Small radii at both internal and external component corners are desirable.

Holes in the direction of pressing are acceptable. The minimum diameter of the holes is 1.5 mm. Holes at right angles to the direction of press cannot be achieved through this process. It is preferable to press blind holes of 6.3 mm diameter or more (unless they are shallow).

Undercuts cannot be achieved with this process because of problems in ejecting the component from the die.

The molding of inserts into the compact is not recommended. Trying to incorporate inserts increases production costs and adversely affects production rates.

Figure 6.12 shows recommendations for reducing a weak punch.

6.2.6 Design guidelines for fine-blanked parts (Fischlin, 1998)

Corners in fine-blanked parts must be well rounded. The combination of the corner angle, material thickness, and type of material determines the minimum radius required. Broad recommendations are as follows (Figure 6.13):

1. For corner angles <90°, radius = 25–30% of the material thickness.
2. For corner angles of 90°, radius = 10–15% of the material thickness.

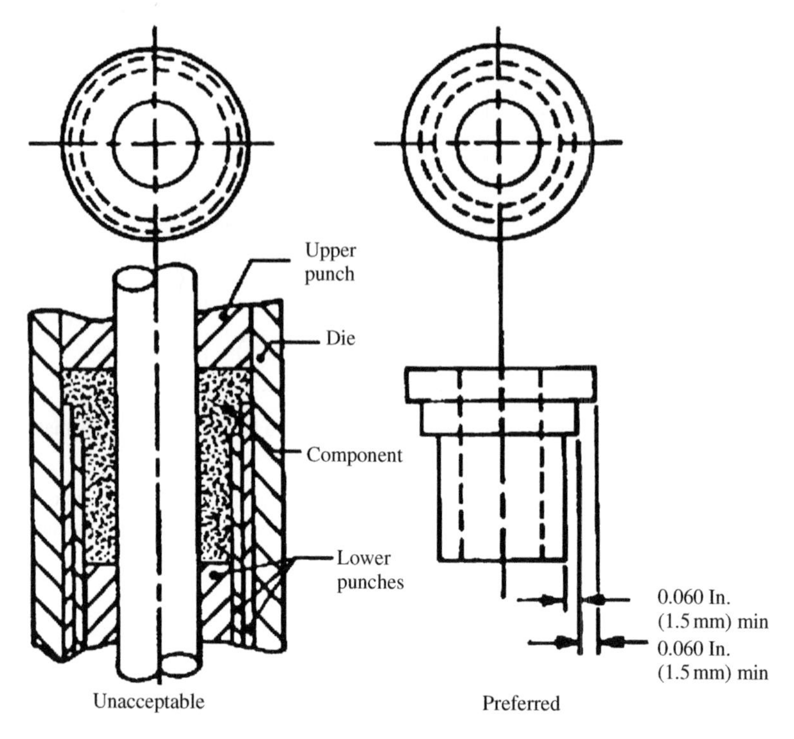

Figure 6.11 Design recommendations for minimum part widths.

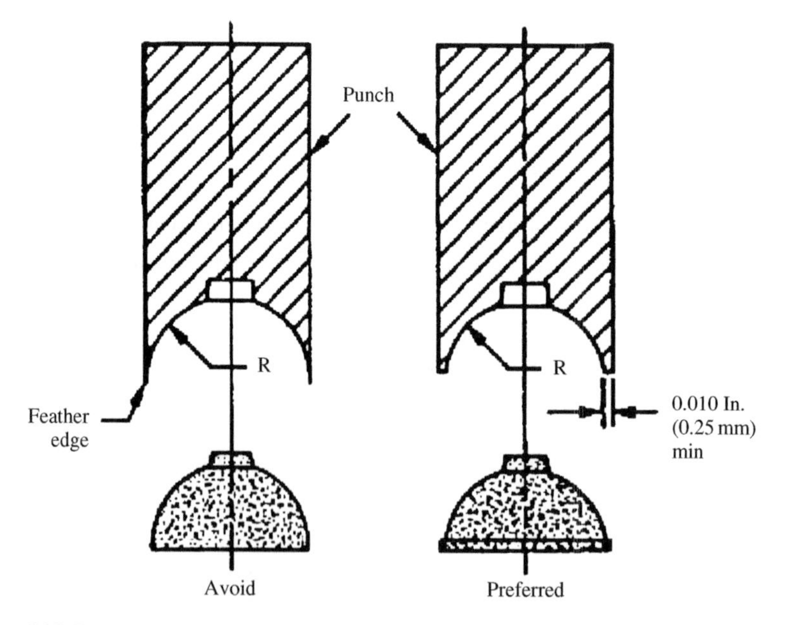

Figure 6.12 Design recommendations to reduce a weak punch.

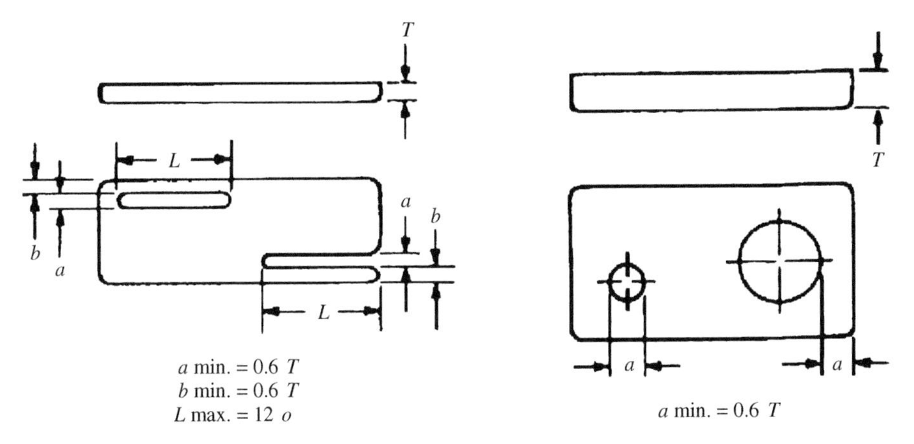

Figure 6.13 Design recommendations for slots and holes in fine-blanked parts.
Source: Adapted from Stein and Strasse (1998).

3. For corner angles >90°, radius = 5–10% of the material thickness.
4. For internal angles, radius = 66% of the external angle values.

Holes in material 1–4 mm thick can be blanked with the width of the sections from inner to outer form corresponding to approximately 60–65% of the material thickness. Gears, spurs, ratchets, and the like can be fine blanked if the width of the teeth on the pitch circle radius is 60% of the material thickness or more. Countersinks and chamfers of 90° can be introduced to depths of one-third of the material thickness without appreciable material deformation (only up to a material thickness of 3 mm). However, the volume of the material to be compressed should not exceed the volume of one-third the material thickness at 90° when countersinks of increased or decreased angles are desired.

6.2.7 Design guidelines for machined parts (Bralla, 1998; DeGarmo et al., 1984)

6.2.7.1 Standardization

If possible, parts should be designed such that they do not need machining. If this is not possible, then minimize the amount of machining required. In general, a low-cost product is achieved through the use of net shape processes, such as precision casting, closed die forging, or plastic molding, or near net shape processes such as impression die forging.

Machined parts should be designed such that the features can be achieved with standard cutting tools (Figure 6.14). Utilize standard preshaped workpieces to the maximum extent.

6.2.7.2 Raw material

Choose raw materials that will result in minimum component costs, without sacrificing any absolute functional requirements.

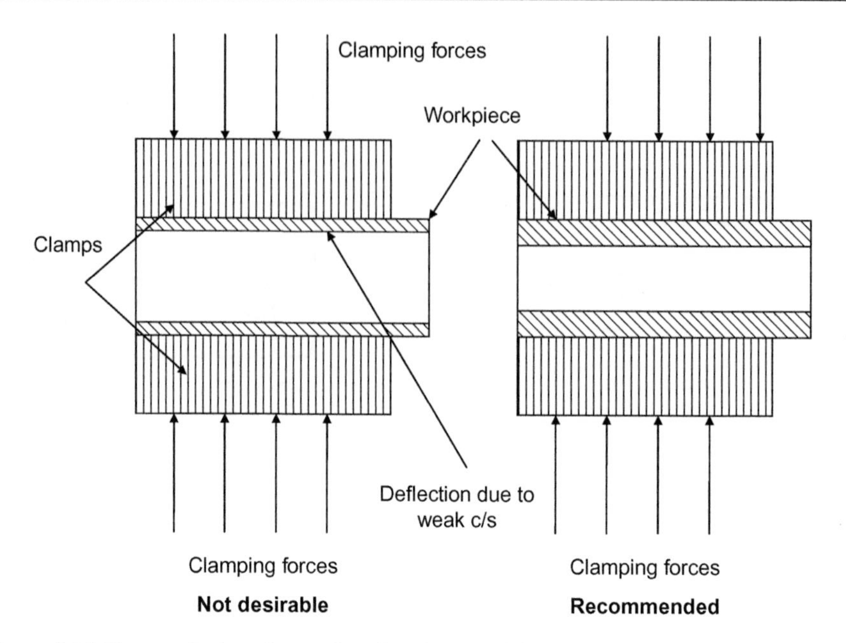

Figure 6.14 Use standard cutting tools rather than special tools.

Machined parts should be designed so that they can be produced from standard available stock. Also, use stock dimensions whenever possible, if doing so will eliminate a machining operation or the need for machining additional surface.

Materials with good machinability must be selected for better cutting speed and hence higher production rates.

6.2.7.3 Component design (general)

Try to design components so that they can be machined with only one machine tool.

Tolerances should be specified to satisfy functional needs, but the capabilities of a process should also be considered. Excessively close tolerances add cost but may not add value to the parts.

The surface finish should also be specified to meet functional needs or esthetic requirements.

Machined features such as sharp corners, edges, and points should be avoided, as they are difficult to accomplish. Deep holes that must be bored should be avoided.

Parts should be designed rigid enough to withstand the forces of cutting and work holder clamping (Figure 6.15).

Undercuts should be avoided, as they often require additional setups and operations or special tooling (Figure 6.16). Undercuts could also be points of stress concentration in components.

Avoid tapers, bent holes, and contours as much as possible in favor of rectangular shapes, which permit simple tooling and setup.

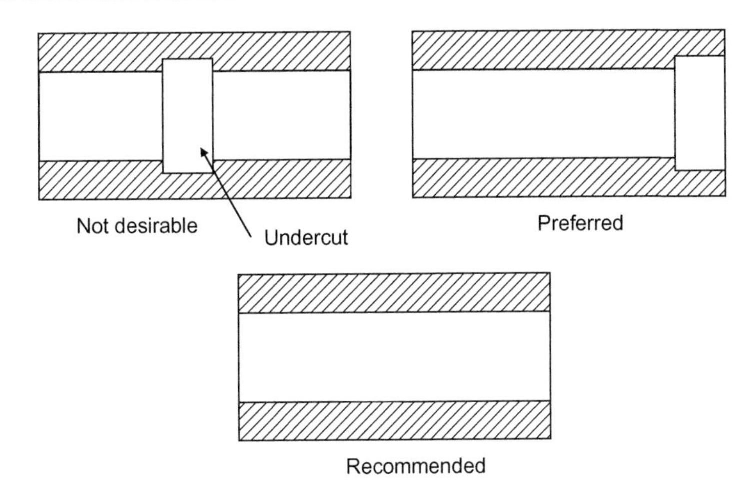

Figure 6.15 Design parts to be rigid enough to withstand clamping and cutting forces.

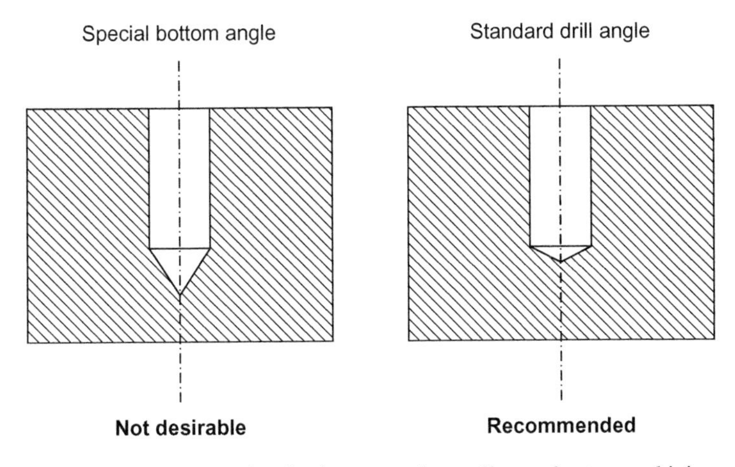

Figure 6.16 Avoid undercuts, as they lead to expensive tooling and extra machining operation.

Reduce the number and size of shoulders, because these usually require additional steps in operation and additional material.

Consider the possibility of substituting a stamping for the machined component.

Avoid using hardened or difficult-to-machine materials unless their special functional properties are essential to the part being machined.

For thin, flat pieces that require surface machining, allow sufficient stock for both rough and finish machining. In some cases, stress relieving between rough and finish cuts may be advisable.

It is preferable to put machined surfaces in the same plane or, if they are cylindrical, with the same diameter to reduce the number of operations required.

Provide access room for cutters, bushings, and fixture elements.

Avoid having parting lines or draft surfaces serve as clamping or locating surfaces.

Burr formation is an inherent result of machining operations. The designer should expect burrs and therefore provide relief space, if possible, and furnish means for easy burr removal.

6.2.7.4 Rotational component design

Try to ensure that cylindrical surfaces are concentric and plane surfaces are normal to the component axis.

Ensure that the diameters of external features increase from the exposed face of the workpiece. Conversely, ensure that diameters of internal features decrease from the exposed surface of the workpiece.

For internal corners, specify radii equal to the radius of a standard rounded tool corner.

Avoid internal features for long components. Also avoid components with a very large or very small length–diameter ratio.

6.2.7.5 Nonrotational component design

Provide a base for work holing and reference.

Ensure that the exposed surface of the component consists of a series of mutually perpendicular plane surfaces parallel and normal to the base.

Ensure that internal corners are normal to the base. Also, ensure that, for machined pockets, the internal corners normal to the base have as large a radius as possible.

If possible, restrict plane surface machining (slots, grooves, etc.) to one surface of the component.

Ensure that, in flat or cubic components, main bores are normal to the base and consist of cylindrical surfaces decreasing in diameter from the exposed face of the workpiece.

Avoid blind bores in large cubic components.

Avoid internal machined features in cubic boxlike components.

6.2.7.6 Assembly design

Ensure that assembly is possible.

Ensure that each operating machined surface on a component has a corresponding machined surface on the mating component.

Ensure that internal corners do not interfere with a corresponding external corner on the mating component, that is, design adequate clearances.

6.2.8 Design guidelines for screw machine parts (Lewis, 1998)

Design components such that the largest diameter of the component is the same as that of the bar stock. Standard sizes and shapes of bar stock should be used in preference to special diameters and shapes.

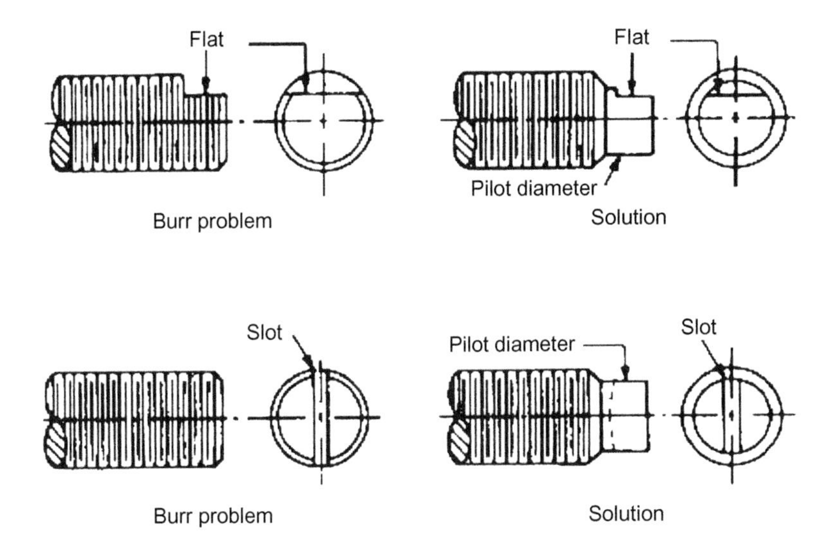

Figure 6.17 Guidelines to avoid burr problem in threaded parts.
Source: Adapted from Engineering Staff, Teledyne Landis Machine (1998).

The design of screw machine parts should be kept as simple as possible, so that standard tools, standard size holes, screw threads, slots, knurls, and the like can be readily machined with available tools.

Avoid secondary operations by designing parts such that components are completed on cutoff from the bar material.

The external length of formed areas should not exceed two and a half times the minimum diameter of the workpiece. Sidewalls of grooves and other surfaces perpendicular to the axis of the workpiece should have a slight draft. The minimum recommended draft is a half degree.

External or internal angular undercuts are not recommended, as they are difficult to machine and should be avoided.

The bottoms of blind holes should have standard angles. Although deep, narrow holes can be provided if necessary, it is better to limit the depth of blind holes to three to four times the diameter of the work.

Rolled screw threads are preferable to cut threads in screw machine products. Providing chamfers and drafts reduces burr problems (Figure 6.17).

Knurled areas should be kept narrow. A knurl's width should not exceed its diameter.

Sharp corners in the design of screw machine parts must be avoided (Figure 6.18). Sharp corners, both internal and external, cause weakness and more costly fabrication of form tools. It is preferable to provide either a chamfer at the corner or an undercut.

When a spherical end is required on a screw machine part, it is better to design the radius of the spherical end to be larger than the radius of the adjoining cylindrical portion. Hence, as a rule of thumb, the end spherical radius is designed greater than $D/2$ where D is the diameter of the part.

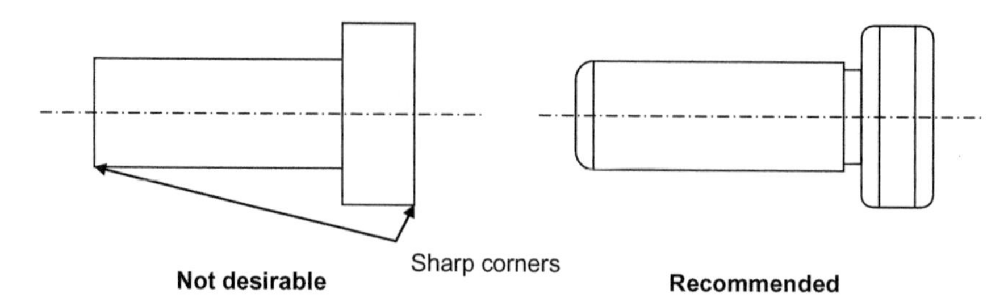

Figure 6.18 Avoid sharp corners by providing chamfers and undercuts.

6.2.9 Design guidelines for milling (Judson, 1998)

Components should be designed such that standard cutter shapes and sizes can be used. Slot widths, radii, chamfers, corner shapes, and overall forms should conform to those of the cutters available rather than ones that would require special fabrication.

Product design should permit manufacturing preference as much as possible to determine the radius where two milled surfaces intersect or where profile milling is involved.

When a small, flat surface is required, as for bearing surface or a bolt-head seat perpendicular to the hole, the product design should permit the use of spot facing, which is quicker and more economical than face milling. When spot faces or other small milled surfaces are specified for casings, it is good practice to design a low boss for the surface to be milled.

When outside surfaces intersect and a sharp corner is not desirable, the component design should allow a bevel or chamfer rather than rounding.

When form milling or machining rails, it is best not to attempt to blend the formed surface to an existing milled surface, because exact blending is difficult to achieve.

Keyway designs should permit the keyway cutter to travel parallel to the center axis of the shaft and form its own radius at the end.

A design that requires the milling of surfaces adjacent to a shoulder or flange should provide clearance for the cutter path. It is recommended that the component be designed such that milling of parting lines, flash areas, and weldments generally will extend the cutter's life. The component design should provide clearance to allow the use of larger size cutters rather than small cutters to permit high material removal rates.

In case of end-milling slots in mild steel, the depth should not exceed the diameter of the cutter.

6.2.10 Design guidelines for planing and shaping (Bralla, 1998)

Parts should be designed so that they can be easily clamped to the worktable and are sturdy enough to withstand deflections during machining.

It is preferable to put machined surfaces in the same plane to reduce the number of operations required.

Avoid multiple surfaces that are not parallel to the direction of the reciprocating motion of the cutting tool.

Allow a relieved portion at the end of the machined surface, because shapers and planers can cut up to only 6 mm of the obstruction or the end of a blind hole.

The minimum size of holes in which a keyway or slot can be machined with a slotter or a shaper is about 1 in.

Due to the lack of rigidity of long cutting tool extensions, it is not feasible to machine slots longer than four times the diameter of the hole.

6.2.11 Design guidelines for screw threads (Engineering Staff, Teledyne Landis Machine, 1998)

External threads should be designed such that they do not terminate too close to the shoulder or adjoining larger diameter. The width of this relief depends on the size of the thread, the coarseness of the thread, and the throat angle of the threading tool. Internal threads should have a similar relief or undercut.

In cases where high thread strength is not required, use of a reduced height thread form is recommended.

It is recommended to have short thread lengths consistent with functional requirements.

All threaded products should have chamfers at the ends of external threads and a countersink at the end of the internal threads.

The surface at the starting end of the screw thread should be flat and square with the thread's center axis.

The use of standard thread forms and sizes is economical and recommended.

Tubular parts must have a wall heavy enough to withstand the pressure of the cutting or forming action.

Threads to be ground should not be specified to have sharp corners at the root.

Centerless ground thread should have a length–diameter ratio of at least 1:1, although a length longer than the diameter is recommended.

Parts for thread rolling have similar requirements of roundness, straightness, and freedom from taper and burrs.

Except for those of the largest size, coarse threads are slightly more economical to produce than fine threads.

6.2.12 Design guidelines for injection molding

Wall thickness must be uniform wherever possible and a general rule of thumb should be that the thickest wall should be less than two times the thinnest wall. Wall thickness should be controlled. Thick cross sections result in slower binder removal and increased injection-molding cycle, leading to surface depressions. Thick sections take longer to cool than thin ones. During the cooling process, if walls are an inconsistent thickness, the thinner walls will cool first while the thick walls are still solidifying. As the thick section cools, it shrinks around the already solid thinner section. This causes warping, twisting, or cracking to occur where the two sections meet. To avoid this problem, try to design with completely uniform walls throughout the part. When uniform walls are not possible, then the change in thickness should be as gradual as possible. Wall thickness variations should not exceed 10% in high mold shrinkage plastics. Thickness transitions should be made gradually, on the order of 3–1. This gradual transition avoids stress concentrations and abrupt cooling differences.

One way to avoid sink marks is to core out the solid sections of the part to reduce thick areas. If the strength of a solid part is required, try using crosshatched rib patterns inside the cored out area to increase strength and avoid sink. As a rule of thumb, make sure that all bosses and locating/support ribs are no more than 60% of the thickness of the nominal wall. Also, textures can be used to hide minor sink marks.

Parts should have a draft angle on sidewalls to ensure easy part removal from the mold. Angles of ½–2° per side are recommended for both inside and outside walls. Larger draft angles should be used for deep sections, complex configurations, and when an inside core or die section is used, the part will tend to shrink around it. Follow the following recommendations:

Use at least 1 degree of draft on all "vertical" faces
1½ degrees of draft is required for light texture
2 degrees of draft works very well in most situations
3 degrees of draft is a minimum for a shutoff (metal sliding on metal)
3 degrees of draft is required for medium texture.

Generous radii and filets should be used as much as possible at all corners. Rounded corners aid material flow in molds and reduce stress concentrations in molds or in the part. A suggested rule is to use a corner radius of one-half the adjacent wall thickness and no <0.4 mm (0.015 in.).

Bosses and ribs are feasible and often desirable in MIM part design. Ribs enable a part to be designed to be strong and rigid even when wall thickness and mass must be reduced. Since these details can be produced by the tooling, they can be incorporated into the design at minimal unit cost.

A "parting line" is the line of separation on the part where the two halves of the mold meet. The line actually indicates the parting "plane" that passes through the part. While on simple parts this plane can be a simple, flat surface, it is often a complex form that traces the perimeter of the part around the various features that make up the part's outer "silhouette." Part lines can also occur where any two pieces of a mold meet. This can include side action pins, tool inserts, and shutoffs. Parting lines cannot be avoided; every part has them. Keep in mind when designing your part that the melt will always flow toward the parting line because it is the easiest place for the displaced air to escape or "vent" (Figures 6.19 and 6.20).

6.3 Manufacturing technology decisions

Advances in technology have had the greatest impact on process design decisions. Technological advances have enabled companies to produce products faster, with better quality, at a cheaper rate. Many processes that were not imaginable only a few years ago have been made possible through the use of technology.

A production process consists of activities that are required in transforming an input set (human resources, raw materials, energy, capital, information, etc.) to

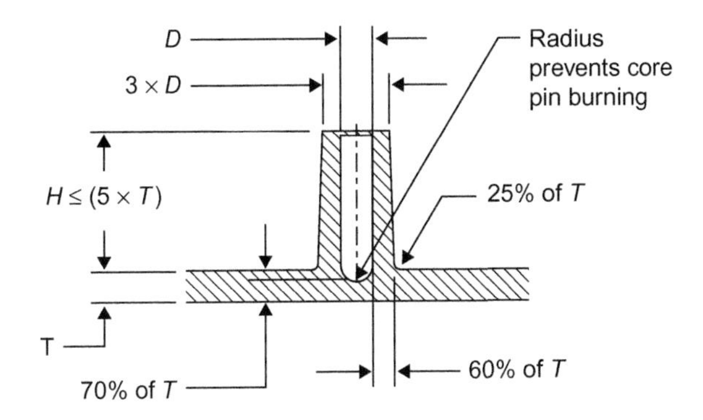

Figure 6.19 Suggested design guidelines for bosses on metal injection molded parts.

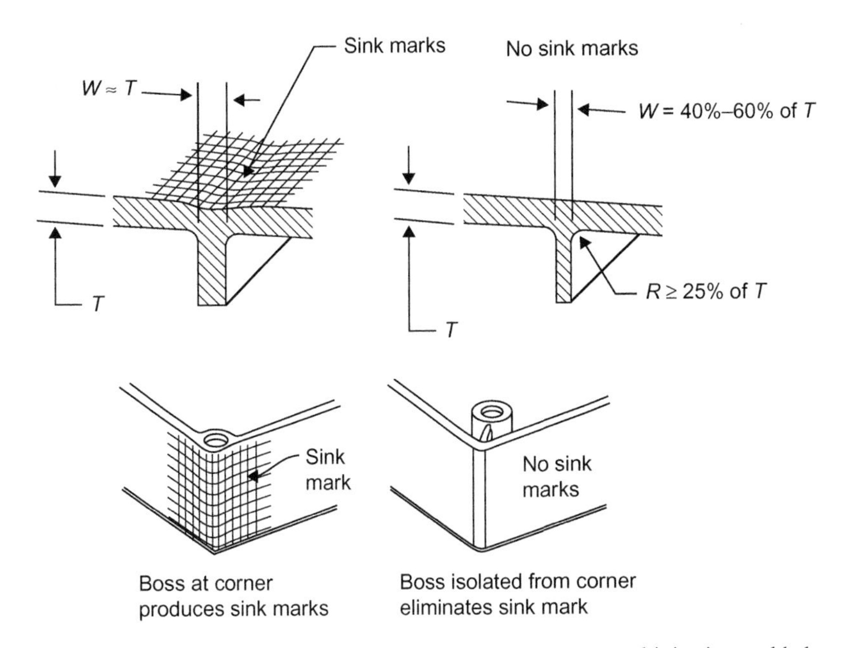

Figure 6.20 Suggested design guidelines for bosses and ribs on metal injection molded parts.

valuable outputs with the help of processes. Table 6.9 shows the two types of production processes based on the machinery used.

Production process selection, to a great extent, depends on the level of manufacturing technology. In recent decades, technologies that have influenced production process decision making are:

- Automation
- Automated material handling:
  - Automated guided vehicles (AGV)
  - Automated storage and retrieval systems (AS/RS)

Table 6.9 Differences between continuous and intermittent systems

#	Decision	Intermittent production systems	Continuous production systems
1	Nature of product	Custom orders (not for stocking)	Based on demand forecast (for stocking)
2	Flexibility of process	Flexible	Not flexible (standardized)
3	Scale of production	Small scale	Large scale
4	Per unit cost	High	Low
5	Range of products	Wide range	One particular type
6	Instructions	Detailed instructions matching customer specifications	Single set of instruction
7	Location change	Easy	Difficult
8	Capital invested	Small	High
9	Product variety	Large	Small
10	Degree of standardization	Low	High
11	Path through facility	Varied pattern	Line flow
12	Critical resource	Labor	Equipment
13	Importance of work skills	High	Low
14	Type of equipment	General purpose	Specialized
15	Degree of automation	Low	High
16	Throughput time	Longer	Shorter
17	Work-in-process inventory	More	Less

- Computer-aided design (CAD) software
- Robotics and numerically controlled (NC) equipment
- Flexible manufacturing systems (FMS)
- Computer-integrated manufacturing (CIM).

The Association for Manufacturing Technology has a broader view of the technology that determines production process selection. Table 6.10 gives this broader view and the means of achieving the technology.

6.4 A typical part drawing and routing sheet

Figures 6.21 and 6.22 show a typical part drawing and a routing sheet, respectively.

Table 6.10 Manufacturing technologies and means to achieve

Technology application	Application medium
Software	Computer-aided design (CAD), computer-aided manufacturing (CAM), computer numerical control (CNC), direct numerical control (DNC), programmable logic control (PLC), numerical control (NC), program optimization software, and systems integration software
Material removal	Turning, milling, drilling, grinding, tapping, electrical discharge machines (EDM), broaching, sawing, water jet cutting equipment, and laser process equipment
Material forming	Stamping, bending, joining, hydroforming, presses, shearing, cold and hot forming equipment
Additive processes	3D printing, laser sintering, and rapid prototyping equipment
Workholding	Chucks, fixtures, clamps, blocks, angle plates, and tooling columns
Tooling	Drills, taps, reamers, boring bars, dies, punches, and grinding wheels
Material handling	Conveyors, automated wire guided vehicles, die handling equipment, robots, pallet changers, and bar feed equipment
Automated systems	Transfer machines, assembly systems, automated systems and cells, and flexible manufacturing systems (FMS)
Biomanufacturing	Use of a biological organism, or part of one, in an artificial manner to produce a product such as developing drugs and medical compounds

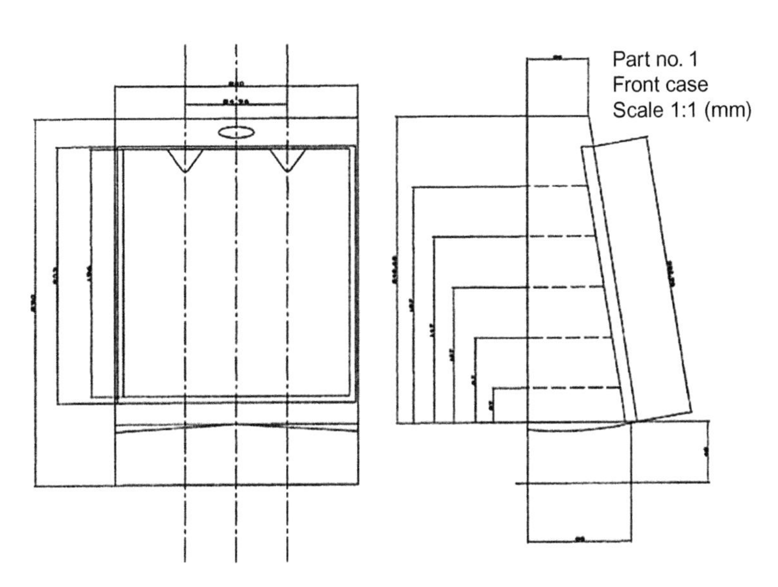

Figure 6.21 Part drawing with some basic dimensions.

PART NO.	P1250-2335	EFFECTIVE DATE 3/23/2005	SHEET 1 OF 1	
PART NAME	FRONT OUTER CASING		PLANNER A.SMITH	
DEPT. NO.	OPERATION	DESCRIPTION	MACHINE TOOL	STANDARD TIME (secs)
27	10	Obtain thermoplastic granules from stock area	Manual	-
102	20	Load granules into hopper feeder	Injection Molding #2ACBX	4.7
30	30	Activate machine	Manual	5
30	40	Process time	Injection Molding #2ACBX	30
56	50	Cool parts and then eject	Injection Molding #2ACBX	30
50	60	Inspect parts as per QC-NA 11-1230	Manual	10
29	70	Place parts in bin	Manual	10

Figure 6.22 Typical routing sheet for the part shown in Figure 6.21.

References

Bralla, J.G. (Ed.), 1998. Design for Manufacturability Handbook, second ed. McGraw-Hill, New York, NY.

DeGarmo, E.P., Black, J.T., Kohser, R.A., 1984. Materials and Processes in Manufacturing, sixth ed. Macmillan, New York, NY.

Engineering Staff, Teledyne Landis Machine, 1998. Screw threads Design for Manufacturability Handbook, second ed. McGraw-Hill, New York, NY.

Fischlin, J.K., 1998. Fine-blanked parts Design for Manufacturability Handbook, second ed. McGraw-Hill, New York, NY.

Groover, M.P., 1996. Fundamentals of Modern Manufacturing: Materials, Processes, and Systems. Prentice Hall, Upper Saddle River, NJ.

Heilman, M.H., Guichelaar, P.J., 1998. Forgings Design for Manufacturability Handbook, second ed. McGraw-Hill, New York, NY.

Judson, T.W., 1998. Parts produced on milling machines Design for Manufacturability Handbook, second ed. McGraw-Hill, New York, NY.

Lewis, F.W., 1998. Screw-machine products Design for Manufacturability Handbook, second ed. McGraw-Hill, New York, NY.

Stein, J., Strasse, F., 1998. Metal stampings Design for Manufacturability Handbook, second ed. McGraw-Hill, New York, NY.

Swan, B.H., Powell, C.J., 1998. Powder metallurgy parts Design for Manufacturability Handbook, second ed. McGraw-Hill, New York, NY.

Zuppann, E.C., 1998. Castings made in sand molds Design for Manufacturability Handbook, second ed. McGraw-Hill, New York, NY.

Designing for Assembly and Disassembly

7

7.1 Introduction

7.1.1 Definition and importance of the assembly process

A consumer product often is an assemblage of several individual components. Each component has been planned, designed, and manufactured separately. However, by themselves, there is very little use to component parts. Only after they are assembled into the final product can they effectively perform their intended function.

Assembly of a product is a function of design parameters that are both intensive (material properties) and extensive (physical attributes) in nature. Examples of such design parameters include, but are not limited to, shape, size, material compatibility, flexibility, and thermal conductivity. It is easy to see that, when individual components are manufactured with ease of assembly in mind, the result is a significant reduction in assembly lead times. This leads to savings in resources (both material and human). Designers have grappled with the problem of designing products for assembly since at least the beginning of the Industrial Revolution.

The importance of designing for ease of assembly cannot be overemphasized. The case of designing for easy and efficient (in terms of time as the singular metric) assembly has been made numerous times by researchers. This is obvious in light of the fact that a product more often than not is an assemblage of various individual components. The spatial alignment between functionally important components is what makes the product function. Given this background, it is imperative that each component be designed in such a way as to align and mate efficiently. This entails the design and processing of the component in a specific manner with respect to shape, size, tolerances, and surface finish. A component designed for assembly leads to a substantial reduction in assembly time as well as cost.

7.1.2 Definition and importance of the disassembly process

In an engineering context, disassembly is the organized process of taking apart a systematically assembled product (assembly of components). Products may be disassembled to enable maintenance, enhance serviceability, and/or to achieve end-of-life (EOL) objectives, such as product reuse, remanufacture, and recycling.

Counterintuitively, disassembly is not necessarily the opposite of assembly. In many ways, components need to be designed for disassembly so that the process can be effected without damage to the parts' intensive and extensive properties. Disassembly has begun to gain in importance as a process only comparatively recently. This can

Product Development. DOI: http://dx.doi.org/10.1016/B978-0-12-799945-6.00007-7

be attributed to growing scarcity of natural resources, increased processing costs for virgin materials (such as mining iron ore for steel manufacturing), and environmental legislation to make manufacturers more responsible with regard to waste disposal.

7.2 Design for assembly

7.2.1 Definition

Design for assembly (DFA) seeks to simplify the product so that the cost of assembly is reduced. Consequently, applications of DFA principles to product design usually result in improved quality and reliability and a reduction in production equipment and part inventory. It has been repeatedly observed that these secondary benefits often outweigh the cost reductions in assembly.

DFA, in principle, recognizes the need to analyze the design of both the part and the whole product for any assembly problems early in the process to cut costs during the entire product cycle. *DFA* may be defined as a process for improving product design for easy and low-cost assembly, which is achieved by means of concurrent focus on the dual aspects of functionality and ease of assembly.

The practice of DFA as a distinct feature of design is a relatively recent development, but many companies have been doing DFA for a long time. For instance, General Electric published an internal manufacturing producibility handbook in the 1960s, which was meant to serve as a set of guidelines and manufacturing data for designers. These guidelines included many of the principles of DFA as we know it today, without ever using that particular nomenclature or distinguishing it from the rest of the product development process.

7.2.2 Different methods of assembly

The different methods of assembly are as follows:

- **Manual assembly**: Manual assembly is a process characterized by operations performed manually, with or without the aid of simple, general-purpose tools, such as screwdrivers and pliers. The cost per unit is constant, and the process requires little initial investment. Manual assembly involves parts that are transferred to workbenches, where the assembly of individual components into the final product takes place. Hand tools generally are used to aid the worker for easy assembly. Although this is the most versatile and adaptable assembly method, there usually is an upper limit to the production volume, and labor costs (including benefits, workers compensation due to fatigue and injury, and overhead for maintaining a clean and healthy environment) are higher.
- **Automatic assembly**: Often referred to as *fixed automation*, this method uses either synchronous indexing machines and part feeders or nonsynchronous machines, where parts are handled by a free transfer device. The system generally is built for a single product, and the cost per unit decreases with increasing volume of production.
- **Fixed or hard automation**: Fixed or hard automation characteristically involves a custom-built machine that assembles only one specific product and entails a large capital investment. As production volume increases, the fraction of the capital investment compared to

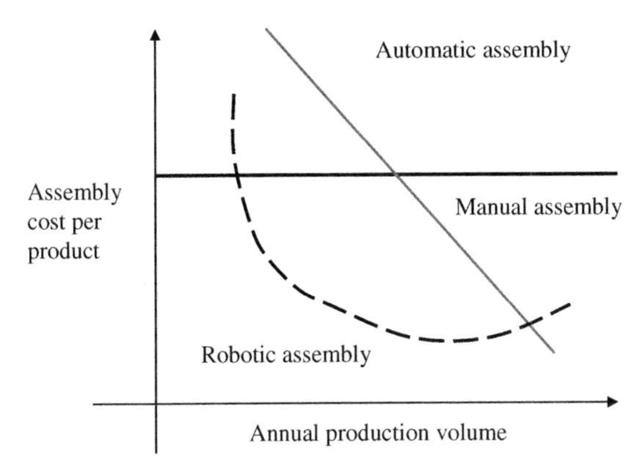

Figure 7.1 Comparing different methods of assembly on the basis of type, production volume, and cost.

the total manufacturing cost decreases. Indexing tables, parts feeders, and automatic controls typify this inherently rigid assembly method. In some instances, automatic assembly is also referred to as *Detroit-type* assembly.
- **Robotic assembly**: This form of assembly is best suited for those products whose production volume lies between the volumes for manual and automatic assembly methods. This method of product assembly can achieve volumes closer to the automatic assembly methods. Soft automation or robotic assembly incorporates the use of robotic assembly systems. This can take the form of a single robot or a multistation robotic assembly cell with all activities simultaneously controlled and coordinated by a PLC or computer. Although this type of assembly method can have large capital costs, its flexibility often helps offset the expense across many different products.

Figure 7.1 draws a comparison between the relative costs of the different methods of assembly by type as well as production volume. Figure 7.2 depicts the production ranges for each type of assembly.

7.3 Design guidelines for different modes of assembly

7.3.1 Manual assembly

The following design guidelines may be incorporated into product design when designing for manual assembly:

- Eliminate the need for decision making by the worker, including making final adjustments. Ensuring this step removes all subjectivity from the decision-making process, thereby improving the accuracy of the assembly process.
- Ensure good product accessibility as well as visibility.
- Eliminate the need for assembly tools or special gauges by designing individual components to be self-aligning and self-locating. Parts that snap and fit together eliminate the need for separate fasteners. This results in speedy as well as more economical assembly.

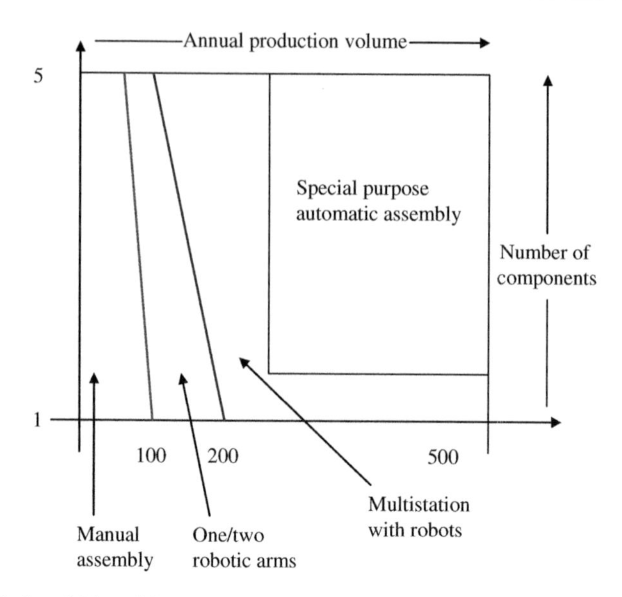

Figure 7.2 Distinguishing different assembly methods based on production ranges.

- Minimize the total number of individual parts, if possible. To facilitate this objective, multipurpose components may be used.
- Eliminate excess parts and combine two or more parts into one, if functionally possible.
- Avoid or minimize the need to reorient the part during the assembly process. Ensure that all insertion processes are simple. Avoid the need for rotation, releasing, and re-gripping. Vertical insertion always is preferable, since it utilizes gravity to accomplish the task.

The process of manual assembly entails extensive component handling by the operators. As such, components need to be designed with a view to minimizing the need for extensive handling to make the process faster and more accurate. The following are some guidelines to accomplish this objective:

- Design parts that have end-to-end symmetry and rotational symmetry about the axis of rotation. If this is not possible, the design should incorporate maximum symmetry.
- When it is impossible to incorporate symmetry into product design, obvious asymmetry should be used to facilitate ease of orientation and insertion.
- Provide features that prevent jamming and entanglement of parts.
- Design parts so there is little or no resistance to insertion. This can be facilitated by providing chamfers to guide insertion of two mating parts.
- Design for a pyramidal method of assembly, if possible. Provision needs to be made for progressive assembly about one axis of reference. In general, it is best to assemble from above.
- Avoid holding down the part. If this is unavoidable, the part should be designed so that it is secured as soon as possible after insertion.
- Design parts to facilitate location before release. Release of a part before location is a potential assembly problem.
- Use of common fasteners increases assembly cost in the following order: snap fit, plastic bending, riveting, and screwing. Bear this factor in mind when designing securing and fastening methods for holding together two or more parts.

7.3.2 Automatic assembly

A clear distinction was drawn between manual assembly and automatic assembly in the preceding section. Due to the inherently varying natures of the two assembly methods, the design guidelines for automatic assembly are significantly different from those for manual assembly. These guidelines are as follows:

- Self-aligning and self-locating features need to be incorporated into the design to facilitate assembly. Considerable improvement can be achieved by using chamfers, guide pins, dimples, and cone and oval screws.
- Use the largest, most rigid part of the assembly as a base or fixture, where other parts are stack assembled vertically to take advantage of gravity. This, in turn, eliminates the need to use an assembly fixture. The best assembly operation is performed in a layered fashion. If this is not possible, the assembly should be divided into subassemblies and plugged together at a later stage.
- As with all other design for X principles, use a high percentage of standard parts. Employing the concept of group technology, begin with fasteners and washers. Use standard modules and subassemblies.
- Avoid the possibility of parts tangling, nesting, or shingling during feeding, since this can complicate and unduly delay the assembly process.
- Avoid flexible, fragile, and abrasive parts and ensure that the parts have sufficient strength and rigidity to withstand the forces exerted on them during feeding, assembly, and use.
- Avoid reorienting assemblies, as such moves may require a separate workstation or machine, thereby increasing costs.
- Design parts to ease automation by presenting or admitting parts to the assembly machine in the right orientation after the minimum possible time in the feeder. Parts that are symmetrical or clearly asymmetrical can be oriented easily.
- Design parts with a low center of gravity, thereby imparting in them a natural tendency to be fed.

7.3.3 Robotic assembly

Guidelines are as follows:

- Reduce part count as well as part type. Many robot manipulators have poor repeatability; therefore, features such as lips, leads, and chamfers assume a great deal of importance.
- Ensure that parts that are not secured immediately on insertion are self-locating in assembly.
- Design components such that all can be gripped and inserted using the same robot gripper. Gripper and tool changes are a major source of inefficiency.
- Design products to allow vertical assembly directly from above.
- Design components to do away with the need for reorientation. Design parts so that they can be presented to the robot arm in an orientation suitable for gripping.
- Design individual components to promote ease of handling from bulk. It is important to avoid parts that nest or tangle; are thin, heavy, or exceedingly large or small; or are slippery.

7.4 Methods for evaluating DFA

Various researchers have proposed methods for evaluating the efficiency of a product design from the perspective of product assembly. Comparison of two different product designs using one of these methodologies can enable one to ascertain the better design. As such, these methods concentrate on an objective design evaluation. Note that several methods for assembly evaluation exist such as

- The Hitachi assembly evaluation method
- The Lucas DFA method
- The Fujitsu productivity evaluation system
- The Boothroyd-Dewhurst DFA method
- The AT&T DFA method
- The Sony DFA method
- SAPPHIRE (a software package used to analyze ease of product assembly).

The three techniques discussed in this chapter are

1. **The Hitachi assembly evaluation method**: This method aims to facilitate design improvements by identifying weaknesses in the design at the earliest stage in the process by using an assemblability evaluation score and an assembly cost ratio.
2. **The Lucas DFA method**: Analysis is carried out in three sequential stages—functional, feeding, and fitting.
3. **The Boothroyd-Dewhurst DFA method**: This method seeks to reduce the number of parts by consideration of manual handling and manual insertion times.

7.4.1 The Hitachi assemblability evaluation method

The objective of the Hitachi assemblability evaluation method (AEM) is to facilitate design improvements by identifying weaknesses in product design at the earliest possible stage. This is achieved using two principal indicators: an assemblability evaluation score ratio (E), which assesses design quality by determining the difficulty of operations, and an assembly cost ratio (K), which projects elements of assembly cost.

The Hitachi method considers both cost and quality important. This means that a low-cost design is not necessarily the best; alternatively, a good design may be too expensive. This is the only evaluation method that takes product design economics into account and hence is not purely technical in nature. Figure 7.3 illustrates the flow of logic in making design decisions using the Hitachi evaluation method.

The following is Hitachi's stepwise procedure for general design evaluation:

- The general universe of assembly operations is categorized into 20 elemental assembly tasks. Each task is assigned a symbol indicating the task content. Each task relates specifically to insertion and fastening processes and not to parts handling.
- Each of the elemental tasks is subject to a penalty score that reflects the degree of difficulty of the task. The penalty scores are obtained from analysis of shopfloor data and are revised constantly to reflect changes in technology and methods. The penalty scores then are ranked, and all are compared to the elemental task with the lowest penalty score. For instance, the simple task of placing an object on top of another object without requiring much accuracy is assigned a penalty score of 0 due to the inherent ease with which the task can be performed. Conversely, the more complicated task of soldering (assembly method) is assigned the much higher penalty score of 20.
- Factors that influence elemental tasks are extracted as coefficients and the penalty scores are modified accordingly.

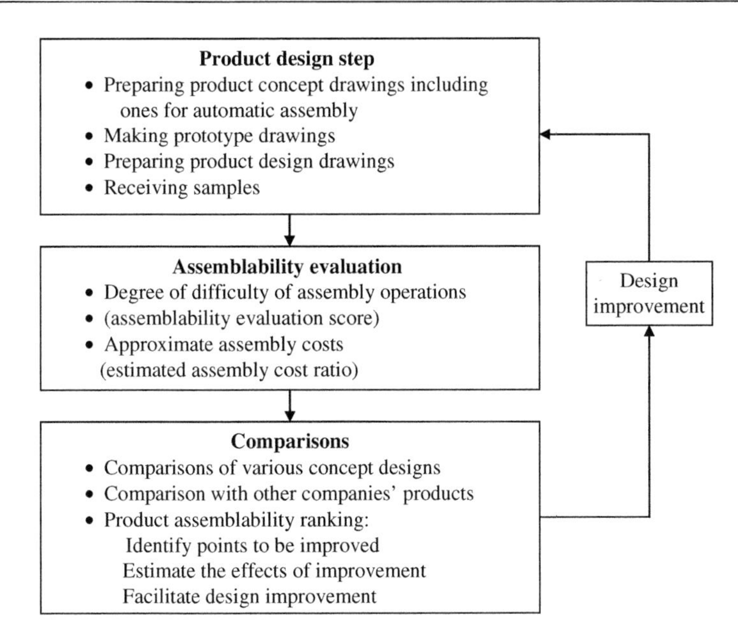

Figure 7.3 Assemblability evaluation and design improvement flow diagram for the Hitachi evaluation method.

- Attaching (contacting) conditions appropriate for each part are expressed using further AEM symbols.
- The total of the various penalty scores for an individual component are modified by the contacting coefficients (as described previously) and subtracted from the best possible score (100) to give the assemblability evaluation score for the part.
- The total score for the product is defined as the sum of the assemblability scores for individual tasks divided by the total number of tasks.
- Generally speaking, a score of 80 or above and a K value of 0.7 (implying savings of 30%) or lower are acceptable.

7.4.2 Lucas DFA evaluation method

The Lucas DFA method was developed in the early 1980s by the Lucas Corporation in the United Kingdom. The Lucas method is based on a point scale that gives a relative measure of the difficulty associated with assembly. This method is based on three separate and sequential analyses, which are described by means of the assembly sequence flowchart. Figure 7.4 depicts the Lucas design for mechanical assembly procedure.

The procedure follows the steps below.

1. Product design specification
2. Product analysis
3. Functional analysis (first Lucas analysis); loop back to step 2 if the analysis yields problems
4. Feeding analysis (second Lucas analysis)

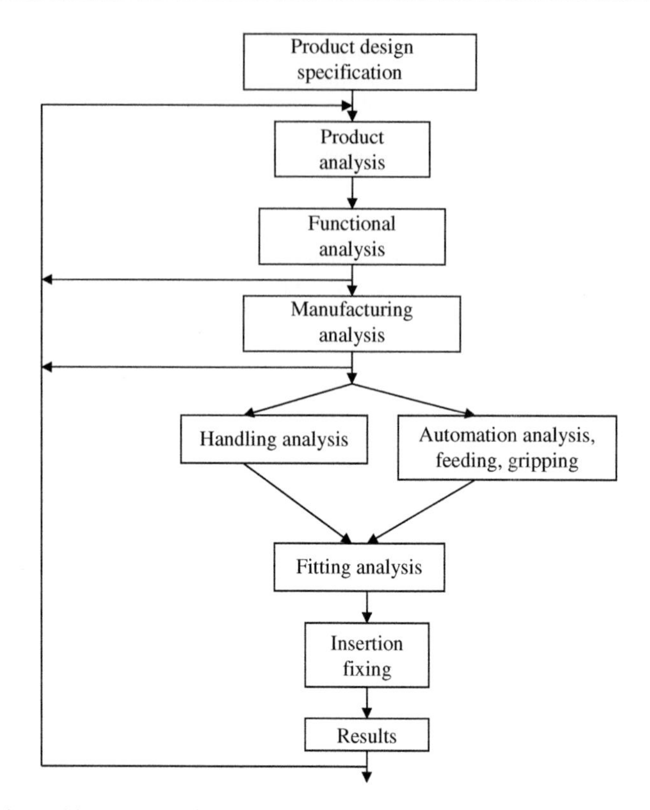

Figure 7.4 Assembly sequence flowchart for the Lucas DFA evaluation model.

5. Fitting analysis (third Lucas analysis)
6. Assessment
7. Return to step 2 if the analyses identify problems.

The functional analysis forms the first part of this evaluation system. Components are divided into two groups. The first group includes components that perform a primary function, and therefore exist for fundamental reasons. These components are considered essential, or A, parts. The second group, B components, are nonessentials, such as fasteners and locators. The design efficiency (DE) is computed using the formula:

$$\text{DE} = A/(A + B) \times 100$$

The target efficiency is at least 60%.

The feeding analysis forms the second part of this evaluation system. This analysis is concerned with problems associated with handling components and subassemblies before they are admitted to the assembly system. By answering a group of questions regarding the size, weight, handling difficulties, and orientation of a part,

Table 7.1 **Lucas manual handling analysis (Handling Index = $A + B + C + D$)**

	Score
A. Size and weight of part	
Very small, requires tools	1.5
Convenient, hands only	1
Large and/or heavy, requires more than one hand	1.5
Large and/or heavy, requires hoist or two people	3
B. Handling difficulties	
Delicate	0.4
Flexible	0.6
Sticky	0.5
Tangible	0.8
Severely nesting	0.7
Sharp or abrasive	0.3
Untouchable	0.5
Gripping problem, slippery	0.2
No handling difficulties	0
C. Orientation of part	
Symmetrical, no orientation required	0
End to end, easy to see	0.1
End to end, not visible	0.5
D. Rotational orientation of part	
Rotational symmetry	0
Rotational orientation, easy to see	0.2
Rotational orientation, hard to see	0.4

its feeding/handling index can be calculated. The feeding/handling ratio is computed as follows:

$$F/H \text{ ratio} = (\text{Feeding/handling index})/\text{Number of essential components}$$

The target value is 2.5.

The fitting analysis is similar to the feeding analysis. A fitting index of 1.5 is a goal value for each assembly. However, note that there usually is greater variance in the fitting indices than in the feeding indices. Again, an overall fitting ration of 2.5 is desired.

$$\text{Fitting ratio} = (\text{Total fitting index})/(\text{Number of essential components})$$

Table 7.1 depicts the manual handling analysis used in the Lucas method, while Table 7.2 depicts the manual fitting analysis for this method.

Table 7.2 **Lucas manual fitting analysis (Fitting Index = $A + B + C + D + E + F$)**

	Score
A. Part placing and fastening	
Self-holding orientation	1.0
Requires holding	2.0
Plus one of the following:	
Self-securing (i.e., snaps)	1.3
Screwing	4.0
Riveting	4.0
B. Process direction	
Straight line from above	0
Straight line not from above	0.1
Not a straight line	1.6
Bending	4.0
C. Insertion	
Single insertion	0
Multiple insertions	0.7
Simultaneous multiple insertions	1.2
D. Access and/or vision	
Direct	0
E. Alignment	
Easy to align	0
Difficult to align	0.7
F. Insertion force	
No resistance to insertion	0
Resistance to insertion	0.6
Restricted	1.5

The last part of the Lucas method is to calculate the cost of manufacturing each component. This manufacturing cost can influence the choice of material and the process by which the part is made. Although not a true costing of the part, this method helps guide designers by giving a relative measure of manufacturing cost. Values of each of the following coefficients are derived from detailed tables developed for the purpose. The part manufacturing cost index is

$$M_i = R_c P_c + M_c$$

where

$R_c = C_c C_{\mathrm{mp}} C_s$ (C_t or C_f) is the relative cost;
C_c = complexity factor;
C_{mp} = material factor;
C_s = minimum section;
C_t = tolerance factor, or C_f =finish factor (whichever is greater);
P_c = processing cost;
$M_c = VC_{\mathrm{mt}}$, W_c is the material cost;
i = volume (mm^3);
C_{mt} = material cost;
W_c = Waste coefficient.

7.4.3 The Boothroyd-Dewhurst DFA evaluation method

The Boothroyd-Dewhurst method of assembly evaluation is based on two principles: the application of criteria to each part to determine if it should be separate from all other parts, and the estimation of the handling and assembly costs for each part using the appropriate assembly process.

The Boothroyd-Dewhurst method relies on an existing design, which is iteratively evaluated and improved. The process follows the following steps:

1. Select an assembly method for each part
2. Analyze the parts for the given assembly methods
3. Refine the design in response to shortcomings identified by the analysis
4. Refer back to step 2 until the analysis yields a satisfactory design.

The analysis generally is performed using a specific worksheet. Tables and charts are used to estimate the part *handling* and part *insertion* time. Each table is based on a two-digit code, which in turn is based on a part's size, weight, and geometric characteristics. Handling and insertion times are a function of the following component parameters. Each of these parameters directly affects the assembly process by simplifying or complicating it:

- Component size
- Component thickness
- Component weight
- Tendency of the component to nesting
- Tendency of the component to tangling
- Component fragility
- Component flexibility
- Component slipperiness
- Component stickiness
- Necessity of using two hands to effect assembly
- Necessity of using specialized grasping tools to effect assembly
- Necessity of optical magnification to effect assembly
- Necessity of mechanical assistance to effect assembly.

Nonassembly operations also are included in the worksheet. For example, extra time is allocated for each time the assembly is reoriented.

Next, all parts are evaluated on the basis of whether each part is really necessary in the assembly by asking the following questions:

- Does the part move relative to another part?
- Are the material properties of the part necessary?
- Does the part need to be a separate entity for the sake of assembly?

The list of all parts then is evaluated to obtain the minimum number of theoretically needed parts, denoted N_m. Table 7.3 depicts the table commonly used for assembly evaluation using the Boothroyd-Dewhurst method.

In column I, the number 1 is used to represent that a part is essential, and 0 to represent that a part is not essential. The method assumes that the ideal assembly time for a part is 3 s. Given that assumption, the DE can be calculated as $(3\,\text{s} \times N_m)/T_m$.

It is clear from this discussion that the method can be quite time consuming, owing to the amount of intricate detail involved in the analysis procedure. A software package has been developed to accelerate the application of this process. Table 7.4 shows an example of estimated times required for manual handling of components for product assembly.

The leftmost column of Table 7.4 specifies the part's symmetry. Alpha symmetry depends on the angle through which a part must be rotated about an axis perpendicular to the axis of rotation, to repeat its orientation. Beta symmetry depends on the angle through which a part must be rotated about the axis of insertion, to repeat its orientation. Parts are categorized, in the rows of the table, by the total degrees of these angles.

Columns 0–3 list parts of nominal size and weight that are easy to grasp and manipulate with one hand without the aid of tools. The parts in columns 4–7 require grasping tools due to their size. Column 8 has parts that severely nest or tangle in bulk. Column 9 has parts that require two hands, two people, or mechanical assistance for handling.

Groups 1 and 2 are further subdivided into categories representing the amount of orientation required based on part symmetry.

The second digit of the handling code is based on flexibility, slipperiness, stickiness, fragility, and nesting characteristics of the part. This digit also depends on the group divisions of the first digit as follows:

1. For columns 0–3, the second digit classifies the size and thickness of a part.
2. For columns 4–7, the second digit classifies the part thickness, type of tool required for handling, and the necessity for optical magnification during the handling process.
3. For column 8, the second digit classifies the size and symmetry of a part.
4. For column 9, the second digit classifies the symmetry, weight, and interlocking characteristics of parts in bulk.

The first digit is divided into three main groups:

1. A first digit of 0–2 means the part is not secured immediately after insertion.
2. A first digit of 3–5 means the part secures itself or another immediately after insertion.
3. A first digit of 9 means that the process involves parts already in place.

Table 7.3 **Boothroyd-Dewhurst method to evaluate DFA**

A	B	C	D	E	F	G	H	I	Name of Assembly
Part ID	Number of consecutive identical operations	2-digit handling code	Manual handling time per part	2-digit insertion code	Manual insertion time per part	Operation time (BD +F)	Operation cost	Essential part?	
					Total	$T_m =$	$C_m =$	$N_m =$	

Table 7.4 **Classification, coding, and database for part features affecting manual handling time in seconds (for parts that can be grasped and manipulated by one hand without the aid of grasping tools)**

		Parts easy to grasp and manipulate					**Parts with handling difficulties**				
		Thickness > 2 mm			**Thickness ≤ 2 mm**		**Thickness > 2 mm**			**Thickness ≤ 2 mm**	
		Size > 15 mm	**Size 6–15 mm**	**Size < 6 mm**	**Size > 6 mm**	**Size ≤ 6 mm**	**Size > 15 mm**	**Size 6–15 mm**	**Size < 6 mm**	**Size > 6 mm**	**Size ≤ 6 mm**
		0	**1**	**2**	**3**	**4**	**5**	**6**	**7**	**8**	**9**
$(\alpha+\beta)<360$	0	1.13	1.43	1.88	1.69	2.18	1.84	2.17	2.65	2.45	2.98
$360\leq(\alpha+\beta)\leq540$	1	1.5	1.8	2.25	2.06	2.55	2.25	2.57	3.06	3	3.38
$540\leq(\alpha+\beta)\leq720$	2	1.8	2.1	2.55	2.36	2.85	2.57	2.9	3.38	3.18	3.7
$(\alpha+\beta)=720$	3	1.95	2.25	2.7	2.51	3	2.73	3.06	3.55	3.34	4

Groups 1 and 2 are further subdivided into classes that consider the effect of obstructed access or restricted vision on assembly time.

The second digit of the assembly code is based on the following group divisions of the first digit:

1. For a first digit of 0–2, the second digit classifies the ease of engagement of parts and whether holding down is required to maintain orientation or location.
2. For a first digit of 3–5, the second digit classifies the ease of engagement of parts and whether the fastening operation involves a simple snap fit, screwing operation, or a plastic deformation process.
3. For a first digit of 9, the second digit classifies mechanical, metallurgical, and chemical processes.

The Boothroyd-Dewhurst method has been known to reduce the total number of individual components in an assembly. However, this has often been achieved at the cost of part complexity. Part complexity does not lend itself easily to manufacturability. Similarly, complex parts are inherently more difficult to disassemble.

7.5 A DFA method based on MTM standards

An improved assembly methodology takes into consideration numerous factors, such as the weight, size, and shape of components being assembled; frequency of assembly tasks (based on number of similar products being assembled within a particular time frame); personnel requirements; postural requirements; material handling requirements; and need for component preparation. A number of human factors, in addition to design and economic factors, merit consideration due to labor intensity of the assembly process.

The most commonly used assembly operations are recorded and described in sufficient detail. Every assembly operation is subdivided into basic elemental tasks. Only a fraction of these tasks actually perform the assembly. The remaining tasks constitute such actions as reaching for and grasping tools. A methods time measurement (MTM)-based index for assembly is presented in Table 7.5. The simplest assembly task—inserting an easily grasped object without the exertion of much force by hand by a trained worker under average conditions—is considered as the basic assembly task. A score of 73 Time Measurement Units was assigned to this task, which corresponds to time duration of approximately 2 s. Subsequent scores were assigned based on detailed study of the most commonly encountered assembly operations.

Figure 7.5 depicts a system of measures that can be utilized to enhance ease of assembly of product architectures. Figure 7.6 depicts a method to enable design improvement for product assembly. It should be noted that these tables and figures will be referenced in a subsequent chapter on designing for maintenance.

Table 7.5 Evaluation system for the numeric analysis of assembly

Design attribute/ feature	Design parameters	Score	Interpretation
Assembly force			
Straight line motion, no exertion of pressure	Push operations by hand	0.5	Little effort required
		1	Moderate effort required
		2	Large effort required
Straight line and twisting motion without pressure	Twisting and push operations by hand	1	Little effort required
		2	Moderate effort required
		4	Large amount of effort required
Straight line motion with exertion of pressure	Intersurface friction or wedging	2	Little effort required
		2.5	Moderate effort required
		4	Large effort required
Straight line and twisting motions with exertion of pressure	Intersurface friction or wedging	2.5	Little effort required
		3	Moderate effort required
		5	Large effort required
Twisting motions with pressure exertion	Material stiffness	2.5	Little effort required
		4	Moderate effort required
		6	Large effort required
Material handling			
Component/fastener size	Component dimensions (very large or very small)	2	Easily grasped
		3.5	Moderately difficult to grasp
		4	Difficult to grasp
	Magnitude of weight	2	Light (<7.5 lb)
		2.5	Moderately heavy (<17.5 lb)
		3	Very heavy (<27.5 lb)
Component/fastener symmetry	Symmetric components easy to handle	0.8	Light and symmetric
		1.2	Light and semisymmetric
		1.4	Light and asymmetric
		2	Moderately heavy, symmetric
		2.2	Moderately heavy, semisymmetric
		2.4	Moderately heavy, asymmetric
		4.4	Heavy and symmetric
		4.6	Heavy and semisymmetric
		5	Heavy and asymmetric
Requirement of tools for assembly			
Exertion of force		1	No tools required
		2	Common tools required
		3	Specialized tools required
Exertion of torque		1	No tools required
		2	Common tools required
		3	Specialized tools required

(*Continued*)

Table 7.5 (Continued)

Design attribute/ feature	Design parameters	Score	Interpretation
Accessibility of joints/grooves			
Dimensions	Length, breadth, depth, radius, angle made with surface	1	Shallow, broad fastener recesses; large, readily visible slot or recess in case of snap fits
		1.6	Deep, narrow fastener recesses, obscure slot or recess in case of snap fits
		2	Very deep, very narrow fastener recesses, slot for prying open snap fits difficult to locate
Location	On plane surface	1	Groove location allows easy access.
	On angular surface	1.6	Groove location difficult to access; some manipulation required
	In a slot	2	Groove location very difficult to access
	On vertical surface	1.5	Some manipulation required against gravity
	On horizontal surface	1	Groove location allows easy access
Positioning			
Level of accuracy required to position the tool	Symmetry	1.2	No accuracy required
		2	Some accuracy required
		5	High accuracy required
	Asymmetry	1.5	No accuracy required
		2	Some accuracy required
		5	High accuracy required

7.6 A DFA case study

It is clear that once design anomalies are identified, alterations can be undertaken to improve product design from an assembly perspective. Once corresponding design decisions have been made, the cost of manufacturing those components can be taken into consideration to optimize the manufacturing process and maximize profit potential.

Figure 7.6 illustrates the practical application of the DFA method. It deals with the assembly operation of a computer monitor (Table 7.6 presents the various components that constitute a computer monitor) and is presented in tabular form in Table 7.7.

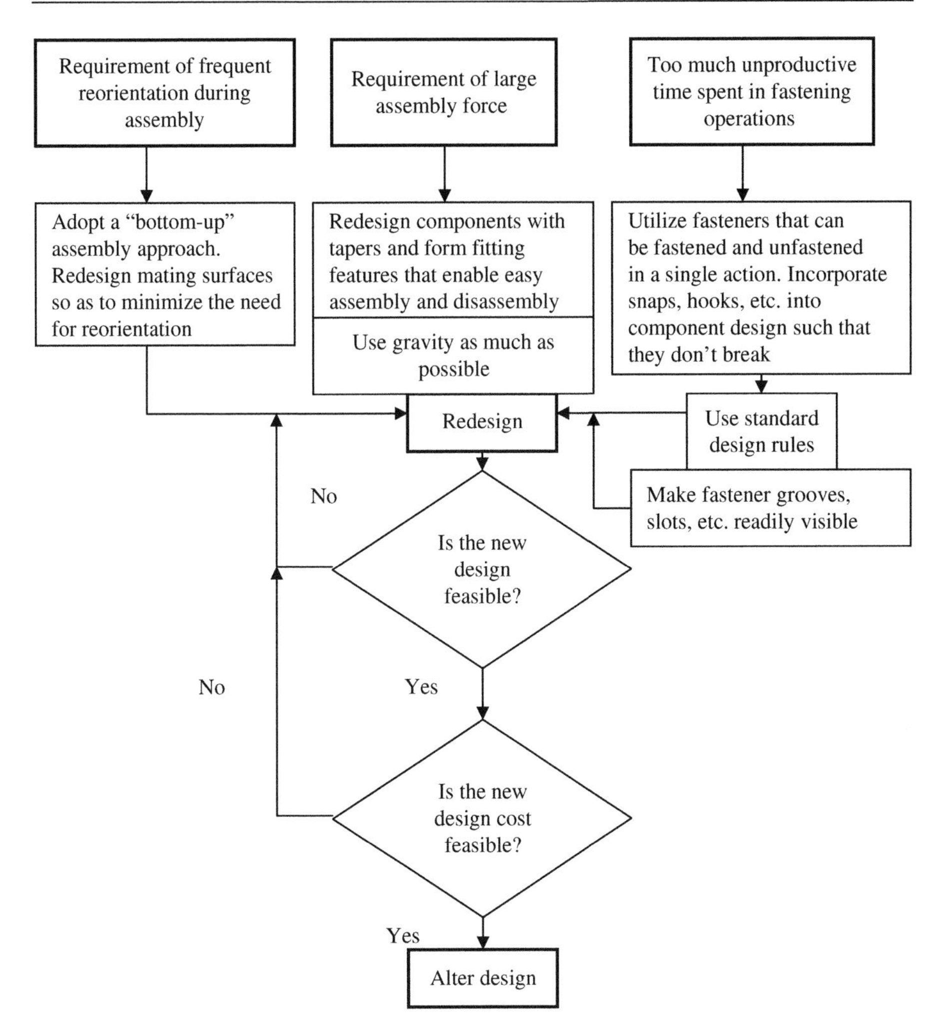

Figure 7.5 Enhancement of product design features to enable ease of assembly.

The total amount of time taken to assemble a typical computer monitor is about 3.378 min. Fixing screws and bending lugs are two of the most time-consuming tasks to be addressed from a design perspective. Simplifying these tasks through improvements in product design can cut assembly time as well as related costs.

The remainder of this chapter focuses on the issue of product disassembly: its importance in the context of growing environmental concerns, its cost implications, and some methods to enable product designers to incorporate disassembly as one of the Xs in the DFX methodology.

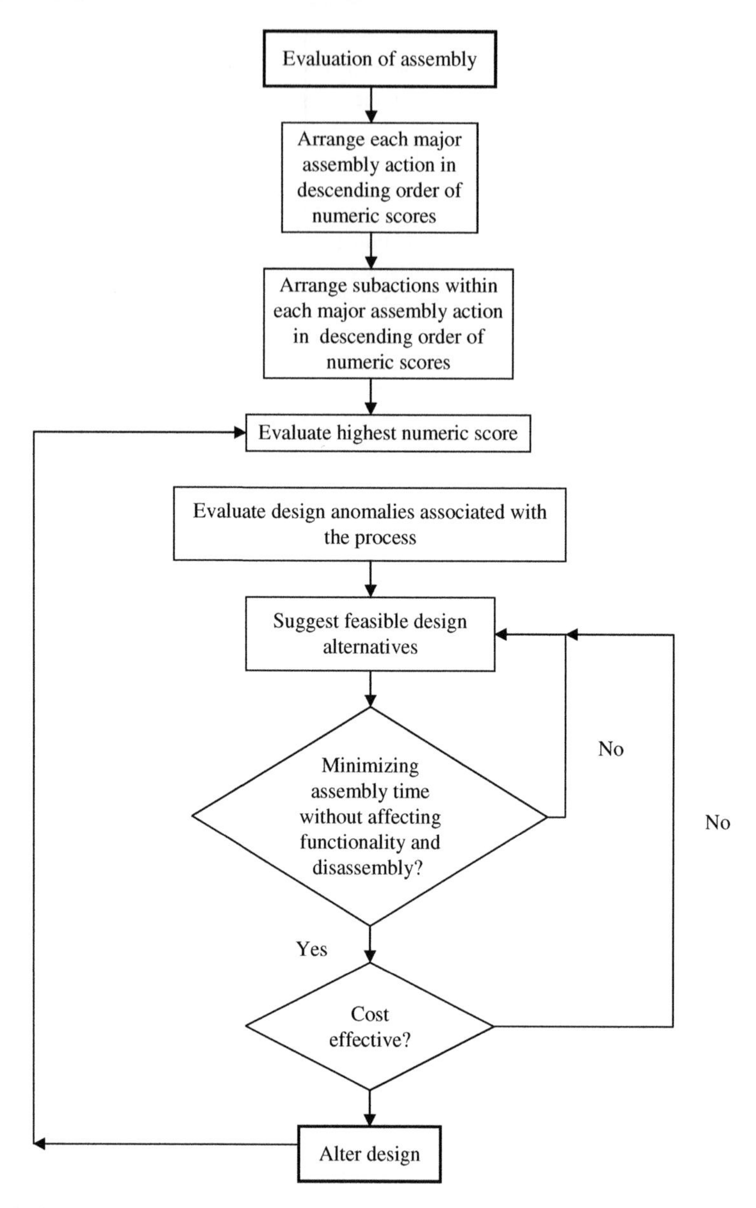

Figure 7.6 Methodology to enable design improvement for product assembly.

7.7 Design for disassembly

7.7.1 Definition

In the present era of environmental awareness, EOL objectives, such as component reuse (components from a retired product used without upgrading in a new product), remanufacture (components from a retired product used in a new product after a

Table 7.6 **Components of a computer monitor**

No.	Component name	Component material	Quantity
1	Back screw	Copper	4
2	PCB screws	Copper	2
3	CRT screws	Copper	4
4	CRT/PCB assembly	Mixed	1
5	Back cover	Plastic	1
6	Swivel base	Plastic	1
7	Pivot	Plastic	1
8	Yoke assembly	Mixed	1
9	Deflection wire lead	Mixed	1
10	Retainer screws	Copper	2
11	Main wire lead	Copper	1
12	Adjusting knobs	Plastic	4
13	PCB retainer screw	Copper	1
14	Retaining lugs	Aluminum	4
15	PCB assembly	Mixed	1
16	Rear board	Plastic	1
17	CRT	Mixed	1

technological upgrade), and recycling (reuse at the material level, such as recycling of plastics), constitute some of the most important reasons for disassembling products. This can be attributed to the staggering impact of industrial and domestic waste on the environment. Widespread diffusion of consumer goods and shorter product life cycles have led to an unprecedented number of used products being discarded. For example, in 2005, there was an average of one computer per family in the United States. In 1991, Carnegie Mellon University estimated that some 150 million obsolete PCs, none with readily recoverable materials, required more than 8 million cubic meters of landfill space at a cost of around $400 million (Lee et al., 2001). However, the number of potential landfill sites for nonhazardous solid wastes has seen an exponential decrease. In the United States alone, landfill sites diminished from 18,000 in 1985 to 9000 in 1989. According to one study, the United States lost more than 70% of its landfill sites by 1997 (Zhang et al., 1997), with landfills in many states reaching their permitted capacities at an alarming rate. EOL products contain extensive amounts of reusable material that is too expensive to dispose of; retrieval of this material would benefit the manufacturer as well as the environment.

Depending on the extent of disassembly, nondestructive disassembly can be further classified into two categories:

- **Total disassembly**: The entire product is disassembled into its constituent components. This may not be economically feasible due to the imposition of external constraints, such as time, economic factors, and presence of hazardous materials.
- **Selective disassembly**: Selective disassembly is the reversible dismantling of complex products into less complex subassemblies or single parts (Lambert, 1999). It involves the systematic removal of desirable constituent parts from an assembly while ensuring that there is no impairment of parts due to the process (Brennan et al., 1994).

Table 7.7 **Assembly operation of a computer monitor**

Task description	Task total	Assembly force			Material handling		
		Inter-surface friction	Inter-surface wedging	Material stiffness	Compo-nent size	Compo-nent weight	Compo-nent symmetry
1. Assemble rear board							
a. Place rear board in place	10.92	–	2	–	2	2	1.2
b. Bend first retaining lug	22	–	–	6	3	2	1.4
c. Bend second retaining lug	22	–	–	6	3	2	1.4
d. Bend third retaining lug	22	–	–	6	3	2	1.4
e. Bend fourth retaining lug	22	–	–	6	3	2	1.4
2. Assemble PCB							
a. Fit PCB in place	11.55	–	2	–	2	2	0.8
b. Bend first retaining lug	24.53	–	–	6	4	2	0.8
c. Bend second retaining lug	24.53	–	–	6	4	2	0.8
d. Bend third retaining lug	24.53	–	–	6	4	2	0.8
e. Bend fourth retaining lug	24.53	–	–	6	4	2	0.8
f. Fit PCB retaining screw	18.27	5	–	–	4	2	0.8
3. Fit CRT/PCB assembly							
a. Screw first PCB screw	15.12	2	–	–	2	2	0.8
b. Screw second PCB screw	15.12	2	–	–	2	2	0.8
c. Screw first PCB screw	15.12	2	–	–	2	2	0.8
d. Screw second PCB screw	15.12	2	–	–	2	2	0.8
e. Screw third PCB screw	15.12	2	–	–	2	2	0.8
f. Screw fourth PCB screw	15.12	2	–	–	2	2	0.8
g. Fit yoke assembly	13.13	–	3	–	2	2	1.4
h. Fit deflection wire lead	13.02	–	3	–	2	2	0.8
4. Fit main wire lead							
a. Fit main wire lead	17.77	–	3	–	4	2.5	2.2
b. Fix first retainer screw	15.65	2	–	–	2	2	0.8
c. Fix second retainer screw	15.65	2	–	–	2	2	0.8
5. Assemble back cover							
a. Remove back cover	15.44	–	3	–	3.5	2	1.2
b. Screw first back screw	15.12	2	–	–	2	2	0.8
c. Screw second back screw	15.12	2	–	–	2	2	0.8
d. Screw third back screw	15.12	2	–	–	2	2	0.8
e. Screw fourth back screw	15.12	2	–	–	2	2	0.8
6. Assemble swivel pivot							
a. Fit swivel pivot	18.16	–	–	4	3.5	2	1.2
b. Fit swivel support	10.92	–	2	–	2	2	1.2
7. Assemble swivel base							
a. Fit swivel base	10.50	–	2	–	2	2	1.2
b. Rotate swivel base about pivot	13.54	–	4	–	2	2	1.2
8. Assemble adjusting knobs							
a. Fit first adjusting knob	10.5	–	1.5	–	2	2	0.8
b. Fit second adjusting knob	10.5	–	1.5	–	2	2	0.8
c. Fit third adjusting knob	10.5	–	1.5	–	2	2	0.8
d. Fit fourth adjusting knob	10.5	–	1.5	–	2	2	0.8
Total score	**563**						
Total time for assembly operation: 5630 TMUs = 3.378 min.							

Tooling		Accessibility and positioning			Allowances			
Force exertion	Torque exertion	Dimensions	Location	Accuracy of tool placement	Posture allowance	Motions allowance	Personnel allowance	Visual fatigue allowance
1	–	1	–	1.2	–	–	–	5%
3	–	1.6	1	2	–	–	–	10%
3	–	1.6	1	2	–	–	–	10%
3	–	1.6	1	2	–	–	–	10%
3	–	1.6	1	2	–	–	–	10%
1	–	1	1	1.2	–	–	–	5%
3	–	2	2	2.5	–	–	–	10%
3	–	2	2	2.5	–	–	–	10%
3	–	2	2	2.5	–	–	–	10%
3	–	2	2	2.5	–	–	–	10%
1	–	1.6	1	2	–	–	–	5%
–	2	1.6	2	2	–	–	–	5%
–	2	1.6	2	2	–	–	–	5%
–	2	1.6	2	2	–	–	–	5%
–	2	1.6	2	2	–	–	–	5%
–	2	1.6	2	2	–	–	–	5%
–	2	1.6	2	2	–	–	–	5%
1	–	1	1	1.6	–	–	–	1%
1	–	1	1	1.6	–	–	–	1%
1	–	1.6	1	2	–	–	–	1%
–	2	1.6	2	2	–	–	–	5%
–	2	1.6	2	2	–	–	–	5%
–	1	1	1	2	–	–	–	5%
–	2	1.6	2	2	–	–	–	5%
–	2	1.6	2	2	–	–	–	5%
–	2	1.6	2	2	–	–	–	5%
–	2	1.6	2	2	–	–	–	5%
2	–	1.6	1	2	–	–	–	5%
1	–	1	–	1.2	–	–	–	5%
1	–	1	1	1.2	–	–	–	1%
1	–	1	1	1.2	–	–	–	1%
1	–	1	1	1.2	–	–	–	–
1	–	1	1	1.2	–	–	–	–
1	–	1	1	1.2	–	–	–	–
1	–	1	1	1.2	–	–	–	–

Products are selectively disassembled to realize the following objectives:

1. Enabling maintenance and repair (serviceability). This enables the ease of performing all service-related operations, including maintenance, malfunction, diagnosis, and repair. Benefits of designing for serviceability include reduced warranty costs (higher earnings before income tax), enhanced customer appeal, and lengthened service life due to the product's ability to be serviced economically.
2. Availability of subassemblies as service parts or for assembly in new products. Components that are robust enough for extended use in another product are reused to achieve economies of scale by component reuse (cost optimization).
3. Removal of parts prior to setting free other desired parts.
4. Availability of parts intended for material reuse (recycling).
5. Increased purity of materials by removal of contaminants.
6. Complying with regulations that prescribe removal of definite parts, materials, and substances for environmental and safety reasons, such as removal of working fluids such as engine oils and lubricants.

7.7.2 Disassembly process planning

If the process of disassembly is to be included in the product at the design stage, comprehensive disassembly process planning needs to be carried out. Figure 7.7 illustrates the general concept of a disassembly process plan.

It is clear from Figure 7.7 that a disassembly process plan begins with a given product group to be disassembled. Every product group has certain characteristics in common that can be aggregated into categories. A disassembly process plan consists of four essential stages as follows:

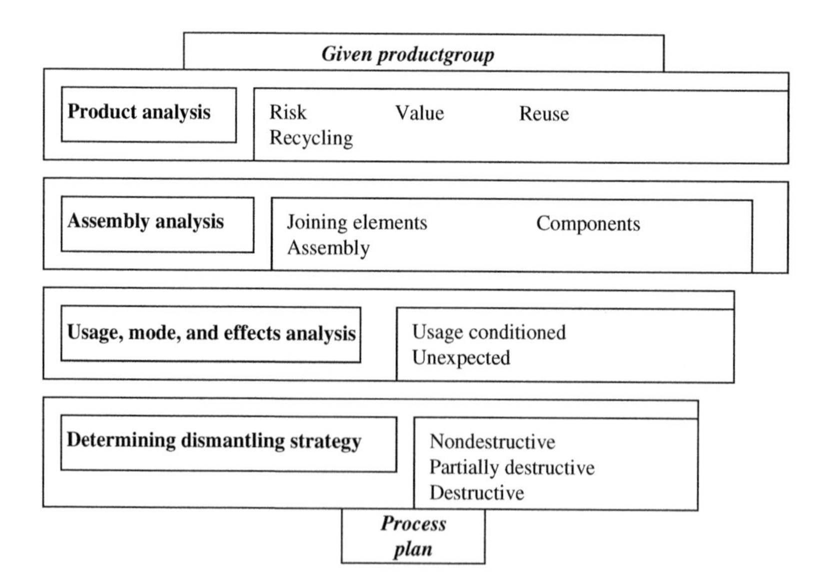

Figure 7.7 Disassembly process plan.

1. **Product analysis**: Product analysis consists of assessing the end value to be realized by disassembling the product. Disassembly value comprises the potential of the product to be reused, its value to be recovered, and risk potentials, as well as existing recycling technology. It is necessary to ascertain the state of recycling technology, since all subsequent recycling depends on this factor.
2. **Assembly analysis**: It is essential to understand how a product has been put together in order to take it apart. For example, if the designer is aware of the kinds of fits, adhesives, bonding members, grooves, slots, and so forth holding the product together, it is easy to plan for disassembly. This planning consists of determining the tooling requirements, magnitude of force, time, and personnel necessary to effect disassembly, as well as knowledge of functionally more valuable components. Assembly analysis involves an analysis of joining elements, component hierarchy, and assembly sequence.
3. **Usage, mode, and effects analysis**: Since most products are disassembled after they have been put to actual use (for either maintenance or EOL purposes), they have been subjected to considerable wear and tear. The way products have been used determines to a large extent what kind of unintended modifications may have been introduced in their intensive and extensive properties. These modifications were not incorporated in the design by the designer. Knowledge of these conditions helps disassembly planners incorporate any contingencies that may arise in the future.
4. **Determination of dismantling strategy**: A determination of dismantling strategy is in keeping with the basic reason for disassembling the product. For example, if the product is to be recycled (at the material level), disassembly need not be performed carefully. Since only the intensive properties of components are important, destructive disassembly may be used to quicken the process. On the other hand, if certain components of the product are to be reused (often the case with functionally important and valuable components), one has to be very careful in separating that component from the product structure to maintain both its intensive and extensive characteristics. In this case, nondestructive disassembly may be used to significant advantage.

7.8 Design for disassembly guidelines

A plethora of literature discusses disassembly planning and design in terms of disassembly design guidelines. The gist of most of these guidelines is to keep product variety to a minimum, use modular product construction, and cluster similar materials together. Some of the more important disassembly guidelines follow.

To minimize assembly work,

1. If possible, similar elements need to be combined in a group.
2. Material variability should be minimized to predict disassembly procedures with a degree of certainty.
3. As far as possible, compatible materials should be used to facilitate disassembly.
4. Any harmful materials, if functionally important, should be grouped together into subassemblies for fast disposal.
5. Any valuable, reusable, and harmful parts need to be easily accessible. This saves a lot of time and effort trying to reach the part in question.

To achieve a predictable product configuration,

1. Aging and corrosive material combinations need to be avoided, since disassembling them cleanly and efficiently (due to their tendency to corrode, spread corrosion, and break off inside the product) often is difficult.
2. What is said in the preceding point holds equally true as far as protecting subassemblies from corrosion, the reasons being the same.

To achieve easy disassembly,

1. Drainage points need to be easily accessible.
2. Fasteners need to be easy to remove or destroy.
3. The number of fasteners must be minimized to save time and effort.
4. Easy access to disjoining, fracture, and cutting points must be provided.
5. Generally speaking, the disassembly path needs to be a simple and straightforward route along which most components are removed. To that end, multiple directions and complex movements for disassembly need to be avoided.
6. Metal inserts in plastic parts should be avoided, since this increases material variety and part complexity and necessitates multiple directions and complex movements in disassembly.

To achieve easy handling,

1. At least one surface needs to be left available for grasping.
2. Nonrigid parts are to be avoided, since they can move, bend, twist, and create problems in disassembly.
3. Any toxic substances, if necessary, need to be placed in sealed units to minimize health hazards.

To achieve easy separation,

1. Any secondary coating processes, such as painting, are to be avoided, since they inhibit access to and removal of components.
2. To separate different materials, they need to be marked accordingly to minimize confusion while disassembling the product.
3. Any parts and materials that are likely to damage machinery need to be avoided.

To reduce variability,

1. Standard subassemblies and parts need to be used (modular product construction).
2. A minimum variety of fasteners should be used, since a large variety of fasteners requires a large variety of tools, skills, surface preparation, and working postures.

7.9 Disassembly algorithms

The bulk of research conducted on disassembly examines such issues as disassembly sequence planning, disassembly evaluation and analysis, and product recovery. This section reviews various approaches addressing these and related issues.

7.9.1 Product recovery approach

Thierry et al. (1995) proposed a product recovery management approach where returned products can be recovered at four levels: product, module, part, and material

Table 7.8 **Options for product recovery after disassembly (Thierry et al., 1995)**

Option	Objective	Level of disassembly	Result
Repair	Restore to working condition	Product level (limited disassembly and fixing)	Some parts repaired
Refurbishing	Improve to quality level, though not like new	Module level (some technological upgrading)	Some modules repaired or replaced
Remanufacturing	Restore to quality level, as new	Part level	Used and new parts in new products
Cannibalization	Limited recovery	Selective disassembly and inspection of potentially reusable parts	Parts reused, recycled, or disposed of
Recycling	Reuse materials only	Material level	Materials used in new products

(in that order). Product recovery options achievable by disassembly may be classified into the categories listed in Table 7.8.

The objective of this method is to recover as much as possible of the economic as well as ecological value of products, components, and materials, so as to minimize the ultimate quantities of waste. One drawback of this approach is the obvious entanglement among the disposal options, which can be quite confusing. For example, the distinction between refurbishing and remanufacturing is too subtle to be implemented from the design perspective.

Krikke et al. (1998) considered the problem at the tactical management level to determine an optimal product recovery and disposal strategy. The model used the disassembly tree as the starting point to describe the disassembly process for the return product. Retrievable parts, modules, and subparts were identified and represented at various sublevels. Products as well as retrievable components are called *assemblies*. The aim of disassembly is to make separate recovery or disposal possible for every single subassembly. In this particular approach, materials are not considered as the lowest disassembly level. Assemblies can be disassembled into subassemblies, and materials can be separated from assemblies. Reuse and recycling options are classified depending on the quality level of the end products. Reuse is classified into three distinct categories: upgrade, restore, and downgrade. Similarly, recycling can be classified as: high-grade material recycling, low-grade material recycling, and alternative material recycling. Disposal is either incineration or landfill. The optimization problem is solved using the stochastic dynamic programming approach.

7.9.2 Optimal disassembly sequence planning for product recovery

A disassembly sequence plan (DSP) is a program of tasks that begins with a product to be disassembled and terminates when all the desired parts of the product are disconnected (Gungor and Gupta, 1998). A DSP aims to optimize product recovery through the minimization of cost, maximization of material recovered, and minimization of disassembly time using mathematical techniques such as linear programming, dynamic programming, and graphical tools.

Navin-Chandra (1994) described product recovery using a CAD tool (ReStar) to find a recovery plan that balances the amount of effort put into recovery and the amount of effort saved by reusing parts and materials through the use of breakeven analysis. All feasible sequences of disassembly are ascertained using an AND/OR graph. An OR relationship between two components, c1 and c2, exists with respect to c3 if either c1 or c2 has to be removed prior to the removal of c3. An AND relationship between c1 and c2 exists with respect to c3 if both c1 and c2 have to be removed prior to removing c3. The sequence of disassembly operations depends on the spatial constraints between components. The traveling salesman methodology is used to solve the problem with products in a certain state of disassembly being analogous to cities and disassembly steps being analogous to connections between cities.

The problem of dealing with future uncertainties of recycling options was addressed by Zussman et al. (1994) using the utility theory. Attributes such as technology refinement, prices, and dumping fees are bound to change with time (dynamic) and, as such, involve uncertainty for the designer and policy maker. The method incorporates the optimization of disassembly sequences to achieve objectives such as maximization of net profit, maximization of parts reuse, or minimization of waste headed toward landfills. A recovery graph is constructed using the AND/OR logic as just described. Values of different entities, such as disassembly sequences, recycling options, and related costs, are used to plot the recovery graph.

Two algorithms were presented by Gungor and Gupta (1997) to obtain a scheme for disassembling multiple product structures having common parts. The core algorithm determines the number of root items to be disassembled in order to minimize disassembly cost. A *root* is defined as a major subassembly, which is further composed of minor subassemblies that can be subdivided into individual components. The allocation algorithm is used to determine a disassembly schedule for the roots and subassemblies by allocating the disassembly requirements over the planning horizon.

Gungor and Gupta (1997) presented an evaluation method to choose the best disassembly process from among several alternatives. Disassembly sequences are generated heuristically by considering the following factors: precedence relationship of components of the product under consideration and average difficulty ratings for each component of the product. The total time of disassembly, a metric affected by attributes such as joint type and direction of disassembly changes, is chosen for evaluating a particular disassembly sequence.

Kuo (2000) divided disassembly planning into four distinct stages: geometric assembly representation, cut-vertex search analysis, disassembly precedence matrix analysis, and disassembly sequence and plan generation for obtaining the disassembly sequence of electromechanical products during the design stage. Disassembly cost is classified as target disassembly, full disassembly, or optimal disassembly. The disassembly tree analysis is the basic idea behind disassembly sequence generation. The component–fastener relationship graph and disassembly precedence matrices represent interrelationships among components. The disassembly process is simplified by forming disassembly modules. The cut-vertex search analysis enables the decomposition of a module into submodules. A cut-vertex in a component–fastener graph is a vertex (component) whose removal disconnects the graph. In a real component, the cut-vertex is the main part that is connected to other components. In a module, the cut-vertex is the connection component between two other components. Three distinct types of geometric assemblies can be defined as:

- Type I. An assembly having a main component to which other components or subassemblies are directly or indirectly assembled.
- Type II. An assembly having no main component. All components are assembled with others. This can be disassembled only as a single component.
- Type III. An assembly that is a combination of both these types. This can be disassembled as further components.

Three kinds of sequences, driven by disassembly cost, are analyzed on formation of the disassembly tree:

- Type I (target disassembly sequence). Specific components are disassembled to remove valuable components.
- Type II (optimal disassembly sequence). Disassembly is stopped when marginal return on the operation becomes uneconomical.
- Type III (complete disassembly sequence). Complete disassembly of the product.

Penev and de Ron (1994) proposed an algorithm for designing processes and systems based on the detection and removal of preferred components. This involves the identification of a poisonous component that needs to be released from the product. All feasible sequences of disassembly are ascertained and depicted using the AND/OR graph technique. Depending on the various options and constraints, the most efficient method of releasing the component or part is determined. Thereafter, the remaining parts are considered for disassembly. This process continues as long as it remains profitable to do so. The dynamic programming approach is used to optimize the problem. Lambert (1999) addressed the problem of selective disassembly sequence generation using a linear programming approach. An economic optimization model with environmental constraints was presented. The disassembly process involves a sequence of single steps or actions, such as the removal of a part or separation into two separate subassemblies. Sequences that combine maximum net revenue with environmental requirements are selected. The profitability of each step is calculated. Locally unprofitable actions should be carried out to enable subsequent profitable steps or to comply with regulations. In this approach, possible disassembly sequences are represented by a graph, which is transformed into a linear programming model. Two sets, S (subassemblies) and A (actions or process steps) are defined. Set S contains all feasible combinations of parts, the original complete product as well as single components. A disassembly sequence is described by a sequence of actions. The optimal disassembly sequence is the one that generates maximum net revenue, subject to specific constraints.

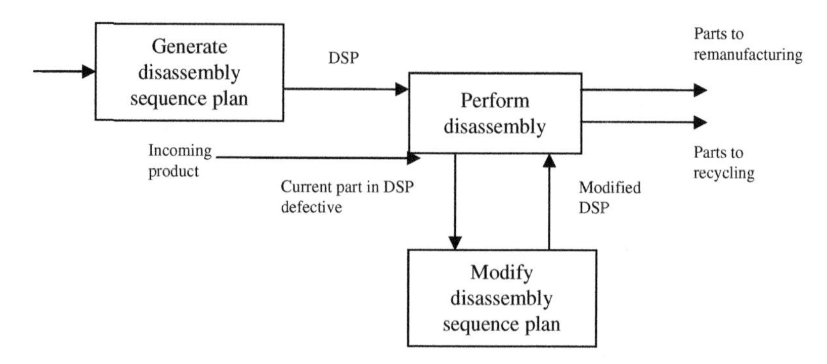

Figure 7.8 Handling uncertainty in disassembly sequence planning.
Source: Adapted from Gungor and Gupta, 1998.

7.9.3 Disassembly sequence planning for a product with defective parts

A high degree of uncertainty is introduced into the disassembly process by the upgrading or downgrading of a product during the course of its use by customers and defects occurring in use or during disassembly. Gungor and Gupta (1998) proposed a methodology for the disassembly sequence planning of products with defective parts. Changes in DSPs have to be incorporated to handle factors leading to uncertainties. This is depicted in Figure 7.8.

Availability of original CAD drawings and an unchanged product structure have been assumed. The physical relationships among components is represented using the disassembly precedence matrix developed from the original CAD model of the product using the AND, OR, and AND/OR relationships. Next, an optimum DSP is generated. Finally, the actual disassembly process is performed. An unexpected situation is dealt with by appropriate modification of the DSP.

7.9.4 Evaluation of disassembly planning based on economic criteria

The issue of disassembly costs was touched on by Feldmann et al. (1999). Disassembly costs must be justified by the economic advantages of recycling. Recycling costs and benefits differ for specific fractions of recovered materials. The more important economic considerations to be taken into account during the disassembly process include (de Ron and Penev, 1995) such factors as value added to products and materials during manufacturing, disassembly cost and revenue per operation, and the penalty if poisonous materials are not completely removed. Operating costs continue to be one of the most daunting concerns for manufacturers. The EOL economic value of components can be computed using the costing technique suggested by Lee et al. (2001). This technique employs conventional costing practices in addition to specifying a miscellaneous cost (the summation of collecting cost and processing cost), which in turn is used to calculate values of other entities, such as reuse value, remanufacture value, primary and secondary recycling values, incineration value, and landfill cost.

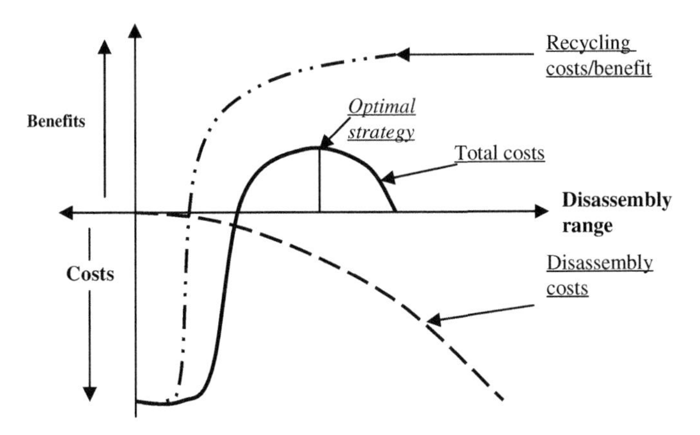

Figure 7.9 Determination of optimal recycling and disassembly strategy.
Source: Modified from Feldmann et al. (1999).

Table 7.9 **Type of recycling according to component composition**

Type of recycling	Definition	Component composition
Primary recycling	Recycling on a comparable quality level	No alloy present in the component Polymer content in the component
Secondary recycling	Recycling on a lower quality level, down cycling	Presence of an alloy in the component No polymer content Ceramic content Elastomer or composite material
Tertiary recycling	Decomposition	
Quaternary recycling	Incineration with energy retrieval	No polymer content Ceramic content

The role played by economic factors in determining an optimal recycling and disassembly strategy is illustrated in Figure 7.9.

A disassembly strategy based on the economic validity of the process was devised by de Ron and Penev (1995). The strategy involves determining the value of the abandoned product as the first step. This value is expressed as the difference between the summations of (i) revenue from disassembled parts and recycling remaining materials and (ii) transportation and miscellaneous costs. The second step involves ascertaining the product's state in its life cycle when discarded. If a product has not completely finished its operational life cycle, it can be brought back into operation with minimum effort by appropriate servicing. This is recommended if the product is in the second phase of its life cycle. Servicing is recommended if the potential value of the product after incurring the costs of disassembly exceeds the revenue expected from disassembly. Lee et al. (2001) proposed guidelines for determining feasible EOL options including the economic value of products and their components (Table 7.9).

Several objectives have been evaluated using economic guidelines. These include the minimization of environmental impact, the minimization of deficit or maximization of surplus, and the minimization of the time for disassembly. Disassembly is stopped when one or more of the preset criteria are met. These criteria include attainment of the highest rate of return or environmental impact and incurring greatest positive net cost.

7.9.5 Geometric models and CAD algorithms to analyze disassembly planning

The issue of disassembling a geometrically constrained assembly of components was addressed by Srinivasan and Gadh (2002). Two types of constraints were considered: spatial constraints due to three-dimensional geometric interactions between components and user-defined constraints imposed by clustering components into subassemblies. The problem was solved utilizing an algorithm that focuses on removing components simultaneously instead of sequentially (the global selective disassembly algorithm).

A geometric algorithm to selectively disassemble a product by determining the disassembly sequence and minimum number of removals, given the component to be disassembled, was presented by Srinivasan and Gadh (1998). The algorithm utilizes the principle of wave propagation by analyzing the assembly from the component outward and ordering the components for disassembly.

Another CAD-based selective disassembly algorithm was presented by Srinivasan et al. (1999). After the disassembly sequences have been determined using the wave propagation algorithm, they are evaluated using an objective function, such as minimization of cost. Disassembly design decisions then are based on this evaluated disassembly sequence.

Shyamsundar and Gadh (1999) addressed the problem of determining a valid assembly for which at least one disassembly sequence exists. This approach contradicts those considered earlier, since it determines whether a disassembly sequence exists for a particular product. Two methodologies were presented: an assembly topology graph, whose nodes represent the components in a product, and a set of boundary components representing components that intersect the boundary of an assembly (components most easily accessible for disassembly).

7.9.6 Automation of disassembly technology and predicting future trends

The automation of disassembly and recycling technology has been dealt with in a recent Delphi study. Obsolescence of EOL technology motivated this study. Since electronic products and automobiles have an average lifespan of 15–25 years, EOL technologies will have changed by the time these products are finally discarded. The conclusions of the study point out that the main obstacles facing future disassembly and recycling technology are more economic than technological in nature (Boks and Templemon, 1998). Automatic disassembly (for a limited product variety only) is

Table 7.10 **Technical feasibility and economic attractiveness of automated disassembly technology for electrical and electronic goods and automotive products (in parentheses)**

Year	Technical feasibility (%)			Commercial feasibility (%)		
	Full	Partial	Limited	Full	Partial	Limited
By 1998	2 (6)	7 (6)	29 (22)	4 (0)	2 (0)	12 (16)
By 2000	7 (2)	28 (14)	40 (35)	5 (0)	23 (4)	25 (16)
By 2005	26 (4)	39 (33)	24 (22)	11 (2)	25 (24)	39 (35)
By 2010	30 (28)	14 (29)	3 (18)	14 (15)	28 (31)	14 (16)
By 2015	11 (17)	11 (12)	2 (0)	23 (17)	5 (12)	2 (4)
By 2020	9 (15)	2 (4)	2 (2)	12 (17)	5 (10)	5 (4)
Later or never	16 (28)	0 (2)	0 (0)	32 (49)	14 (18)	4 (8)

Source: Modified from Boks and Templemon (1998).

expected to gain importance. Results for the automation of electrical and electronic consumer goods disassembly and recycling technology are tabulated in Table 7.10.

Some of the main obstacles believed to be preventing automated disassembly from becoming a commercially successful activity range from large product variety (most important) to variations in returned products (only partially important) to a highly damaged product (least important). Active research disassembly on oriented life cycle analyses is being conducted at the Computer-Integrated Manufacturing Institute at the Georgia Institute of Technology. This research evaluated the recyclability of a product in the light of possible future trends in the development of recycling technology and economy (Kuo, 2000). In addition to the preceding methodologies, a number of researchers have explored to recyclability of materials. For instance, Brinkley et al. (1996) dealt with the life cycle inventory of PVC.

7.10 A proactive design for disassembly method based on MTM standards

A review of the literature demonstrates that most research pertaining to designing for disassembly has focused on mathematical algorithms that seek to optimize the disassembly sequence. These algorithms tend to be highly theoretical in nature and cannot be readily applied to product design. They also tend to be reactive in approach and too limited in their scope to affect the disassembly process in any meaningful way. Over 70% of product life cycle costs are ascertained at the design stage itself. This makes it important to have a method that seeks to accomplish proactive product design for disassembly. Such a method has been the pioneering work of the authors and is presented in this section.

The disassembly methodology takes into consideration numerous factors, such as the weight, size, and shape of the components being disassembled; frequency of disassembly tasks (based on the number of similar products being disassembled within a particular time frame); requirements for personnel; postural requirements; material handling requirements; and the need for component preparation, such as cleaning and degreasing. A number of human factors must be considered, as the disassembly work is highly labor intensive. These factors directly affect the disassembly process and, hitherto, have been neglected in the formulation of both disassembly algorithms and design for disassembly methods.

The most widely used disassembly operations are recorded and described in sufficient detail. Every disassembly operation then is subdivided into basic elemental tasks. Only a fraction of the tasks in the disassembly operation actually are responsible for performing disassembly. The remaining tasks constitute such actions as reaching for tools, grasping tools, and cleaning components prior to disassembly. For example, the following case study considers a simple unscrew operation that may be subdivided into several elemental tasks.

7.11 A design for disassembly case study

The elemental tasks in a simple unscrew operation are

1. Constrain the product to prevent motion during disassembly
2. Reach for tool (power screwdriver)
3. Grasp the tool
4. Position the tool (accessibility of fastener)
5. Align the tool for commencement of operation (accessibility of fastener)
6. Perform disassembly (unscrew operation: force exertion in case of manual unscrew operation)
7. Put away the tool
8. Remove screws and place them in a bin
9. Remove the component and put it in a bin.

As is evident from this sequence of operations, tasks 4, 5, and 6 actually effect disassembly; the remaining tasks are more auxiliary in nature. Tasks 1, 2, and 3 are preparatory. Altering these tasks would have little or no effect on the efficiency of the disassembly process. Assuming constancy of all other conditions, such as operator dexterity and speed of operation, weight and size of tool, and workplace conditions, the efficiency of the disassembly process can be directly attributed to tasks 4, 5, 6, and 9. Examination of these tasks reveals that they are directly affected by the design configuration of the product. For example, some designs allow easy access to components for disassembly, while others may not. Accessibility of components and fasteners is a design attribute that enables effective positioning and alignment of a tool for disassembly purposes. Similarly, task 9 can also be shown to be directly affected by product design. Component removal is influenced by design attributes such as the size, shape, weight, and material of the component. Large, unsymmetrical, and heavy components as well as minute and sharp components are difficult to manipulate and result in a decline in disassembly efficiency. Similarly, all these tasks require the adoption of a particular posture during the disassembly process. If a large number of such operations is to be performed during the work shift (frequency of operations) and the worker is forced to adopt an unnatural posture, resulting in the onset of static fatigue, the long-term effects can be devastating, not only for the worker but for the organization as well.

Meaningful disassembly evaluation criteria therefore should include all these factors, since they relate directly to product design. Other factors that affect the disassembly process include the weight and size of the tool (large, heavy, and unsymmetrical tools are difficult to operate) and preparation operations such as cleaning and degreasing, which can be minimized through the identification of potential dirt traps prior to disassembly.

The proposed method consists of the following distinct elements: a numeric disassemblability evaluation index and systematic application of design for disassembly methods.

The numeric disassemblability evaluation index is a function of several design parameters that directly or indirectly affect the process of consumer product disassembly. Numeric scores are assigned each of these parameters depending on the ease with which they can be attained. The following parameters have been addressed:

- **Degree of accessibility of components and fasteners**: Easy access is a prerequisite for quick and efficient disassembly operation. The less accessible a component or fastener is, the higher numeric score it receives.
- **Amount of force (or torque) required for disengaging components (in case of snap fits) or unfastening fasteners**: The less the amount of force required, the better is the design. The amount of effort required is directly proportional to the value of numeric score received.
- **Postural requirements for performing disassembly tasks**: The disassembly process still is predominantly labor intensive and, according to a recent Delphi study, is expected to remain so in the foreseeable future. As a result, disassembly operations that require workers to assume unnatural postures would be highly detrimental to the operator performing those operations. An unnatural posture is one responsible for the onset of static muscular fatigue. This issue assumes even greater importance in light of the high frequency of disassembly tasks. A provision for including additional allowances in the disassembly score based on this category has been made in the method.
- **Identification of dirt traps**: This factor is important for obvious reasons. A product that has been in regular use is bound to accumulate internal dirt over a period of time. From a disassembly perspective, components that accumulate dirt need to be cleaned and degreased before disassembly and therefore involve prior preparation. This activity is time and labor intensive. Empirical data can enable easy identification of dirt traps at the design stage. This can enable component redesign to facilitate disassembly.
- **Design factors such as the weight, shape, and size of components being disassembled**: This can be a crucial consideration in product disassembly, especially since it involves the use of special fixtures and apparatuses or simply more workers. For example, the CRT of a 25″ television set can be quite heavy and large for a single person to manipulate efficiently. These factors have been addressed through the introduction of an additional multiplier for material handling.

Any product configuration may be considered to be composed of two distinct entities: functional elements and fastening elements. Functional elements are directly responsible for jointly performing the primary, secondary, and tertiary functions of the product. For example, the CRT of a television set performs the primary function of displaying the picture, whereas the speakers are responsible for emitting sound. The secondary function of protecting the internal components of a TV set is performed by the cabinet.

Fasteners are used to securely link various functional elements. Various fastener configurations may be employed, depending on such factors as nature of current technology, component design modified according to EOL options, human factors, and various economic factors. The key to solving the problem of designing products for easy disassembly is to choose appropriate fasteners and joining methods that enable quick, easy, and economic dismantling of an assemblage of functionally important components (with corresponding design modifications of the functional components). The method described here addresses this situation in two parts.

Identification of design anomalies helps optimize component design from the disassembly perspective. Several design factors, such as accessibility, mating surface condition, corrosion, size, weight, and shape, play an important role in disassembly. It is imperative, therefore, that they should be addressed in detail before a particular component design is finalized. The application of the disassembly evaluation criteria to a product results in numerical indices for various categories of evaluation. These scores can be multiplied by additional allowances prespecified for worker posture, motion, and visual fatigue and personnel requirements. The higher the score an evaluation category obtains, the greater is the chance of detecting a design flaw within that category. For example, in the simple unscrew operation, if the category accessibility receives a higher numeric score, this implies that the product design hampers easy accessibility of fasteners and an opportunity for product redesign in that realm exists.

Once design anomalies have been identified, the second part of the method may be applied to achieve design optimization. Discrete EOL options are assigned various components after a thorough economic analysis. Depending on these options, appropriate fastener configurations are chosen to enhance disassembly. The components now can be (re)designed based on the previous factors. Taking the example as a case in point, design optimization may be achieved by redesigning the component to enable optimal placement of fasteners (if more than one), so that all fasteners are removed before the tool finally is put away. This not only entails optimal placement but also the use of uniform standardized fasteners. The design for disassembly methodology can be put to practical use to actually work for the designer. In most cases, accessibility is directly proportional to the need for assuming an unnatural posture. Thus, addressing and optimizing one evaluation category could result in a solution for another category as well.

A time-based numeric index for disassembly is presented in Table 7.11. The simplest disassembly task, removing an easily grasped object without the exertion of much force by hand by a trained worker under average conditions, has been considered as the basic disassembly task. A score of 73 TMUs was assigned this task, which corresponds to a time duration of approximately 2 s. Subsequent scores were assigned based on the detailed study of most commonly encountered disassembly operations.

Allowances for various attributes affecting the dismantling process have been presented in Table 7.12. Relevant allowances include those made for posture, fatigue, and types of motions of the worker and personnel needs. The allowances section of the index remains unchanged for disassembly, maintenance, and assembly procedures.

Table 7.13 indicates the various components used in a computer CRT and Table 7.14 shows the case study for the computer monitor. Figure 7.10 is a hierarchical representation of the DFD methodology being discussed.

Table 7.11 Evaluation system for a numeric analysis of disassembly

Design attribute	Design feature	Design parameters	Score	Interpretation
Disassembly force	Straight line motion without exertion of pressure	Push–pull operations with hand	0.5	Little effort required
			1	Moderate effort required
			3	Large amount of effort required
	Straight line and twisting motion without pressure	Twisting and push–pull operations with hand	1	Little effort required
			2	Moderate effort required
			4	Large amount of effort required
	Straight line motion with exertion of pressure	Intersurface friction or wedging	2.5	Little effort required
			3	Moderate effort required
			5	Large amount of effort required
	Straight line and twisting motions with exertion of pressure	Intersurface friction or wedging	3	Little effort required
			3.5	Moderate effort required
			5.5	Large amount of effort required
	Twisting motions with pressure exertion	Material stiffness	3	Little effort required
			4.5	Moderate effort required
			6.5	Large amount of effort required
Material handling	Component size	Component dimensions (very large or very small)	2	Easily grasped
			3.5	Moderately difficult to grasp
			4	Difficult to grasp
		Magnitude of weight	2	Light (<7.5 lb)
			2.5	Moderately heavy (<17.5 lb)
			3	Very heavy (<27.5 lb)
	Component symmetry	Symmetric components easy to handle	0.8	Light, symmetric
			1.2	Light, semisymmetric
			1.4	Light, asymmetric
			2	Moderately heavy, symmetric
			2.2	Moderately heavy, semisymmetric
			2.4	Moderately heavy, asymmetric
			4.4	Heavy, symmetric
			4.6	Heavy, semisymmetric
			5	Heavy, asymmetric

(*Continued*)

Table 7.11 **(Continued)**

Design attribute	Design feature	Design parameters	Score	Interpretation
Requirement of tools for disassembly	Exertion of force		1	No tools required
			2	Common tools required
			3	Specialized tools required
	Exertion of torque		1	No tools required
			2	Common tools required
			3	Specialized tools required
Accessibility of joints and grooves	Dimensions	Length, breadth, depth, radius, angle made with surface	1	Shallow, broad fastener recesses; large, readily visible slot/recess in snap fits
			1.6	Deep, narrow fastener recesses; obscure slot/recess in snap fits
			2	Very deep, very narrow fastener recesses; slot for prying open snap fits difficult to locate
	Location	On plane surface	1	Groove location allows easy access
		On angular surface	1.6	Groove location is difficult to access; some manipulation required
		In a slot	2	Groove location very difficult to access
Positioning	Level of accuracy required to position the tool	Symmetry	1.2	No accuracy required
			2	Some accuracy required
			5	High accuracy required
		Asymmetry	1.6	No accuracy required
			2.5	Some accuracy required
			5.5	High accuracy required

Table 7.12 **Allowances for disassembly, maintenance, and assembly procedures**

	Percentage multipliers
Posture allowances	
Sitting down	0%
Standing up	2%
Bending down	5%
Lying down	3%
Crouching	5%
Stretching	8%
Squatting	8%
Motions allowances	
Normal motions	0%
Limited motions	5%
Awkward motions	5%
Motions with confined limbs	10%
Motions with confined body	10%
Personnel allowances	
One extra worker	100%
Two extra workers	200%
Each additional worker	$(200 + 100x)\%$, x =number of additional workers
Visual fatigue allowances	
Intermittent attention	1%
Continuous attention	5%
Fixed focus	10%

Table 7.13 **Components of a computer monitor**

Component	Material	Quantity
1. Back screw	Copper	4
2. PCB screws	Copper	2
3. CRT screws	Copper	4
4. CRT/PCB assembly	Mixed	1
5. Back cover	Plastic	1
6. Swivel base	Plastic	1
7. Pivot	Plastic	1
8. Yoke assembly	Mixed	1
9. Deflection wire lead	Mixed	1
10. Retainer screws	Copper	2
11. Main wire lead	Copper	1
12. Adjusting knobs	Plastic	4
13. PCB retainer screw	Copper	1
14. Retaining lugs	Aluminum	4
15. PCB assembly	Mixed	1
16. Rear board	Plastic	1
17. CRT	Mixed	1

Table 7.14 **Case study to analyze disassembly of a computer monitor**

Task description	Task total	Disassembly force			Material handling		
		Inter-surface friction	Inter-surface wedging	Material stiffness	Component size	Component weight	Component symmetry
1. Disassemble rear board							
a. Rotate swivel base about pivot	13.54	–	4	–	2	2	1.2
b. Pull out swivel base	11.51	–	2	–	2	2	1.2
2. Disassemble back cover							
a. Unscrew first back screw	15.65	2.5	–	–	2	2	0.8
b. Unscrew second back screw	15.65	2.5	–	–	2	2	0.8
c. Unscrew third back screw	15.65	2.5	–	–	2	2	0.8
d. Unscrew fourth back screw	15.65	2.5	–	–	2	2	0.8
e. Remove back cover	15.44	–	3	–	3.5	2	1.2
3. Disassemble CRT/PCB assembly							
a. Unscrew first PCB screw	15.65	2.5	–	–	2	2	0.8
b. Unscrew second PCB screw	15.65	2.5	–	–	2	2	0.8
c. Unscrew first CRT screw	15.65	2.5	–	–	2	2	0.8
d. Unscrew second CRT screw	15.65	2.5	–	–	2	2	0.8
e. Unscrew third CRT screw	15.65	2.5	–	–	2	2	0.8
f. Unscrew fourth CRT screw	15.65	2.5	–	–	2	2	0.8
g. Remove CRT/PCB assembly	18.90	–	3	–	4	2.5	2.4
4. Remove CRT							
a. Remove yoke assembly	13.13	–	3	–	2	2	1.4
b. Remove deflection wire lead	13.02	–	3	–	2	2	0.8
c. Remove CRT	18.06	–	3	–	4	2.5	2.2
5. Remove main wire lead							
a. Remove first retainer screw	15.65	2.5	–	–	2	2	0.8
b. Remove second retainer screw	15.65	2.5	–	–	2	2	0.8
c. Remove main wire lead	13.13	–	3	–	4	2.5	2.2

Tooling		Accessibility/positioning			Allowances			
Force exertion	**Torque exertion**	**Dimen-sions**	**Location**	**Accuracy of tool placement**	**Posture allowance**	**Motions allowance**	**Personnel allowance**	**Visual fatigue allowance**
1	–	1	1	1.2	–	–	–	1%
1	–	1	1	1.2	–	–	–	1%
–	2	1.6	2	2	–	–	–	5%
–	2	1.6	2	2	–	–	–	5%
–	2	1.6	2	2	–	–	–	5%
–	2	1.6	2	2	–	–	–	5%
–	1	1	1	2	–	–	–	5%
–	2	1.6	2	2	–	–	–	5%
–	2	1.6	2	2	–	–	–	5%
–	2	1.6	2	2	–	–	–	5%
–	2	1.6	2	2	–	–	–	5%
–	2	1.6	2	2	–	–	–	5%
–	2	1.6	2	2	–	–	–	5%
1	–	1.6	1	2.5	–	–	–	5%
1	–	1	1	1.6	–	–	–	1%
1	–	1	1	1.6	–	–	–	1%
1	–	1.6	1	2	–	–	–	1%
–	2	1.6	2	2	–	–	–	5%
–	2	1.6	2	2	–	–	–	5%
1	–	1.6	1	2	–	–	–	1%

(*Continued*)

Table 7.14 **(Continued)**

Task description	Task total	Disassembly force			Material handling		
		Inter-surface friction	Inter-surface wedging	Material stiffness	Component size	Component weight	Component symmetry
6. Remove adjusting knobs							
a. Remove first adjusting knob	11	–	2	–	2	2	0.8
b. Remove second adjusting knob	11	–	2	–	2	2	0.8
c. Remove third adjusting knob	11	–	2	–	2	2	0.8
d. Remove fourth adjusting knob	11	–	2	–	2	2	0.8
7. Remove PCB assembly							
a. Remove PCB retainer screw	18.27	5	–	–	4	2	0.8
b. Bend first retaining lug	25.08	–	–	6.5	4	2	0.8
c. Bend second retaining lug	25.08	–	–	6.5	4	2	0.8
d. Bend third retaining lug	25.08	–	–	6.5	4	2	0.8
e. Bend fourth retaining lug	25.08	–	–	6.5	4	2	0.8
f. Remove PCB assembly	11.55	–	2	–	2	2	0.8
8. Remove swivel pivot							
a. Pry out swivel support	18.7	–	–	4.5	3.5	2	1.2
b. Remove swivel support	10.92	–	2	–	2	2	1.2
9. Remove rear board							
a. Bend first retaining lug	23.65	–	–	6.5	3.5	2	1.4
b. Bend second retaining lug	23.65	–	–	6.5	3.5	2	1.4
c. Bend third retaining lug	23.65	–	–	6.5	3.5	2	1.4
d. Bend fouth retaining lug	23.65	–	–	6.5	3.5	2	1.4
e. Remove rear board	10.92	–	2	–	2	2	1.2
Total score	**698**						

Notes: Total time for disassembly = 6980; TMUs = 4.188 min.Total maintenance time = 4.188 min.Task 1 for disassembly analysis: Remove PCB assembly. Most feasible cost effective design solution: Replace four retaining lugs with two or use snap fits to hold PCB assembly in place.Conclusion: Most amount of time is spent in bending the retaining lugs. Too many lugs hamper disassembly.

Tooling		Accessibility/positioning			Allowances			
Force exertion	Torque exertion	Dimen-sions	Location	Accuracy of tool placement	Posture allowance	Motions allowance	Personnel allowance	Visual fatigue allowance
1	–	1	1	1.2	–	–	–	–
1	–	1	1	1.2	–	–	–	–
1	–	1	1	1.2	–	–	–	–
1	–	1	1	1.2	–	–	–	–
1	–	1.6	1	2	–	–	–	5%
3	–	2	2	2.5	–	–	–	10%
3	–	2	2	2.5	–	–	–	10%
3	–	2	2	2.5	–	–	–	10%
3	–	2	2	2.5	–	–	–	10%
1	–	1	1	1.2	–	–	–	5%
2	–	1.6	1	2	–	–	–	5%
1	–	1	–	1.2	–	–	–	5%
3	–	1.6	1	2.5	–	–	–	10%
3	–	1.6	1	2.5	–	–	–	10%
3	–	1.6	1	2.5	–	–	–	10%
3	–	1.6	1	2.5	–	–	–	10%
1	–	1	–	1.2	–	–	–	5%

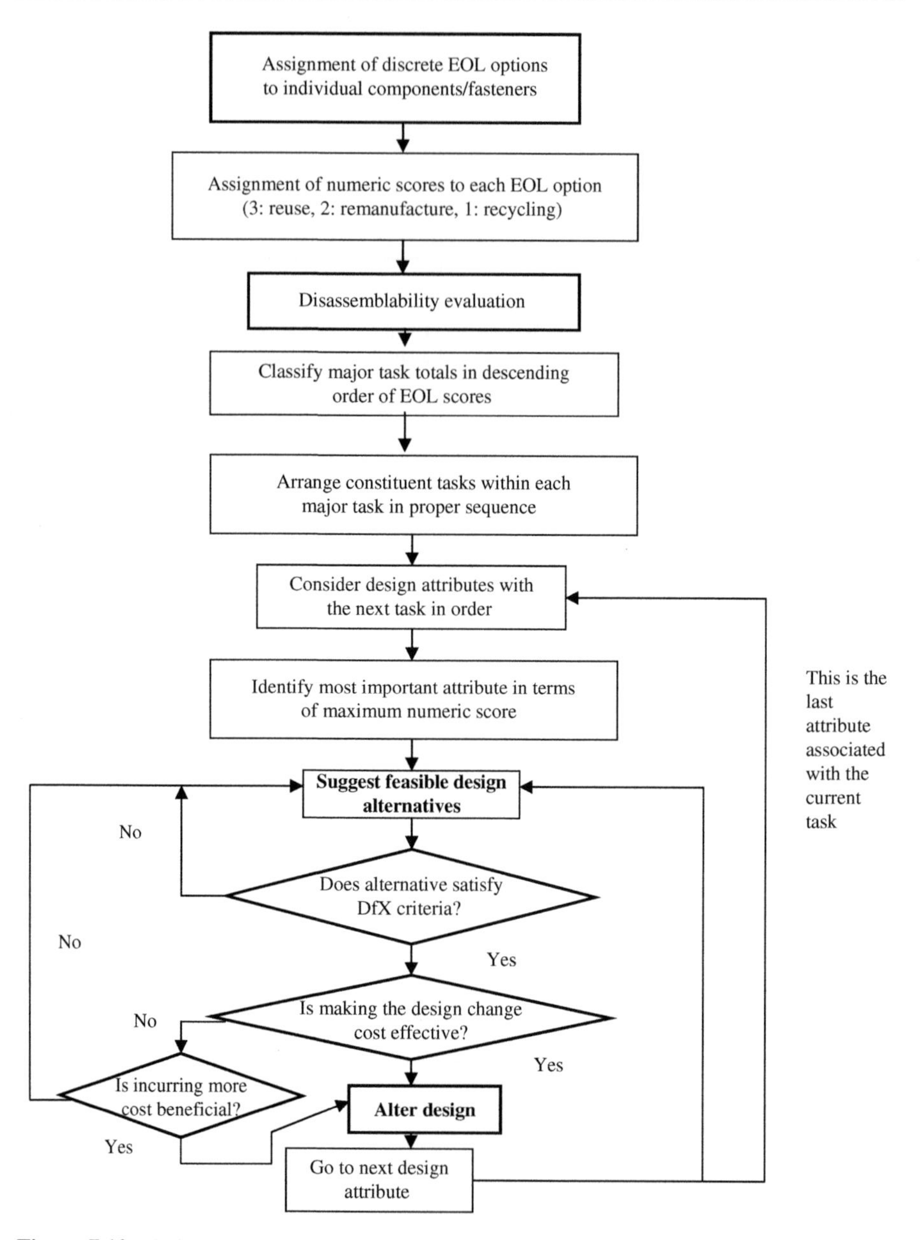

Figure 7.10 Methodology for design improvement for disassembly.

7.12 Concluding remarks

This chapter addressed the twin issues of product assembly and disassembly. It discussed the importance of both processes and presented an overview of several design methods to facilitate each process. Contrary to widespread opinion, one process is not always the exact opposite of the other. In other words, the assembly process may not be the exact opposite of disassembly, and vice versa, due to the presence of several product and process parameters that might hamper the ease with which either process is performed. The designer should always consider the pros and cons of both assembly and disassembly simultaneously when considering specific design options.

Time is only one metric that needs to be considered in evaluating the ease with which either assembly or disassembly can be carried out. Other metrics that can be used in this evaluation include effort, cost, use of fixtures and equipment, and workforce needs. It is up to the designer to use the metric that suits the condition under consideration.

References

Boks, C., Templemon, E., 1998. Future disassembly and recycling technology. Results of a Delphi study. Futures 30 (5), 425–442.

Brennan, L., Gupta, S.M., Taleb, K.N., 1994. Operations planning issues in an assembly/disassembly environment. Int. J. Oper. Prod. Manag. 14 (9), 57–67.

Brinkley, A., Kirby, J.R., Wadehra, I.L., 1996. Life cycle inventory of PVC: manufacturing and fabrication processes Proceedings of the 1996 IEEE International Symposium on Electronics and the Environment. IEEE Press, Piscataway, NJ.

Navin-Chandra, N.-D., 1994. The recovery problem in production design. J. Eng. Design 5 (1), 65–86.

de Ron, A., Penev, K., 1995. Disassembly and recycling of electronic consumer products: an overview. Technovation 15 (6), 363–374.

Feldmann, K., Traunter, S., Meedt, Otto, 1999. Innovative disassembly strategies based on flexible partial destructive tools. Ann. Rev. Control. 23, 159–164.

Gungor, A., Gupta, S.M., 1997. An evaluation methodology for disassembly process. Comput. Ind. Eng. 33 (1–2), 329–332.

Gungor, A., Gupta, S.M., 1998. Disassembly sequence planning for products with defective parts in product recovery. Comput. Ind. Eng. 35 (1–2), 161–164.

Krikke, H.R., van Harten, A., Schuur, P.C., 1998. On a medium term recovery and disposal strategy for durable assembly products. Int. J. Prod. Res. 36 (1), 111–139.

Kuo, T.C., 2000. Disassembly sequence and cost analysis for electromechanical products. Robotics Comput. Integrated Manuf. 16, 43–54.

Lambert, A.J.D., 1999. Linear programming in disassembly/clustering sequence generation. Comput. Ind. Eng. 36, 723–738.

Lee, S.G., Lye, S.W., Khoo., M.K., 2001. A multi-objective methodology for evaluating product end-of-life options and disassembly. Int. J. Adv. Manuf. Technol. 18, 148–156.

Penev, K.D., de Ron, A.J., 1994. Determination of a disassembly strategy. Int. J. Prod. Res. 34 (2), 495–506.

Shyamsundar, N., Gadh R., 1999. Geometric abstractions to support disassembly analysis. IIE Trans. 31 (10), 935–946.

Srinivasan, H., Gadh, R., 1998. A geometric algorithm for single selective disassembly using the wave propagation abstraction. Comput. Aided Des. 30 (8), 603–613.

Srinivasan, H., Gadh, R., 2002. A non-interfering selective disassembly sequence for components with geometric constraints. IIE Trans. 34 (4), 349–361.

Srinivasan, H., Figueroa, R., Gadh, R., 1999. Selective disassembly for virtual prototyping as applied to demanufacturing. Robotics Comput. Integrated Manuf. 15 (3), 231–245.

Thierry, M., Salomon, M., Van Nunen, J., Van Wassenhowe, L., 1995. Strategic issues in product recovery management. Calif. Manag. Rev. 37 (2), 114–135.

Zhang, H.C., Kuo, T.C., Lu, H., Huang, Samuel H., 1997. Environmentally conscious design and manufacturing: a state-of-the-art surge. J. Manuf. Syst. 16 (5), 352–371.

Zussman, E., Kriwet, A., Seliger, G., 1994. Disassembly-oriented assessment methodology to support design for recycling. Ann. CIRP 43 (1), 9–14.

Designing for Maintenance

8.1 Introduction

A system or product is said to be maintainable or repairable if, when it fails to perform as required, it can be maintained by a suitable methodology, be it repair, overhaul, or replacement, either manually or by an automated action (Reiche, 1994). Modern complex systems and products involve a major load on maintenance and support resources, in terms of both personnel and cost. It is important, therefore, that every effort be made to reduce maintenance requirements for newly introduced systems and equipment. Maintenance analysis during the design, acquisition, and selection phases ensures that maintenance requirements are minimized in the future.

The ability of a product to work successfully over a prolonged period of time is referred to as *reliability*. While the concept of reliability has grabbed the attention of engineers to the point of becoming an obsession, no worthwhile results have come out of that fascination. Achieving 100% reliability all the time is nothing more than an imagined fallacy. However, maintaining products periodically by adhering to a strict maintenance regimen can not only help prolong the life of equipment but can also ensure that it works smoothly in the future without breakdown.

Note that significant reference has been made to the terms *equipment* and *systems* throughout this chapter. This has been done to include the vast array of products that are interlinked to perform a specific function. This chapter deals with the ease of maintenance of single products as well as an assemblage of interlinked products (systems). All maintenance concepts, as well as procedures, that deal with equipment or systems are equally applicable to single consumer products.

8.1.1 *Importance of designing for maintenance*

Maintainability can be defined as "the degree of facility with which an equipment or system is capable of being retained in, or restored to, serviceable operation. It is a function of parts accessibility, interval configuration, use and repair environment and the time, tools and training required to effect maintenance" (Morgan et al., 1963). The U.S. Department of Defense defines *maintainability* as "a characteristic of design and installation which is expressed as the probability that an item will conform to specified conditions within a given period of time when maintenance action is performed in accordance with prescribed procedures and resources" (Harring and Greenman, 1965). Given the ongoing discussion regarding the importance of ease of product maintenance, it is clear that designing for maintenance assumes paramount

Product Development. DOI: http://dx.doi.org/10.1016/B978-0-12-799945-6.00008-9

importance in ensuring reliable equipment operation. To that end, reliability actually follows effective maintenance, instead of it being the other way around.

Maintainability is applicable to commercial equipment as well as military systems and equipment. If a commercial product cannot be maintained in or returned to usable condition within a reasonable period of time and at an advantageous cost, it cannot survive long in a competitive market. As far as military systems are concerned, this competition is among nations. National survival is attained through deterrence of aggression, if possible, or through victory, if the former option is not possible. Given these alternatives, the defense industry has assumed leadership in promoting maintainability as an important contributor to matériel readiness (Harring and Greenman, 1965).

It is a fact that individual components of a machine or product assembly eventually will break down as a result of fatigue and wear (sometimes also as a result of improper use). Similarly, no amount of redundancy built into the assembly will yield consistent performance over an extended operational horizon unless periodic maintenance is performed.

8.1.2 Factors affecting ease of maintenance

The rapidly evolving complexity of products has kept pace with evolving technology. Improvements in reliability techniques, however, have been unable to keep pace with the growing degree of product complexity (Crawford and Altman, 1972; Morgan et al., 1963; Oborne, 1981). New problems in equipment downtime have been proliferating, and the concept of maintenance as a tool to reduce downtime has assumed growing importance (Imrhan, 1991).

As far as designing equipment for maintenance is concerned, it has been practiced more as an art than as a science, to the extent that it has evolved more as a result of common sense than by means of scientific investigation (Oborne, 1981). It is worth noting in this context that maintenance is perhaps the most expensive of all human–machine system activities. This is because of the increasing need to perform maintenance activities and the high and ever increasing cost of human labor. An estimation of the cost of human labor is extremely important, since maintenance may be the only field of operation that relies solely on human capital and human skill.

Some examples from the military aircraft industry corroborate this claim:

- Aircraft maintenance costs in the United States have been estimated to amount to approximately 35% of life cycle costs of military systems (McDaniel and Askrein, 1985).
- The technical complexity of modern aircraft has compounded the problem of quick, cost-effective maintenance even more. Adding to the degree of complexity is the wide array of hardware made possible by computer-aided design systems (Adams and Patterson, 1988).
- In 1970, the U.S. Department of Defense allocated one quarter of its budget to maintenance costs (Smith et al., 1970). This fraction has grown with the increasing level of complexity of components and machine assemblies.

From this discussion, it is clear that machines and products designed with a view to enhancing ease of maintenance lend themselves more easily to that particular function. This results in maintenance operations being performed at a fraction of their regular costs and in a fraction of the time required otherwise. In this context, the importance of designing for maintenance cannot be overemphasized. However, as this chapter points out, very little research really has been performed with a view to enhancing the maintainability of products and machines. Before getting to that section, a distinction needs to be drawn between functional design (design for operability, in this context) and design for maintenance.

Designing for maintenance is more difficult than designing for operability, for the following reasons:

- Environmentally speaking, the maintenance workplace is much more variable and less predictable. Maintenance technicians often are forced to work in limited and cramped workspaces, which in turn have not been designed to allow maintenance operations.
- The degree of variability inherent in equipment is staggering. This is truer in the case of consumer products such as consumer electronics. The problem is further compounded by the rapid pace at which equipment becomes obsolete. This, in turn, underlines another problem: training and retraining of the maintenance crew. However, this point is not within the scope of this chapter, hence it is not appropriate to discuss it here.
- A crucial point of difference is the obviously conflicting goals as far as maintenance and operation of equipment is concerned. For instance, clearance within a machine and between machine parts is crucial to enable maintenance. However, the extra space may not always enhance operability of the product. The present trend is toward miniaturization (Tichauer, 1978), driven by the need to lower cost of production, ease the manipulation of machines during operation and transportation, and satisfy the demands of a gradually shrinking workplace site. The issue of designing a worksite, while relevant to the general scope of maintenance, is not directly related to machine design for maintenance. Hence, it is not discussed in detail in this chapter.

Maintenance friendliness is important in terms of both production and safety. Also, machines that are difficult to maintain routinely are less likely to receive the required standard of maintenance (Ferguson et al., 1985). For example, according to Johnson (1988), breakdown on some machines was often found to be associated with or was a direct result of lack of maintenance or abuse of equipment rather than just poor engineering. This brings us back to the point made previously. It is clear that, although there have been significant recent improvements in reliability, in all likelihood, developing a totally reliable machine or product will not be cost feasible. For this reason, good maintainability always will be important.

The importance of the maintenance process is undeniable. A methodology that seeks to approach the maintenance process from the design perspective remains sorely lacking. Researchers have tried to approach this issue in a variety of ways. Since maintenance is largely a manual process, design methodology is bound to draw on ergonomic data. However, the ergonomics data available in texts (Van Cott and Kinkade, 1972) do not readily help designers make the trade-off between ergonomics and engineering issues. To this end, it is worth noting, a holistic methodology that offers new concepts as well as builds on previous research works has not evolved.

This chapter examines current research on the topic of designing for maintenance. To enable this scrutiny, the following approach is adopted sequentially:

- Study of maintenance elements and concepts
- Study of mathematical models for maintenance
- Critical study of design for maintenance algorithms.

8.2 Maintenance elements and concepts

8.2.1 Maintenance elements

Maintenance elements describe the maintenance concepts and requirements for any system. This includes the analysis and verification of customer requirements. The priority selection of each element depends on particular requirements. Figure 8.1 depicts these elements as well as the interconnections among them. A study of these elements is necessary to achieve effective maintenance once a system has been conceptualized. To that end, various maintenance elements must be fully integrated and form part of the initial tasks to be performed. Each of these elements must be controlled and incorporated into system design (Reiche, 1994). It is necessary to realize that the implementation of these elements must be timely and not lag behind the system design.

The International Electrotechnical Commission has been promoting the idea of customer satisfaction as a measure of reliability and maintainability. Given this background, the maintenance parameter may be depicted in the form presented in Figure 8.1. A subclassification of various maintenance elements is presented in Figure 8.2.

To maintain a product with minimum downtime, it is often necessary to carry out corrective or preventive maintenance, making use of minimal maintenance resources. Examples of such resources include but are not limited to personnel, tools, test equipment, technical expertise, and materials. We next outline some basic concepts related

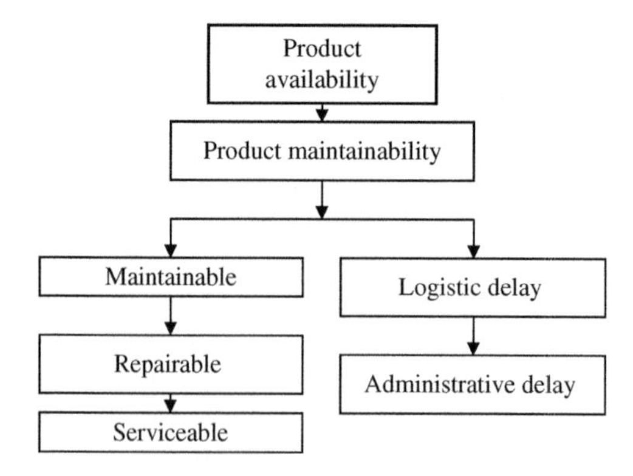

Figure 8.1 Relationship between maintenance and customer satisfaction. *Source*: Modified from Reiche (1994).

to designing for maintenance. It should be noted that many of these concepts essentially are maintenance philosophies in themselves, which can be built upon to form a cohesive design for maintenance methodology. A critical examination of various designs for maintenance methods is covered in the following section.

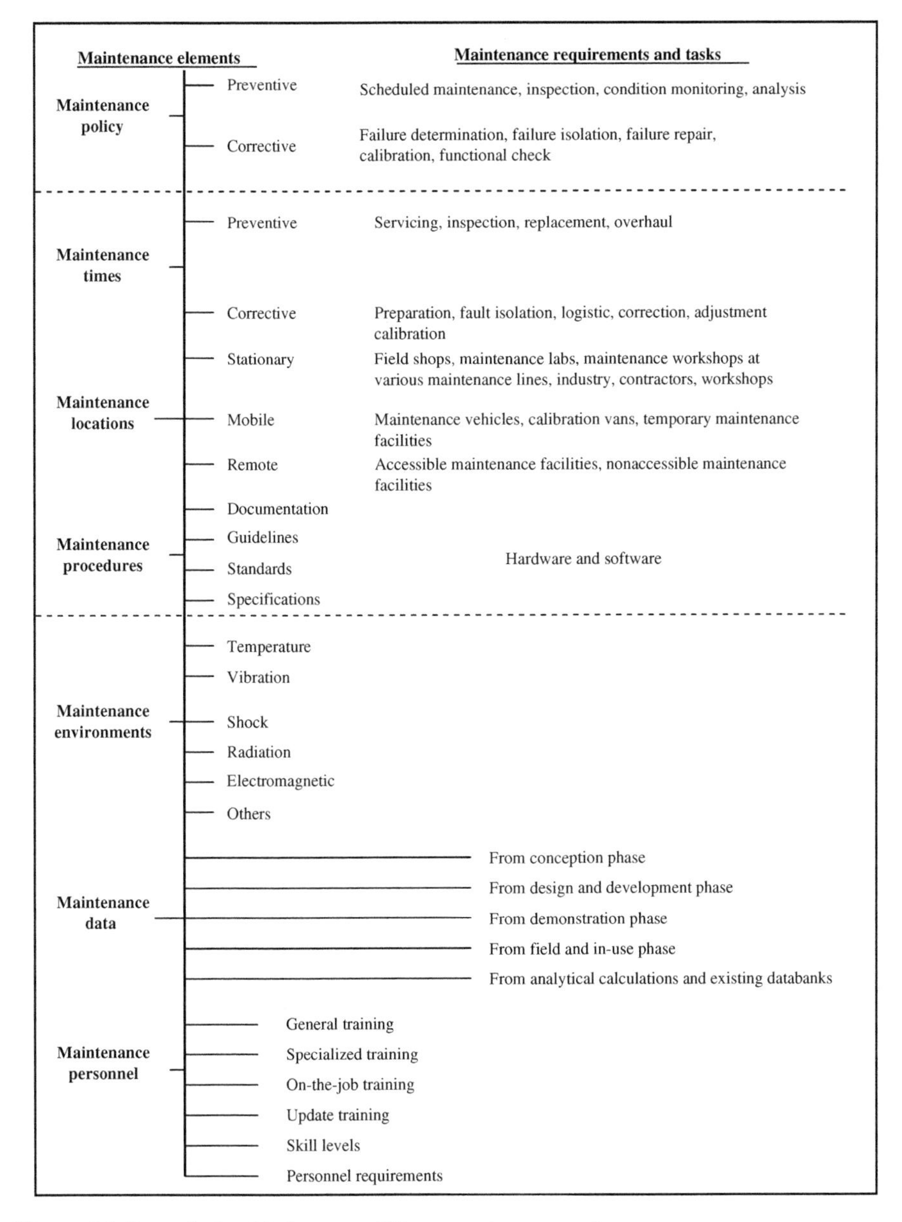

Figure 8.2 Interrelationship between different maintenance elements.
Source: Modified from Reiche (1994).

8.2.2 Maintenance concepts

8.2.2.1 Corrective (reactive) maintenance

Corrective maintenance is reactive in nature. Every time a product or system fails, repair or restoration must follow to restore its operability. The following steps constitute corrective maintenance:

- Once the failure has been detected, it must be confirmed. If the failure is not confirmed, the item generally is returned to service. This no-fault-found problem leads to a considerable waste of time at significant cost. It also entails carrying an unnecessarily large inventory all the time.
- If the failure is confirmed, the item is prepared for maintenance and the failure report is completed.
- Localization and isolation of a failed part in the assembly is the natural next step in corrective maintenance.
- The failed part is removed for disposal or repair. If disposed of, a new part is installed in its place. Examples of repairable parts and connections include broken connections, an open circuit board on a PCB, or a poor solder.
- The item may be reassembled, realigned, and adjusted after repair. It is checked before being put back to use.

The chief disadvantage of this maintenance procedure is the inherent amount of uncertainty associated with it. Similarly, the procedure is extremely reactive in nature, capable of shutting down an entire operation because of a single failure in a single machine under extreme conditions (often leading to a severe bottleneck and lost productivity). As a result of its drawbacks, another, more proactive maintenance method (recognizing that equipment needs periodic maintenance to function smoothly, which should be provided before a breakdown occurs) was developed.

8.2.2.2 Preventive (and predictive) maintenance

As its name implies, preventive maintenance is carried out to minimize the probability of a failure. Preventive maintenance often is referred to as *use-based maintenance* (Swanson, 2001). It comprises maintenance activities undertaken after a specific amount of time or equipment use (Gits, 1992; Herbaty, 1990). This type of maintenance relies on the estimated probability of equipment failure in the given interval of time. Preventive maintenance tasks may include equipment lubrication, parts replacement, cleaning, and adjustment (e.g., tightening or slackening). Equipment also may be checked for telltale signs of deterioration during preventive maintenance.

Due to its inherent nature, preventive maintenance must follow maintenance schedules to be fully effective. To that end, preventive maintenance schedules are published for many systems and pieces of equipment. For new designs, however, schedules must be established by or on the basis of information available from the manufacturer. It is worth noting that corrective maintenance experience exerts the greatest influence on decisions concerning preventive maintenance schedules and procedures (Reiche, 1994). Primary or periodic maintenance inspections may have to be planned to carry out preventive maintenance effectively.

To prepare a preventive maintenance plan, the objectives of the plan should be clear. Examples of such objectives include the following: attempting to maintain system design reliability and availability, reducing corrective maintenance actions, increasing planned maintenance work, and improving the effectiveness of maintenance.

These goals can be accomplished effectively by predicting maintenance actions, applying diagnostic procedures to detect system deterioration prior to failure, performing regular inspections and calibrations, monitoring system performance, and making repairs and overhauls based on test results (Reiche, 1994).

The advantages of preventive maintenance have already been outlined. However, to effect preventive maintenance, equipment has to be taken off-line. The resulting downtime is one of the chief disadvantages of this maintenance philosophy.

Predictive maintenance is an adaptation of the preventive maintenance procedure. It is based on essentially the same principles, except it employs different criteria to determine the need for specific maintenance actions. Diagnostic equipment measures the physical condition of equipment for such conditions as abnormal temperature, vibration, noise, corrosion, and need for lubrication (Eade, 1997). In other words, these attributes are not related to inherent material properties. When any one or more of the indicators reaches a specified level, the system is taken off-line to rectify the problem.

A chief advantage of predictive maintenance over preventive maintenance is that equipment is taken off-line only when the need to do so is imminent, not after a passage of time, as is the case with preventive maintenance (Herbaty, 1990; Nakajima, 1989).

To summarize, preventive maintenance is performed routinely to accomplish the following three goals (Smith and Hinchcliffe, 2004):

- Prevent or mitigate failure.
- Detect the onset of failure. Doing this can enable the maintenance engineer to take precautionary actions before a catastrophic failure occurs.
- Discover a hidden failure.

8.2.2.3 Maintenance of a degrading system

Most systems operate with some sort of degradation occurring throughout their useful lives. To enable the maintenance of such systems, a review has to be done periodically to determine what actions need to be taken. To optimize the maintenance schedule, it has been suggested that the level of degradation be monitored instead of time. This approach enables the addition of factors such as maintenance costs and distribution of degradation (Reiche, 1994). After each monitoring period, the amount of degradation is measured. Maintenance is carried out when degradation passes a specified point. The amount of degradation is assumed to be a nonnegative, continuous random variable, and for each monitoring period, it is the same and independently distributed. An optimal maintenance plan obviously depends on cost factors. As such, the costs of overhauling and operating a system must be included in the evaluation. A maintenance model was suggested by Sivakian (1989) to this end. It seems that this approach reduces long-term discounted costs.

8.2.2.4 Aggressive maintenance

It is clear from its nomenclature that aggressive maintenance implies a much more aggressive and far-seeking maintenance philosophy than preventive maintenance. An aggressive maintenance strategy seeks to improve overall equipment operation, drawing on the concept of total productive maintenance (TPM). Hence, it is essential to understand the concept of TPM to fully realize the benefits of aggressive maintenance.

TPM may be defined as a partnership approach to maintenance (Maggard and Rhyne, 1992). It is a philosophy that chiefly deals with maintenance management designed to complement the implementation of just-in-time systems in Japanese plants (Swanson, 2001). TPM activities seek to eliminate the "six major losses" related to equipment maintenance: equipment failure, setup and adjustment time, idling and minor stoppages, reduced speed, defects in process, and reduced yield (Macaulay, 1988). Under TPM, small groups or teams create a cooperative relationship between maintenance and production that ultimately aids in the accomplishment of maintenance tasks. Also, given the team nature of work, production workers are involved in performing maintenance work, thereby allowing them a role in equipment monitoring and upkeep. This consequently raises the skill of production workers and their efficiency in maintaining equipment.

Maintenance prevention teams work to improve equipment performance through improved equipment design (Swanson, 2001). To this end, the maintenance department works cohesively with the engineering department during the early stages of design. The result is equipment that is easy to operate and maintain (Adair-Heeley, 1989).

The chief advantage of TPM (and, hence, aggressive maintenance) is the obvious improvement in equipment availability and reduction in maintenance costs. This further leads to better maintenance efficiency and reduced repair time.

8.2.3 Design review for maintainability: planning for maintenance and its management

The emphasis on maintainability does not mean that it should be the only issue on the agenda. As such, it should not be dealt with alone. Other design factors have to be included to arrive at a comprehensive design methodology. It should be clearly understood that maintainability is an integral part of the product design process.

The design review is one of the most important means of achieving good maintainability and reliability. It may be defined as "the quantitative and qualitative examination of a proposed design to ensure that it is safe and has optimal performance with respect to maintainability, reliability and performance variables needed to specify the equipment" (Thompson, 1999). It is useful and necessary to undertake a review at four principal levels of deign:

- Design specification review, including market need in product design
- System review
- Equipment (functional unit) evaluation
- Component analysis.

Nominally, subsystems should be included in the system-level review. Similarly, subassemblies should be included in the equipment review. As far as this classification is concerned, generally a commonsense approach is needed. Generally speaking, these four levels of classification should be sufficient. This recognition of distinct levels facilitates the selection of appropriate review methods for different tasks and adoption of a systematic approach for an efficient and effective design review. A comprehensive design review may be characterized by distinct stages, as presented in Table 8.1. A brief description of each activity follows.

8.2.3.1 Review of design specifications

The objective of the design specifications review is to make certain that all parts and specifications are understood at the outset and the importance of different statements is appreciated. At this stage, the client and design team (either in-house or contracted) should discuss the salient features of the specifications to eliminate any misunderstandings. The specifications are the most common reference point in contractual

Table 8.1 Structured design review procedure

Stage and activity	Purpose	Timing
1. Review of design specifications	To ensure that the significance of all points contained within the design specifications is understood	Prior to the commencement of any design activity
2. Activity systems level review	To identify critical areas of the design that may affect plant availability and communicate to the detail design teams the necessity to pay particular attention to these areas To comment on the advisability of pursuing projects with a high-risk content	Prior to the start of equipment design
	To examine equipment groups to maximize uniformity and stability To maximize the reliability systems formed by manufacturing and process considerations	After the completion of the first equipment designs
3. Equipment (functional unit) evaluation	To evaluate quantitatively critical items of equipment To undertake qualitative reviews of equipment	After the completion of the first detailed designs
4. Component analysis	To check that certain important sets of components will not give rise to maintainability or reliability problems in service	After the completion of the first detailed design

Source: Modified from Thompson (1999).

disputes. Hence, it is in the interest of all to be clear in terms of definitions and requirements. The following specifications are of particular significance in the context of maintainability:

- Maintainability and reliability objectives that are quantitative in nature. This helps avoid any discrepancies in perception.
- A consideration of environmental conditions that may affect maintainability and reliability.
- Particular maintainability requirements need to be addressed in detail, such as the necessity for modular construction, restrictions on the skill level of maintenance workers, and designs that entail multiskill working.
- That the equipment can be effectively and reliably maintained should be demonstrable and acceptance criteria explicitly specified.

8.2.3.2 System review

The first system review is done prior to forming detailed designs of the product or equipment. As such, it is necessary to review the parameters of the manufacturing plant in terms of part availability, inventories, buffer capacities, and the like. This is where the issue of what is called *maintenance management* emerges. It is clear that this is a system review, not a review of equipment design. The objective at this stage is not to undertake a precise quantitative reliability analysis (yielding system failure rate predictions), since the equipment is yet to be designed. This stage of the design review identifies critical areas that, if a breakdown occurs, may cause a total plant shutdown. This review is accomplished by utilizing information concerning nominal production rates, buffer capacities, operational contingencies, and so forth.

This stage of the design review should make certain that the appropriate equipment design teams are made fully aware of the presence of any critical areas of the plant (Thompson, 1999).

The second stage of the design review enables the designers to complement the initial design by examining equipment groups that have commonalities with seemingly different groups. These are equipment groups that cut across conventional system boundaries. For example, a review of pumps to be used in a plant will reveal whether there is a substantially large diversity of manufacturers (leading to the need for more spares). Keeping this principle in mind, equipment groups should be defined and analyzed to maximize uniformity to reduce spares. Avoiding diverse products enables maintenance teams to more readily build up knowledge and competence in maintenance design practice (Thompson, 1999). Figure 8.3 depicts the role played by system review in the design process.

8.2.3.3 Equipment evaluation

Different items of equipment require different evaluation techniques. The design team has the opportunity to evaluate a design quantitatively at this stage of the design review. Evaluation methods proposed by researchers include the concept evaluation technique, the device performance index (DPI), and the parameter profile analysis.

The concept evaluation technique, proposed by Pugh (1991), involves quantitative evaluation in which design concepts are compared to a reference design concept. The

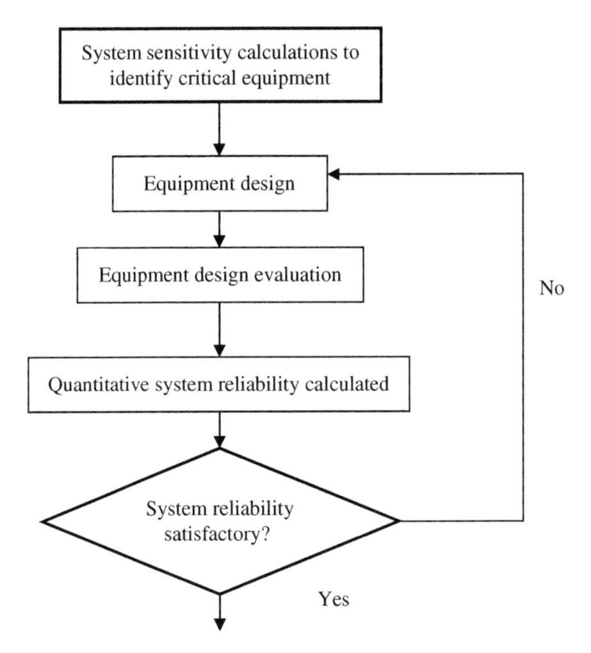

Figure 8.3 Interaction between system and equipment design levels in a design review. *Source*: Modified from Thompson (1999).

reference concept usually is a standard design or a design considered just acceptable. In some cases, it could even be one of the proposed concepts that appeared to be the best on first inspection. However, this method of choosing the standard design is rare.

An evaluation matrix is constructed with concepts (1 to m) arranged against the evaluation criteria (1 to n). To make things easier to understand, a small sketch of each concept could be made on the grid. Each proposed concept is compared to the reference concept, which is chosen as the reference or datum level. If a concept is better than the datum with respect to a particular criterion, a score of (+) is assigned to the concept for that criterion. Similarly, if the proposed concept is worse than the reference for a particular criterion, a score of (−) is assigned to that concept for the particular criterion. If no judgment can be made, an s is assigned, which is equivalent to a score of 0. The scores for each concept are totaled and that with the highest score generally is chosen. The chosen concept then is evaluated to find out if the design can be modified to improve on the negative and null scores. This system of choice caters readily to maintenance criteria early during the design stage.

One of the chief drawbacks of this process is that it does not distinguish among the relative importance of various criteria, which would involve assigning successively higher numerical weights to successively more important criteria. Doing this would enable designers to reach a more balanced decision as far as choice of designs is concerned.

Figure 8.4 depicts the sequential process of generating ideas and concepts for design review. The process is not described in detail here, but the figure is clear enough for readers to understand the process.

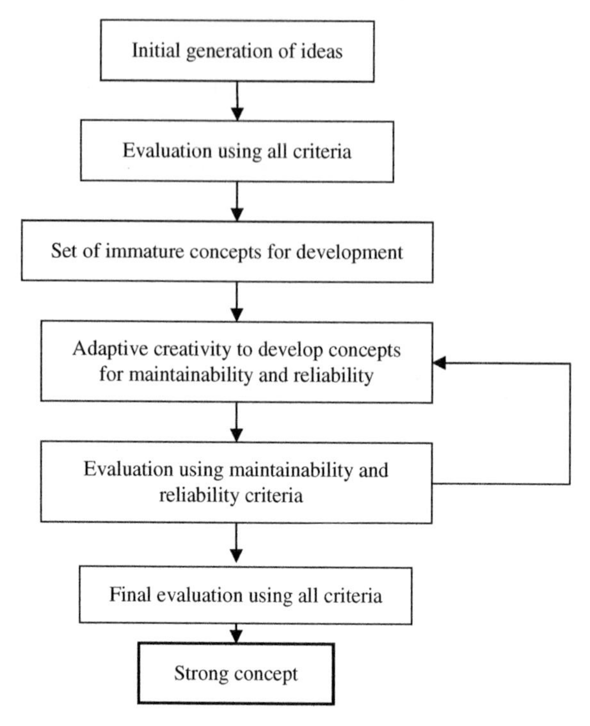

Figure 8.4 Development of good concepts from initial ideas.
Source: Modified from Thompson (1999).

The DPI evaluates equipment that has been designed in detail or compares alternative proposals. It compares quantitative assessments with respect to different performance parameters, including maintainability and reliability. It can also incorporate subjective value judgments.

The DPI is based on an inverse method of combining individual value scores of all criteria for each design concept. The overall value is found by calculating the DPI as follows:

$$\text{DPI} = n \times [(1/u_1) + (1/u_2) + \cdots + (1/u_n)]^{-1}$$

where u_i are value scores for each criterion and n is the number of criteria.

This method has a significant advantage over other methods, in that it utilizes only a simple addition of scores. For instance, if there is a low score with respect to one criterion, then the value of the numerator also is small, since it is the multiple of all individual score values (Thompson, 1999). Hence, if a design scores low with respect to maintainability (one of the criteria for evaluation), then the DPI will be equally small. This in turn reduces the chances of that design being selected in the final evaluation (since the value of DPI is directly proportional to the probability of success).

The parameter profile analysis evaluates equipment performance as well as system characteristics. Research by Moss and Strutt (1993) indicated that the expensive systems made of many relatively low-cost items often are subjected to superficial design reviews. This analysis method is suited for just such systems. The aim of the evaluation is to identify weak points in the system and highlight areas where system performance is near its limit. The performance parameters that define a system are described in a matrix with respect to the items of equipment. When an operating performance requirement moves beyond the performance limit of an item of equipment (e.g., operating pressure exceeds the pressure limit of a valve), the system fails. A set of data points can be obtained for each item with respect to performance parameters that are relevant to that item. Maintainability applies to all items of equipment and is included in the matrix of data points. Maintenance performance is measured by calculating the mean corrective repair time of an item (Thompson, 1999).

8.2.3.4 Component analysis

Component evaluation is clearly different from equipment and system evaluation, because components usually are constituents of a larger system. The question is one of scale; for example, a component may be a bearing, a motor, a gasket, or a rivet. From the perspective of maintainability, it is not practical to consider a general survey of components in a manufacturing plant. In this case, certain component classes need to be identified to facilitate detailed analysis. Examples include components that are functionally important (seals in fluid containers and welded joints, for instance). Experience is important to identify such component classes.

Figure 8.5 depicts different maintainability design features. A study of the maintainability universe would serve well to impart an introductory idea as to the composition of the maintenance occupation.

8.3 Mathematical models for maintainability

Numerous mathematical models seek to address the problem of effective maintenance through objective, numerical problem formulation. However, there is a basic challenge as far as this approach is concerned. The success of this approach, which is strictly a branch of applied mathematics or statistics, can be measured only in terms of its impact upon the solution of real maintenance problems (Scarf, 1997). It should be pointed out up front that a major problem exists with this approach. While new theories keep appearing at an unprecedented rate (Cho and Parlar, 1991), too little attention is paid to data collection and the consideration of the usefulness of models for solving real problems through model fitting and validation (Ascher and Feingold, 1984). This section covers some of the more important mathematical algorithms for design for maintenance.

8.3.1 Simple models

When speaking in terms of mathematical models, simple models contain a small number of unknown parameters. An example of a simple model would be that proposed

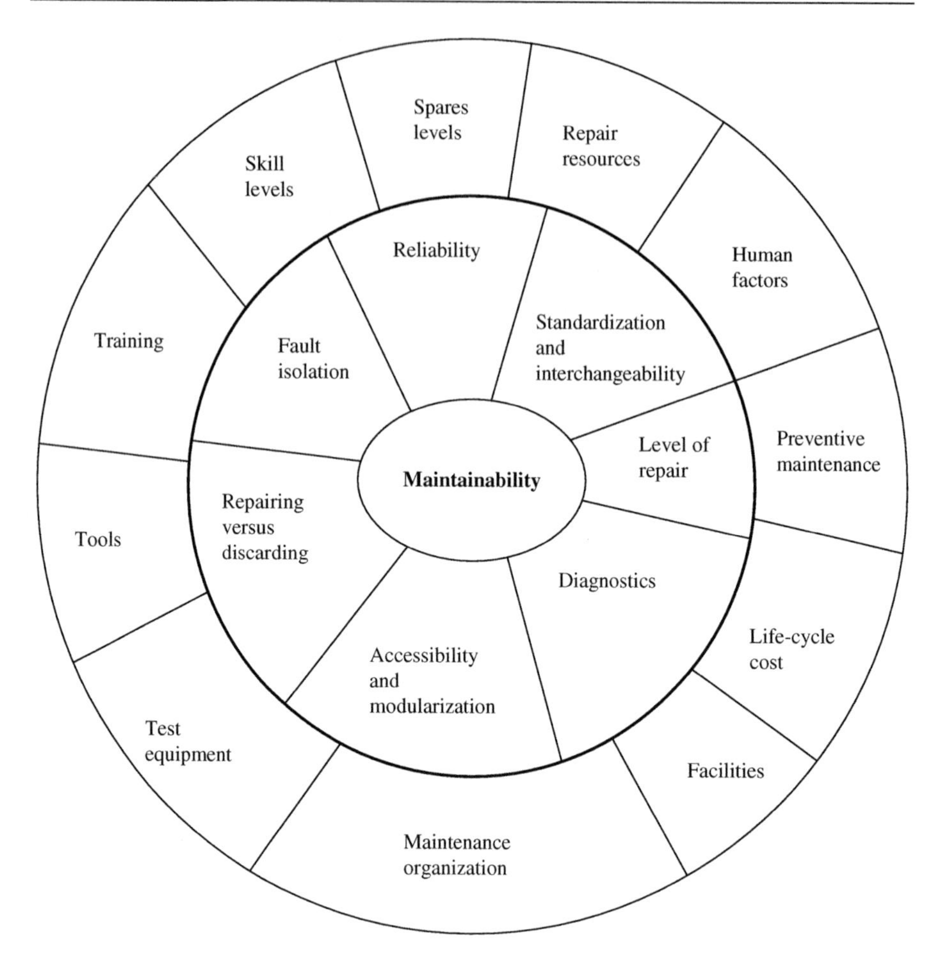

Figure 8.5 The maintainability universe: inherent and secondary design features.
Source: Modified from Ebeling (1997).

by Barlow and Hunter (1960), an age-based replacement model for a component. Components are replaced using a two-parameter Weibull time-to-failure distribution. In this model, according to Baker and Scarf (1995), only a small number of observations of time to failure (approximately 10) are required to enable determination of the optimal or near optimal value of the critical age at which preventive maintenance should be carried out. One of the chief disadvantages with this model is the obvious scarcity of real-life examples that require a data set of such minuscule dimensions. As a result, the practical validity of this model is highly suspect. Other examples of simple age-based replacement models were proposed by Christer and Keddie (1985) and Vanneste and Van Wassenhove (1995). Scarf (1997) pointed out in this case that, while component test data might be sufficient in numbers, environmental factors affecting the maintenance process may be quite different than those assumed in the problem formulation.

Other models are more complex in their problem formulation with respect to the number of unknown parameters. A chief drawback of these models is that the degree of correlation among different parameter estimates is high. In other words, the parameter estimates tend to overlap each other to a large extent. This means that the proposed model is unable to distinguish clearly between different parameter combinations, which leads one to be quite certain about the inefficacy and invalidity of problem formulation in the first place. Complex models often lack the necessary and sufficient data required to arrive at a solution. Another drawback with these models is that they present a very complicated solution and are unable to make accurate (or even feasible) predictions.

Finally, management and engineering are looking for simple, straightforward, and transparent models to solve what essentially is a very practical problem. Theoretical solutions arrived at by solving highly complicated mathematical equations (often with very little relevant data available) offer very little by way of a real solution to a real problem.

8.3.2 An integrated approach to maintenance

The integrated approach to maintenance involves the qualitative as well as quantitative aspects of model formulation for maintenance. The sequence is as follows:

- Recognizing a problem
- Collecting data and designing an exercise for collection of data
- Designing systems for future data collection
- Feasible and effective modeling and problem formulation using collected data
- Comparing results with other techniques
- Formulating a revised, alternative maintenance policy based on the results
- Training maintenance managers in the new technique
- Calculating economic gains from implementation of the new maintenance model.

The problem recognition phase of this technique is based on traditional industrial engineering tools, such as quality management, Pareto analysis, and cause and effect diagrams (Vanneste and Van Wassenhove, 1995). Some researchers (Christer and Whitelaw, 1983) refer to this technique as *snapshot modeling*.

The integrated approach to maintenance incorporates many of the practical, real-life aspects of the maintenance process in mathematical modeling. It tends to present a rather holistic picture of the process. However, it still provides a mathematical solution that is proactive. It seeks to solve problems after they are created and does not try to avoid problems in the first place. In fact, it would be difficult (if not impossible) for any mathematical model to present a proactive solution to a maintenance design problem, because all mathematical models need data with which to work. For that to happen, a problem needs to be present.

8.3.3 Capital replacement modeling

Capital replacement modeling is an area of maintenance considered by some (Pintelon and Gelders, 1992) as strategic or long-term maintenance. It is considered to be a part of strategic planning. Strategic planning consists of providing resources to safeguard an organization's future competitiveness. Hsu (1988) pointed out that, while technological and economic factors may be the principal drivers for equipment replacement, maintenance costs and unavailability are just as important. Quite a few researchers proposed models to solve the capital replacement problem (Christer and Scarf, 1994; Eilon et al., 1996; Hastings, 1969; Jardine et al., 1976; Scarf and Bouamra, 1995; Simms et al., 1984). The models proposed by these researchers generally are simple and offer little opportunity for mathematical exploration (Scarf, 1997). From the ongoing discussion, it is clear that mathematicians have had the most influence on problem formulation and attempts to solve them.

8.3.4 Inspection maintenance

Inspection maintenance has been a topic of extensive study in the past and still holds its importance among the research community. Quite a few researchers have addressed the issue of inspection maintenance from a mathematical viewpoint (Baker and Christer, 1994; Baker and Wang, 1991, 1993; Christer et al., 1995; Day and Walter, 1984). A chief concern in this modeling, as with other mathematical models, is the need to keep the model simple. A case in point is the model proposed by Christer and Walter (1984), a two-parameter delay-time model (a Poisson process of defect arrivals with rate α, exponentially distributed delay times with mean $1/\gamma$, with perfect inspection for faults). The maximum likelihood estimate in this case is quite easy to compute when inspections are evenly spaced (occurring at regular intervals, Δt time intervals apart). For a component observed over $(0, T)$, the maximum likelihood estimates satisfy the condition

$$\alpha = n/T$$

$$[(n-k)\gamma\Delta/(e^{\gamma\Delta}-1)] + [\sum \gamma t_i/(e^{\gamma\Delta}-1)] = (n-k)$$

where k failures are observed at times t_i ($i = 1$–k) from the last inspection and $(n-k)$ defects are found at inspections.

8.3.5 Condition-based maintenance

Condition monitoring techniques have gained importance over the years in an effort to tackle problems such as the following: rising requirements for production performance, increasing cost and complexity of manufacturing plants, and a drastic decrease in the downtime available for routine maintenance.

Recently, mathematical techniques to tackle the problem of conditional maintenance have proliferated. These techniques focus on tracking a condition-related variable, X, over time. Repair and maintenance activities are initiated when X exceeds some preset level, c (Scarf, 1997). Most researchers have tried to determine the appropriate variable(s) to monitor (Chen et al., 1994), and to design systems to enable condition monitoring of data acquisition (Drake et al., 1995) and condition monitoring data diagnosis (Harrison, 1995; Li and Li, 1995).

Numerous drawbacks are associated with these models. For instance, no substantial research has been conducted to determine the optimal level of the variable *c*. The critical level, *c*, is chosen subjectively and on the recommendations of the supplier and monitoring equipment manufacturers (Scarf, 1997). Similarly, no cost considerations are used in the decision-making process.

From the ongoing discussion regarding the mathematical modeling of maintenance problems, the following glaring anomalies make themselves evident:

- The issue of mathematical modeling takes precedence over actual real-life problem solving. Models are hardly successful, if at all, in solving problems faced by maintenance engineers and managers.
- Even if mathematical models are developed to address maintenance problems, they often suffer from a lack of the relevant data necessary to obtain a clear solution.
- Some of the models have problems in themselves, such as an inability to distinguish between parameters or parameter combinations. This leads to an extremely complicated solution, in direct contrast to real-life situations, which seek unambiguous, easy-to-follow, practically applicable solutions.
- Since all mathematical problem formulations require concrete data sets to obtain solutions, clearly most mathematical modeling is reactive in nature. As such, it does nothing to try to prevent problems from occurring in the first place.
- Maintenance is a highly practical problem. All mathematical research that fails to address this important characteristic is bound to be practically inapplicable.

8.3.6 Maintenance management information systems

The development of condition monitoring coupled with decision models puts new demands on maintenance management information systems (MMIS). Currently, a substantial number of systems are available with the goal of managing maintenance (Kobaccy et al., 1995). The specific objectives of such systems are as follows:

- Track specific components through the maintenance cycle
- Provide logistic support to plant managers, maintenance engineers, and maintenance personnel. An example of logistic support includes providing information on and tracking the spare parts inventory
- Record, maintain, and provide an equipment maintenance history
- Alert personnel to impending predetermined maintenance activity
- Produce management reports to enable strategic actions such as aggressive maintenance.

While a large majority of MMIS are able to accomplish these, a small number of systems also can perform the following auxiliary functions: analyze the maintenance history and determine the optimal policy for components and subsystems. This is tantamount to optimizing control of the spare parts inventory, helping to reduce spare parts overhead costs substantially.

In large, complex systems with many subsystems and component interactions, MMIS must provide solutions in the following areas:

- Incorporate expert opinion in a knowledge base
- Include subjective data from experts in the maintenance field and related fields

- Draw up a schedule of maintenance activities, especially from the perspective of preventive and aggressive maintenance
- Update maintenance schedules with the occurrence of operational events, such as component or system failures and unscheduled replacements
- Inventory plan resources
- Measure the effectiveness of maintenance activities. Note that the formulation of an objective index is most helpful in this regard.

Research conducted by Dekker (1995) deserves special mention in the context of combining maintenance activities into schedules. Dekker restricted attention specifically to those maintenance activities for which the next execution moment was determined from the previous one. However, a problem with this approach is that it essentially is a static combination of maintenance activities. As such, it may not necessarily yield optimal results in case of failure maintenance, and condition-based maintenance activities have to be carried out independently.

The proactive approach to maintenance is necessarily a design issue. The next section provides an overview of the different design approaches used by researchers to design products and systems for ease of maintenance.

8.4 Prediction models for maintenance

Prediction procedures for maintainability enable the designer to forecast the effects of design on system repair. The findings of maintainability prediction indicate the extent to which design contributes to ease of support. As a result, it is easy to estimate what additional maintainability features are required (Harring and Greenman, 1965). Prediction models indicate the downtime to be expected from a system prior to its operation in real-life situations. They also point to which of the system's features are likely to cause serious trouble. Maintainability predictions complement qualitative design parameters. This is important, since the maintenance engineer often is concerned with system availability and must resort to quantitative criteria to measure the effects of qualitative design features. This section discusses some of the most commonly used prediction models for maintainability. It is interesting to note that all the models being presented in this section are based on the concept of preventive maintenance. This should stress the importance of the process to industry, and hence underline what was said in Section 8.3.

8.4.1 The RCA method

The RCA method utilizes support time as the criterion of maintainability. It is a technique that essentially uses a checklist of physical features of product design. Support time is regarded as a function of the following attributes: physical design features, support requirements, and personnel requirements essential to effect efficient maintenance. Design features play a pivotal role in evaluating the physical aspects of a system. They also are utilized to determine the effects of layout, accessibility, and packaging on support time. The physical design of a product is evaluated on the basis of 15 sets of questions. Each question is assigned a value in terms of the impact of physical design on repair time.

Table 8.2 Partial representation of checklist A for the RCA method: physical design features

Physical design features	Score
1. Access	
a. Access adequate for both visual and manipulative tasks (electrical and mechanical)	4
b. Access adequate for visual but not for manipulative tasks	2
c. Access inadequate for visual and manipulative tasks	0
2. Latches, fasteners, and connectors	
a. External latches, fasteners, and connectors are captive, need no special tools, and require only a partial turn for release	4
b. External latches, fasteners, and connectors meet one of these three criteria	2
c. External latches, fasteners, and connectors meet none of these three criteria	0

Source: Modified from Harring and Greenman, 1965.

A linear equation is developed for support time by regression analysis of the empirical data (produced by more than 100 support incidents occurring in the operation of ground electronic equipment). This is presented as follows:

$$Z = 3.54651 - 0.02512A - 0.03055B - 0.01093C$$

where A, B, and C represent measures of the three parameters.

As far as the three elements affecting support time are concerned, the group of features listed is regarded as the most significant. Table 8.2 depicts a partial representation of the checklist under discussion.

Clearly, a higher score achieved on the scale translates into better maintainability. A major disadvantage of checklist A is that the scoring system is not time based. As such, complex regression is required to arrive at a meaningful matrix for maintainability. Similarly, the term *adequate access* is too subjective. Since maintenance is primarily a manual activity, sufficient accessibility for one person may not be so for another. The scoring system does not take these factors into consideration. Further, the justification for assigning the points the way they are (e.g., 4 points for adequate access for all kinds of jobs) is not available. The scoring system is based on empirical data, which need not have a scientific basis.

Table 8.3 depicts the scoring system for support items that are dictated by system design. It consists of a set of seven questions, each of which is assigned a numeric value. From Table 8.3, if a task can be accomplished without external test equipment, it can be accomplished in less time and with less effort. This makes the task simpler from the maintainability perspective. Again, as in Table 8.2, the system of scoring uses a base number of 4. Each successive task with increasing difficulty receives a numeric score equal to half that of the previous one. Checklist B is not exhaustive as far as analysis of certain key equipment such as, say, external test equipment is

Table 8.3 Partial representation of checklist B for the RCA method: design-dictated facilities

Design-dictated facilities	Score
1. External test equipment	
a. Task does not require the use of external test equipment	4
b. One item of test equipment is needed	2
c. Two or three items of test equipment are needed	1
d. Four or more items are required	0
2. Assistance (technical personnel)	
a. Task requires only one technician	4
b. Two technicians are required	2
c. More than two technicians are required	0

Source: Modified from Harring and Greenman (1965).

Table 8.4 Representation of checklist C for the RCA method: maintenance skills

Skills required	Score
1. Arm, leg, and back strength	
2. Endurance and energy	
3. Eye–hand coordination, manual dexterity, and neatness	
4. Visual activity	
5. Logical analysis	
6. Memory for things and ideas	
7. Planning capability and resourcefulness	
8. Alertness, caution, and accuracy	
9. Concentration, persistence, and patience	
10. Initiative and incisiveness	

Source: Modified from Harring and Greenman (1965).

concerned. Such equipment is not described in sufficient detail. The feature "assistance" (not depicted in Table 8.3) is extremely subjective. Expressions such as *some assistance needed* or *considerable assistance needed* are highly vague. Similarly, as far as the "assistance" subsection is concerned, more objectivity needs to be introduced in terms of physical ability of a healthy maintenance worker working under normal conditions.

Table 8.4 depicts checklist C, which consists of a series of support personnel requirements imposed by the system design. Each is scored from 0 to 4.

The scores for each of the elements featured in checklist C are assigned by moderators or supervisors or by the workers actually performing maintenance activities. As such, the element of subjectivity clearly is present. Similarly, the individual terms such as *arm*, *leg*, and *back strength* are extremely vague as far as their application value is concerned. An inclusion of "postural dynamics" would have been more relevant in this case. This would have enabled the designer to more fully understand the effects of unnatural postures on the musculoskeletal system (which, in turn, affects worker efficiency). Other elements, such as visual activity or logical analysis, have similar drawbacks.

As we have seen throughout this discussion, the RCA method is vague in representation, subjective in analysis, and incomplete in coverage. Also, that it is not time based essentially diminishes its utility. It is obvious that there is substantial room for improvement as far as this method is concerned.

8.4.2 The Federal Electric method

The Federal Electric method analyzes complex maintenance tasks and applies time analysis to gauge the maintainability of equipment. The four major steps are as follows:

- Identification of principal parts
- Determination of the failure rate of each part
- Determination of the time required for the maintenance of each part
- Computation of the expected maintenance time for the equipment by utilizing the information obtained in the first three steps.

While the first two steps are concerned with routine maintenance, ascertaining the time required to maintain each part forms the heart of this methodology. To ascertain maintenance time for each part of the equipment, the following seven actions are recognized. These actions are essential to restore broken equipment to working condition.

1. **Localization**: The first step concerns pinpointing the location of the malfunction without using auxiliary test equipment.
2. **Isolation**: This step concerns determining the location of the malfunction by the use of appropriate auxiliary test equipment, built-in test points, and the like.
3. **Disassembly**: Disassembly (full or partial) is essential to remove or replace a defective part from a machine. Clearly, factors such as accessibility and ease of component removal play their usual roles here, too.
4. **Interchange**: This process involves the substitution of a sound part in working condition for one or more that have failed.
5. **Reassembly**: As the name implies, this process involves restoring equipment to its original condition after disassembly.
6. **Alignment**: The various steps involved in this process are making adjustments, calibrations, and other checks and changes that have been made necessary due to the repair.
7. **Checkout**: This step involves verification of the desired level of performance. It makes certain that the equipment indeed has been restored to its initial condition or any other condition that was destined for it.

Standard repair time charts have been developed for each of these actions, based on more than 300 repair tasks. However, before times are predicted, a functional level analysis is carried out by analyzing the equipment under consideration. This is done by breaking down the equipment under study into a hierarchical arrangement of functional levels in order of complexity (part, subassembly, assembly, unit, group, and equipment). This breakdown is put into functional level diagram form, an abbreviated example of which is depicted in Figure 8.6.

The breakdown of equipment, as just described, enables sharper estimates of repair time than otherwise would be possible. For instance, the level of repair to be accomplished directly affects disassembly, reassembly, alignment, and checkout times.

The greatest advantage of using this technique is that equipment repair time can be predicted with some degree of accuracy. However, the system does not provide much utility as far as designing equipment for maintenance is concerned. It is reactive in nature. Also, the time measures used are based on empirical studies with sample sizes that are more or less insignificant (300 in this case). As a result, there are chances for a substantial margin of error in repair-time estimation. This methodology has room for improvement in the sense that various alternative system (product) hierarchies can be examined at the product design stage itself. This can be coupled with various design and human factors to evolve a somewhat holistic design methodology. As such, there is definite value in the Federal Electric method from the design perspective, but the method needs to be modified substantially to harness its potential.

8.4.3 The Martin method: TEAM

The technique for evaluation and analysis of maintainability (TEAM) is a method of prediction that does not rely on prior experience to design for maintenance. This method is based on the graphic representation of a troubleshooting scheme.

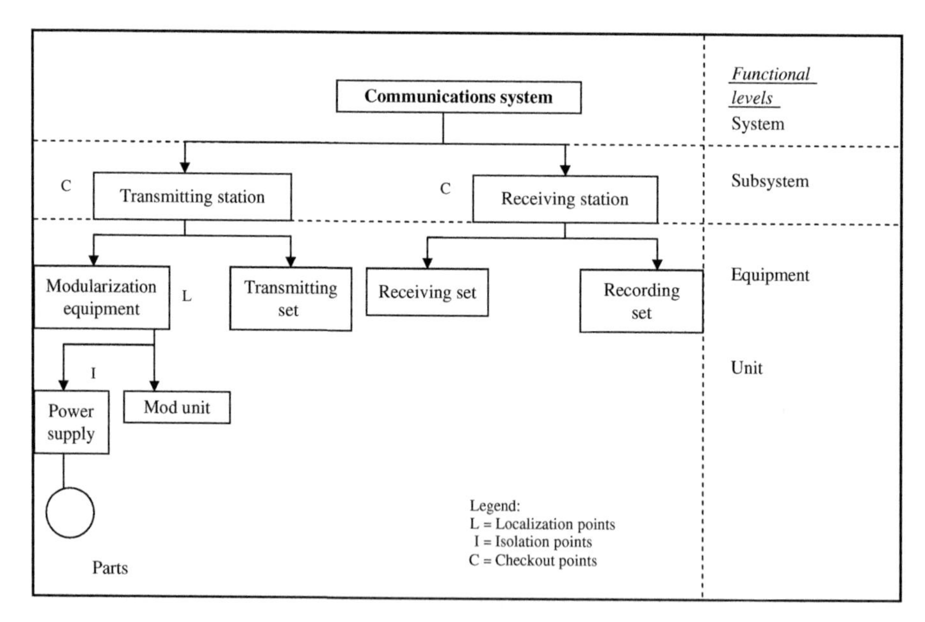

Figure 8.6 Abbreviated functional level diagram of a communications system: the Federal Electric method.
Source: Modified from Harring and Greenman (1965).

The representation begins with a symptom of a fault and works logically toward a solution to rectify that fault. Since the repair process, in essence, can be composed of several substages, the time required for each stage of the repair process (thus traced) is estimated to predict maintenance time (Harring and Greenman, 1965).

TEAM has some factors in common with conventional design for maintenance methodologies. For instance, it provides accessibility criteria based on reliability data, establishes requirements for test points, as well as other testing features. Further, it provides guides that facilitate the development of logical packaging schemes.

TEAM relies on PERT-type graphical representation as the foundation for estimating maintenance requirements as well as maintenance time. The chart depicts a chain of sequences. This sequence starts with a symptom of failure identified during checking. All items that could produce a given symptom are included in the chain, which begins with that particular symptom. All actions required to correct a fault (e.g., dismantling, removal, replacement, and repair) are entered on the chart. This is followed by an in-depth evaluation of each action to facilitate estimation of time required to perform the repair or maintenance task.

The failure rate for each replaceable component (F_r) in the system is estimated and entered in the PERT diagram. Failure rates in conjunction with repair times for various items are the principal determinants of the order in which replaceable items appear in the chain of sequence. For instance, a motor having the highest failure rate of all items in a chain invariably is placed at the head of the list, because any potential failure is likely to be traced back to this component. This, in turn, means that the part with the highest failure rate should have the most accessibility.

Once the TEAM diagram for a given symptom has been completed, the next step is an estimation of the mean time to repair (MTTR) necessary to correct the fault. This is obtained by adding the times for all the steps in the chain that lead to successful elimination of the particular fault. For any area of a system, the relevant data is entered on a worksheet similar to the one depicted in Table 8.5.

The MTTR is calculated as follows:

$$\text{MTTR} = (\Sigma F_r R_t) \,/\, \Sigma F_r$$

Table 8.5 A worksheet for TEAM analysis

Path no.	**Replaceable item**	**Repair time, R_t (min)**	**Failure rate/1000, F_r (h)**	$F_r \times R_t$	**MTTR for chain**	**MTTR for unit (min)**
1	Transmitter	13.5	0.315	4.2525	13.5	
2	Power supply	17.5	0.102	1.785	17.5	
3	Encoder	16.75	0.195	3.26625	16.75	
4	RF chassis	27.25	0.360	9.81	27.25	
5	Audio compressor	19.25	0.103	1.98	19.22	
	Total	–	1.075	21.09	–	19.61

Source: Modified from Harring and Greenman, 1965.

Calculation of the MTTR for each chain points out the relationship with the greatest potential for improving maintainability. This, in PERT terminology, is referred to as the *critical path*. In most instances, the longest troubleshooting paths are critical in nature, and hence should be shortened.

The TEAM method is quite adept at sequencing a maintenance plan of action. It does not provide any kind of design guidelines that can be effectively utilized to design equipment for maintenance, however. As such, it does not in any way take into account any design variables or human variables, which play such a crucial role in the maintenance process as a whole. Also, as far as maintenance time is concerned, at best, the TEAM can provide an estimate of MTTR. While the estimate is based on a good estimation of variables, such as failure rates and repair times for component parts, it still is not sufficient for implementation early on during the design stage. This makes the TEAM, like other methods, a reactive method of maintenance.

8.4.4 The RCM method: maintenance management

Reliability-centered maintenance (RCM) was developed in the aviation industry to determine scheduled maintenance policies for civil aircraft. RCM emphasizes the role of reliability in focusing preventive maintenance activities on certain aspects. These aspects enable retention of the equipment's inherent design reliability. Clearly, this maintenance technique centers on reliability technology. The RCM philosophy was a result of efforts by industry, especially the airlines industry (United Airlines in particular) in the 1960s to undertake a complete reevaluation of preventive maintenance strategy. Since then, the importance of RCM has grown by leaps and bounds. For example, RCM specifications have been developed (U.S. Air Force, 1985), a course in RCM is offered by the Air Force Institute of Technology, and the Navy published a handbook on RCM (U.S. Navy, 1983).

The RCM methodology is a special case of Pareto analysis, where resources are focused on solving the few yet vital problems that could cause serious system malfunction. It can be described completely by four unique features (Smith and Hinchcliffe, 2004):

1. Preserve functions
2. Identify failure modes that can defeat the functions
3. Set priorities based on function need (via failure modes)
4. Select applicable and effective preventive maintenance tasks for high-priority failure modes.

The first feature of RCM is to preserve the function of the component or system. Doing this enables the system to perform well in the future. Unlike other methods, which seek to preserve the component, RCM seeks to preserve the function of a component (components, in most cases, are designed exclusively for their functions). This method of thinking enables the designer to isolate functionally superior parts (primary functional parts) from functionally inferior parts.

The second feature of RCM is to identify specific failure modes that could cause the unwanted failures. Since preservation of function constitutes the first step of RCM, it is obvious that the next step would try to seek out failure modes that could cause the loss of that intended function. In the past, failure modes were identified using one of the many industrial engineering tools available specifically for that purpose, such as FMEA (failure mode effects analysis). Research identified six failure patterns, which show the conditional probability of failure against operating age. These patterns are exhaustive in their coverage and applicable to a wide variety of electrical and mechanical items. A list of the patterns (Knezevic, 1997) follows:

Pattern 1, bathtub curve pattern.

Pattern 2, a pattern that demonstrates constant or slowly increasing failure, probability with age, ending in a wear-out zone.

Pattern 3, a pattern that indicates a slow increase in the probability of failure.

Pattern 4, a pattern that shows a low failure probability when the item is new, followed by a rapid increase until a plateau is reached.

Pattern 5, a pattern that exhibits a constant probability of failure at all ages; a random failure pattern.

Pattern 6, a pattern that starts with a burn-in and eventually drops to a constant or very slowly increasing probability of failure.

Identification of possible failure modes enables designers to take that possibility into account early during the design stage itself. This way, the component can be designed and built to resist failure. Similarly, additional redundancy can be designed into the equipment to ensure smooth functioning even in the case of future failure.

The third feature of RCM is to set priorities based on the importance of failure modes. In other words, the third feature enables designers and product planners to concentrate their efforts (time, resources, and finances) on the most significant failure modes. This means that components that are functionally more important than others need to be guarded against failure (since failure in this case may cause system breakdown). Priorities set in this way can be used to develop a priority assignment rationale (Figure 8.7).

The fourth feature of RCM, for the first time, deals with actual preventive maintenance. Once the component has been identified, its probable cause of failure ascertained, and the priority sequence is in place, the next logical step is to perform preventive maintenance. Each potential PM task has to be judged as applicable and effective. *Applicability* refers to the ability of the PM task to accomplish one of the three reasons for doing it (prevent or mitigate failure, detect onset of a potential failure, or discover a hidden failure) (Smith and Hinchcliffe, 2004). *Effective* refers to the willingness of management to spend resources to perform PM.

The RCM methodology obviously is a maintenance management method. It does very little as far as design issues are concerned. Mere pinpointing of probable failure modes and components is insufficient by itself. Assigning priorities to failure modes and components would not be very useful unless supported by a sound design philosophy. This would corroborate the strengths of the RCM methodology.

8.4.5 Design attributes for enhancing maintainability

Several rules of thumb facilitate design for maintenance. However, design teams routinely tend to ignore them, often at their own expense. While these rules do not form any particular methodology, they nevertheless are important to effective and efficient

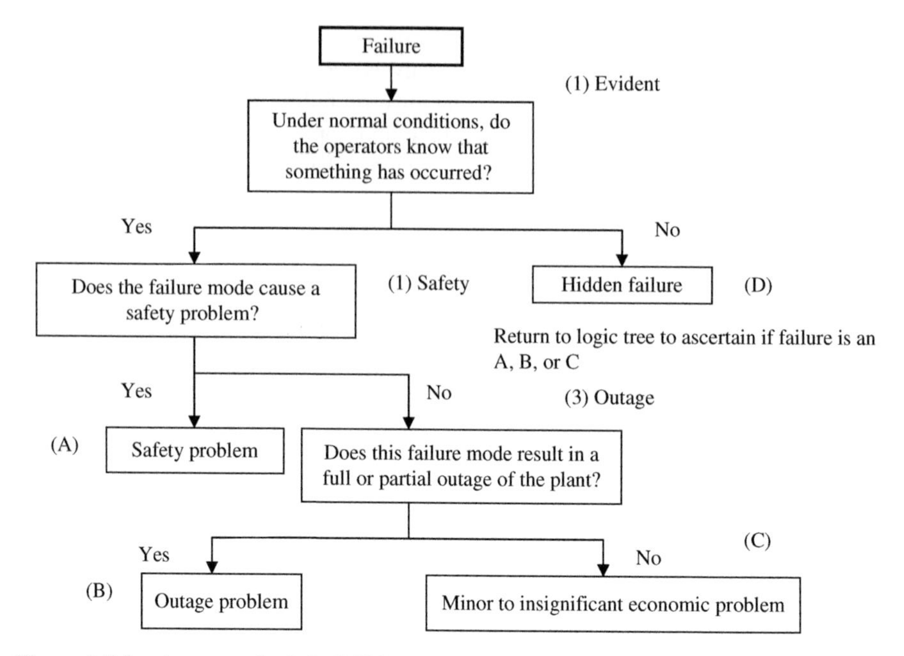

Figure 8.7 Logic tree analysis in RCM to assign priorities of resources to each failure mode. *Source*: Modified from Smith and Hinchcliffe (2004).

design. Knezevic (1997) provided an overview of design principles applicable to design for maintenance:

1. **Accessibility**: All equipment and subassemblies that require routine inspection should be located such that they can be accessed readily and easily. They should also be fitted with parts that can be connected rapidly for all mechanical, air, electric, and electronic connections. The TGV train of France is an example of this principle. The design of the train is such that the roof panels can be rapidly dismounted and lateral access panels and numerous inspection points allow for progressive inspections in a short span of time. Similarly, the auxiliary equipment in the power cars and passenger cars is located such that they allow work positions for maintenance staff to be as ergonomically sound as possible. As far as practically possible, it should not be necessary to remove other items to gain access to those items that require maintenance. Similarly, it should be easy to replace or top up items such as lubricants without requiring disassembly (Knezevic, 1997).
2. **Modularity**: The greater the degree of modularity introduced in a design, the easier it is to replace a component. Modularity is a design system in which functionally similar parts are grouped together into subassemblies, which in turn can be put together to form the product. However, effective modularization can be achieved only if interface equipment is standard (such as standard couplings, joints, fits, and the like). Modularity, by its inherent nature, ensures that no further readjustments are required once the modules are put into place. An example of effective modularity is the SAAB Gripen's (aircraft) RM 12 engine. The engine design is modular, enabling ease and quickness of inspection. Also, replacement entails replacing only the faulty module. It is not necessary to dig down into individual component parts (Knezevic, 1997).

3. **Simplicity**: It is a matter of common sense that a simpler design is inherently easier to maintain. Simplicity can be achieved by undertaking measures such as reducing the number of different parts or part varieties. It is a surprising yet true fact that no tools are required to open and close the service panels on the SAAB Gripen aircraft. Here is a case of an exceedingly simple design that perfected the disassembly process. All control lights and switches needed during turnaround time are positioned in the same area. These are placed together with connections for communication with the pilot and those for refueling.
4. **Standardization**: There are several advantages to using standard fasteners, connectors, test equipment, materials, and so on, when designing a product. Standardization allows for easy replacement of faulty components. It also assures designers of a certain level of quality associated with the component in question. This avoids nasty surprises later on during the design process. Cost effectiveness is yet another advantage of using standardized components, because of their ready availability (due to manufacturing on a wide scale).
5. **Foolproofing**: Items that appear to be similar but are not usable in more than one application should be designed to prevent fitting to the wrong assembly (Knezevic, 1997). Incorrect assembly should be obvious immediately during the manufacturing process, not later. Some of the measures that can be undertaken to enhance foolproofing are
   - If an item is secured with three or more fasteners, their spacing should be staggered.
   - Ensure that shafts that are not symmetrical about all axes cannot be wrongly fitted, either end to end or rotationally.
   - Whenever shafts of similar lengths are used, ensure that they cannot be used interchangeably. This means that their diameters need to be varied.
   - With pipes, avoid using two or more pipe fittings close together with the same end diameters and fittings.
   - Flat plates should have their top and bottom faces marked if they need to be installed with a particular orientation.
   - Springs of different rates or lengths within one unit also should have different diameters (Knezevic, 1997).
6. **Inspectability**: Whenever possible, create a design that can be subjected to a full, nondestructive, functional check, unlike, say, a fuse, which needs to be destructively tested for the test to be effective. The ability to inspect important dimensions, joints, seals, surface finishes, and other nonfunctional attributes is an important characteristic of maintainable design. The term *inspectability* often is used interchangeably with *testability*.

8.4.6 The SAE maintainability standard

The Society of Automotive Engineers (SAE) formulated a design for maintenance standard to be utilized early during the design stage of a new product, system, or machine (SAE J817-1, 1976). The SAE information report established a hierarchy of product effectiveness, defined *serviceability*, *maintainability*, *reparability*, and *diagnostics* and related these attributes to product effectiveness. Figure 8.8 depicts the hierarchy of product effectiveness as defined by the SAE standard.

Figure 8.8 graphically depicts the important role played by maintainability and serviceability in enhancing overall product effectiveness. This is even more important when maintenance decisions are taken at the design stage to build high quality into the product.

The standard under discussion establishes a numerical value to rate an existing machine or a new conceptual machine. Maintenance is the primary criterion for

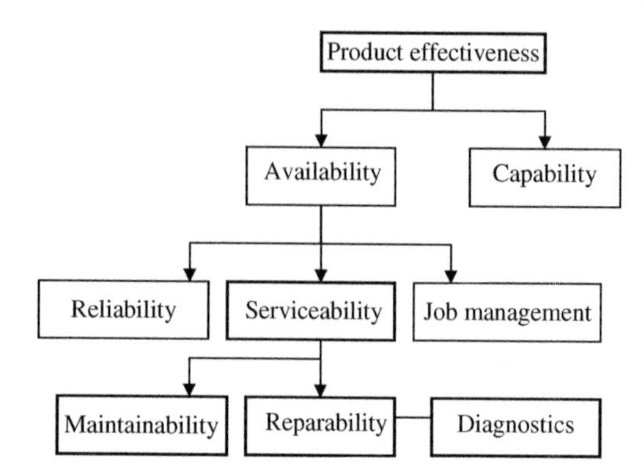

Figure 8.8 Hierarchy of product effectiveness.
Source: Modified from SAE Information Report, SAE J817.

evaluation. Based on certain requirements, the SAE index assigns point values to lubrication and maintenance items. Lubrication and maintenance operations can be subject to preset conditions such as location, accessibility, operation, and other miscellaneous factors. These requirements are clearly defined (as will be elaborated in the following pages). Each maintenance operation is described in detail and evaluated using conventional task analysis procedures. Each suboperation requiring the use of features such as location and accessibility is noted. Scores are assigned based on a preconceived system. The higher the score, the lower the maintainability of the machine, and vice versa. Each requirement enumerated in the original standard is reproduced in the following sections briefly. Note that the degree of ease with which a requirement can be accomplished translates to a higher or lower score assignment.

8.4.6.1 Location

Location refers to the position in which maintenance personnel should be positioned to perform the task. This section of the index assumes that only one operator is required. If more than one operation can be accomplished from the given position, the first operation is assigned the points applicable to that location and each subsequent operation is assigned 1 point. Table 8.6 depicts the numeric scores attributed to the design feature location.

It is clear from Table 8.6 that tasks requiring substantial moving around or special accessories, such as ladders, take longer to perform. As such, they are assigned a higher score. Similarly, tasks that require the assumption of unnatural postures, such as bending or kneeling, are difficult (not very natural) to perform. This, in turn, adds to the amount of time necessary to complete the task, which leads to a higher score on the index.

Location is a design attribute that dictates the primary posture requirement necessary to perform a task. This is followed by access features that facilitate (or complicate) the maintenance operation. This is described next.

Table 8.6 Locations and their respective point values in the SAE index

Positions	Points
1. Ground level, working within normal reach	1
2. Ground level, bending or stretching outside normal reach	2
3. Ground level, squatting, kneeling, or lying (except under the machine)	3
4. Mount machine, normal reach	10
5. Mount machine, bending, stretching, or squatting	15
6. Any position (other than upright) under or within the confines of the machine	25
7. Must climb into position without handrails, steps, or platforms provided	50

Source: Modified from SAE Information Report, SAE J817.

Table 8.7 Accessibility parameters and point scores of the SAE index

Accessibility parameters	Points
1. Exposed	1
2. Exposed through opening	2
3. Flip up cover or flap	3
4. Door or cover, hand operated	4
5. Door or cover, single fastener	10
6. Door or cover, multiple fasteners	15
7. Hood removal	35
8. Multiple covers, multiple fasteners	50
9. Radiator guard removal	50
10. Tilt cab	75
11. Crankcase or drive train guard removal, hinged and bolted	75
12. Crankcase or drive train guard removal, bolted only	100

Source: Modified from SAE Information Report, SAE J817.

8.4.6.2 Access

Access refers to the ease of reaching a lubrication or maintenance point. Here again, if multiple operations can be accomplished using the same access facilities, the first operation is assigned points applicable to the access. Each subsequent operation is assigned 1 point. Accessibility considerations and their respective point values are presented in Table 8.7.

Accessibility is a measure of the ease with which a maintenance point can be reached. Obviously, a maintenance point that is exposed can be reached easily and gets the lowest score. A point that is exposed but flanked by an opening gets a higher score due to the constraints imposed by the opening. Any access point that incorporates design features that impose constraints on its accessibility gets a higher score.

For instance, accessibility that requires the removal of a hood gets 35 points. This is because the hood, first, is an obstruction and, second, may be heavy or unwieldy and its removal may take time based on its design features.

This, however, does not mean that all access points must be exposed. Sometimes, functionality dictates the incorporation of additional design features, such as the ones featured in Table 8.7. The crucially important factor is to be able to reach a design compromise so that the maintenance point is easy to access and functional at the same time.

The accessibility feature is followed by the most important component of this standard: operation, which is the basic objective for its formulation. Location and access are merely two design features that facilitate or compound the ease with which the main maintenance operation can be performed. They can be viewed as stepping stones to the actual maintenance operation.

8.4.6.3 Operation

Operation refers to the action required to service the listed items. The SAE index was formulated to cater to the maintenance requirements of heavy machinery, such as off-road work machines. The operations section of the index makes this clear, inasmuch as the major operations categories have been designed with the service of heavy machinery in mind. Table 8.8 presents the various operation categories along with the point scores assigned to each.

Table 8.8 presents a short version of the operations section of the SAE index. As before, a task that can be accomplished easily gets a lower score, and vice versa. Design features that facilitate the performance of that task are scored accordingly. For instance, it is clear in Table 8.8 that a visual check of a liquid compartment is easier and less time consuming than that requiring a dipstick or that entails using multiple screw caps and an unfastening tool.

8.4.6.4 Miscellaneous considerations

The miscellaneous section of the SAE index includes requirements that are undesirable but required by functional or other design or occupational constraints. The point values listed alongside each item in fact are punitive, penalty points. An abridged version of this section is presented in Table 8.9.

Of special significance in Table 8.9 is the inclusion of operations and positions requiring caution. These items are obviously dangerous from the maintenance as well as the operational perspective. For this reason, they are assigned a score of 100 penalty points each. Another factor that deserves attention is the need for a special tool to perform an operation. It is clear by now that operations that can be done by hand are the most feasible from the maintenance perspective. Any operation that requires a standard tool is acceptable. However, a score of 4 penalty points is assigned because of the obvious necessity of skill required to operate the tool. Note that such skills may not always be readily available. This compounds the problem further.

In other words, all avoidable and undesirable operations (caused by particular design features) that result in compounding the problem receive a higher penalty

Table 8.8 Operations considerations and point scores for SAE index, abridged version

Operation considerations	Points
1. Compartment checking (liquid)	
Visual check	1
Dipstick	3
Screw cap, hand removable	4
Multiple screw caps, hand removable	6
Screw cap or plug requiring tool	8
Multiple screw caps or plugs requiring tool	10
2. Component checking	
Visual check	1
Hand check of belt tension	2
Nonprecision tool (includes tire pressure check or torque wrench)	5
Precision tool	10
3. Draining	
Drain valve, hand operable	1
Drain valve, tool required	3
Horizontal plug	6
Vertical plug	8
Cover plate	10
Multiple plugs or covers	15

Source: Modified from SAE Information Report, SAE J817.

Table 8.9 Abridged version of miscellaneous considerations from the SAE index

Miscellaneous considerations	Points
1. Bleeding required	3
2. Priming required	3
3. Special tool required	4
4. Need for special instruction	10
5. Inadequate clearance for required operation	20
6. Operation requiring caution	100
7. Position requiring caution	100

Source: Modified from SAE Information Report, SAE J817.

score. This is because the aim of the index is to introduce simplicity in design from the maintenance perspective.

All maintenance operations are repeated over a duration of time. This important practical fact has been incorporated into the SAE index by means of the frequency multiplier.

Table 8.10 **Frequency multipliers of the SAE standard**

Maintenance interval	Frequency multiplier
1. 1000 h, semiannually or greater	1
2. 500 h, quarterly or as required	2
3. 250 h, monthly	4
4. 100 h, semimonthly	10
5. 50 h, weekly	20
6. 10 h, daily	50

Source: Modified from SAE Information Report, SAE J817.

8.4.6.5 Frequency multiplier

This part of the index does not take into account one-time maintenance or that which requires less than a hundred hours of maintenance work to be performed. Table 8.10 presents an overview of the frequency multiplier assigned to different maintenance schedules. The maintenance hour intervals listed conform to SAE recommended practice J753. If intervals other than those shown are to be used, the frequency multiplier of the nearest SAE interval is applied and a penalty of 2 points added. Each lubrication and maintenance item is assigned a frequency multiplier once, the most frequent interval performed.

The SAE index is one of the most comprehensive attempts to quantify the maintenance occupation in terms of equipment design. However, a few anomalies exist that, if rectified, could improve the index substantially:

1. The index is not time based. Mere objectivity can impart a numeric score that can be used for objective comparison. However, if this objectivity could be linked to time indices, it would be able to pinpoint actions and design anomalies that are not maintenance friendly and obstruct the maintenance procedure.
2. The index needs more flexibility to take care of complicated maintenance tasks.
3. The SAE standard seeks to address maintenance requirements of off-road heavy machinery. This curtails its universality in terms of field of application.
4. There is no arrangement in the index to allocate resources to specific areas of machine design based on maintenance requirements, design features, and functionality. Incorporating this element would make it more "intelligent." This would enhance its appeal to both maintenance engineers and maintenance managers.

8.4.7 The Bretby maintainability index

The Bretby maintainability index was formulated as a substantial improvement over the SAE index, in that it sought to quantify the maintainability of products and machines. This section describes the different parts of the Bretby index, explains their highlights, and comments specifically on the chief drawbacks of the index, to enable further improvement and restructuring.

8.4.7.1 Description

The Bretby maintainability index has been described in detail by Mason (1990). It is an evaluation index that assigns time-based scores to various maintenance tasks and procedures. Researchers initially sought to modify the SAE index, described in the previous section, and evaluate its compatibility with a time-based system of scoring maintenance tasks. However, certain anomalies were found in the SAE index. These anomalies are (Mason, 1990):

- The SAE index produced a figure of merit for a particular task as opposed to a time estimate and is extremely limited in its area of application.
- The SAE index takes no account of any preparation needed prior to maintenance, nor does it take into account the weight of components to be handled, size or position of access apertures, or restricted access for tools necessary to effect appropriate maintenance.
- Developers of the Bretby maintainability index noted that, if the maintenance tasks had any degree of added difficulty, the SAE system, which was relatively simple, was incapable of satisfactorily handling operational difficulties beyond the basic maintenance task.

As far as the structure of the Bretby index is concerned, it is essentially classified into two distinct sections: gaining access to the job and the maintenance operations themselves.

8.4.7.2 Access section

The access section of the index is subdivided into two sections. The first subsection concerns the removal and replacement of hatches and covers. This means it deals directly with gaining access to the machine from outside. The second subsection deals with the space inside openings and apertures. However, just obtaining access to apertures, hatches, and covers is insufficient to effect maintenance. A good maintenance methodology should also address other equally important and practically applicable factors, such as surface or component preparation and manual activities such as carrying and lifting. A consideration of manual activities further entails the inclusion of related factors, such as energy expenditure estimates and postural difficulty (important from the viewpoint of musculoskeletal disorders). Table 8.11 summarizes some of the more important attributes covered by the access section of the Bretby index.

To this end, the difference between the Bretby and SAE indices is quite prominent. The Bretby index addresses in detail quite a few practically important points that the SAE index fails to even consider.

A similar section for component location has been added to the Bretby index to make it more comprehensive. The location section assigns scores to machine components based on how easy they are to reach. Ergonomically speaking, the components most within reach and those that do not entail the adoption of awkward, unnatural postures receive the lowest score. It should be remembered that this is a linear scale of scoring. Each score is further converted into a time metric. The lower the score, the more time is needed to perform the operation, and vice versa. Table 8.12 depicts the location subsection of the access part of the method.

Table 8.11 Abridged version of miscellaneous considerations from the SAE index

Description	Point score
1. Flip up cover or flap, no fasteners	3 per cover
2. Door or cover, hand-operated fasteners	4 per cover
3. Door or cover, single fastener, tool operated	5 per cover
4. Door or cover, multiple fasteners, tool operated	10 per cover
5. Lift-off or lift-up panel, easy to handle, <12 kg	2 per cover
12–24 kg	4 per cover
25–35 kg	6 per cover
>35 kg	10 per cover

Source: Modified from SAE Information Report, SAE J817.

Table 8.12 Location subsection of the access section of the Bretby maintainability index

Description	Point score
1. Ground level, working upright, within normal reach	1
2. Ground level, bending or squatting, outside normal reach	2
3. Ground level, squatting, kneeling, or lying (not under machine)	3
4. Mount machine, normal reach	6
5. Mount machine, bending, stretching, or squatting	8 (S)
6. On machine, subsequent operations within normal reach	1 each
Subsequent operations bending or stretching	2 each
Subsequent operations, squatting or kneeling	3 each
7. Any position (other than upright) under or within confines of machine	10 (S)
8. Enter driver or operator cab	3

Source: Modified from Mason, 1990.

As is evident from Table 8.12, the Bretby index takes into account the need for assuming awkward postures to perform maintenance procedures. This inclusion of postural requirements addresses the concern of many professionals that such postures may lead to the onset of musculoskeletal disorders. It is clear from the table that the simplest, most natural postures receive the lowest scores, which automatically means that they are less time consuming. A lower score also means that they are the most ideal postures on the list. Consequently, machine components, fasteners, and the like that need more complicated and unnatural postures are pinpointed accurately for design modifications to improve their degree of maintainability.

8.4.7.3 Operations section

The operations section of the index is divided into 12 sections. The more important ones deal with component removal or replacement, component carrying and lifting,

Table 8.13 Removal or replacement subsection of the operations section of the Bretby maintainability index

Description	Point score
1. Spin-on fastener	1
2. Single fastener, not requiring tool	3
3. Single fastener, requiring tool	4
4. Additional fasteners, not requiring tool	2 each
5. Additional fasteners, requiring tool	3 each

Source: Modified from Mason (1990).

Table 8.14 Slackening or tightening section of the Bretby maintainability index

Description	Point score
Fastener type	
1. Single fastener, not requiring tool	1
2. Single fastener, requiring tool	2
3. Additional fasteners	1 each
Fastener force requirements	
4. Slackening fastener, high forces needed	1 (H), (S)
Requiring impact	1–8 (S)
5. Tighten to unspecified torque	2

Source: Modified from Mason (1990).

and component preparation. Component removal or replacement is further modified by way of a subsection on operations that do not involve complete removal of a component or fastener. Oftentimes in industry, it is necessary only to slacken fasteners to effectively perform maintenance operations. Similarly, the converse is equally true. Slackened fasteners need to be retightened after maintenance to ensure smooth operation of machinery. The clear subclassification of this process indeed is unique to the Bretby method and adds much-needed flexibility as well as practicality to the index. An example of the removal or replacement index is presented in Table 8.13. The slackening or tightening index is presented in Table 8.14.

Table 8.13 deals only with the removal or replacement of fasteners. Bear in mind that machine components may not need fasteners to be held in place. Conversely, allowances have to be made for handling machine component weight (especially those that are heavy for the average worker to handle comfortably) once the fasteners are removed. The Bretby index makes allowances for handling unusually heavy components. For example, components that are easy to handle (weighing <12 kg) are assigned a score of 2 points per component. This is necessary since maintenance is largely a manual activity. As such, handling machine components during maintenance (lifting, moving, and refitting) is a time-consuming process. The lighter the components, the better the operation is from the maintenance perspective.

While the Bretby index takes into account the weight of individual components, it fails to assign weight scores to awkwardly shaped components (given that part variety in products and machines is staggering). Components that are irregularly shaped, have sharp edges, are made of fragile materials, or have an eccentric center of gravity, for example, need a separate scoring system as far as part handling is concerned. The Bretby index fails to take this into consideration.

Additionally, the data presented in Table 8.13 takes into consideration fasteners based on two criteria: those that need tools and those that do not. This is in addition to the typical spin-on type of fasteners. However, no distinction is made between those spin-on fasteners that require tools and those that do not. Similarly, no distinction is made between fasteners and components that need such extreme measures as the use of a pry bar, for example. Here is an example of a situation that entails the use of a tool with the exertion of force and requires substantial clearance within and around the machine (depending on location of the fastener or component). A consideration of such situations would make the index even more valuable from the practical viewpoint.

8.4.7.4 Other features

The Bretby index has numerous salient features that underscore its importance as a leading index on maintainability. These features include carrying and lifting tasks, preparation tasks, and inclusion of important practical factors.

Consideration is given specifically to carrying and lifting activities, especially important in the case of large machines with heavy components. Within the carrying and lifting category, allowances have been made for frequency of lifting as well as machine design from the perspective of provision of headroom to enable satisfactory maintenance and lifting. Special consideration also is given to a one-person lifting task as against a two-person task (depending predominantly on the weight of components).

It is assumed that one person can satisfactorily perform all lifting and carrying tasks for all objects weighing up to 35 kg. This is too random an assumption, especially in the case of machines that do not allow the requisite clearance in terms of either headroom or other clearances. Two people may be required for heavier objects (as is often the case in typical push–pull activities). A special allowance needs to be made for a second person in such cases. To ensure that this is incorporated effectively in the index, the carrying and lifting index needs to be split to incorporate allowance for the inclusion of an additional person. Each additional person performing the task in less maintenance-friendly conditions (such as insufficient clearances or headroom) needs to be assigned successively higher values to reflect obvious anomalies in machine design from the maintenance perspective. An additional allowances section is included in the index, but it gets too confusing to couple the carrying index, as is, along with the allowances. A simpler formulation is possible and would be helpful to practitioners.

Most maintenance operations entail one or more preparatory tasks before the actual maintenance operations can be carried out. The Bretby index does a good job of including an entire section on preparation tasks to be performed prior to maintenance. To that end, specific points have been allotted to discrete preparation tasks. For example, the task of cleaning around unions, fasteners, and the like has been allotted four points. Jacking up and chocking the machine prior to maintenance has been allotted 20 points. Similarly, donning protective equipment such as gloves or goggles (standard equipment) has been allotted two points, since it is quick and habitual to don standard personal protective equipment (PPE) and can be performed quickly. The process of donning nonstandard PPE, on the other hand, has been allotted a more generous five points due to more time spent in the process.

While the Bretby index includes most preparation tasks satisfactorily, special mention needs to be made about abrasive cleaning solutions, such as acids and alkalis, necessary to effectively complete preparation for maintenance. The use of such solutions entails the donning of nonstandard PPE (especially to protect the worker from noxious fumes). It also entails the use of concentrated chemicals that may take some time to complete the cleaning action before the machine may be accessed for maintenance (as is often the case in cleaning tough grease and grime). This means that the worker essentially has to wait for sometime before it is safe to commence further operations. The index could be modified to include this very important and widely utilized method of preparation.

Similarly, the index makes mention of cleaning small and extensive areas of the machine. This is very subjective, since machines come in all shapes and sizes. A modification could include affected surface area as a function (percentage) of total principal surface area. To this end, the parameter "surface area" could be classified as primary (essential functionally) and secondary. The point system could be modified to take this into account.

Additionally, as far as cleaning is concerned, the formulators left out an important variable: cleaning in hard-to-reach, inaccessible, and barely accessible areas. This action is most certainly time consuming and may require unnatural postures and abrasive cleaning products.

The Bretby index scores positive points as far as inclusion of important practical factors, such as component checking, lubrication, and draining. It gives due consideration to tool access parameters to effect maintenance. For example, a two to three flats access for wrenches and Allen wrenches is considered sufficient clearance and awarded one point per fastener that affords this kind of clearance. The point score increases in inverse proportion to clearance. The index also includes several miscellaneous items, such as energy output, frequency of operations, and visual fatigue.

A chief drawback of the energy output multiplier is that it takes into consideration only underground conditions and is vague as far as quantification is concerned. Similarly, as far as visual fatigue is concerned, the index has no provisions to take into account lighting conditions while checking as well as performing the maintenance operation.

8.4.7.5 Using the index

To use the index on a machine, it is necessary to obtain a list of all maintenance tasks to be performed as well as their frequency. Similarly, each task has to be described in sufficient detail (task analysis) for the necessary features of the index to be accessed. This description may be obtained from observations on the machine or discussion with experienced engineers and fitters (Mason, 1990).

8.4.7.6 General observations about the index

Clearly, the Bretby index approaches the maintenance procedures well by breaking down the process into easy-to-understand, sequential subprocesses. However, numerous applicable variables have been left out, as pointed out in the preceding discussion. Similarly, the index has been structured only for large machines (such as mining machines). It cannot be flexibly modified to include smaller machines or consumer products. As such, the Bretby index addresses only one specific section of the maintenance industry and is not universally applicable. There is definite scope for an index formulated within a more flexible framework and appendices that can adapt to product and machine variety as well as maintenance situations. Adaptability introduced in this way, in essence, would enhance its universal appeal.

Another important aspect that cannot be overlooked is the lack of a scheme by which the firm's resources can be effectively utilized toward maintenance operations (a system of priorities is lacking). The Bretby index could use an addendum by means of which maintenance issues can be managed as well as designed (because maintenance is as much a management issue as a design issue).

8.5 A comprehensive design for a maintenance methodology based on methods time measurement

We developed a new, comprehensive design for maintenance methodology to address anomalies existing in current research on designing for maintenance. A list of these anomalies follows:

1. **Anomalies related to reactive methods**: All maintenance-related methods are reactive in nature. This means that they seek to solve a problem after it has occurred. Study of current as well as past work practices indicates that equipment fatigue and failure is a very real problem that consumes precious resources: financial, material, human, and time. Given this background, it is easy to see that current methods do not serve a proactive purpose. They do not seek to prevent problems. Some of the more important consequences of such a line of thought, action, and design are:
   - Significant costs in terms of repair costs and labor costs.
   - Frequent maintenance means that relevant spare parts have to be kept on hand at all times. This further leads to an escalation in inventory costs as well as storage space.
   - Maintenance, in most cases, cannot be done effectively while the equipment is in operation. For effective maintenance, equipment needs to be taken off-line. This leads to lost time and money, leading to decreased efficiency.
   - Taking equipment off-line for maintenance implies significant equipment downtime, which can have serious repercussions, such as manufacturing bottlenecks and related economic consequences such as low rate of return.

2. **Anomalies related to mathematical and partial methods**: Most maintenance-related methods are strictly mathematical in nature, involving the quantitative analysis of a very real and practical problem. They do not consider real maintenance-related subjective issues, such as equipment condition, requirement of tools and labor, different types of equipment, various maintenance procedures, and accumulation of grime, for example. A consideration of these factors is essential to the formulation of a relevant and effective method to design equipment for maintenance. Some of the more important consequences of purely quantitative approaches to maintenance are:
   - The chief drawback of a mathematical methodology is that it cannot be used in design. All mathematical methodologies rely on past failure and scheduling data, which predisposes them to be reactive and, therefore, of little practical significance in forming a proactive design method.
   - Since most maintenance methods tend to rely to a large extent on mathematics, the practitioner has to be well versed in operations research fundamentals and the ability to use these concepts in practice.
   - Failure to consider subjective maintenance-related factors renders current methodologies ineffective, since they address only a part of the problem and fail to see that there is more to maintenance than mere scheduling.
3. **Current methods not based on time**: Most current maintenance methods are not time based. This is an important oversight, since it is common knowledge that maintenance is essentially time based, in that it costs the enterprise not only in terms of financial and labor resources but also in terms of time. In this context, time can be taken to mean both equipment downtime and time lost in terms of productivity.
4. **Current methods not based on human factors**: Current maintenance methodologies also are not based on considerations of labor. Maintenance is predominantly a manual activity and has yet to be automated on a large scale. It therefore follows that any maintenance methodology that seeks to build ease of maintenance into product design needs to build on human factors, work standards, and ergonomic considerations. Doing so would help alleviate worker stress. It would also go a long way toward automating the process.

8.5.1 A numeric index to gauge the ease of maintenance

The development of a numeric index for evaluating the ease of maintenance is crucial to this design effort. A brief description of how this most important and primary of objectives is achieved follows.

The most widely used maintenance operations are recorded and described in complete detail. Every maintenance operation then is subdivided into basic elemental tasks. Only a fraction of the tasks in the maintenance operation involve actual maintenance. The remaining tasks constitute such actions as reaching for tools, grasping tools, and cleaning components prior to maintenance. For example, consider a simple lubrication operation that may be subdivided into the following elemental tasks:

1. Isolate the component to be lubricated.
2. Constrain the product to avoid displacement during maintenance.
3. Locate the component to be lubricated (location of component).
   a. Visual location.
   b. Tactile location.
   c. Visual and tactile location.

4. Access the component to be lubricated (accessing the component): Tactile access.
   a. Access with tool and accessories: A difficult maintenance operation may involve component unfastening, slackening, or removal for maintenance (time consuming).
   b. Access without tool or accessories: This is an easier maintenance operation and may involve only on-site lubrication (less time consuming).
5. Check the component.
   a. Visual check: This is the easiest and least time consuming.
   b. Tactile check: This is easy and quick.
   c. Check requiring tool: This is more time consuming than visual and tactile checks.
   d. Check requiring precision tool: This is complicated and may take the most time.
6. Cleaning for maintenance.
   a. Cleaning around unions and fasteners to facilitate maintenance: This takes the least time, depending on the number of fasteners and unions.
   b. Cleaning small areas on machine.
   c. Cleaning large areas on machine: This may be the most time consuming, depending on the amount of cleaning required.
7. Perform maintenance operation.
   a. Lubricate on site: This is the easiest and least time consuming.
   b. Slacken fasteners and perform lubrication, refasten the component.
   c. Remove fasteners and remove component, perform lubrication, and put the component back in place followed by refastening it.

It is clear from this sequence that the maintenance operation is not limited solely to maintenance. The entire spectrum of tasks, beginning with isolation and location of the component(s) and ending with cleaning prior to actual maintenance, comprises an indispensable gamut of operations to be performed to achieve effective maintenance.

Each of these tasks can be further expanded to accommodate practical issues related to maintenance. For example, tasks 3 and 4 can be expanded to take into account postural requirements and ergonomic issues such as bending, stretching, stooping, and visual fatigue while performing maintenance. Similarly, as far as the maintenance operation itself is concerned, task 7c, which involves lifting a component, can be further expanded to include the component's physical and chemical properties.

Examples of physical properties include such parameters as weight, physical dimensions, shape, and the nature of the surface finish.

Examples of chemical properties include toxicity of component material, malleability and ductility of component material, and brittleness. Each of these parameters directly affects maintenance and worker safety and hence plays an integral role in evolving an effective design for maintenance algorithm.

A detailed representation of factors that directly and indirectly affect the maintenance operation is depicted in Figure 8.9. The figure is a graphic representation of the factors discussed in the preceding paragraphs. Certain elements that deserve elaboration (as explained before) are exploded as per specific needs. For instance, the entire lifting operation is described in detail in terms of component attributes. Similarly, other operations such as cleaning, checking, and locating are addressed in relevant detail.

Figure 8.9 depicts the stages in a typical sequential maintenance process. More important, it depicts variables that directly affect each step of the process. For instance, maintenance may be performed on site or require removal of the component.

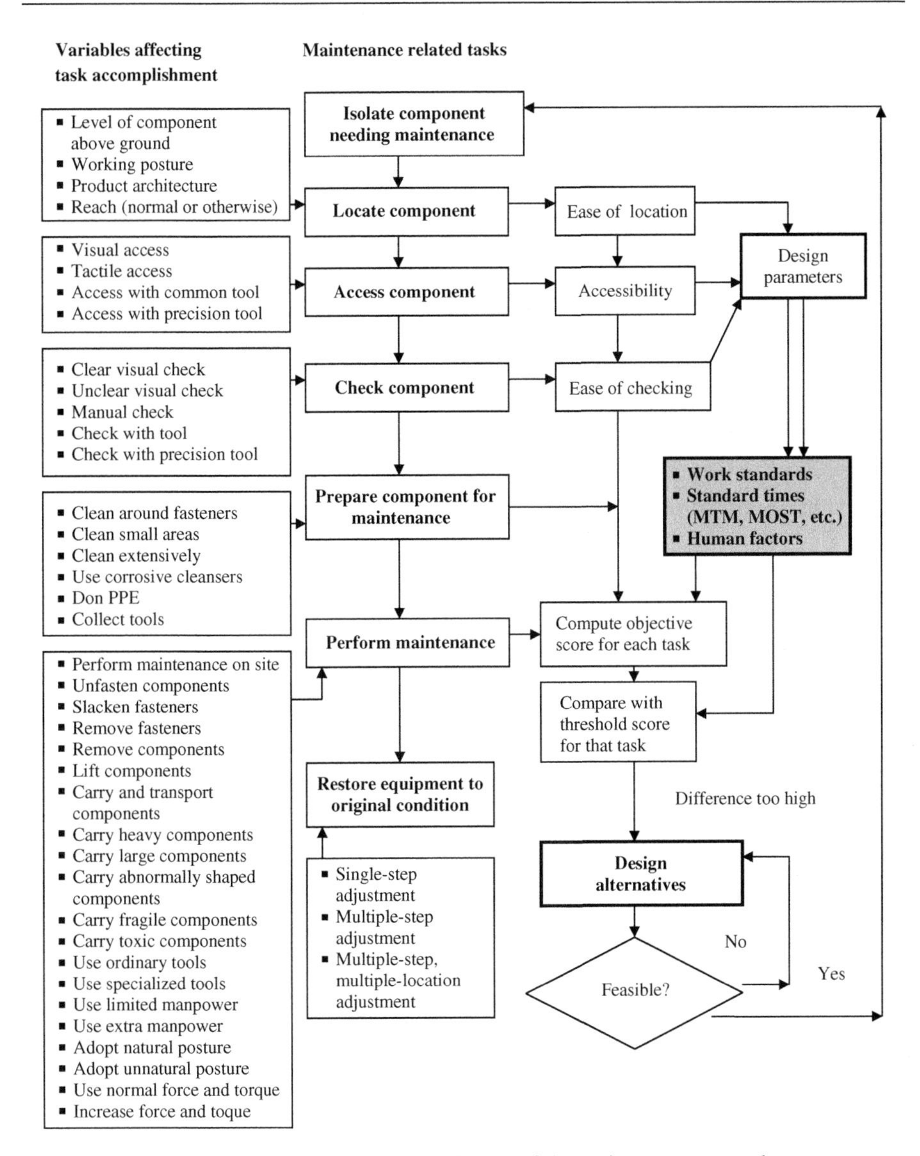

Figure 8.9 Schematic of factors affecting each step of the maintenance procedure.

If the component needs to be removed, one or more fasteners may need to be removed using ordinary tools or special tools. Other related factors, such as the necessity for exertion of normal or abnormal force, can be explained on similar lines.

It is easy to see from this explanation that each of the variables affecting a task is a direct function of design variables. For instance, nonstandard fasteners may require specialized tools for slackening and unfastening as well as refastening (for restoration). Physical features of a component, such as weight, shape, and size, also dictate the need for extra workers. This clearly is not maintenance friendly, since it entails extra labor requirements every time a maintenance operation on the component is to be performed.

On a similar note, lifting and carrying a component is a function of component design variables such as shape, size, weight, and composition. Whenever a component is made of a toxic substance, such as asbestos, additional precautions are necessary to protect the worker(s). In such instances, it may be necessary to don protective equipment, which adds to the total maintenance time.

As far as ergonomic factors are concerned, it is well established that, under normal conditions, a man can exert more force than a woman. Given this situation, it is essential to take into consideration the requirement for adoption of unnatural postures for a long period of time. This issue gains even more importance when one considers that adoption of unnatural postures for a long period of time can lead to the early onset of musculoskeletal disorders.

8.5.2 Role of work standards and standard times

Standard time data, such as that obtained from methods time measurement (MTM) or Maynard Operations Sequence Technique (MOST) systems, provide a ready reference regarding standard times for a variety of industrial operations. *Standard times* is defined as the amount of time taken by an ordinary worker to perform a task under normal conditions. MTM data enable the designer to estimate with a high degree of accuracy the amount of time necessary to perform a specific task. Using MTM data also does away with stopwatch time studies and other time measurements that are inherently subject to a high degree of error, based on the skill level of worker, accuracy of the time-measuring instrument, experience of the observer, and so forth.

Standard times are widely used and well regarded in industry as well as in academia for research purposes. Figure 8.9 depicts MTM and MOST data as input to computing objective scores for each maintenance task. The total score for each task is obtained as the sum of each attribute affecting the task (as depicted in "Variables affecting task accomplishment").

Consideration of work standards and human factors is necessary on account of the reliance on manual labor required for most industrial maintenance operations.

8.5.3 Common maintenance procedures and the parameters affecting them

Most industrial equipments that involve moving parts or components that exhibit relative motion in all planes (vertical, horizontal, and rotational) are subject to wear with the passage of time. Friction inherent in all pairs of mating surfaces is directly responsible for surface degradation. Once surface breakdown exceeds a threshold value, it interferes with equipment performance. Components no longer function in a symbiotic and synchronous manner, leading to eventual equipment breakdown.

Restoration of faulty equipment back to full working condition is accomplished by means of various maintenance procedures. Some commonly used maintenance procedures include processes such as lubrication, cleaning, filling, draining, repair, and replacement.

As pointed out earlier, most maintenance procedures are manual in nature. This means that factors affecting safe and efficient work practices directly affect the efficient performance of most maintenance work as well. Work practices are furthermore the direct function of several design parameters of components, such as weight, shape, and size. The relationship between different maintenance procedures and equipment design parameters is depicted in Figure 8.10, which represents the most commonly performed maintenance procedures and how they relate to different product design parameters.

Most common maintenance actions are a direct function of design parameters. Note that factors such as energy requirements, personnel requirements, and general posture requirements need to be explained in more detail. In Figure 8.10, "extra allowances" is mentioned alongside each of these factors. This is explained in the following section.

8.5.4 Provision for additional allowances for posture, motion, energy, and personnel requirements

Maintenance workers, in most cases, need to adopt a particular posture and expend a requisite amount of energy to accomplish a particular maintenance task. It is clear from Figure 8.10 that postures, for instance, can be classified into several categories, depending on the orientation of the worker with the work surface. The simplest and easiest task to perform is one that can be performed at desk level while the worker is sitting down. This entails the least expenditure of energy (in terms of posture requirements) and is the most natural working position. In most instances, postures such as bending down, crouching, squatting, and the like are uncomfortable. Obviously, an effective design methodology needs to take into account the different postures adopted by workers performing maintenance tasks, because the same results can be obtained in two different ways. A worker who needs to bend continuously or, worse, perform a crouch to perform a job obviously expends more energy due to that posture.

As far as personnel requirements are concerned, design parameters often dictate the number of workers needed to perform a maintenance task effectively and quickly. Examples of such design parameters include heavy components, a large number of joints or fasteners, and objects that are unwieldy to carry. In such cases, more than one worker is necessary to complete the task satisfactorily. Any maintenance task requiring more than one worker usually is considered detrimental to general work practice as well as to company overhead. To incorporate maintenance tasks requiring more than one person, personnel allowances need to be included in the design methodology.

8.5.5 Design parameters affecting premaintenance operations

Premaintenance operations are those that need to be performed to enable effective maintenance, such as the following:

- Slackening or removal of fasteners to remove a component for repair or replacement.
- Slackening or removal of fasteners to lubricate joints (maintenance).
- Cleaning joints or surfaces prior to maintenance.

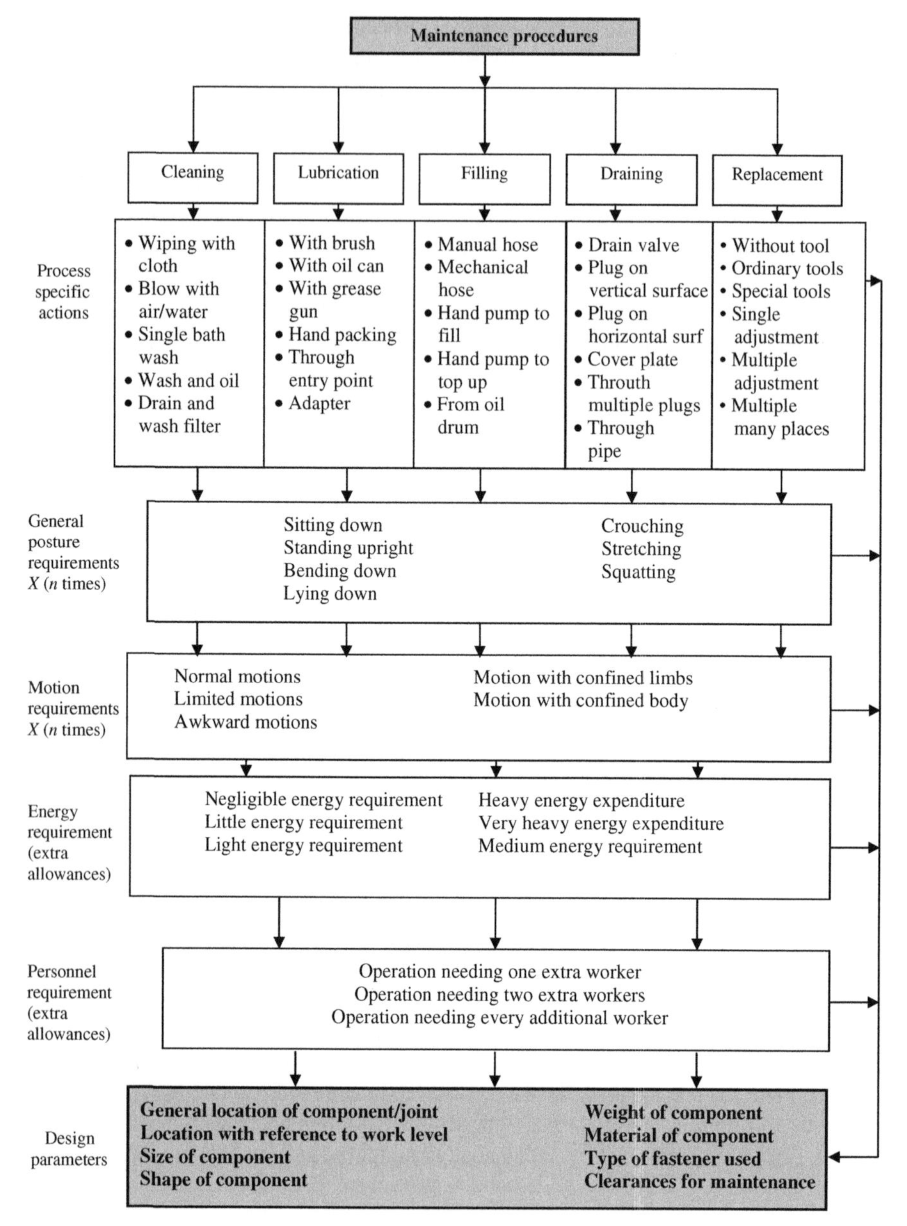

Figure 8.10 Relationship between maintenance procedures and equipment design parameters.

- (Re)alignment of joints or components to perform maintenance.
- Removing and carrying components to perform maintenance of bearings and related surfaces (such as treating or replacing corroded surfaces).
- Removing and carrying components to perform maintenance.

Note that, in all of these cases, certain product design parameters directly influence effective task performance. Examples of such design parameters include the following:

- **Physical design parameters, weight, shape, and size of components**: A consideration of physical design parameters is essential from the point of view of enhancing maintenance. A component that is abnormally heavy (requiring more than one worker to lift, carry, or adjust in place prior to replacement) obviously features a design anomaly, since it prolongs the maintenance operation and increases personnel requirements. Similarly, a component that is unwieldy in shape is a design liability from the perspective of maintenance. This is the reason the design literature often stresses the use of components of standard physical dimensions and characteristics.
- **Types of fasteners used, standard or nonstandard**: The type of fasteners used is important from the point of view of ascertaining specific tool requirements. For example, fasteners requiring precision tools for slackening, tightening, removal, or replacement often incorporate a higher design score (penalty) in the proposed methodology. This is because nonstandard and precision tools require specialized skills that may not be readily available. On the other hand, ordinary tools can be handled skillfully by the ordinary worker and are readily available.
- **Clearances in equipment design**: Clearances are an essential component of equipment design. They enable performance of maintenance tasks by enhancing worker accessibility. Ideally, clearances should facilitate not only easy and ready access to joints, fasteners, and components but sufficient room for manipulation of tools.
- **Material of components**: Component material can be classified into many categories, depending on criteria of evaluation. Examples include corrosive or noncorrosive, brittle or nonbrittle, and toxic or nontoxic. Asbestos, which has been proven to be carcinogenic, is an example of a toxic material used in industrial design. The objective of the methodology is to discourage the use of materials that would prolong the performance of maintenance or adversely affect worker health. For instance, corrosive materials often deteriorate with age due to corrosion. This not only poses hazards to maintenance workers (cuts, bruises) but prolongs the maintenance operation (by jamming fasteners or components, which in some cases need to be pried out or broken for removal). Corroded surfaces also entail the exertion of a greater amount of disengaging force in removal.
- **Amount of force necessary to enable maintenance**: Generally speaking, maintenance should require the least amount of exertion, the fewest tools, and a posture that causes the least physical stress. However, reality is far from this ideal situation. Often, in industry, it is essential to exert a large amount of force to enable maintenance. This feature is a design liability, since it stresses the worker, may require use of specialized tools, and requires additional clearance to manipulate tools. It also is more time consuming. Note that exertion actually is a function of the design variables.

The objective of this design method is to isolate product design parameters that reduce maintainability of equipment. Doing so enables the redesign of components by altering the design parameters in question. Product redesign, in turn, enhances equipment maintainability. This process is depicted in Figure 8.11. The relationship between premaintenance operations and design parameters is depicted in Figure 8.12.

8.5.6 Structure of the index

The principal objective of this methodology is to develop a composite numeric index to quantitatively evaluate maintainability of equipment based on quantification of

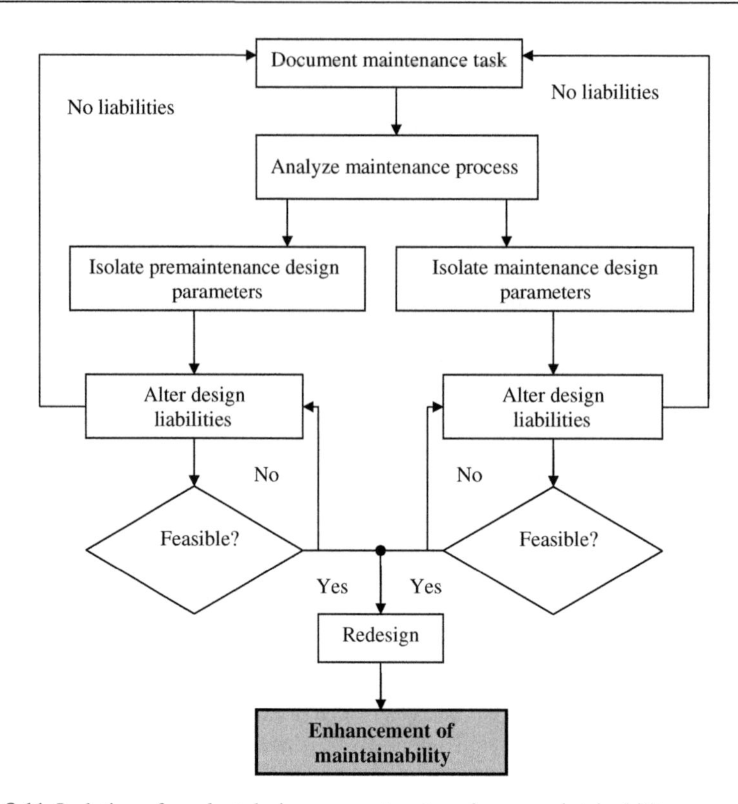

Figure 8.11 Isolation of product design parameters to enhance maintainability.

design variables that directly or indirectly affect the process of maintenance. The preceding section had two principal objectives. First, it sought to explain in detail the different maintenance practices used in industry. It sought to outline the different design variables involved in maintenance planning and establish a relationship between maintenance tasks and design variables. Second, it established a clear demarcation between principal maintenance activities and premaintenance activities. Premaintenance activities essentially are those that need to be performed to enable actual maintenance. The relationship between premaintenance activities and product design parameters also was established. The next logical step is the formation of a structured composite index to enable quantitative evaluation of the various design parameters that affect equipment maintainability.

The index essentially comprises three sections. The first section includes premaintenance actions, such as slackening, tightening, removing, lifting, and cleaning. The design parameters affecting these actions are to be evaluated. The second section of the index focuses on evaluating design variables directly affecting actual maintenance processes, such as lubrication, fitting, and replacement. The third section takes into account all allowances as described earlier (such as allowances for posture requirements, personnel requirements, types of motions, and energy expenditure).

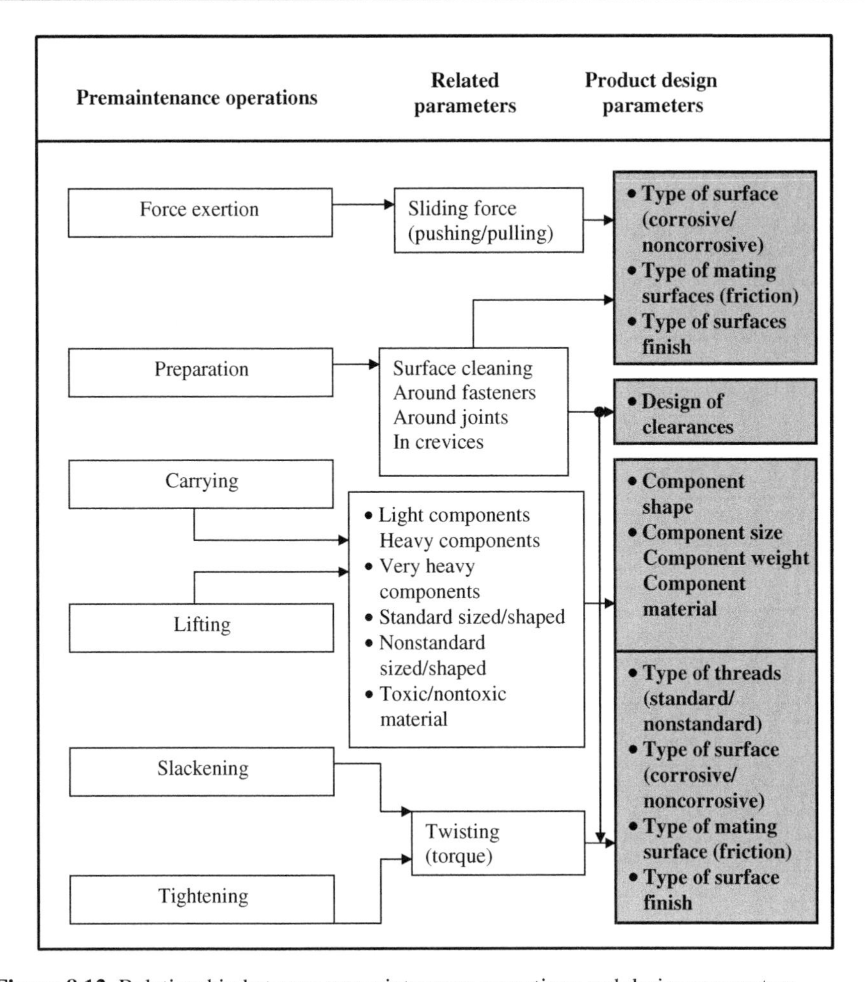

Figure 8.12 Relationship between premaintenance operations and design parameters.

8.5.6.1 Gaining access to components

Access to components is most easily gained in the absence of any interfering equipment parts. Examples of such parts include covers, handles, and any functional or nonfunctional equipment. However, practically speaking, this is not always possible. For functional purposes, it often is necessary to incorporate design features that obstruct direct access to components. The simplest way to gauge scores for such obstructions is to estimate the amount of time necessary to overcome the obstruction, which may be a function of the size, weight, or material of the obstruction.

Similarly, it is a function of any fasteners and amount of force needed to unfasten any locked covers. These features have been addressed in detail already. The Bretby index classified this section separately. However, when tasks are reclassified into premaintenance and maintenance, all tasks pertaining to gaining access are counted in

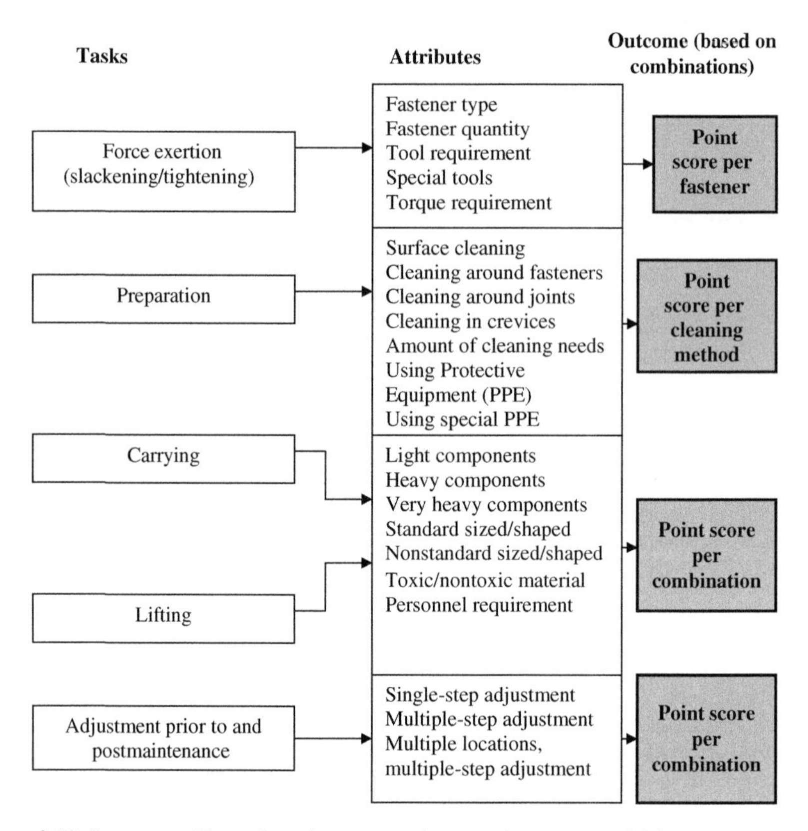

Figure 8.13 Structure of index based on pre- and postmaintenance activities.

the former category, which does away with the need for an additional section focusing on accessibility. Consider that the time and score for each action is calculated in the same way as other premaintenance activities.

8.5.6.2 Pre- and postmaintenance activities after access

The structure of premaintenance activities is depicted in Figure 8.13. Each action is evaluated on a scale covering all possible combinations (such as little force required for slackening to very high force required for slackening) using MTM.

Each score as detailed in Figure 8.13 is essentially a conversion of the standard time necessary to perform that function. For instance, if operation A requires 20 TMUs for completion, the score assigned to the operation is 2. If the time required happens to be a fraction, such as 75.5, it is converted to the closest integer by rounding up.

8.5.6.3 Maintenance activities

Major maintenance activities were identified already. Examples of such activities include lubrication, cleaning, filling, fitting, and replacement. The major difference

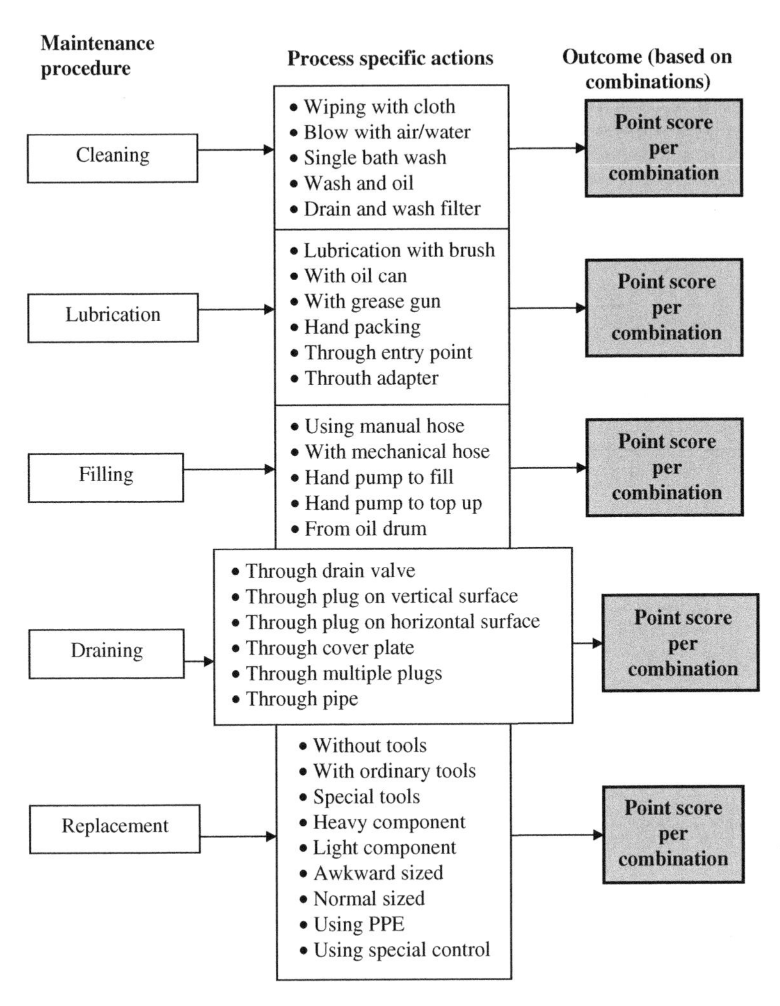

Figure 8.14 Structure of the index based on maintenance activities.

between maintenance activities and pre- or postmaintenance activities is quite subtle. As the definition implies, the former constitute actual reinstatement of equipment to working conditions. The latter, on the other hand, constitute either preparation of equipment conditions for maintenance or "top off" actual maintenance activities. Adjustment at single or multiple locations and steps is an example of a postmaintenance activity.

Another important characteristic of maintenance activities is that the ease afforded by equipment to enable effective maintenance is a direct function of features outlined earlier, such as accessibility, force requirements, and personnel requirements. As explained in the previous section, these features can be modified to a large extent by modifying the underlying product design. This is where maintenance activities overlap with their counterparts. As has been observed time and again, the ease of maintenance finally boils down to equipment design features. This is depicted in Figure 8.14.

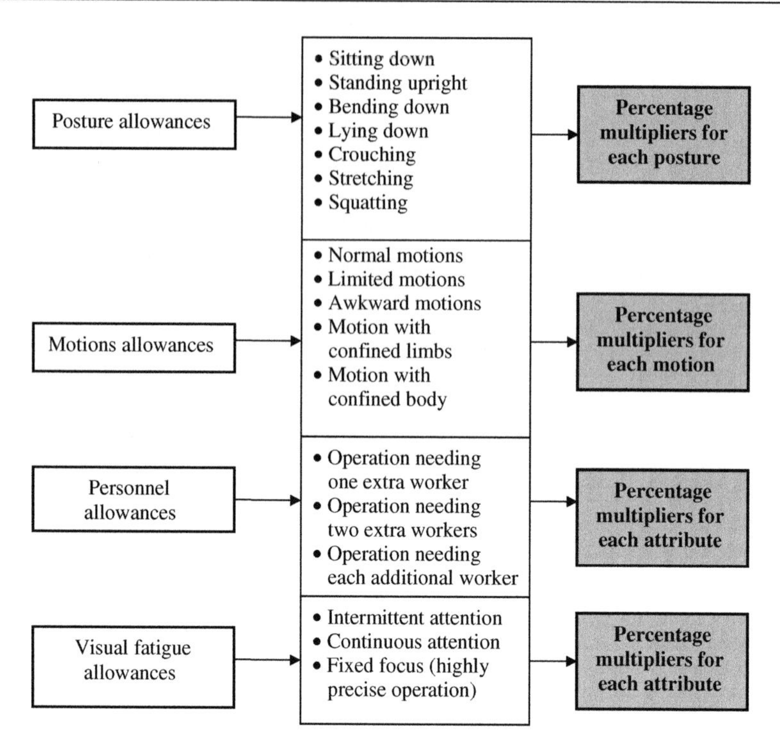

Figure 8.15 Additional allowances for maintenance procedures.

8.5.6.4 Maintenance allowances

The preceding discussion dealt primarily with the correlation between maintenance tasks and product design features and the structure of the maintenance index based purely on task analysis. To make the index comprehensive and to resemble reality, certain external yet related allowances must be included in the big picture. These allowances are outlined in Figure 8.15 and are explained further as follows:

- **Posture requirements**: A specific posture needs to be adopted to perform a certain task. Postures can be classified into two categories: natural postures, which are the most comfortable way of performing a task, and unnatural postures, which place a certain amount of strain and stress on the worker. It is easy to deduce that unnatural postures require a greater amount of energy and lower productivity. All other factors being equal (such as working conditions and skill of the worker), product design features often require unnatural postures. The inclusion of posture allowances in the index is necessary to take into account the greater amount of energy expended in performing a task as well as the lower productivity. The formation of an allowance is necessary because these effects are not necessarily discrete but continuous and can be observed only after a certain amount of time has passed. This means they cannot really be counted off one by one.
- **Motion requirements**: Motion requirements are a follow-up to posture requirements. Ideally, a maintenance task needs to be performed using normal unhindered motions on equipment designed with sufficient clearance. However, in practice, this is not the case most

of the time. Motions often are hindered due to a variety of reasons, including suboptimal product design and insufficient clearances. This occurs when equipment is not designed with ease of maintenance in mind. As in the case of posture requirements, awkward motions can result in lower productivity, more time spent in performing a particular task, and more far-reaching physical disorders such as injuries and musculoskeletal disorders.

- **Personnel requirements**: Most maintenance operations should be accomplished with minimum personnel, ideally one worker. However, when equipment is not designed for maintenance or due to a variety of other functional reasons, additional workers often are required. In such instances, the organization needs to either hire extra workers or diverted them from their normal jobs. Other related factors, such as additional training requirements and additional overhead, are further undesirable hindrances.
- **Visual fatigue**: All maintenance tasks require a certain amount of visual attention throughout or during specific portions of the task. The portion of the maintenance task requiring the most visual attention is either highly precise or the result of faulty design. Concentrated visual attention on a task throughout its duration can result in significant worker fatigue and affect productivity. Hence, the incorporation of allowances for visual fatigue assumes importance.

Since these factors are intended to be incorporated into the index primarily in the form of allowances, they are expressed in the form of percentages. The most ideal manner of incorporating allowances into task analysis is to arrive at a complete score for a task based on the numerous factors outlined previously and multiply the subtask (or entire task) affected by allowances by the appropriate percentage value.

Doing so gives the index significant flexibility, since the element of algebraic multiplication is introduced for a much fairer and objective evaluation of design features affecting maintenance.

For instance, assume that a task is composed of three subsections: A, B, and C. Only subsection C requires unnatural postures and subsection B can be performed only with awkward motions. Let the allowance be 4% for lying down and that for awkward motions be 3%. In this case, the composite score after taking into account all these allowances would be $A + 1.03B + 1.04C$. The incorporation of maintenance allowances is depicted in Figure 8.15.

8.5.7 Using the index

The index is used as outlined in Figure 8.16. Due to the methodical and stepwise approach to be utilized in the formation of the index, its usability lends itself to spreadsheet friendliness. Several features that make spreadsheet software appealing to users (such as automatic multiplier functions, addition functions, and repeatability functions) can be incorporated into the index as well. This makes using the index extremely user friendly.

8.5.8 Priority criteria for design evaluation

Once maintainability scores for each maintenance task have been developed, the next step is to evaluate design attributes associated with each task. As we have seen countless times before, the efficiency of a maintenance task is a direct function of the

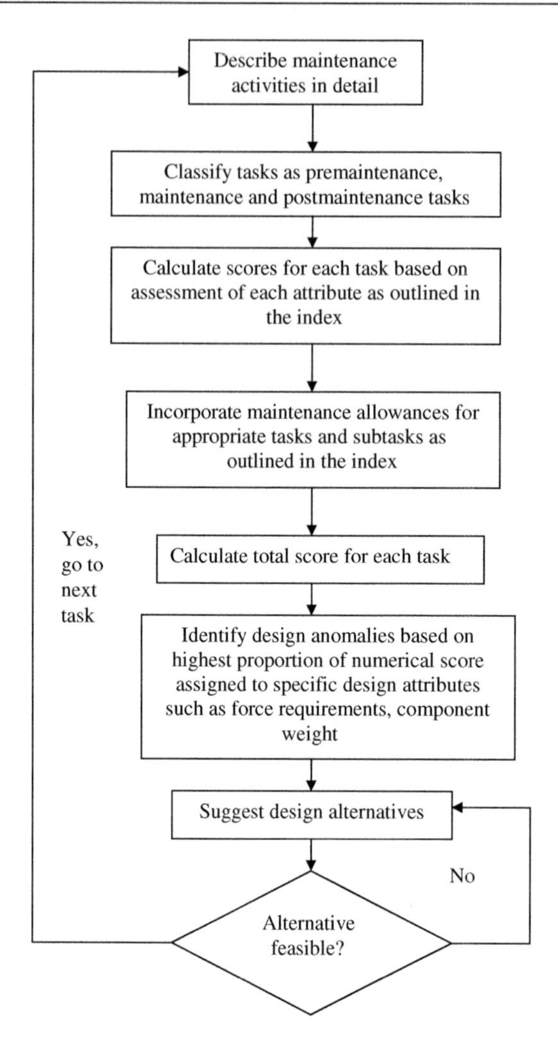

Figure 8.16 Implementing the index in practice.

component's design parameters. However, when faced with the challenging situation of choosing among different tasks for design modification, how does one go about setting priorities?

This section addresses exactly this tricky situation by developing a sequence of actions aimed at choosing priorities intelligently. Focusing on the most important components for design modification results in savings of time, effort, and money.

Design choice is primarily a function of functional importance and maintenance frequency:

- **Functional importance**: Generally speaking, the more important a component is to the functioning of the product or equipment, the more vital it is. Therefore, it is of paramount

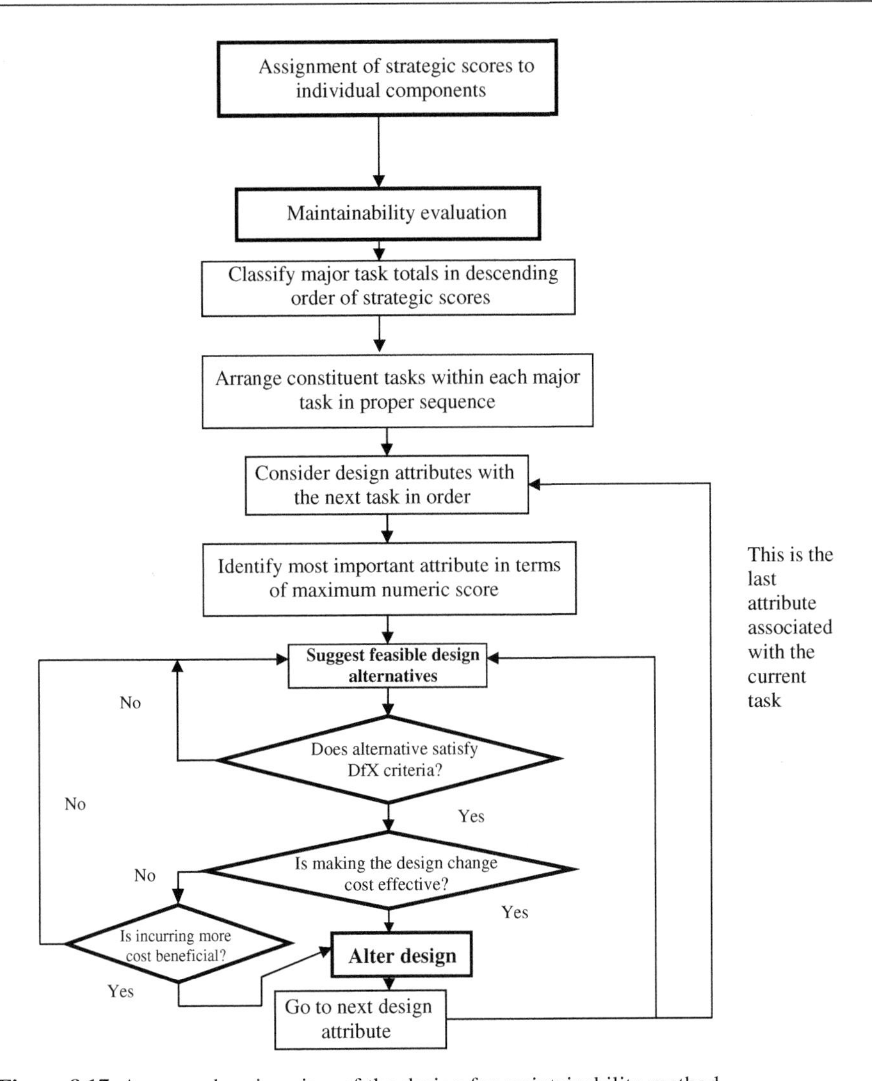

Figure 8.17 A comprehensive view of the design for maintainability method.

importance to maintain it in working condition all the time. A case in point is that of an automobile engine or a television CRT.

- **Maintenance frequency**: Maintenance frequency refers to the number of times a particular component needs to be maintained during a specific period of time. The greater the maintenance frequency, the higher is the probability that the component in question will need to be accessed, cleaned, lifted, and adjusted to repair it or restore it to working condition.

Based on a consideration of these factors, a series of objective scores is developed that establish the overall importance of individual components. Once this scoring matrix is developed, the method can be fully implemented in practice. The sequence of operations for using the method is depicted in Figure 8.17.

8.6 Developing and evaluating an index

Almost all maintenance operations comprise three major tasks: disassembly, maintenance, and reassembly. Note that disassembly may be partial or total and is done solely to gain access to the component being maintained. As such, product architectures that can do away with the need to disassemble a product prior to maintenance result in the best maintenance times and associated costs. The index for evaluating ease of maintenance is subdivided into three major sections, each section dealing with one of these major tasks.

This section deals with developing an index for product disassembly, maintenance, and reassembly. Also, systematic methods for evaluation of the index are developed. Each index for disassembly, maintenance, and reassembly is developed using the MTM system as outlined in the preceding section. The advantages of using such a system have already been elaborated.

Finally, a number of practical applications of the index are demonstrated using real-life situations. A case study is presented in the following section.

8.6.1 *Numeric index and design method for disassembly and reassembly*

The time-based numeric index and method for product disassembly and assembly was presented in Chapter 7. Refer Section 7.8 on disassembly and Section 7.4 on assembly, for the design methods.

8.6.2 *Numeric index and method for maintenance*

A time-based numeric index for maintenance is presented in Table 8.15. The most common maintenance operations include cleaning, lubrication, draining, and filling. The index addresses each of these tasks and evaluates the ease with which each can be performed by assigning it a numeric value representative of its design features.

8.6.3 *Priority criteria for maintenance*

In any product architecture, a few parts are vital to the product's functionality, whereas a large majority of the remaining components serve auxiliary and support functions. Most maintenance functions serve to maintain functionally important components in working condition. An effort needs to be made to set priorities on maintenance operations based on functional importance of the components as well as the frequency of maintenance. Any component that requires frequent maintenance, by its nature, constitutes a design anomaly, and that factor needs to be given due consideration.

In keeping with the preceding discussion, priority criteria for maintenance ranks design alterations based on a combination of product functionality and maintenance frequency. A product (multiplication) of functionality (higher functionality equates to a higher score) and maintenance frequency (higher frequency equates to a higher score) is arrived at for each component (fasteners not included). Each maintenance

Table 8.15 Numeric index for maintenance operations

Maintenance task	Design feature	Score	Interpretation
Cleaning			
Wiping with cloth	Cleaning the machine:	2/surface	Minor cleaning on surface
Wiping in crevices	around fasteners,	3.5/crevice	Minor cleaning in inaccessible areas
Blow with air/water	in crevices, on surfaces,	2	Major cleaning in inaccessible areas
Single bath wash	etc.	5	Cleaning needs washing with solvent
Wash and oil		10/surface	Washing followed by oiling to prevent corrosion
Drain and wash filter		10	Filter needs to be washed and drained
Multiple washes		15/surface	Requires more than one wash/surface
Lubrication			
With brush	Lubrication between mating	2/location	Minor lubrication on surface or crevice
With oil can	surfaces to minimize	2/location	Minor lubrication in inaccessible areas
With grease gun	friction and prevent material	2/location	Minor greasing in inaccessible areas, slots, etc.
Hand packing	loss	15/location	Major lubrication to compensate for material loss, etc.
Through access point		1/location	Simple lubrication through access point
Through adapter		3/location	Simple lubrication requiring special equipment
Filling			
Using manual hose	Filling to replenish stock of	4/filling	Simple filling operation using manual hose
With mechanical hose	lubricant (oil)	8/filling	Simple filling operation with mechanical hose
Hand pump to fill		8/gallon	Hand pump requires additional effort
From oil drum		6/gallon	Simple filling operation, direct from drum

(Continued)

Table 8.15 **(Continued)**

Maintenance task	Design feature	Score	Interpretation
Draining			
Through drain valve Through plug on vertical surface Through plug on horizontal surface Through multiple plugs Through pipe	Removing and draining used lubricant to replace it for future use	1/valve 7/operation 9/operation 7/plug 4/operation	Simple, quick draining operation Vertical drain requires extra care Horizontal drain requires more time Proportional to number of plugs
Replacement			
Combination of disassembly, assembly, and cleaning			

operation then is classified first based on this product and the methodology depicted in Figure 8.18.

A holistic approach to maintenance involves improving product design for disassembly to facilitate access to the component being maintained or to enhance overall accessibility for comprehensive maintenance, and reassembly after the maintenance operation. Figure 8.19 depicts a holistic methodology for maintainability.

8.6.4 A holistic method for maintainability

The method strategically combines elements from disassembly, assembly, and maintenance methods presented earlier in this chapter. It shortens the task sequence necessary for performing a maintenance operation by optimizing the three processes that constitute maintenance. An illustration of the logical thought process is depicted in Figure 8.19.

8.6.5 Design modifications and measures to enhance ease of maintenance

Table 8.16 lists design alternatives to overcome design anomalies and improve component or product design from the maintenance, assembly, and disassembly perspective. A careful study of these measures would enable the designer to design components and mating parts right the first time, thereby eliminating the need to redesign in the future.

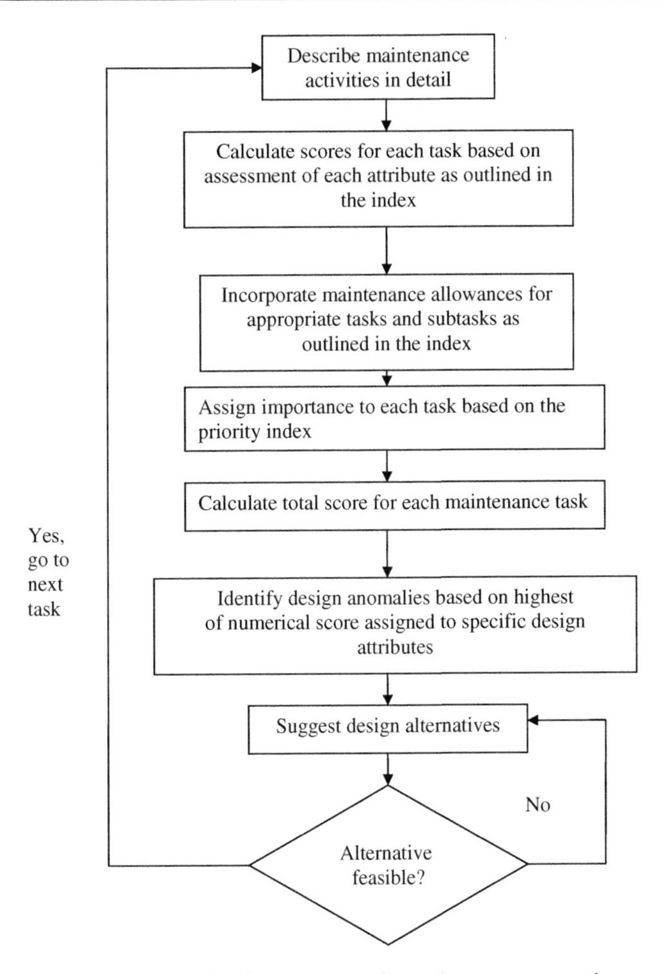

Figure 8.18 Method to enable design improvement for maintenance operations.

8.7 Design for maintenance case study

This section presents a case study to illustrate the maintenance (lubrication) of a handheld drill rotor. It involves a relatively simple maintenance operation that does not require the adoption of unnatural postures (most of the maintenance tasks are done sitting down). Also, the disassembly and reassembly operations are straightforward. A list of individual components to be disassembled, serviced, and reassembled is presented in Table 8.17. The entire maintenance operation is presented in Table 8.18.

The individual elemental scores and time elements for specific suboperations as well as the entire lubrication operation are based on the scoring matrix developed and presented earlier in the chapter. It is clear from Table 8.18 that operations that entail complicated sequences of motions, handling of nonstandard components, or require

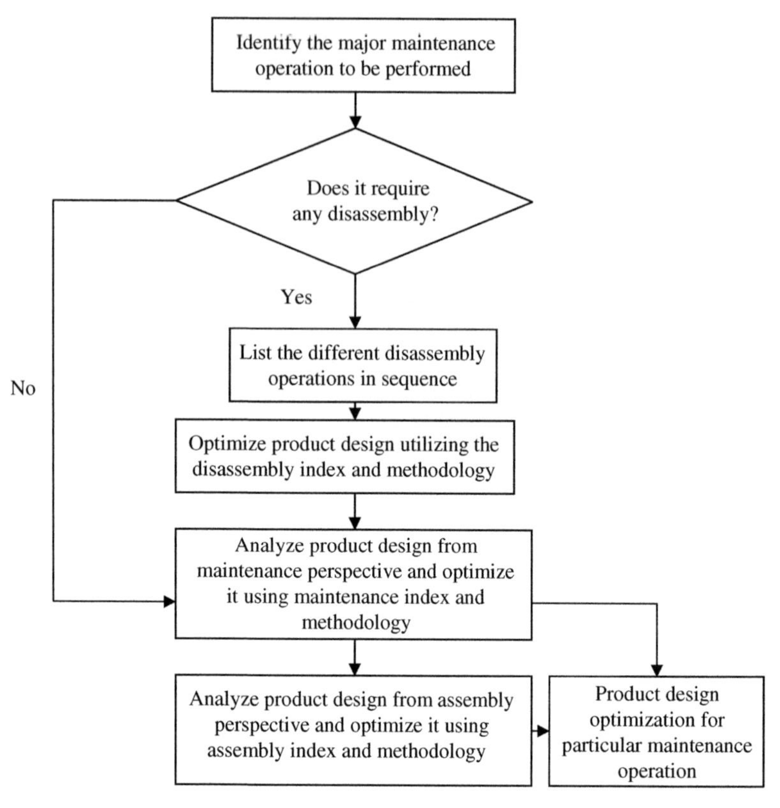

Figure 8.19 A holistic methodology for maintainability.

Table 8.16 Possible remedial measures to enhance the ease of maintenance

Design attribute/ feature	Remedial measures	Component redesign required?
Accessibility		
Deep fastener recesses	Redesign recess to facilitate tool access	Yes
Narrow fastener recesses	Redesign recess to facilitate tool access	Yes
Small fastener head	Increase fastener head size	No
Obscure fastener	Choose standard fastener sizes	No
	Increase fastener size	No
Deformed fastener	Improve fastener rigidity to withstand stresses during operation	No
Deformed component	Improve component rigidity to withstand stresses during operation	Yes
Deformed bearing surface of component	Improve component rigidity to withstand stresses during operation	Yes

(*Continued*)

Table 8.16 **(Continued)**

Design attribute/ feature	Remedial measures	Component redesign required?
Need for cleaning before access	Redesign component/fastener interface	Yes
Obscuring components	Redesign assembly sequence based on disassembly priority of components	No
Insufficient clearance for effective tool manipulation	Redesign component recesses, slots, or fasteners	Yes
Force exertion		
Moderate to large force required	Select appropriate materials for component bearing surfaces or fastener to reduce intersurface friction	Yes
	Redesign holding surfaces in component (e.g., bores in case of screwed in fits)	Yes
Wedges while disengaging	Redesign holding surfaces in component (e.g., bores in screwed in fits) or mating surface of fastener (e.g., provide taper in screw shanks); improve rigidity of material(s) if bearing surfaces are deformed; provide appropriate tolerances on mating surfaces to improve mutual fit	Yes
Tight snap fits	Redesign components to provide adequate clearance and taper to allow easy dismantling of snap fits	Yes
Occurrence of corrosion at component interfaces	Select appropriate (noncorrosive) materials for component bearing surfaces or fastener	Yes
Coarse threads on fastener	Select fasteners with finer threads or greater thread pitch	No
Positioning		
Moderate to high degree of precision required to place tool	Redesign access path, modify component bearing surfaces or fastener	Yes
Component weight	Redesign component to reduce weight	Yes
Component size	Use standard sizes	Yes
	Optimize component size for functionality and material handling	Yes
Component shape	Use standard, symmetric shapes	Yes
	Avoid protrusions in component shape design	Yes
Component material	Use nonhazardous materials	Yes

(Continued)

Table 8.16 (Continued)

Design attribute/ feature	Remedial measures	Component redesign required?
Mating surface condition		
Either or both surfaces corroded	Select appropriate (noncorrosive) materials for component bearing surfaces or fastener	Yes
Either or both surfaces deformed	Select appropriately rigid materials to withstand forces during operation/disassembly	Yes
	Redesign component bearing surfaces to allow for appropriate clearances	Yes
	Redesign fastener holding surfaces	No

Table 8.18 Lubrication operation of drill rotor

Task description	Task total	Inter-surface friction	Inter-surface wedging	Material stiffness	Compo-nent size	Compo-nent weight	Compo-nent symmetry
		Disassembly force			Material handling		
1. Remove upper housing							
a. Unscrew 1st front/back screw	15–65	2.5	–	–	2	2	0.8
b. Unscrew 2nd front/back screw	15.65	2.5	–	–	2	2	0.8
c. Unscrew 3rd front/back screw	15.65	2.5	–	–	2	2	0.8
d. Unscrew 4th front/back screw	15.65	2.5	–	–	2	2	0.8
e. Unscrew 5th front/back screw	15.65	2.5	–	–	2	2	0.8
f. Unscrew 6th front/back screw	15.65	2.5	–	–	2	2	0.8
g. Unscrew 1st middle screw	15.65	2.5	–	–	2	2	0.8
h. Unscrew 2nd middle screw	15.65	2.5	–	–	2	2	0.8
i. Pull out upper housing	8.7	–	1	–	3.5	2	1.4

Table 8.17 **Individual components of a handheld drill for maintenance**

Component name	Component material	Quantity
1. Front/back screw	Copper	6
2. Middle screw	Copper	2
3. Bushing	Brass	1
4. Insulating washer	Plastic	1
5. Upper housing	Plastic	1
6. Lower housing	Plastic	1
7. Rotor	Mixed	1
8. Wire lead	Copper/plastic insulation	1

Force exertion	Torque exertion	Dimen-sions	Loca-tion	Accuracy of tool placement	Posture allowance	Motions allowance	Personnel allowance	Visual fatigue allowance
Tooling		**Accessibility and positioning**			**Allowances**			
–	2	1.6	2	2	–	–	–	5%
–	2	1.6	2	2	–	–	–	5%
–	2	1.6	2	2	–	–	–	5%
–	2	1.6	2	2	–	–	–	5%
–	2	1.6	2	2	–	–	–	5%
–	2	1.6	2	2	–	–	–	5%
–	2	1.6	2	2	–	–	–	5%
–	2	1.6	2	2	–	–	–	5%
1	–	1	1.6	1.6	–	–	–	1%

(Continued)

Table 8.18 (Continued)

Task description	Task total	Inter-surface friction	Inter-surface wedging	Material stiffness	Component size	Component weight	Component symmetry
		Disassembly force			Material handling		
2. Access drill rotor							
a. Pull out bushing	10.5	1	–	–	2	2	0.8
b. Pull out insulating washer	10.5	1	–	–	2	2	0.8
3. Lubricate rotor and clean housing							
a. Lubrication: 2/ location × 2 locations	4.2	–	–	–	–	–	–
b. Clean upper housing: 2	2.02	–	–	–	–	–	–
4. Reassemble drill							
a. Reinstall washer	10.5	1	–	–	2	2	0.8
b. Reinstall bushing	10.5	1	–	–	2	2	0.8
c. Refit upper housing	8.7	–	1	–	3.5	2	1.4
d. Screw 1st middle screw	15.65	2.5	–	–	2	2	0.8
e. Screw 2nd middle screw	15.65	2.5	–	–	2	2	0.8
f. Screw 1st front/ back screw	15.65	2.5	–	–	2	2	0.8
g. Screw 2nd front/ back screw	15.65	2.5	–	–	2	2	0.8
h. Screw 3rd front/ back screw	15.65	2.5	–	–	2	2	0.8
i. Screw 4th front/ back screw	15.65	2.5	–	–	2	2	0.8
j. Screw 5th front/ back screw	15.65	2.5	–	–	2	2	0.8
k. Screw 6th front/ back screw	15.65	2.5	–	–	2	2	0.8
Total	294						

Notes: Total time for maintenance operation = 2940; TMUs = 1.764 min. Task 1 for maintenance analysis: Unscrewing/screwing back of various screws of upper housing. Most feasible, cost-effective design solution: Use toggle type snap fits for upper housing in place of screws to reduce maintenance time. Total disassembly time for 100% disassembly = 2.044 min (maintenance requires almost complete disassembly of drill). Task 1 for disassembly analysis: Disassembly of rotor-bushing subassembly. Most important design anomaly for disassembly: Force required wedging out subassembly of rotor and bushings. Task 1 for assembly analysis: inserting trigger assembly. Total assembly time = 1.30 min. *Conclusion*: Most amount of time is spent in *accessing* the maintenance area. Too many fasteners hamper disassembly and assembly.

Force exertion	Torque exertion	Dimen-sions	Loca-tion	Accuracy of tool placement	Posture allowance	Motions allowance	Personnel allowance	Visual fatigue allowance
Tooling		**Accessibility and positioning**			**Allowances**			
1	–	1	1	1.2	–	–	–	5%
1	–	1	1	1.2	–	–	–	5%
–	–	–	–	–	–	–	–	5%
–	–	–	–	–	–	–	–	1%
1	–	1	1	1.2	–	–	–	5%
1	–	1	1	1.2	–	–	–	5%
1	–	1	1.6	1.6	–	–	–	1%
–	2	1.6	2	2	–	–	–	5%
–	2	1.6	2	2	–	–	–	5%
–	2	1.6	2	2	–	–	–	5%
–	2	1.6	2	2	–	–	–	5%
–	2	1.6	2	2	–	–	–	5%
–	2	1.6	2	2	–	–	–	5%
–	2	1.6	2	2	–	–	–	5%
–	2	1.6	2	2	–	–	–	5%

the unnatural postures tend to take more time as well as effort. This is reflected in higher elemental times, which in turn can be pinpointed as specific design anomalies that need rectification.

Since maintenance is largely a holistic operation, consisting of a variety of suboperations, it is imperative that the design and fault detection processes be construed in a mutually synergistic manner. This point is driven home when one considers the case of assembly, disassembly, and maintenance of a handheld drill.

8.8 Concluding remarks

This chapter addressed the issue of product maintenance in detail. It examined the need for product maintenance, the variety of commercially available consumer products, and the distinction among different methods of product maintenance. Several terms associated with the general topic of product maintenance were presented.

Several maintenance methods were presented to inform the reader of the variety of approaches adopted to facilitate product design for ease of maintenance. The practical utility of each methodology was scrutinized to examine its value in dealing with real-world situations when the product, equipment, and systems are operating in the field.

References

Adair-Heeley, C., 1989. The JIT challenge for maintenance. Prod. Invent. Manag. Rev. APICS News 9, 34–35.

Adams, S.K., Patterson, P.J., 1988. Maximum voluntary handgrip torque for circular electrical connectors. Hum. Factors 30 (6), 733–745.

Ascher, H.E., Feingold, H., 1984. Repairable Systems Reliability. Marcel Dekker, New York, NY.

Baker, R.D., Christer, A.H., 1994. Review of delay-time or modeling of engineering aspects of maintenance. Eur. J. Oper. Res. 73, 407–422.

Baker, R.D., Scarf, P.A., 1995. Can models fitted to small data samples lead to maintenance policies with near optimum costs? I.M.A. J. Math. Appl. Bus. Ind. 6, 3–12.

Baker, R.D., Wang, W., 1991. Estimating the delay-time distribution of faults in repairable machinery from failure data. I.M.A. J. Math. Appl. Bus. Ind. 3, 259–282.

Baker, R.D., Wang, W., 1993. Developing and testing the delay-time model. J. Oper. Res. Soc. 44, 361–374.

Barlow, R.E., Hunter, L.C., 1960. Optimum preventive replacement policies. Oper. Res. 8, 90–100.

Chen, W., Meher-Homji, C.B., Mistree, F., 1994. COMPROMISE: an effective approach for condition-based maintenance management of gas turbines. Eng. Optim. 22, 185–201.

Cho, D.I., Parlar, M., 1991. A survey of maintenance models for multi-unit systems. Eur. J. Oper. Res. 51, 1–23.

Christer, A.H., Keddie, E., 1985. Experience with a stochastic replacement policy. J. Oper. Res. Soc. 36, 25–34.

Christer, A.H., Scarf, P.A., 1994. A robust replacement model with application to medical equipment. J. Oper. Res. Soc. 45, 261–275.

Christer, A.H., Walter, W.M., 1984. Reducing production downtime using delay-time analysis. J. Oper. Res. Soc. 35, 499–512.
Christer, A.H., Whitelaw, J., 1983. An operational research approach to breakdown maintenance: problem recognition. J. Oper. Res. Soc. 34, 1041–1052.
Christer, A.H., Wang, W., Baker, R.D., Sharp, J., 1995. Modeling maintenance practice of production plant using the delay-time model. I.M.A. J. Math. Appl. Bus. Ind. 6, 67–84.
Crawford, B.M., Altman, J.W., 1972. Design for maintainability. In: Vancott, H.P., Kinkade, R.G. (Eds.), Human Engineering Guide to Equipment Design U.S. Department of Defense, Washington, DC, pp. 585–631.
Day, N.E., Walter, S.D., 1984. Simplified models of screening for chronic disease: estimation procedures for mass screening programs. Biometrics 40, 1–14.
Dekker, R., 1995. Integrating optimization, priority setting, planning and combining of maintenance activities. Eur. J. Oper. Res. 82, 225–240.
Drake, P.R., Jennings, A.D., Grosvenor, R.I., Whittleton., D., 1995. A data acquisition system for machine tool condition monitoring. Qual. Reliab. Eng. Int. 11, 15–26.
Eade, R., 1997. The importance of predictive maintenance. Iron Age New Steel 13 (9), 68–72.
Ebeling, C.E., 1997. An Introduction to Reliability and Maintainability Engineering. McGraw-Hill, New York, NY.
Eilon, S., King, J.R., Hutchinson, D.E., 1996. A study in equipment replacement. Oper. Res. Q. 17, 59–71.
Ferguson, C.A., Mason, S., Collier, S.G., Golding, D., Graveling, R.A., Morris, L.A., et al., 1985. The Ergonomics of the Maintenance of Mining Equipment Including Ergonomics Principles in Designing for Maintainability. Final report on CEC contract 7247/12/008. Edinburgh: Institute of Occupational Medicine IOM Report TM 85/12.
Gits, C., 1992. Design of maintenance concepts. Int. J. Prod. Econ. 24 (3), 217–226.
Harring, M.G., Greenman., L.R., 1965. Maintainability Engineering. Duke University, Durham, NC.
Harrison, N., 1995. Oil condition monitoring for the railway business. Insight 37, 278–283.
Hastings, N.A.J., 1969. The repair limit replacement method. Oper. Res. Q. 20, 337–349.
Herbaty, F., 1990. Handbook of Maintenance Management Cost Effective Practices, second ed. Noyes Publication, Park Ridge, NJ.
Hsu, J.S., 1988. Equipment replacement policy—a survey. Prod. Inventory Manag. J. 29, 23–27.
Imrhan, S.N., 1991. Workspace design for maintenance. In: Mital, A., Karwowski, W. (Eds.), Workspace, Equipment and Tool Design Elsevier, Amsterdam, pp. 149–174.
Jardine, A.K.S., Goldrick, T.S., Stender., J., 1976. The use of annual maintenance cost limits for vehicle fleet replacement. Proc. Inst. Mech. Eng. 190, 71–80.
Johnson, A.D., 1988. The design of modern road headers. Colliery Guardian.
Knezevic, J., 1997. Systems Maintainability Analysis, Engineering and Management. Chapman and Hall, London.
Kobaccy, K.A.H., Proudlove, N.C., Harper, M.A., 1995. Towards an intelligent maintenance optimization system. J. Oper. Res. Soc. 47, 831–853.
Li, C.J., Li, S.Y., 1995. Acoustic emission analysis for bearing condition monitoring. Wear 185, 67–74.
Macaulay, S., 1988. Amazing things can happen if you … 'Keep it clean.'. Production 100 (5), 72–74.
Maggard, B., Rhyne, D., 1992. Total productive maintenance: a timely integration of production and maintenance. Prod. Inventory Manag. J. 33 (4), 6–10.
Mason, S., 1990. Improving plant and machinery maintainability. Appl. Ergon. 21 (1), 15–24.

McDaniel, J.W., Askrein, W.B., 1985. Report submitted to U.S. Air Force Aerospace Medical Research Laboratory. Wright Patterson Air Force Base, Fairborn, OH.

Morgan, C.T., Cook III, J.S., Chapanis, A., Lund., M.W., 1963. Human Engineering Guide to Equipment Design. McGraw-Hill, New York, NY.

Moss, T.R., Strutt, J.E., 1993. Data Sources for Reliability Analysis Proceedings of the Institution of Mechanical Engineers, Part E. J. Process Mech. Eng. 207, 13–19.

Nakajima, S., 1989. Total Productive Maintenance Development Program: Implementing Total Productive Maintenance. Productivity Press, Cambridge, MA.

Oborne, D.J., 1981. Ergonomics at Work. John Wiley & Sons, Inc., New York, NY.

Pintelon, L.M., Gelders., L.F., 1992. Maintenance management decision making. Eur. J. Oper. Res. 58, 301–317.

Pugh, S., 1991. Total Design. Addison-Wesley, Reading, MA.

Reiche, H., 1994. Maintenance Minimization for Competitive Advantage. Routledge, Amsterdam.

Scarf, P.A., 1997. On the application of mathematical models in maintenance. Eur. J. Oper. Res. 99, 493–506.

Scarf, P.A., Bouamra, O., 1995. On the application of a capital replacement model for a mixed fleet. I.M.A. J. Math. Appl. Bus. Ind. 6, 39–52.

Simms, B.W., Lamarre, B.G., Jardine, A.K.S., Boudreau, A., 1984. Optimal buy, operate and sell policies for fleets of vehicles. Eur. J. Oper. Res. 15, 183–185.

Sivakian, B.D., 1989. Optimum scheduling for a new maintenance program under stochastic degradation. Microelectron. Reliab. J. 29 (1).

Smith, A.M., Hinchcliffe, G.R., 2004. RCM—Gateway to World-Class Maintenance. Elsevier Butterworth-Heinemann.

Smith, R.L., Westland, R.A., Crawford, B.M., 1970. The status of maintainability models: a critical review. Hum. Factors 12, 271–282.

Swanson, L., 2001. Linking maintenance strategies to performance. Int. J. Prod. Econ. 70, 237–244.

Thompson, G., 1999. Improving Maintainability and Reliability through Design. Professional Engineering Publishing Limited, London and Bury St. Edmunds, UK.

Tichauer, E.R., 1978. The Biomechanical Basis of Ergonomics. John Wiley & Sons, Inc., New York, NY.

U.S. Air Force, February 8, 1985. Reliability Centered Maintenance for Aircraft Engines and Equipment. MIL STD 1843 USAF.

U.S. Navy, January 1983. Reliability Centered Maintenance Handbook, revised. Department of the Navy, Naval Sea Systems Command, S 9081-AB-GIB-010/MAINT.

Van Cott, H.P., Kinkade, R.G., 1972. Human Engineering Guide to Equipment Design. U.S. Department of Defense, Washington, DC, 1966.

Vanneste, S.G., Van Wassenhove., L.N., 1995. An integrated approach to improve maintenance. Eur. J. Oper. Res. 82, 241–257.

Designing for Functionality

9.1 Introduction

9.1.1 Definition and importance of functionality

It is well recognized that functional design plays a central role in ensuring design quality and product innovation; products with problems in their main functions do not sell well, no matter how sophisticated their details. Numerous examples exist of products marketed and sold as sophisticated in the features they provide customers but that routinely fail to perform the intended functions or do so in a very unsatisfactory manner. For instance, the Eastman Kodak company's disc camera was marketed as a camera with nearly 40 usability features.

However, due to excessive noise in the output signal and its related negative effect on the quality of picture the camera took, the Kodak disc camera was considered a failure—the camera failed to provide the intended function, to take good pictures. Another example is the ubiquitous can opener found on supermarket shelves. To remove the lid, the cutting edge of the can opener has to progress around the lid and sever lids completely and cleanly without leaving slivers of metal behind. However, this seldom is the case with most can openers (mechanical devices). In addition to not performing the main function, most can openers jiggle the lid and cause it to splatter or submerge in the liquid as the cutter progresses around the can.

9.1.2 Factors affecting functionality

Historically, a variety of factors, both internal and external to a company, have influenced its product design goals. For instance, the mass production paradigm pioneered by Henry Ford resulted in the concepts of building products on assembly lines, using of interchangeable parts, and standardization of parts and components with a view toward reducing product cost (Bralla, 1996; Cross, 1989; Green, 1966; Lacey, 1986; Ziemke and Spann, 1993). Customers' demand for high quality products prompted manufacturing companies to consider quality as their key product design goal (Akiyama, 1991; Taguchi et al., 1989). The establishment of the U.S. Consumer Product Safety Commission in 1972 prompted manufacturers to project product safety as their key design goal (Brauer, 1990; Hammer, 1980; Mital and Anand, 1992). The advent of the computer screen and the resulting digital interface may be the primary reason for companies projecting product usability as their prime product design goal (Nielsen, 1993). Similarly, the need for product manufacturers to reduce assembly time and cost prompted product designs built from design for assembly processes (Bakerjian, 1992; Boothroyd, 1994; Boothroyd and Dewhurst, 1983; General Electric Company, 1960; Gupta and Nau, 1995; Kusiak and He, 1997; Miyakawa and Ohashi, 1986a,b;

Product Development. DOI: http://dx.doi.org/10.1016/B978-0-12-799945-6.00009-0

Miyakawa et al., 1990a,b; Nof et al., 1997; Runciman and Swift, 1985; Taylor, 1997). Recent regulations from the U.S. Environmental Agency have prompted companies to project design for environmental friendliness or "green design" as an important product design goal (Billatos and Nevrekar, 1994; Hermann, 1994; Hundal, 1994; Van Hemel and Keldmann, 1996). The Ford Motor Company initiated a setup for disassembling used cars and selling used parts (an Internet junkyard), the profitability of which depends on designing products for disassembly (Wall Street Journal, 1999). Simultaneous optimization of a number of design goals (design for X, where X could stand for assembly, manufacturability, safety, reliability, or any of the other design goals) is the latest on the research agenda (Asiedu and Gu, 1998; Bralla, 1996; Chu and Holm, 1994; Gupta et al., 1997; Huang, 1996; Huang and Mak, 1998; Jansson et al., 1990; Nevins and Whitney, 1989; Priest, 1990; Sanchez et al., 1997; Ullman, 1997). While all these design goals have gained recognition and acceptance, product performance (or what is broadly known as *product functionality*), as a design goal, often has been taken for granted by designers. Indeed, the provision of functionality in a product often is considered the purpose of design. It is possible that, even though product functionality (functionality being performance of the intended function) may have been an important product design goal for designers when product design was formalized, the necessity to accord other design goals, such as safety and usability, higher priority (due to the demands placed on the designer from time to time with different market needs, such as quality, safety, and usability) than functionality may have relegated the task of ensuring functionality in a product to a distant second place.

Keeping pace with the constantly changing product design goals, product design processes themselves have changed from the times of artisan production to the modern day concurrent engineering process. During the Middle Ages, a single craftsman could design and manufacture a complete product with no formal drawing or modeling of the product before making it. For example, a potter could make a pot with no drawings of the pot. For a larger product, one craftsman possessed enough knowledge of the engineering principles, the materials to use, and manufacturing processes to manage all aspects of the design and manufacture of a product such as a ship.

9.2 Concurrent engineering in product design

By the middle of the twentieth century, design and manufacturing processes had become complex. It was impossible for one person, however skilled, to focus on all aspects of an ever-evolving product. Different groups of people became responsible for different activities, such as product marketing, product design, product manufacturing, product sales, and overall project management. This evolution led to what is now commonly known as the *over-the-wall* design process. This process involves a one-way communication between different groups of people, represented as information that is "thrown over the wall." In this design process, engineering interprets the request, develops concepts, and refines the best concept into manufacturing specifications. These manufacturing specifications are "thrown over the wall" to be produced. Manufacturing then interprets the information passed to it (from design) and builds what it thinks the designer wants (Cross, 1989; Ullman, 1997). The traditional sequential path to product manufacture (marketing–design–manufacturing–sales) does not entail dialog between design and the downstream processes except through a series of standard, tedious engineering change orders.

With the recognition that design decisions made early in the product development cycle can have a significant effect on the manufacturability, quality, cost, product introduction time, and thus on the ultimate marketplace success of the product, and with the prohibitive corrective cost of engineering change orders, concurrent engineering recently has been recognized as a viable design approach (Jo et al., 1993). In concurrent, or simultaneous, engineering, due to the simultaneous design of the product and all its related processes in a manufacturing system, there is a greater possibility of ensuring a good match between a product's structure with functional requirements and the associated manufacturing implications.

An understanding of the key elements involved in the design and manufacturing (for functionality, see Figure 9.1) of consumer products and the tools used to model these elements should help shed light on why functionality is not ensured in products. Is it the design process? Or is it manufacturing? Or is it a lack of close correspondence between design and manufacturing? Are the current criteria for product functionality adequate? Are there problems in translating customer expectations into product functions? Is the definition of functionality adequate?

The following section examines some design concepts and methods central to effectively designing products for functionality. A thorough examination of these concepts is critical to fully understanding the chief drawbacks that exist in current design information and possibly outlook.

Following an in-depth examination of current methodologies, we conclude that some anomalies and deficiencies exist that could be overcome by the development of a new design methodology for product functionality. This methodology is presented in the final section of this chapter.

9.2.1 Functionality in design

This section reviews functionality in design, with a special emphasis on the different definitions of *function* as well as the models and tools used to represent function in a product.

Designs exist to satisfy some purpose or function. Knowledge of functionality is essential in a wide variety of design-related activities, such as generation and modification of designs; comparison, evaluation, and selection of designs; and their diagnosis as well as repair.

Beyond agreement among researchers and designers that function is an important concept in determining a product's fundamental characteristics, there is no clear, uniform, objective, and widely accepted definition of *functionality*. Function has been interpreted historically in a variety of ways. Examples of this interpretation are

- As an abstraction of the intended behavior of a design
- As an index of its intended behavior

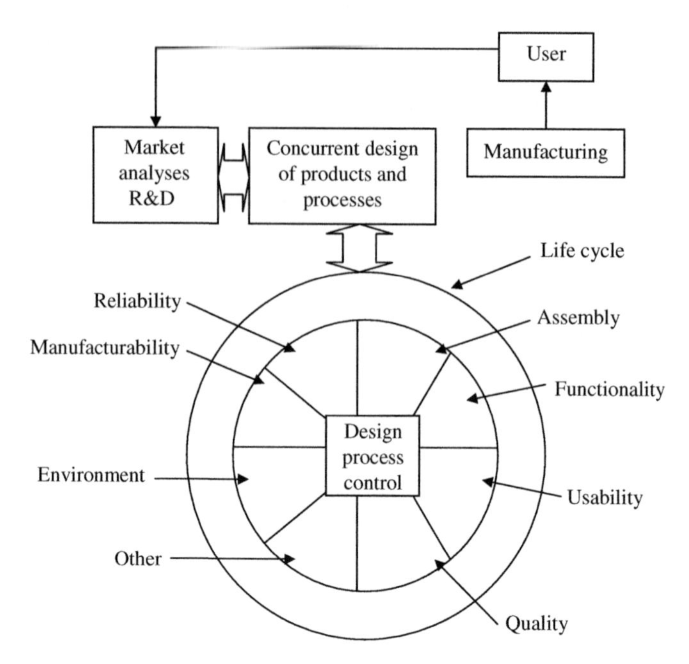

Figure 9.1 Concurrent engineering in product design.

- As a relationship between a design and its environment
- As the external behavior of a design
- As the internal behavior of a design (Umeda and Tomiyama, 1997).

The definition of *function* also has been influenced by design methodologies in use. For instance, if the designer follows the traditional conceptual design methodology, the entire function is first determined by analyzing the specifications of the product to be designed and built. This function is then divided recursively into subfunctions, a process that produces a functional structure. For each subfunction, the next step is to use a catalog to look up the most appropriate functional element—a component or a set of components that perform a function. Finally, the designer composes a design solution from the selected elements. Since the results of the design process using the traditional conceptual design methodology depend entirely on the efficacy of the decomposition of the function, the role of functionality is critical in using such a methodology (Pahl and Beitz, 1988).

A number of new models for abstracting and representing function, in addition to numerous computer-aided design tools for managing the modeling of function in a product, emerged recently. A conceptual or theoretical model represents concepts and ideas in the form of diagrams and other conventional representation methods. Any well-developed device with a physical form that can be used in real life to perform a design activity can be classified as a tool. For instance, a software program used to perform a certain design activity can be considered as a tool, whereas an algorithm powering the software can be classified as a model.

9.2.2 Function and functional representations: definitions

Working action or the action of something is the literary definition of *function*. This definition encompasses all the specific roles possessed by each of the mutually interacting elements constituting a whole.

While functionality is considered an intuitive concept dependent on the designer's intention, traditionally there have been three approaches to representing function in a design:

1. Representing function in the form of verb–noun pairs (Miles, 1961). An example would be the function of a shaft. Its function is represented by two words: "transmit torque."
2. Input–output flow transformations, where the inputs and outputs can be energy, materials, or information (Rodenacker, 1971).
3. Transformations between input–output situations and states. The essential difference between this and the preceding definitions is the type of input and output. For instance, if the product is a household buzzer, the function "to make a sound" can be represented by two behavior states, state 1 representing an upward clapper movement and state 2 representing a downward clapper movement (Goel and Stroulia, 1996; Hubka and Eder, 1992).

Miles (1961) developed the function analysis method of expressing a function as a verb and direct object (a noun or an adjective). The motivating idea for this definition is that any useful product or service has a prime function. This function usually can be described by a two-word definition, such as provide light, pump water (for a domestic water pump), or indicate time (as a clock). In addition to primary functions, secondary functions may be involved in a product. For instance, if the primary function of a light source is to provide light, a secondary function could be that the light source is required to resist shock; a pump for domestic use, with pumping water as the primary function, may have to operate at a low noise level. Although this definition of a function is general, owing to the lack of a clear description of the relationships between product function and product structure, this representation is not considered powerful enough for design applications. Miles's definition of *function* has been used primarily in value engineering, representing a function in the form of "to do something" as well as by comparing the value of function with respect to product cost.

Rodenacker (1971) defined *function* as the transformation between the input and output of material, energy, and information. A coffee mill is an example using Rodenacker's definition. In this case, the input can be conceptualized to consist of the coffee beans, energy, and information to the user in the form of electrical signals. The coffee mill is the black box where the transformation of coffee beans into ground coffee occurs. The output is ground coffee, heat, and information to the user in the form of electrical signals. Although this definition is widely accepted in design research, it has its limitations. For instance, some functions do not strictly involve transformation between input and output.

Umeda et al. (1990) proposed the FBS (function behavior state) diagram to model a system with its functional descriptions. An example is depicted in Figure 9.2.

According to this definition, function is a description of behavior abstracted by the human through the recognition of the behavior to utilize that behavior. The underlying concept in this definition is that it is difficult to distinguish clearly between function and human behavior. It is also not meaningful to represent function independent of

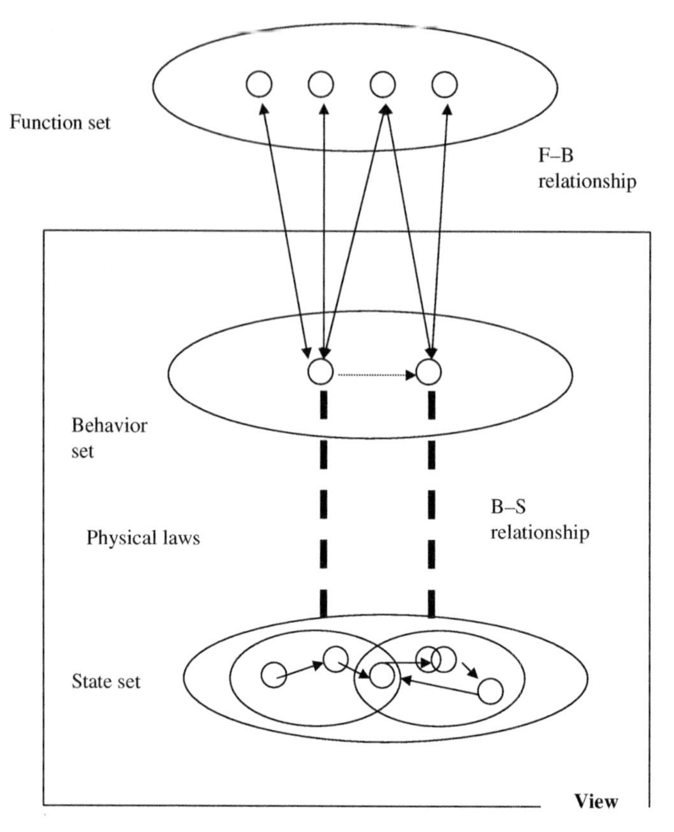

Figure 9.2 Relationship between function, behavior, and state.
Source: Modified from Umeda et al. (1990).

the behavior from which it has been abstracted. Function, in the FBS diagram, is represented as an association of two concepts: the symbol of a function, represented in the form of "to do something," and a set of behaviors that exhibit that function. For instance, some behaviors, such as "ringing a bell" and "oscillating a string," may be used to realize the function "making a sound." Although the concept of symbolic information is meaningful only to a human, this information, associated with its behavior, has been found to be essential to support design. Examples of this include the reuse of design results and clarification of specifications. It is easy to see that function and behavior have a subjective, many-to-many correspondence in their relationship. On the other hand, the representation of behavior of an entity can be determined more objectively based on physical principles. The FBS diagram is intended to assist the designer in the synthetic as well as analytic aspects of conceptual design.

According to Sturges et al. (1990, 1996), *function* is defined as the domain-independent characteristics or behavior of elements or groups of elements. Function-logic methods are modified by Sturges et al. for the development and use of function block diagrams. The concept behind this definition of *function* is that the designer should be able to describe the intended function, expand it into the required subfunctions, and map the

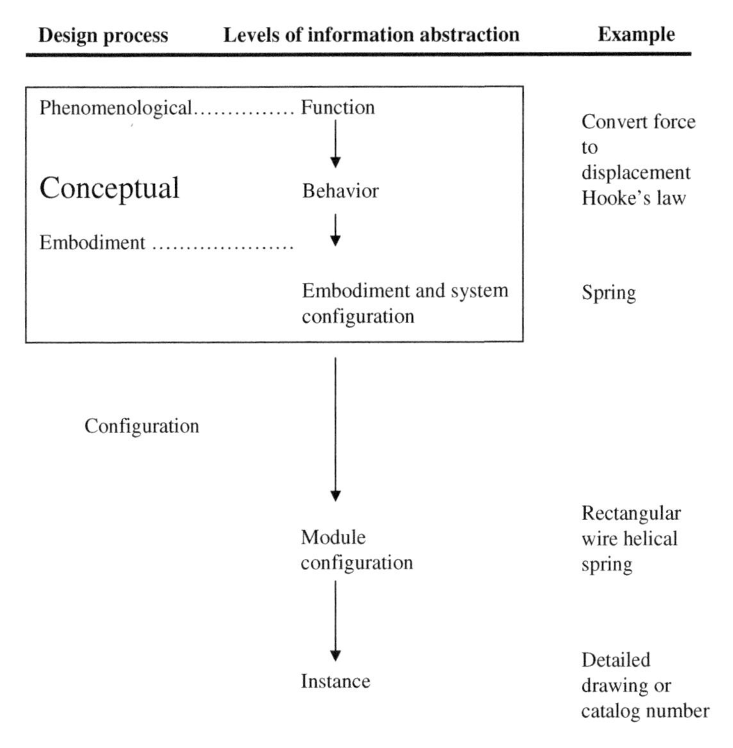

Figure 9.3 Classification of design information and processes.
Source: Modified from Welch and Dixon (1992).

subfunctions into components capable of fulfilling them. The designer is assisted by the computer in this process in terms of systematic identification of functions, allocation of constraints to each function, the interrelationship between functions, and functional evaluation. The approach supports the designer mainly in the identification, articulation, and evaluation of function structures, rather than the search for design solutions. Therefore, it applies to later stages of task clarification and the early stages of conceptual design.

According to Welch and Dixon (1992, 1994), function is a set of causal relationships between physical parameters, as described by the outward physical action of a device. An example is depicted in Figure 9.3.

Behavior is the detailed description of the internal physical action of a device based on established physical principles and phenomena. Functional design is the transition between the three stages. A design problem is stated in terms of a set of functions that must be met. For instance, the conversion of force to displacement is the description of one such design problem. The functional information is transformed by the phenomenological design process to behavior information based on physical principles and phenomena. If the function is conversion of force to displacement, the physical principles of Hooke's law are used to accomplish the function. The embodiment design process, using behavior graphs, models the required behavior as a guide to select and configure systems of embodiments. An embodiment is an abstraction of a physical artifact, such as a spring, gear pair, or electrical motor, that contains not only behavior information but also constraint and evaluation information. In the conversion of force to displacement, a spring could be used to accomplish the function.

A review of the literature indicates that the use of the computer as a design tool (Bracewell and Sharpe, 1996; Chakrabarti and Blessing, 1996; Chakrabarti and Bligh, 1994, 1996; Qian and Gero, 1996) has not changed the primary definition of *function*, while creating new problems in transforming the design information and evaluating alternative design solutions (Peien and Mingjun, 1993; Peien et al., 1996).

9.3 A generic, guideline-based method for functionality

A method has been developed to design consumer products for functionality. It is depicted in Figure 9.4. This method gains importance in light of the absence of design systems that successfully address product functionality and include the design process. The work is split into two distinct phases: development of criteria for functionality and testing and validation of developed criteria and process.

9.3.1 Phase 1. Development of generic criteria for functionality

Current criteria for product functionality are based mainly on product performance. There is a need to consider, during design, downstream manufacturing materials and process variables to ensure product functionality. The current criteria are extended through the following preliminary activities.

The preliminary activities focused on generating generic functionality criteria for product design and manufacturing. A complete, critical review of the research and practice literature, individual experiences, and user complaints with present consumer products (information regarding product returns was collected from leading stores such as Walmart and Kmart) was carried out.

Since different products have different functions and manufacturing processes, a comprehensive list of design and evaluation guidelines across different consumer products is difficult, if not impossible, to generate and validate. This requires extensive study of a large sample of consumer products. For instance, the main function, the manufacturing processes and process variables, and the materials used for manufacturing a coffeemaker are different from those for a car. But it is possible to develop generic product functionality criteria applicable across different consumer products. For example, "safety of the function" is a product functionality criterion applicable to coffeemakers as well as cars. To overcome this problem, broad generic criteria for consumer products were developed, then extensive product design and evaluation guidelines were developed for each functionality criterion for specific products and product families—e.g., a family of coffeemakers.

Information from design handbooks, data from other sources, such as best design and manufacturing practices, designer interviews, and plant visits, were used to generate a comprehensive list of product-specific guidelines for the consumer product family chosen for this research. In addition to these sources for criteria and detailed guideline development, case studies of transformation of product function into

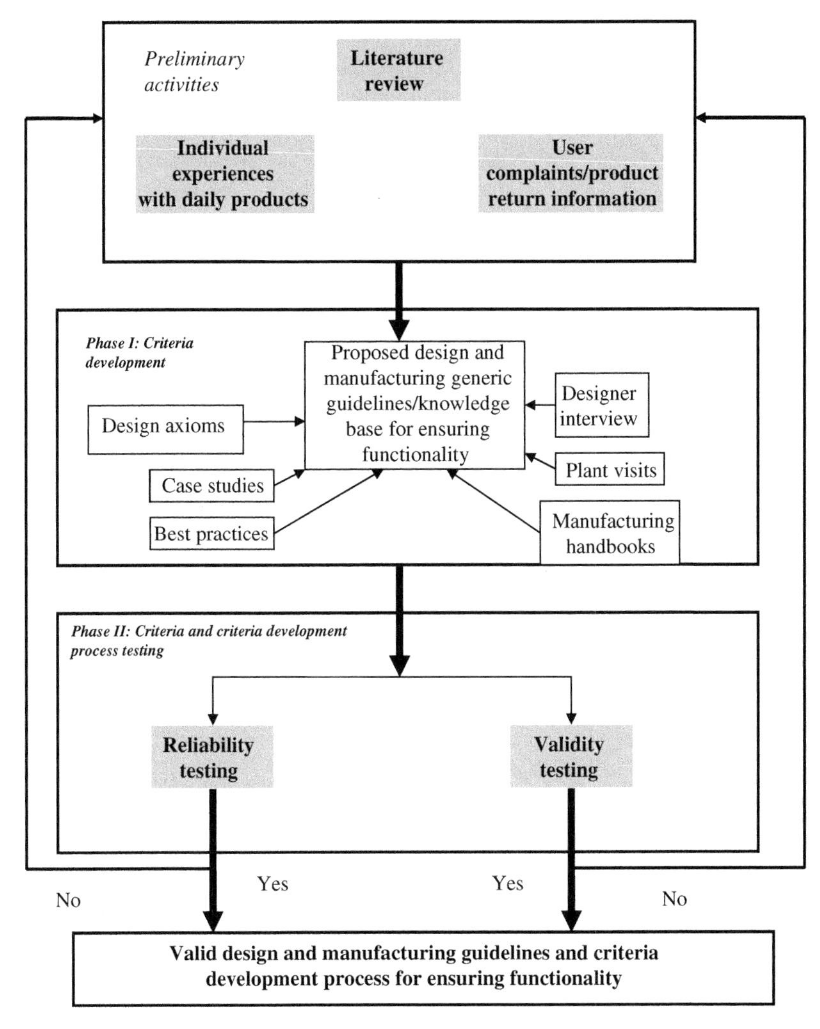

Figure 9.4 The proposed method.

manufacturing process variables were performed for products using transformation matrices (similar to quality function deployment matrices). Sufficient data were generated to deductively reason out generic criteria for functionality for a particular consumer product family. The product family was chosen such that the product was not too simple (such as a can opener) but had a main function and multiple functions supporting the main function. This product family provided a large scope in broadening and extending the traditional definitions of functionality (which is the main objective of this research) to include designer-related and user-related factors, such as reliability of the function and usability and safety of the function.

The systematic process used in developing the functionality criteria and the detailed design and evaluation guidelines for ensuring functionality for a specific product family were modeled. This model can be used by any consumer product designer to ensure functionality for specific products.

An integrated approach to ensure functionality in product design and manufacturing is illustrated in Figure 9.5.

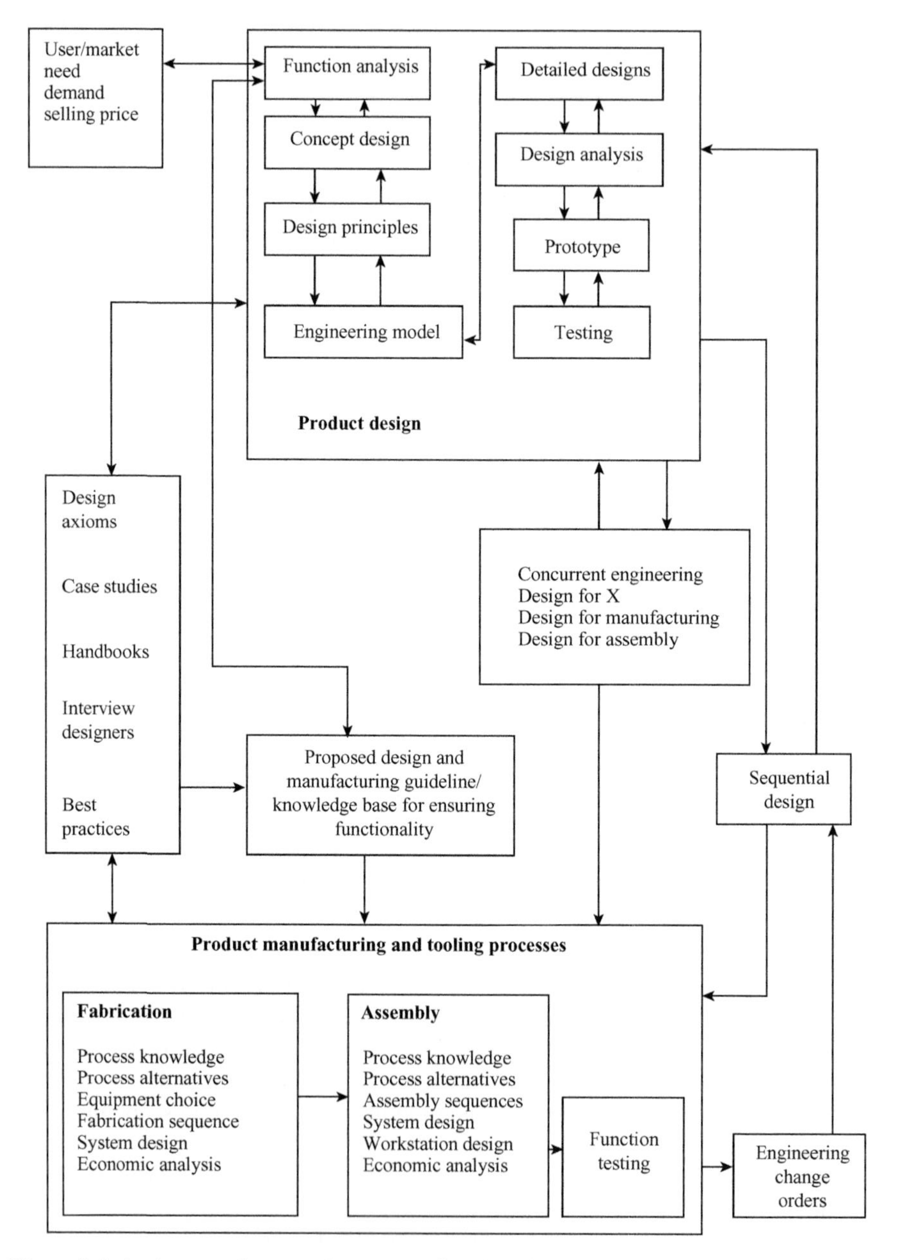

Figure 9.5 An integrated approach to ensure functionality in product design and manufacture.

9.3.2 Phase 2. Validation and testing of developed criteria and processes

The hypothesis of interest in methodology development is whether the new and extended criteria and guidelines, developed as a result of consumer product design and manufacturing information for ensuring functionality, indeed ensure functionality.

Since the final criteria and guidelines were expected to be in the form of design checklists or questions, the goodness of the criteria and guidelines developed as a result of the research were tested using statistical validity and reliability measures. Validity tests how well a technique, instrument, or process measures the particular thing it is supposed to measure. Reliability tests how well or consistently a measuring instrument (or technique or process) measures whatever it is measuring.

For testing the validity of questionnaires, it is standard practice to compare the scores with what is considered standard information. Since the criteria are expected to be new, comparison with a standard to validate the criteria developed in this research was not possible. The validity of the criteria developed in this research for a specific product family was tested by comparing and correlating the overall mean scores for each criterion in the criteria list with the individual item score for each item for each of the functionality criteria.

A high correlation score for the criteria and guidelines for a specific product family implies a high degree of validity for the process used to generate the criteria and guidelines. This developmental process then can be replicated for other products and product families.

Two types of reliability were tested in this research: interitem consistency and interrate consistency. The internal consistency of measures of the homogeneity of the items was tested by computing the Cronbach alpha score for each functionality criteria. Interrate consistency was tested by comparing the responses of designers to specific items in the criteria. Both measures of reliability used the guidelines of product families, and the response from designers of these product families was elicited. If the responses (on a suitable scale such as 1–5 or a Yes/No response) from the designers of different specific products are consistent for the items in the criteria list, then both interitem and interrate consistency (hence, the reliability of the criteria) are expected to be high.

If the reliability and validity of the guidelines are found to be too low, the guidelines are revisited (as indicated by the feedback loop in Figure 9.4) and revised.

9.4 The procedure for guideline development

The main objective was to develop a systematic procedure to generate the design and manufacturing guidelines to ensure product functionality. The major research activities carried out to achieve this objective are as follows:

Step 1 deals with the extension of the definition of *product functionality*. The traditional definitions of *functionality* need considerable extension to include the notion of function in consumer product design. At this point, we use our extended definition as the starting point. We define *function* as "to do something (performance), safely, reliably, in a usable manner, in a high quality manner, with concern for manufacturability and environment friendliness."

Step 2 deals with the development of specific product functionality criteria based on the extended definition of *functionality*. The purpose of this step is to develop a checklist of generic terminology for generating specific product design and manufacturing guidelines. A complete, critical review of the research and practice literature, individual experiences, user complaints with present consumer products, and case studies can achieve this. Since different products have different functions and manufacturing processes, a comprehensive list of specific generic terminologies across different consumer products is difficult to develop. To overcome this problem, in this research, specific generic criteria were developed for each functional criterion, for specific products and product families—e.g., a family of can openers. The important criteria for product functionality were performance, reliability, manufacturability, usability, safety, quality, and environmental friendliness. The assumption was that, in each criterion, we would find important factors (specific functionality criteria) that need to be controlled through design variables, material variables, and manufacturing process variables. Doing so would improve the overall product functionality. Lists of the important factors within each criterion that have links with different design and manufacturing variables that need to be controlled are depicted in Tables 9.1 and 9.2.

Step 3 deals with the classification of each criterion into different design stages (conceptual, embodiment, detailed) for future generation of guidelines. Most DFX tools fail to make a clear distinction as to when and how they should be used but merely provide a list of recommended design rules with little direction on their use. We use the model in Chapter 3 as a descriptor of the product development process. Only the broad phases of task clarification, conceptual design, embodiment design, and detailed design are required for this classification. Some product design guidelines are equally valid over a number of product development phases. These guidelines tend to be more general in nature than those that are applicable during only one phase. However, this methodology focuses on the overall guideline development procedure.

Step 4 deals with the systematic identification of the important design and manufacturing variables that affect product functionality. Design is a natural human activity. The criteria to select such design variables are based on evaluation against functional requirements to determine whether the design variable satisfies the requirements. Some possible design variables that may affect product function include designer experience (novice vs. experienced designer), design tools used (the software and hardware), the type of design (creative vs. adaptive redesign), design budget, and communication mechanisms for parties involved in the design (e.g., the over-the-wall approach vs. concurrent engineering).

Manufacturing variables include both material variables and manufacturing process variables that are closely related. In selecting a material for a product or a component, the primary concern of engineers is to match material properties to the functional requirements of the component. One must know what properties to consider, how these are determined, and what restrictions or limitations should be placed on the application. Some material-related variables that can affect product function

Table 9.1 Factors with links to design and manufacturing variables affecting performance, reliability, usability, and safety

Performance	Reliability	Usability	Safety
Appropriate material	Number of parts	User friendliness	Provision of guards
Function effectiveness	Redundancy	Task simplicity	Avoiding sharp corners
Operating environment	Maintainability	Use mapping	Designing for fail safe
Function performing consistency	Serviceability	Providing feedback	Providing interlocks
Minimum variations in function	Controlling environmental conditions	Good displays	Providing warning devices
Solid base	Diagnosis	Utilizing constraints	Providing safety procedures
Simplicity	Safety factor	Fitting products to users	Abuse by users
Minimal mass/strength ratio	Material strength	Designing controls	Mechanisms to identify source of hazards
Tolerance considerations	Geometric variability	Expecting human errors	Reducing response time
	Wearing out	Avoiding awkward and extreme motions	Providing diagnosis system
	Testability	Reducing learning time	Maintenance and repair considerations
	Protection	Speedy performance	Communication
	Identification of weakest component	Subjective satisfaction	Redundancy
	Loads and capacity	Retention of operative skill over time	Personal protective equipment
	Failure rates	Low rate of errors by users	Rotating/reciprocating parts
	Failure analysis		Flying objects
	Abuse by users		Hazards from gases/liquids
			Explosives
			Understanding designer's response to product liability
			Technical guidelines for safety and safety training
			Continuing responsibility
			Product for children or adults

Table 9.2 Factors with links to design and manufacturing variables affecting quality, manufacturability, and environmental friendliness

Quality	Manufacturability	Environmental friendliness
Consumer wants and needs	Assembly process	Reusable
Product characteristics	Material selection	Recyclables
Critical manufacturing and assembly characteristics	Manufacturing process	Toxic material
Inspection and testing	Standardized designs	Material consumption
Performance data collection	Simplify the design	Energy consumption
Works as it should	Avoid designs requiring machining operations	Manufacturing processes
Lasts a long time	Use materials formulated for ease of manufacturing	Heavy metals
Easy to maintain	Liberal tolerances	Understand DFE principles and design guidelines
Attractive	Manufacturing considerations to avoid sharp corners	Number of parts
Incorporate latest technology	Standardized parts features in minimized numbers	Fasteners
Design and process capability	Using commercially available parts	Disassembly tools consideration
Simplicity	Ease of handling	The number of different materials in a product
Experienced quality control personnel	Avoiding special finishes	
Concurrent design	Designs based on existing products	
	Calculating technical merit	
	State-of-the-art systems and techniques	

significantly include the type of material, material toughness, hardness, and fatigue resistance. The type of material used for a component, in turn, determines the manufacturing process and all manufacturing process dimensions, such as machinability, formability, weldability, and assemblability. Depending on the specific manufacturing process involved in component fabrication, one or more process variables need to be controlled for component and product functionality to be optimal. These variables may include the cutting speed and feed, the depth of cut, the temperature, presence or absence of lubricants, duration of machining, the rate of cooling/heating, current density and voltage, and the type and amount of solvent used.

Design variables are not included because the scope of these variables is too large. The methodology focuses on developing guidelines based only on manufacturing variables.

Step 5 deals with the determination of the links (relationships) between product functionality criteria and design and manufacturing variables. Here we provide a systematic procedure for generating guidelines to ensure product functionality. If the specific generic criteria are obvious, guidelines are introduced to illustrate what needs to be controlled. If the specific generic criteria are not sufficiently obvious, the relationship needs to be illustrated with a different set of variables. Ensuring product functionality is possible only through controlling the design and manufacturing variables within an optimal range. If a relationship between functionality and design attributes and a relationship between the design of a product and its manufacturing attributes can be developed, it is possible to enhance the product's overall functionality. In this step, one case study is presented. This chapter uses the example of a can opener to illustrate the relationship between product functionality on the one hand, and design and manufacturing attributes on the other.

Step 6 deals with the systematic development of the design and manufacturing guidelines incorporating the links identified in step 5, the guidelines for controlling design and manufacturing variables, and the major design activities involved in any design.

The six steps are iterative. We can add new concepts and information at any time and, finally, the guidelines reach an optimal stage. The hypothesis of interest in this procedure is that the new and extended criteria and guidelines, developed as a result of a synthesis of consumer product design and manufacturing information for ensuring functionality, indeed ensure functionality.

9.5 Functionality case study: can opener

This case study illustrates the application of the methodology described in the preceding section to facilitate development of design guidelines for a specific product, a can opener. This case study depicts the links between design/manufacturing factors with various functionality criteria and ways of enhancing them. Functionality transformation matrices have been used to establish such links. The procedure is iterative, and new concepts and information can be added at any time. Development of a comprehensive checklist is not our purpose. We focus on illustration of the methodology.

9.5.1 Can opener architecture

A manual can opener has five main parts: upper handle, lower handle, blade, crank, and the drive sprocket. The upper handle is joined with the crank and drive sprocket to form a subassembly. The lower handle is joined to the blade to form the overall assembly. The upper and lower handles are used for holding the opener and providing gripping force. When mounted properly onto the can and gripped with adequate pressure, the blade pierces the can and the sprocket wheel holds on to the top outside rim of the can. The crank wheel is used to apply torque, which helps the blade cut the can lid and rotate the can until the lid is completely severed.

9.5.2 Can opener manufacturing processes

The principal operations involved in manufacturing a can opener are blanking, piercing, bending, heat treatment, nickel plating, riveting, swaging, and tumbling. The upper handle is cut from SAE 1008 steel strip and the lower handle is cut from SAE 1008 steel wire by a blanking operation. Both handles are individually subjected to stamping operations using progressive dies. The piercing and bending operations produce two holes and a twist in the upper handle. Two protrusions are formed using a die pressing operation on the lower handle. The blade is cut from SAE 1050 steel strip, swaged at the top to produce a sharp cutting edge, bent 90° twice, and pierced to produce a hole at the center and two holes at the bottom using a progressive die in a punch press. The crank is cut from SAE 1008 steel strip and trimmed to achieve the desired shape. These pieces are tumbled to remove the burrs resulting from the stamping operation. The drive sprocket and blade are heat treated. All five parts are nickel plated to promote corrosion resistance and enhance appearance. The drive sprocket and crank are assembled with the upper handle and swaged to manufacture another subassembly. The two subassemblies are riveted together to produce the final assembly.

9.5.3 Guideline development process for the can opener

The generic checklist terminology presented earlier lends itself readily to this case study. It contains some specific generic criteria that can be applied to develop guidelines for the can opener. The following is an example:

- Effectiveness of function (performance)
- Operating environment (performance)
- Redundancy (reliability)
- User friendliness (usability)
- Avoidance of sharp corners (safety)
- Attractiveness (quality)
- Material and process selection (manufacturability)
- Recyclability and avoidance of toxic materials (environmental friendliness).

9.5.4 Identification of important manufacturing variables affecting functionality

Functionality of the can opener can be established based on user requirements and function analysis. The principal functional requirements for a can opener are as follows:

- Smoothness of lid cut
- Slip-free operation
- Ease of cutting
- Safety of handling
- Comfortable grip
- Appearance and durability
- Meet all kitchen can-opening needs
- Be hygienic.

The following functionality features (and the manufacturing processes necessary to achieve them) were determined to be critical:

- **Ease of cutting**: Blade hardness directly influences ease of cutting. Hardness can be achieved by proper heat treatment of the blade as well as appropriate material selection. The temperature and method of quenching constitute the major process variables. Certain parameters normally are associated with heat treatment and can be controlled. For instance, parameters such as decarburization, scaling, cracking, residual stresses, and dimensional changes can be controlled to achieve optimal results.
- **Smoothness of lid cut**: The cutting edge should be sharp enough to rip through the can lid. The sharpness can be achieved by appropriate swaging of the blade's edge. Swaging improves the tensile strength and surface hardness due to improved metal flow and finer grain structure.
- **Slip-free operation**: Slip-free operation is a function of how tightly and smoothly the drive sprocket rolls on the outside of the can rim. The sprocket roll depends on the hardness and wear resistance of the drive sprocket. These properties can be improved by heat treatment and nickel plating.
- **Appearance and durability**: Surface preparation and coating are the main influencing factors as far as appearance and durability are concerned. While the former improves appearance, the latter enhances durability.

9.5.5 Functionality-manufacturing links

Functionality-manufacturing links were obtained by using function transformation matrices (FTMs) similar to quality function deployment matrices and tables. FTMs are used as tools for a structured approach to defining functional requirements and translating them into specific steps to develop the product under consideration. It allows functional requirements to be taken into account throughout all stages of product design.

Transformation matrices use a series of relationship matrices to document and analyze relationships among various factors. While the details of the matrices vary from one stage to the next, the fundamentals remain the same. In the conceptual design stage, functional requirements are identified and translated into design and technical requirements. Product deployment is the second stage of the transformation process. Its purpose is to translate the previously developed design and technical requirements into product specifications and features. During the process deployment stage, various product features are converted into specific manufacturing operations. During the manufacturing deployment stage, various manufacturing processes are related to specific operations and the material variables that control them. It is possible to ensure and control product functionality by adopting the stepwise transformational approach as just detailed.

9.5.5.1 Design and technical requirements deployment

Functional requirements are listed on the horizontal portion of the first stage of the FTM process (Table 9.3). The functional requirements are based on our extended function definition. The can opener manufacturing processes are not complex. Hence,

Table 9.3 Functional requirements: design and technical requirements transformation for a can opener

Functional requirements	Design and technical requirements						
	Structural rigidity	**High force of blade on lid**	**Contact force of sprocket on rim**	**Low handle grip force**	**Low crank torque**	**Smooth surface and edge finish**	**Good rust-proofing**
Performance	S	S	S	M	M		
Reliability	S	M	M	W	W	M	M
Manufacturability	M	S	S			M	M
Safety	M					M	W
Usability	S	S	S	S	S	M	W
Quality	M					S	S
Environmental friendliness		W					

Relationships: Strong (S) =5; Medium (M) =3; Weak (W) =1.

the entire function requirements are transferred in one transformation matrix. For complex components, each definition of functionality is transferred separately.

The functional requirements are translated into vocabulary the organization can use to describe its product for design, processing, and manufacture. The objective of this step is to develop a list of design and technical requirements that should be worked on to satisfy functional requirements.

Relationships between design and technical requirements and functional requirements are established next to identify the relative importance of various design requirements. Every functional requirement in the horizontal portion is compared with design requirements on the vertical portion. The degree of relationship is marked at the intersection; the degree of relationship (strong, moderate, weak) is in accordance with the key as depicted in Table 9.3.

9.5.5.2 Product deployment

Product deployment is the second stage of the transformation process. In this stage, the design and technical requirements taken from the vertical column of the previous stage are listed (Table 9.4). Based on previous design experience, the product features needed to satisfy these design and technical requirements are identified and listed in the vertical column. The degree of relationship is identified as before.

9.5.5.3 Process deployment

Process planning is the third part of the transformation process (Table 9.5). Its purpose is to determine the manufacturing processes that actually produce the product, by relating various product features to specific manufacturing operations. The critical

Table 9.4 **Design and technical requirements: product features transformation for a can opener**

Design and technical requirements	Product features					
	Blade hardness	**Drive sprocket hardness**	**Material**	**Surface and edge regularity**	**Crank surface and edge regularity**	**Surface finish**	**Joints between components**
Structural rigidity		S					S
High cutting force	S	S	S				M
High contact force		S	S				M
Low grip force	S	S	S	W	W		S
Low crank torque	S	S	M	W	W		
Smooth surface and edge finish			M	S	S	S	
Good rust-proofing	M		M	M	M	S	

Relationships: Strong (S) =5; Medium (M) =3; Weak (W) =1.

Table 9.5 **Product features/process features transformation for a can opener**

Product features	Process features															
	Blanking (upper handle)	Stamping (upper handle)	Blanking (lower handle)	Stamping (lower handle)	Blanking (blade)	Swaging (blade)	Heat treatment (blade)	Blanking (crank)	Stamping (crank)	Stamping (drive sprocket)	Heat treatment (sprocket)	Tumbling (edge and surface)	Nickel plating	Swaging (sub-assembly)	Swaging (sub-assembly)	Riveting (overall assembly)
High blade hardness					W	M	S									
High drive sprocket hardness										W	S					
Smooth handle surface and edge	S	M	S	M								S	M			
Smooth crank surface and edge								S	M			S	M			
Good surface finish												W	S			
Strong joint between parts														S	S	S

Relationships: Strong (S) =5; Medium (M) =3; Weak (W) =1.

product feature requirements identified in the previous stage are listed in the horizontal portion of the matrix. The major process elements necessary to develop the product are extracted from the process flow diagram and are shown at the top of the column section of the matrix. The processes that have the most influence in the manufacture of the can opener are the progressive die operations, heat treatment, and inorganic surface coating.

9.5.5.4 Manufacturing deployment

Manufacturing planning is the culmination of the work done in the previous stages. In this particular stage, the various manufacturing techniques necessary to make the product are related to process attributes that affect them (Table 9.6). For instance, the hardness of the blade is affected by the rate of cooling during the heat treatment process. The rate of cooling, in turn, is controlled by properties of the quenching liquid. Although the manufacturing process used for producing the can opener is affected by numerous process variables, only variables that affect the can opener's functionality are considered in this instance. The manufacturing techniques are listed in the horizontal portion and process variables are listed in the vertical portion of the FTM (Govindaraju, 1999). Through an FTM analysis, a clear progression of the relationships linking product functionality features and manufacturing variables that affect manufacturing are established. The FTMs indicate that the overall functionality of a product can be enhanced by adopting an optimum range of values for process variables. Hence, manufacturing variables that affect highly ranked functionality features must be tightly controlled to enhance product functionality.

Based on the final results, design and manufacturing guidelines can be developed, incorporating the links and guidelines for controlling the variables that directly affect the result. For instance, a process feature from Table 9.6 indicates the necessity of the tumbling process. In accordance with this requirement, one guideline could be stated as follows: for upper handle and lower handle surface finishing (tumbling process), care needs to be taken to control rotation speed. If this is the case, only a small fraction of the load is finished at any time, resulting in long process times. Increasing rotation speed improves the processing time, often at the cost of quality. Some of the user checklists and design/manufacturing guidelines for this case study are listed in the following section.

9.5.6 Survey development

A survey was conducted to test the functionality evaluation and design/manufacture guidelines. The survey development was based on the preceding section of this paper. Two sets of questionnaires were developed: one for user evaluation and the other for designer/manufacturing engineering. The responses to the user evaluation survey were obtained through a one-on-one interview with 21 individuals who participated in the study. Each survey lasted approximately 20 min and was conducted at the home of the individual. All individuals owned manual can openers and were very familiar with the main function of the device. The individuals also were asked to rate answers based on a scale of 1–5 (least important to most important).

Table 9.6 **Process features/process variables transformation for a can opener**

Process features	Process variables									
	Pressure	Temperature	Chemical concentration	Duration	Quench solution	Operation speed	Punch-die clearance	Rate of cooling	Rivet-hole clearance	Current density
Blanking	S					M	S			
Stamping	M					S	S			
Swaging	S					S				
Heat treatment		S	S	S	S				S	
Tumbling			M	M		M				
Nickel plating			S	M		W				S
Riveting	S					W			S	

Relationships: Strong (S) =5; Medium (M) =3; Weak (W) =1.

Table 9.7 **User evaluation checklist for performance and reliability**

	Opinion					Comments
	1	2	3	4	5	
Performance						
1. To what extent are you concerned with the force you need to apply to open the can? 2. To what extent are you concerned with the structural rigidity of the can opener when you open the can? 3. To what extent are you concerned with the force needed on the lower/upper handle to make blade pierce the lid? 4. To what extent are you concerned with a comfortable feeling of the grip of the handle bars while opening the can? 5. To what extent are you concerned with the lid or the can having sharp or jagged edges? 6. What is the overall functionality of the can opener in terms of its design for performance in this section?						
Reliability						
1. To what extent are you concerned with the blade failing often to pierce the lid? 2. To what extent are you concerned with the guiding wheel failing often while opening the can? 3. To what extent are you concerned with the joint (rivet) failing (breaking) often? 4. To what extent are you concerned with the joint loosening (loss in structural rigidity) often? 5. To what extent are you concerned with the nickel plating of the can opener wearing out too soon (corrosion resistance and comfortable grip)? 6. What is the overall functionality of the can opener in terms of its design for reliability in this section?						

The design/manufacturing guideline questionnaires were sent to designers and manufacturers by either mail or fax. Background information was collected on each designer using designer profile forms. Completed questionnaires were analyzed to study the results. An exhaustive list of checklists is presented in Tables 9.7–9.18.

9.5.7 Statistical analysis and testing

A hypothesis test was performed to study the interrelation between user evaluation questionnaires and design and manufacturing questionnaires. The purpose of hypothesis testing was to help draw conclusions regarding population parameters based on results observed in random samples. The null hypothesis in this case is that there is no difference between design/manufacturing guidelines and user evaluation

Table 9.8 User evaluation checklist for usability

	Opinion					Comments
	1	2	3	4	5	
1. To what extent are you concerned with the can opener blade easily piercing the lid?						
2. To what extent are you concerned with the can opener's grip for comfort?						
3. To what extent are you concerned that the can opener should perform the function fast and safely?						
4. To what extent are you concerned that the can opener should never leave a sharp or jagged edge on either the lid or the can?						
5. To what extent are you concerned that the upper and lower handle shape should be round, with smooth transitions, and follow the contour of the human hand?						
6. To what extent are you concerned that the can opener should cut smoothly and quietly?						
7. To what extent are you concerned that the can opener should appear small, compact, attractive, and modern?						
8. Do you prefer that the can opener be able to open the entire range of US market cans?						
9. To what extent are you concerned that the can opener should be easy to operate, maintain, clean, and meet all kitchen can-opening needs?						
10. To what extent are you concerned that the cutter wheel or blade never comes in contact with the food?						
11. What is the overall functionality of the can opener in terms of its design for usability?						

checklists. A *t*-test was conducted to compare the means between the average scores of the functionality evaluation questionnaires and the average scores of the design/manufacturing questionnaires. If the design/manufacturing guidelines are useful in developing a more functional product, the outcome of functionality evaluation for different can openers should correspond with the outcome of the design and manufacturing guidelines implementation. On an aggregate level, for a given sample of users and designers, the means should not be significantly different and the hypothesis should not be rejected if any relationship exists. Checklists were returned by 24 users and 16 designers/manufacturers. Statistical analyses and a comparison of analysis were performed. Reliability also was tested by measuring the internal consistency of the score for each questionnaire. The Cronbach alpha values for the seven sections of design/manufacturing checklist as well as the six sections of user evaluation checklist are presented in Table 9.19.

Since a value of 0.4 was found to be adequate, not all design/manufacturing guidelines were found to be reliable. As for the user evaluation guideline criteria, all

Table 9.9 User evaluation checklist for safety

	Opinion					Comments
	1	2	3	4	5	
1. To what extent are you concerned that the burrs on the upper and lower handle can pose a safety problem to the user? 2. To what extent are you concerned that the can opener leaves a sharp and jagged edge on either the can or the lid? 3. To what extent are you concerned that the can opener should provide guards or covers over sharp blades and similar elements? 4. To what extent are you concerned that the can opener should be free from sharp edges, corners, and points that can cause injury? 5. To what extent are you concerned that the can opener should prevent toxic contamination of metal shaving from falling into the food? 6. To what extent are you concerned that the cutter wheel or blade never comes in contact with the food? 7. What is the overall functionality of the can opener in terms of its design for safety?						

Table 9.10 User evaluation checklist for quality

	Opinion					Comments
	1	2	3	4	5	
1. To what extent are you concerned that the can opener should appear small, attractive, and have modern styling? 2. To what extent are you concerned that the can opener should have a good esthetic appearance, bright and easily maintained surface, and a modern and attractive appearance? 3. To what extent are you concerned that the metallic parts of the can opener be rust-proof and corrosion proof to give a feeling of cleanliness, healthiness, and elegance? 4. To what extent are you concerned that the can opener should have considerable long-term value? 5. What is the overall functionality of the can opener in terms of its design for quality?						

Table 9.11 User evaluation checklist for environmental friendliness

	Opinion					Comments
	1	2	3	4	5	
1. To what extent are you concerned that the can opener as much as possible should avoid the use of toxic materials in the product and manufacturing process?						
2. To what extent are you concerned that the can opener should avoid the use of hazardous materials including those that are a hazard when burned, recycled, or discarded?						
3. Do you agree that the can opener should minimize the amount of material in the product?						
4. What is the overall functionality of the can opener in terms of its design for environmental friendliness?						

Table 9.12 Design and manufacturing checklist for performance

	Opinion					Comments
	1	2	3	4	5	
1. Designer of the can opener should select an appropriate material that meets the functional requirements of the users along with various product features.						
2. Design should consider low crank torque and low handle grip force to open the can						
3. For structural rigidity and durability, use two rivets to join the upper handle and lower handle						
4. Design should consider force analysis. The cutting edge of the blade and the crank of the guiding wheel should have a shear stress greater than the maximum material strength of the lid						
5. Design should consider all components strong enough to transmit and resist forces during opening of the can						
6. Removing the burrs from stamping operations on the upper handle, lower handle, blade, and crank is important. Burrs can jam parts, reduce the fatigue life of components, and be a safety hazard						
7. Design of the blade should consider heat treatment so that it never leaves a sharp or jagged edge on either the lid or the can						
8. What is the overall functionality of the can opener in terms of its design for performance?						

Table 9.13 Design and manufacturing checklist for reliability

	Opinion					Comments
	1	2	3	4	5	
1. Design should consider using high-grade metal to ensure product durability						
2. Design of can opener should simplify the design and minimize the number of parts						
3. Design should consider operating conditions by using lower alloy grades that resist corrosion in atmospheric and pure water environments						
4. To ensure reliability of the can opener, heat treatment considerations are method selected, care taken in quenching, and proper choice of quenching method, media, and temperatures						
5. The design of joints of the upper and lower handles requires consideration of the type of loading, such as shear and tension, to which the structure will be subjected and the size and spacing of holes						
6. Design of can opener should consider the use of standard parts and materials						
7. Design should consider impact resistance through the use of materials providing high toughness at temperatures ranging from high to below freezing						
8. What is the overall functionality of the can opener in terms of its design for reliability?						

Table 9.14 Design and manufacturing checklist for manufacturability

	Opinion					Comments
	1	2	3	4	5	
1. Design should consider ease of fabrication, using state-of-the-art techniques for cutting, welding, forming, machining, and fabricating materials						
2. The blade of the can opener should use steel with a carbon content between 0.48% and 0.55%						
3. Design should consider work hardening property of austenitic steel grades that result in a significant strengthening of the material from cold working alone						
4. Design should consider relaxing tolerances in which dimensional deviations do not matter much to decrease manufacturing time and make the process easier without interfering with part functionality						

(*Continued*)

Table 9.14 **(Continued)**

	Opinion					Comments
	1	2	3	4	5	
5. Design should consider the maintenance of specified tolerances vital to the function of subassemblies and interchangeability; excessively tight tolerance and surface finish specifications lead to high manufacturing costs						
6. For upper and lower handle surface finishing, care should be taken to control rotation speed						
7. Compatibility of the fastener material with that of the components to be joined is important. Incompatibility may lead to corrosion						
8. Design of joints of the upper and lower handles should use standard sizes whenever possible						
9. The cutting edge of the can opener has a significant relationship with joint strength and life						
10. When designing the joints of the upper and lower handles, holes should not be located too close to the edges or corners to avoid tearing the material when subjected to external forces						
11. What is the overall functionality of the can opener in terms of its design for manufacturability?						

Table 9.15 **Design and manufacturing checklist for safety**

	Opinion					Comments
	1	2	3	4	5	
1. Burrs resulting from stamping operations on the upper handle, lower handle, blade, and crank should be removed, since they can jam parts, reduce fatigue life, and be a safety hazard						
2. Blade design should consider heat treatment and cutting edge sharpening to never leave a sharp or jagged edge on the can or the lid						
3. Design of can opener should consider providing guards or covers over sharp blades and similar elements						
4. Can opener design should avoid sharp corners, generous radii should be incorporated whenever possible						
5. Design should prevent toxic contamination of food by preventing metal shavings from falling into food while opening the can						
6. Design should consider the depth of cut so the cutter wheel never comes in contact with food						

(*Continued*)

Table 9.15 **(Continued)**

	Opinion					Comments
	1	2	3	4	5	
7. Give careful attention to strength of all parts whose failure might result in user injury. Allow a reasonable factor of safety for stressed or otherwise critical components						
8. What is the overall functionality of the can opener in terms of its design for safety?						

Table 9.16 **Design and manufacturing checklist for quality**

	Opinion					Comments
	1	2	3	4	5	
1. Design of can opener should be small, compact, attractive, and modern						
2. Design should consider esthetic appearance, using bright, easily maintained surfaces of stainless steel to provide a modern and attractive appearance						
3. The metallic parts of the can opener should be corrosion proof to give a feeling of cleanliness, hygiene, and elegance						
4. Design of can opener should consider long-term value, stainless steel is usually the least expensive material option						
5. The heat treatment of blade and guiding wheel should consider avoiding problems such as cracking, distortion, and nonuniform properties throughout the heat treated part						
6. Burrs resulting from stamping operations on the upper handle, lower handle, blade, and crank should be removed						
7. Design should consider all components of the can opener geometrically related in extent and position						
8. Design of can opener should consider using standard parts and materials						
9. Surface finishing operation should be directed to producing a surface that is within tolerances, has the proper roughness and texture, and is free from damage						
10. Before any surface treatment operation is carried out for decorative or protective purposes, a preliminary step is the removal of sand and scales						
11. What is the overall functionality of the can opener in terms of its design for quality?						

Table 9.17 Design and manufacturing checklist for usability

	Opinion					Comments
	1	2	3	4	5	
1. Design should consider usability features such as ease of piercing, comfortable grip, and smooth, quiet, safe, and fast cutting 2. Design of the blade should consider heat treatment so it never leaves a sharp or jagged edge on the lid or can 3. The shape of the upper and lower handles should be round with smooth transitions 4. The can opener should be small, compact, attractive, and modern 5. Can opener should be able to open the entire range of cans on the market 6. Can opener should be easy to operate and maintain 7. Cut should be deep enough to cut cleanly but shallow enough that it never comes in direct contact with the food 8. Design should consider hygiene and should use material such as stainless steel 9. What is the overall functionality of the can opener in terms of its design for usability?						

Table 9.18 Design and manufacturing checklist for environmental friendliness

	Opinion					Comments
	1	2	3	4	5	
1. The use of toxic materials should be avoided to the greatest extent possible in the product as well as the manufacturing processes 2. Use of hazardous materials should be avoided in product design 3. The use of material should be minimized in the product design 4. Process should be designed to minimize manufacturing residue, such as mold scrap and cutting scrap 5. Design the fasteners of the can opener for easy access to aid in disassembly 6. Minimize use of liquids such as acids, alkalis, and solvents during the manufacturing process 7. What is the overall functionality of the can opener in terms of its design for environmental friendliness?						

Table 9.19 Reliability test values for designer and user checklists

Functionality criteria	Cronbach coefficient alpha
Designer/manufacturing checklist	
Performance	0.2259
Reliability	0.5334
Manufacturability	0.7959
Safety	0.3531
Quality	0.7789
Usability	0.2507
Environmental friendliness	0.6407
User checklist	
Performance	0.3619
Reliability	0.6821
Manufacturability	0.5313
Safety	0.6265
Quality	0.5217
Environmental friendliness	0.7826

Table 9.20 Comparison of mean scores between designer/manufacturing checklist and user checklist

Section	Designer/manufacturer checklist ($n = 16$)		User checklist ($n = 24$)		t-test	
	Mean	**SD**	**Mean**	**SD**	t**-value**	p**-value**
Performance	4.4375	0.2684	4.3229	0.4191	0.967	0.34
Reliability	4.2768	0.3439	4.3917	0.5258	−0.769	0.446
Safety	4.4732	0.2486	4.2917	0.4771	1.397	0.171
Quality	4.2306	0.4242	3.8646	0.6030	2.101	0.042
Usability	4.3828	0.2765	4.2083	0.3623	1.634	0.110
Environmental friendliness	4.5417	0.3305	4.3472	0.8193	0.899	0.374

sections are reliable except the performance section. The reasons for unreliability are discussed in the following section.

To test the validity of any questionnaire, a standard questionnaire is generally compared with the test questionnaire under consideration. Here, this facility is not available. Validity, therefore, was tested by comparing the average scores of the questionnaire items with the overall score of that questionnaire. Correlation analysis was performed between the mean score of all the questionnaire items and the overall score of each questionnaire. The correlation coefficients and significance values of the design/manufacturing and user checklists are listed in Table 9.20.

Among the 13 sections of the two checklists, 11 sections have correlation coefficients over 0.4 and 2 sections have a low correlation coefficient, below 0.4. According to the rule of large correlation, a coefficient of 0.19 and below is very low, 0.20–0.39 is low, 0.40–0.69 is modest, 0.70–0.89 is high, and 0.90–1 is very high. The correlation between the overall score and average score in most sections is modest or high, while in the safety section for the designer/manufacturing checklist and the usability section for the user checklist, the correlation between the overall score and average score is low. The correlation between the average score and overall score in designer/manufacturing checklist is highly significant at the 0.01 level as far as reliability and environmental friendliness are concerned, whereas the correlation of performance is significant at the 0.05 level, but attributes such as quality, safety, and usability are not significant. The correlation between the average score and the overall score in the user checklist is highly significant at the 0.01 confidence level for performance, manufacturability, safety, quality, and environmental friendliness; however, the correlation for usability is insignificant ($p = 0.348$).

To put it succinctly, from the perspective of the designer/manufacturing checklist, the performance, reliability, manufacturability, and environmental friendliness sections are valid, while the quality and usability sections fail to test positive for validity due to low correlation. From the perspective of the user checklist, the performance, reliability, safety, quality, and environmental friendliness sections are valid, while usability is not valid, since the correlation is not only low but also insignificant.

9.5.8 Hypothesis test results

It is hypothesized that, if all can openers are analyzed collectively, the mean scores of sections of the user checklist should not be significantly different from the scores of the designer/manufacturing checklist. A *t*-test was performed between the mean score of the two checklists for the same section. Since the manufacturability section is a special questionnaire for the designer/manufacturer, no *t*-test was done on this section. The comparison of mean scores between the designer/manufacturing checklist and the user checklist is evaluated in Table 9.20.

Comparison between the two checklists indicates that the hypothesis could not be rejected ($p > 0.05$) for all attributes except quality. The comparison for quality is significant ($p = 0.042$), hence, the hypothesis is rejected in this case. This means that there is a significant difference between the mean scores of the two checklists. Therefore, we conclude that, except for quality, the general opinion is compatible on all other aspects of functionality.

9.5.9 Discussion of the results

9.5.9.1 Discussion of the reliability test

The Cronbach alpha values for all sections except performance, safety, and usability were over 0.4, considered an acceptable value, and hence reliable. The other sections

Table 9.21 **Item total statistics for the performance section**

Question	Scale mean if item deleted	Alpha if item deleted
1	25.58	0.2104
2	25.58	0.2967
3	25.42	0.5563
4	25.75	0.2411
5	25.58	0.3692
6	25.50	−0.1467
7	25.58	−0.1877

did not meet this standard (for the reason described in considering performance as a case in point). The performance section of the designer checklist scored a value of 0.2259, considered unreliable. To determine means of increasing the reliability of the performance section, a detailed reliability analysis was performed. The test result is presented in Table 9.21.

Table 9.21 illustrates the relationship between individual questions and the composite score. It is evident that, if question 3 is deleted, the alpha value increases to 0.5563, which is an acceptable value to evaluate the reliability of the section. It is possible that question 3 contains some inherent ambiguity, thereby affecting the reliability of the entire section. We conclude, therefore, that to improve reliability of this particular section, the third question needs to be modified.

9.5.9.2 *Discussion of the validity test*

From the preceding sections, it will be seen that the Pearson's correlation values for almost all sections are acceptable ($r > 0.4$). However, they are not acceptable for the safety section of the designer checklist and the usability section of the user checklist. For 9 of the 11 sections, which have acceptable correlation values, the probability that a correlation coefficient of at least 0.4 is obtained when there is no linear association in the population between the overall value and the average value is 0.05. For the quality and usability sections in the designer checklist, the significance is 0.121 and 0.083, respectively. This indicates a probable lack of association between the populations of overall and average values. It cannot be argued that this section is not valid owing to unacceptable correlation values and significance. It does indicate, however, that the overall and average values are uncorrelated. A small percentage of the population being reviewed did not agree with some of the questions on the section, despite its perceived usefulness. In yet another instance, the subjects objected to the entire section. This disjoint is responsible for the lack of coherence between the overall and average values. This indicates that, to illustrate actual validity, explanation of overall value needs to be enhanced.

9.5.9.3 Discussion of the comparison between the two checklists

The means between the average values of the two checklists were compared. Theoretically speaking, the means should not be significantly different if the checklists are related for a given sample of users and designers. The means were not found to be significantly different from each other for each of the categories except quality. This may indicate that different opinions may exist between designers and users as to what constitutes acceptable quality. It can be read as a conflict between business interests and pragmatism. For instance, designers tend to improve can opener quality in an effort to boost sales. However, while some people may consider those very features, other users may prefer performance and usability of the can opener rather than its quality.

9.6 Functionality case study: automotive braking system

In this section, we focus on the detailed manufacturing processes to ensure product functionality. The functionality transformation matrix deals with details at the component level, not the entire product. The reason for this is that, if we focus on the whole product, only higher level conceptual guidelines can be generated. This being a case-based illustration, its purpose is to develop detailed guidelines to ensure product functionality. The product chosen for the purpose is an automotive braking system.

9.6.1 The function of an automotive braking system

Automotive brakes require attention more often than most other units of a vehicle. An understanding of the requirements of braking systems of automotive vehicles requires knowledge of the following:

- The purpose of brakes
- An appreciation of their contribution to safety
- Recognition of the factors controlling the stop
- An understanding of braking action
- An appreciation of possible stopping distances.

The basic functions of a braking system include decelerating a vehicle, including stopping, maintaining vehicle speed during downhill operation, and holding a vehicle stationary on a grade.

The function of a braking system is to enable the user to stop the car whenever necessary, safely, usably (the system has to withstand abuse from the user, depending on the type of driver), reliably, simply, efficiently, and in the least expensive manner possible.

The safe operation of a motor vehicle requires continuous adjustment of its speed to changing traffic conditions. The braking system must perform safely under a variety of operating conditions including slippery, wet, and dry roads; when a vehicle is lightly or fully laden; when braking straight or in a curve; with new or worn brake linings; with wet or dry brakes; when applied by a novice or experienced driver; and when braking on smooth or rough roads or when pulling a trailer.

9.6.2 The components of an automotive braking system

The typical braking system has many subsystems or components, such as the brake pedal, master cylinder, wheel cylinder, hydraulic lines, and flexible hose. We chose one subsystem (wheel cylinder) to further illustrate the potential links between functionality attributes and design and manufacturing variables.

The main components of a wheel cylinder are the cylinder body, piston, cup seal, return spring and cup expanders, dust boot, and bleeder screw (Figure 9.6). We concentrate on the cylinder body manufacturing, since it involves almost all manufacturing processes. The other parts are considered standard and are not dealt with in this chapter.

9.6.3 Wheel cylinder architecture

A typical wheel cylinder (Figure 9.6) has two opposed pistons. Each brake shoe fits into a slot at the outer end of the piston. As the brake pedal is depressed, the master cylinder forces fluid along the brake lines to the wheel cylinders, where it enters between the two pistons. Pressure is exerted between the two pistons, forcing the shoes outward against the drum. Leakage of fluid is prevented by the rubber cups between the piston and the fluid. A spring between the two piston cups sets them firmly against the piston at all times. Each cylinder is provided a bleeder valve to permit the removal of any air in the hydraulic system.

9.6.4 Wheel cylinder manufacturing processes

The process description of the cylinder body was obtained from the manufacturer. The first step is to form the cylinder shape. For cost considerations, function requirements, and manufacturability, casting is the preferred process for this step. The materials can be aluminum AC8B, AC2B, or casting iron, using die casting to manufacture. The next step is boring to enlarge the hole by the previous casting process and provide the working allowance for the final surface finishing. A lathe can then be used to

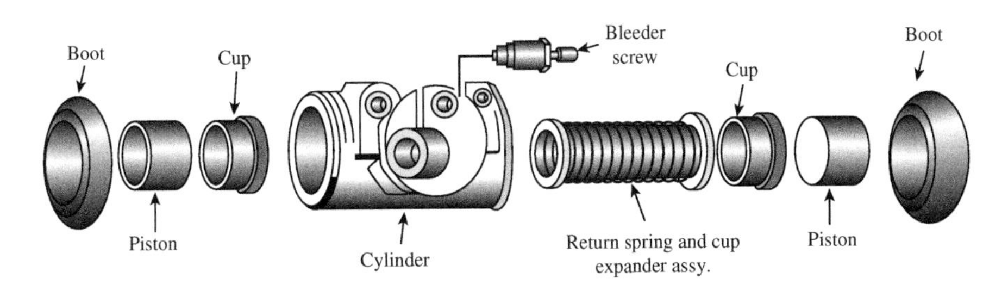

Figure 9.6 Components of a wheel cylinder.

manufacture two side faces and taper for the cylinder body and to produce the slots for installing the dust boot. After this step, the drilling and tapping processes can be used to manufacture the fluid inlet and bleeder on the cylinder body. A drilling machine can be used for drilling holes and tapping. For the cylinder body, the internal cylindrical surface finish is obtained using a honing process. Honing is one of the best operations performed on the surface of the cylinder body. To remove the dust resulting from previous operations, an ultrasonic cleaning process is used. Making all surfaces corrosion resistant is the final stage of the manufacturing process.

9.6.5 Guideline development procedure for the automotive brake system

Step 1 essentially is the same as for the can opener case study. The extended definition of functionality can be used as the starting point for step 1.

In step 2, some specific generic criteria can be applied to develop the guidelines for this case study. The checklist-type generic terminology developed earlier in this chapter is a good starting point. The modifications can be done at any time. The following is a list of potential criteria:

Appropriate material (performance)
Effectiveness of function (performance)
Operating environment (performance)
Tolerance considerations (performance)
Redundancy (reliability)
Maintainability and serviceability (reliability)
Control the environmental conditions for product use (reliability)
Diagnosis (reliability)
Material strength (reliability)
Wear out (reliability)
Protection (reliability)
Identification of the weakest component of the product (reliability)
Loads and capacity (reliability)
Abuse (reliability)
User friendliness (usability)
Providing feedback (usability)
Avoid awkward and extreme motions (usability)
Speedy performance (usability)
Designing for a "fail-safe" mode of operation (safety)
Anticipating the environment in which the product will be used (safety)
Provision of warning devices (safety)
Response time reduction (safety)
Provision of diagnosis system (safety)
Maintenance and repair considerations (safety)
Redundancy (safety)
Customer wants and needs (quality)
Critical manufacturing and assembly characteristics (quality)
Inspection and testing (quality)
Performance data collection (quality)

Satisfying intended function (quality)
Long life (quality)
Latest technology (quality)
Design and process capability (quality)
Assembly process (manufacturability)
Material selection (manufacturability)
Manufacturing process (manufacturability)
Standardized designs (manufacturability)
Avoiding designs requiring secondary machining operations (manufacturability)
Using materials formulated for easy manufacture (manufacturability)
Using liberal tolerances (manufacturability)
Manufacturing considerations to avoid sharp corners (manufacturability)
Using commercially available parts (manufacturability)
Ease of handling (manufacturability)
Avoiding special finishes (manufacturability)
Designs based on existing products (manufacturability)
Calculating technical merit (manufacturability)
State-of-the-art systems and techniques (manufacturability)
Reusability (environmental friendliness)
Recyclability (environmental friendliness)
Toxic materials (environmental friendliness)
Material consumption (environmental friendliness)
Energy consumption (environmental friendliness)
Manufacturing processes (environmental friendliness)
Heavy metals (environmental friendliness)
Understand DFE principles and design guidelines (environmental friendliness)
Disassembly tools consideration (environmental friendliness)
The number of different materials in a product (environmental friendliness).

The procedure to be followed in step 3 is identical to the procedure outlined earlier.

A complete example illustrating step 4, how to identify the important design and manufacturing variables that affect functionality of a system, was presented in the previous case study.

9.6.6 Functionality-manufacturing links

The functional requirements for the cylinder body require that it contain fluid and transfer fluid pressure to the pistons. To accomplish these requirements, other product characteristics, such as dimensional accuracy, long life, manufacturability, avoidance of wear, and ease of assembly, also are required.

9.6.6.1 Design and technical requirements deployment

The purpose of the functional transformation matrix is to identify the relationship between design attributes and manufacturing variables. If only a few manufacturing processes are involved in product manufacture, a single chart for each stage is sufficient. However, for a complex system, there may be a need to transfer one criterion at a time or two criteria simultaneously.

Here, one criterion (reliability) is used to demonstrate transfer from functional requirements to design and technical requirements. The results are depicted in Table 9.22. The criteria that affect the cylinder body reliability are chosen. They are as follows:

Material strength
Wear
Cleaning
Safety
Protection
Maintainability and serviceability
Load and capacity
Identification of the weakest components of the product.

The functional requirements for reliability are listed in the horizontal portion of the first stage on the FTM process. The functional requirements (reliability) were translated into appropriate design and technical requirements as depicted in Table 9.22. The main function of the cylinder body is to contain fluid and transfer fluid pressure. It should have sufficient structural rigidity to resist pressure and be durable as well. The degree of internal surface finish of the cylinder body has a significant effect on wear resistance. It also protects the cup (seal). In general, cleanliness is essential for effective application of metal working fluids, coating and painting, adhesive bonding, welding, brazing, and soldering. Finally, it is necessary to provide a watertight seal to prevent dust, water, and other external elements from entering the wheel cylinder. Table 9.22 illustrates the relationship between functional requirements and design requirements for the cylinder body.

9.6.6.2 Product deployment

During the next stage, the conceptual design of the product is chosen to implement the technical requirements listed previously. This involves the functional mechanism, the technical component subassemblies related to this function, and the product architecture (Table 9.23).

9.6.6.3 Process deployment

The product features developed through the selection of the concept design can be implemented only through the appropriate selection of process features, such as materials, machines, and tools (Table 9.24).

9.6.6.4 Manufacturing deployment

The manufacturing processes for a wheel cylinder are material–casting–boring–face turning–taper turning–slot cutting–drilling–tapping–honing–ultrasonic cleaning–rust prevention and proofing (Table 9.25).

The processes and machine tools used to manufacture the components should be tightly controlled to achieve the desired quality. The desired goal of producing

Table 9.22 **Functional requirements: design and technical requirements transformation matrix for a braking system (cylinder body)**

Functional requirements	Design and technical requirements							
	Structural rigidity	**Smooth inner wall surface**	**High corrosion resistance**	**Good dimensional control**	**Release residual stress**	**Good rust-proofing**	**Smooth surface and edge finish**	**Material selection**
Material strength	S				M			S
Wear		S	S	M		S		S
Cleaning			M			S	M	
Maintainability and serviceability			M	S	W	M		
Safety factor	S							S
Protection			S			S		
Identify weakest product component		M	M			M		M
Loads and capacity	S							S

Relationships: Strong (S) =5; Medium (M) =3; Weak (W) =1.

Table 9.23 **Design and technical requirements: product features transformation for a braking system (cylinder body)**

Design and technical requirements	Product features						
	Material	**Surface/edge regularity**	**Surface treatment**	**Body inner wall surface finish**	**Cylinder body wall strength**	**Dimensional accuracy**	**Seal**
Structural rigidity	S				S		
Smooth inner wall surface			S	S			S
High corrosion resistance	M		S	S			M
Good dimensional control						S	S
Release residual stress	M				S		
Good rust-proofing		S	S				M
Smooth surface and edge finish	M	S		M			

Relationships: Strong (S) =5; Medium (M) =3.

Table 9.24 **Product features: process features transformation for a braking system (cylinder body)**

Product features	Process features											
	Material	**Casting**	**Heat treatment**	**Boring**	**Face turning**	**Taper turning**	**Slot cutting**	**Drilling**	**Tapping**	**Honing**	**Ultrasonic cleaning**	**Rust prevention/ proofing**
Material	S	S	S									M
Surface and edge regularity		M										S
Surface treatment	M										S	S
Cylinder body inner wall surface finish				S						S		
Cylinder body wall strength	S	S	S									
Dimensional accuracy					M	M	S	S	S			
Seal	S	M	M	M		M	M			S	M	S

Relationships: Strong (S) =5; Medium (M) =3.

Table 9.25 Process features: process variables transformation matrix for a braking system (cylinder body)

Process features	Process variables												
	Depth of cut	Cutting speed	Solvent/ quenchant	Rate of cooling	Degree of surface finish	Tolerance	Cutting fluid	Lubricants	Material cost	Material fatigue resistance	Material hardness	Material toughness	Temperature
Material									S	S	S	S	
Casting				S		S			S		S	S	S
Heat treatment			S	S		M							S
Boring		S			S	S	S				M		S
Face turning	S	S				S	S	S					
Taper turning	S	S				S	S	S					
Slot cutting	M	M				S	M	M					
Drilling		S				S	S	S			S		S
Tapping		S				S	S	S					
Honing		S			S		S	S			S		S
Ultrasonic cleaning			S										S
Rust prevention/ proofing			S						M				S

Relationships: Strong (S) =5; Medium (M) =3.

functional parts will not be realized unless the manufacturing process is made robust by controlling the manufacturing variables. Some of the important processes relevant to the wheel cylinder body manufacture follow.

- **Turning**: Turning is a secondary operation for producing cylindrical precision surfaces. Since the turned parts must be rotated during operation, the process imposes limitations on size, weight, and shape. Workpiece size and length inversely affect the dimensional accuracy. The larger these dimensions are, the greater the possible variation after machining. Machine design and construction must provide control over operating disturbances due to factors such as vibration, deflection, thermal distortion, and wear on functional parts of the machine, which may account for piece-to-piece variation. Other factors include part deflection, tool wear, measuring tool accuracy, and operator skill. In addition, the surface finish of a turned part is directly related to the feed rate, tool sharpness, tool geometry, tool material, and workpiece material.
- **Drilling/boring**: Holes are machined in whenever the primary production process does not produce holes or fails to produce them at the necessary size, accuracy, straightness, or surface finish. The accuracy of both the diameter and the straightness of drilled holes depend on the correctness of the drill sharpening, the play and lack of rigidity in the typical drill spindle, thermal expansion of the material to be drilled, workpiece distortion from clamping, and the presence of a drill bushing during the drilling process.
- **Surface treatment**: Plating and polishing operations are to improve the surface finish rather than refine dimensions. The most important functions of polishing operations are to improve appearance; remove burrs; clean a surface for brazing, soldering, or surface finishing; and improve resistance to corrosion. Anodizing, thermal spraying, hard facing, porcelain enameling, hot dipping, chemical vapor deposition, ion vapor deposition, vacuum metalizing, sputtering, ion implantation, electroplating, electroforming, and electrodeless painting are the process variables.
- **Heat treatment**: Heat treatment imparts the following characteristics to a part: lower residual stress, increased surface hardness with improved resistance to wear, microstructure modification for improved mechanical properties, and higher machinability. Tempering, annealing, normalizing, carburizing, nitriding, carbonitriding, chromizing, boronizing, high-frequency resistance hardening, induction hardening, flame hardening, electron-beam hardening, and laser hardening are some heat treatment techniques. Optimal heat treatment depends on the critical time–temperature transformation relationship, composition and condition of the metal to be heat treated, response of the metal to quenching, and the method of quenching used.

In Step 6, we study the wheel cylinder body manufacturing process in detail and use the transformation matrix to identify the important manufacturing variables to generate guidelines that ensure product functionality. At the same time, some important design concepts are generated. For instance, when studying the boring process, the fixture design is very important for dimensional accuracy. Similarly, at the design stage, we need to consider the cylinder body shape for ease of manufacturing. Several examples illustrate the process of generating design and manufacturing guidelines. For instance, in step 2, we have the criterion "redundancy" so we can have the following guidelines:

- Use two or more circuits to transmit braking energy to the wheel brakes. In the event of a circuit failure, partial braking effectiveness is provided.
- To increase reliability, use two independent cylinders, that is, one master cylinder controls two wheel cylinders.

As another example, from step 5, the honing process is included in the process features so we can have the following guidelines:

- Choose two motion speeds (rotation and reciprocation at the ratio 3:1) to give a resulting crosshatch lay pattern with an included angle of 30–60°, which is the best surface lay pattern to resist wear for the internal cylinder body surface.
- Choose appropriate honing speeds based on material, hardness, and bore characteristics (plain, interrupted, etc.). If the honing speed is too high, it is apt to generate more heat, which results in surface damage. Conversely, if the speed is too low, it increases the manufacturing cost as well as time.

9.6.7 Survey development

The automotive braking system is an internal mechanical system. End users of the vehicle have limited contribution in evaluating the functional requirements of the system. A survey was conducted to test the functionality only for design/manufacturing guidelines. The survey was developed based on the specific product functionality criteria (in the generic form) and the results of the FTM. Development of guidelines for manufacturability is depicted in Table 9.26. This is followed by a list of checklists to develop other related product parameters (Tables 9.27–9.32).

9.6.8 Testing and statistical analysis

The procedure for data collection is similar to the can opener example illustrated earlier. The reliability and validity tests are the same. There is no correlation test for the users and the manufacturers, since users are not being surveyed in this case. We use the same method as in the testing of the can opener checklist. Cronbach alpha is used to test the reliability among the questions of each section in the questionnaire, and Pearson's product moment correlation is used to test the validity of the questionnaire. Twenty-one designers/manufacturers returned the questionnaire checklists. Statistical analyses were performed to test the reliability and validity of the questionnaire.

9.6.8.1 Reliability test results

The method used to test the reliability of each section in the braking system functionality questionnaire is the Cronbach alpha method, as in the case of the can opener functionality guideline questionnaire. The Cronbach alpha values for the design/manufacturing guideline checklists are presented in Table 9.33.

Since the Cronbach alpha values of all sections are over 0.8, all sections are highly reliable and to be used in the braking system functionality guidelines.

9.6.8.2 Validity test results

Validity refers to the extent to which the measurement procedures accurately reflect the conceptual variable being measured. For the braking system functionality guidelines, all the sections were tested for the correlation between the overall score and the

Table 9.26 Manufacturability guidelines to enhance the functionality of an automotive braking system

	Opinion disagree–agree					Comments
	1	2	3	4	5	
1. Use standard parts and materials						
2. Simplify the design and minimize the number of parts						
3. Try to improve the drawback of the drum brake system. The rear drum brake has an inherent low cost compared to an equivalent disc brake and the parking brake requirement is built in with very little extra complexity						
4. Finishing must be directed toward producing a surface that is within tolerances, has the proper roughness and texture, and is free of damage and of harmful residual stresses						
5. To ensure sufficient wear life, thermal performance, and low noise, the maximum allowable brake diameter is limited by rim size and, as such, determined by vehicle weight						
6. Honing: The degree of surface finishing required depends on the life of the cup and manufacturing cost; if the quality of the cup is good, degree of surface finishing can be reduced						
7. Gating and risering techniques must ensure smooth, complete filling of the die cavity followed by orderly solidification to prevent sand, skin, and scab in the cylinder body						
8. Use casting process to form the cylinder body shape						
9. Casting: Based on functional requirements, choose appropriate material for cylinder body for easy manufacturing and low cost						
10. Casting: Design of cylinder body should consider provisions for contraction of cast metals by a shrinkage allowance on patterns; other allowances are those for machining stock for finishing and occasionally to compensate for expected distortions						
11. Casting: Sharp corners and fillets should be avoided, as they may cause cracking and tearing during solidification of the metal						
12. Casting: After the casting is removed from the mold, various cleaning, finishing, and inspection operations may be performed						
13. Casting: Heat transfer must be locally controlled to prevent starvation of late solidifying portions of the casting and minimize porosity in the cylinder body						

(Continued)

Table 9.26 **(Continued)**

	Opinion disagree–agree					Comments
	1	2	3	4	5	
14. Honing: For internal surface finish of the wheel cylinder body, use honing; from the degree of surface finish (Ra 0.15–1.5 μm), match the functional requirement; the cost of honing lower than other operations						
15. Honing: Choose appropriate grain size, which depends on the surface finish requirement (material, hardness, type of abrasive, etc.); the wrong will raise manufacturing costs or will not meet the functional requirements						
16. Honing: Provide 0.05 mm internal allowance for honing. This is an optimal range for cost (time) and machining						
17. Boring: Use to enlarge a hole in the cylinder body made by the previous casting process and to cut the allowance for final surface finishing						
18. Boring: For surface finishing the internal cylinder body, if the degree of surface finishing is high, then the machining allowance can be reduced; otherwise it needs to be increased						
19. Boring: The machining allowance for final surface finishing should maintain minimum requirements to reduce further machining cost						
20. Boring: Fixture of workpieces for boring operations is extremely important for dimensional accuracy						
21. Boring: For master or wheel cylinder body, if the diameter/length ratio is too low, use a gun drill instead of boring to avoid vibration and chatter						
22. Boring: A boring bar must be sufficiently stiff (a material with high elastic modulus, such as tungsten carbide) to minimize deflection, avoid vibration and chatter, and maintain dimensional accuracy						
23. Drilling and tapping: Designs should allow holes to be drilled on flat surfaces, perpendicular to the drill motion; otherwise, the drill tends to deflect and the hole may not be located accurately. Exit surfaces for the drill should be flat						
24. Drilling and tapping: Parts should be designed so that all drilling can be done with a minimum of fastening or repositioning the workpiece						
25. Drilling and tapping: Interrupted hole surfaces should be avoided or minimized for better dimensional accuracy						

(*Continued*)

Table 9.26 **(Continued)**

	Opinion disagree–agree					Comments
	1	2	3	4	5	
26. Drilling and tapping: Hole bottoms should match standard drill point angles. Avoid flat bottoms or odd shapes						
27. Drilling and tapping: Generally, the holes produced by drilling are larger than the drill diameter. The amount of oversize depends on the quality of the drill, and equipment and practices used. Depending on their properties, the final hole should be smaller than the drill diameter						
28. Drilling and tapping: Chip removal can be a significant problem, especially in drilling and tapping, and lead to tool breakage. The use of a proper cutting fluid and periodic reversal and removal of the tap from the hole are effective means of chip removal and improving the quality of the hole. Note that tapping is among the most severe processes, requiring effective cutting fluids						
29. Face turning, taper turning, and slot cutting: Primary factors in any turning operation are speed, feed, and depth of cut. Other factors, such as type of material and type of tool, also are important						
30. Face turning, taper turning, and slot cutting: Consider turning process parameters such as tool geometry, material removal rate, relief angles, cutting edge angles, forces in turning, tool materials, feed and cutting speeds, and cutting fluids						
31. Face turning, taper turning, and slot cutting: Reduce vibration and chatter; minimize tool overhang; support workpiece rigidly; use machine tools with high stiffness and damping capacity; when tools begin to vibrate and chatter, modify the process parameters, such as tool geometry, cutting speed, feed rate, depth of cut, or cutting fluid						
32. Face turning, taper turning, and slot cutting: Unless workpiece is held on a mandrel, it must be turned end for end after the first end is completed and the facing operation repeated (if both ends of the work are to be faced)						
33. Face turning, taper turning, and slot cutting: In the facing of castings, the depth of the first cut should be sufficient to penetrate the hard material to avoid excessive tool wear						

(*Continued*)

Table 9.26 **(Continued)**

	Opinion disagree–agree					Comments
	1	2	3	4	5	
34. Face turning, taper turning, and slot cutting: Burrs may interfere with the assembly of parts and cause jamming and misalignment; they also may reduce the fatigue life of components. In the wheel cylinder assembly, the cup may heat and reduce cup life. Pay special attention to deburring. What is your overall functionality of the brake system in terms of its design for manufacturability in this section?						

Table 9.27 **Reliability guidelines to enhance the functionality of an automotive braking system**

	Opinion disagree–agree					Comments
	1	2	3	4	5	
1. Use two or more braking circuits to transmit braking energy to the wheel brakes to design for redundancy 2. Use two independent cylinders: one master cylinder control and two wheel cylinders 3. If a partial failure should occur, try to minimize the effects of that failure on pedal force and pedal travel 4. Decide braking system component sizing based on fatigue loading and overload 5. Design wear in the form of product lining friction coefficient and mechanical pressure 6. If possible, design friction components in the air stream where they are easily cooled to increase brake life and for rapid recovery 7. Design self-adjusting devices in the braking system to ensure easy maintenance and reduce labor requirements 8. Provide a vehicle brake function monitoring system to improve braking system reliability 9. Design should consider system reliability under partial failure condition of braking system. When this occurs, partial braking effectiveness should be maintained under the following conditions: Service system circuit failure Braking effectiveness with partial/complete loss of power assist						

(Continued)

Table 9.27 **(Continued)**

	Opinion disagree–agree					Comments
	1	2	3	4	5	
Braking effectiveness with brakes in thermal fade condition						
Directional stability with diagonal split failure						
Increased pedal travel with service system circuit failure						
Increased pedal force with service system circuit failure						
10. Braking system design should consider heat flux into drum or rotor surface to avoid thermal fade						
11. Importance of brake reliability, life, and maintenance should be stressed in service manuals						
12. Maximum allowable brake diameter should be limited by rim size and determined by vehicle weight						
13. Parts can be minimized by combining with other parts by checking whether:						
The part or subassembly moves relative to its mating parts during normal functioning of the product						
The part or subassembly must be of a different material than its mating part to fulfill its function						
The combination of certain parts would not affect the assembly of other parts						
Field service does not require their disassembly						
14. Casting: Heat treat the cylinder body to relieve stresses and increase toughness of cylinder wall						
15. Casting: Heat transfer must be locally controlled to prevent starvation of late solidifying portions of the casting and to minimize porosity in the cylinder body						
16. Honing: Choose two motion speeds to give a resulting crosshatch lay pattern with an included angle of 30–60°						
17. Honing: Choose appropriate honing speeds based on material, hardness, and bore characteristics						
18. Honing: Choose appropriate cutting fluids, usually a sulfurized mineral oil or lard oil mixed with kerosene or similar light oil						
19. Ultrasonic cleaning: Cleaning is essential for more effective application of metal working fluids, coating and painting, brazing, and soldering						

(*Continued*)

Table 9.27 (Continued)

	Opinion disagree–agree					Comments
	1	2	3	4	5	
20. Drilling and tapping: Chip removal can be a significant problem, especially in drilling and tapping and can lead to tool breakage. Tapping is among the most severe processes, requiring effective cutting fluids, otherwise the operation will damage the cylinder body surface 21. Face turning, taper turning, and slot cutting: Burrs may interfere with assembly of parts and can cause jamming and misalignment; burrs also can reduce the fatigue life of components; pay special attention to deburring What is the overall functionality of the brake system in terms of its design for reliability in this section?						

Table 9.28 Safety guidelines to enhance the functionality of an automotive braking system

	Opinion disagree–agree					Comments
	1	2	3	4	5	
1. Design physical indicators in the braking system to allow the driver to notice brake failure prior to an accident 2. Provide a brake function monitoring system 3. For air brakes, consider application and release time lags; for hydraulic brakes, consider pedal force boost lag 4. The basic functions of a braking system have to be performed under normal operation and to a lesser degree of braking effectiveness, using a brake system failure 5. Use two or more circuits to transmit braking energy 6. Use two independent cylinders to increase safety 7. Redesign may be required to ensure adequate component performance and life 8. Determination of whether the operating environment is adverse to braking system components is important for braking safety effectiveness						

(*Continued*)

Table 9.28 **(Continued)**

	Opinion disagree–agree					Comments
	1	2	3	4	5	
9. Brake design should consider braking effectiveness: Minimum stopping distance without wheel lockup Minimum stopping distance without loss directional control with wheel lockup for dry and cold brakes and cold and heated brakes Minimum stopping distance without wheel lockup while turning						
10. Design should consider partial failure; when this happens, at least partial braking effectiveness should be maintained: Braking effectiveness with service system circuit failure Braking effectiveness with partial or complete loss of power assist Braking effectiveness with brakes in thermal fade condition Directional stability with diagonal split failure Increased pedal travel with service system circuit failure Increased pedal force with service system circuit failure						
11. Design should consider thermal analysis: Heat transfer coefficient for drum or rotor Brake temperature during continued and repeated braking and maximum effectiveness stop Reduced braking effectiveness during thermal faded conditions Thermal stresses to avoid rotor cracking and heat checking Brake fluid temperatures in wheel cylinders to avoid brake fluid vaporization						
12. Design of braking system should consider heat flux into drum or rotor surface						
13. Design should consider horsepower absorption by brake lining or pad						
14. Brake design should consider maximum straight line wheels, unlocked deceleration						
15. Design should consider the effectiveness of the parking brake: Maximum deceleration by application of emergency brake lever on level and sloped roadways						

(*Continued*)

Table 9.28 **(Continued)**

	Opinion disagree–agree					Comments
	1	2	3	4	5	
Maximum grade-holding capacity Determination under what conditions an automatic emergency application should occur 16. Design of components should consider limiting heat transfer into the fluid, else the brake fluid might boil, resulting in vapor lock 17. If a partial failure should occur, try to minimize the effects of partial failure on pedal force and pedal travel 18. Maintenance: Design self-adjusting devices 19. Manufacturer should emphasize the importance of brake maintenance in service manuals 20. Determination of whether certain maintenance practices or lack of maintenance by particular user groups may require redesign to ensure adequate component performance and life 21. Casting: Heat treat the cylinder body to relieve stresses and increase the toughness of cylinder wall 22. Honing: Choose two motion speeds to give a resulting crosshatch lay pattern with an included angle of 30–60° 23. Face turning, taper turning, and slot cutting: Burrs may interfere with assembly of parts and can cause jamming and misalignment; burrs also can reduce the fatigue life of components; pay special attention to deburring						
What is the overall functionality of the brake system in terms of its design for safety in this section?						

Table 9.29 **Quality guidelines to enhance the functionality of an automotive braking system**

	Opinion disagree–agree					Comments
	1	2	3	4	5	
1. Express dimensions and their tolerance in the engineering drawing 2. Emphasize engineering metrology and in-process measurement in quality control						

(Continued)

Table 9.29 **(Continued)**

	Opinion disagree–agree					Comments
	1	2	3	4	5	
3. Do not overtolerance dimensions; use reversed chain tolerance to establish sensible tolerances of components at progressive steps in the chain						
4. Maintenance of specified tolerances is vital to the function of assemblies and makes interchangeability possible; excessively tight tolerance and surface finish specifications lead to excessive manufacturing costs						
5. Using high quality brake fluid, which cannot vaporize at the highest operating temperature, needs a high boiling point and low vaporization pressure						
6. Design should consider brake fluid and rub compatibility: Brake fluid should have properties that do not make the rub expand and soften during use in the defined operation range, else piston block or fluid leak might occur						
7. Design should consider the lube properties of brake fluid for brake system components						
8. Design should consider the stability properties of brake fluid, since the fluid is used for a long period of time						
9. Design should consider the water-acceptable properties of the brake fluid; a good brake fluid should not change properties if water ingestion occurs						
10. Casting: When using cast test bars for testing mechanical properties of the cylinder body, make certain cooling rates are the same as in the casting						
11. Casting: Nondestructive testing techniques are particularly important in detecting internal defects, whether due to solidification shrinkage, internal hot tearing, or porosity						
12. Casting: Design of cylinder body can apply tolerance of $\pm^{1}/_{16}$ in. for dimensions up to 12 in.						
13. Casting: Produce mold cavity in desired shape and size with due allowance for shrinkage of the solidifying metal						
14. Casting: Heat transfer must be locally controlled to prevent starvation of late solidifying portions of the casting and minimize porosity of the cylinder body						
15. Casting: A melting process must be capable of providing molten material not only at the proper temperature but also in the desired quantity at an acceptable quality and reasonable cost						

(Continued)

Table 9.29 **(Continued)**

	Opinion disagree–agree					Comments
	1	2	3	4	5	
16. Casting: Heat treat the cylinder body to relieve the stresses and increase the toughness of the cylinder wall						
17. Honing: The degree of surface finishing required depends on the life of the cup and the manufacturing cost; if the quality of the cup is good, the degree of surface finishing can be reduced						
18. Honing: The machining allowance for honing should consider the piston and hole matching requirements						
19. Boring: A boring bar must be sufficiently stiff, made of a material with a high modulus of elasticity to minimize deflection and chatter to maintain dimensional accuracy						
20. Boring: Fixtures of workpieces for boring operations are critical for dimensional accuracy						
21. Ultrasonic cleaning: Cleaning is essential for more effective application of metal working fluids, coating and painting, brazing, and soldering; clean, reliable parts are important to machinery and assembly operations for wheel cylinder and master cylinder						
22. Drilling and tapping: Avoid or minimize interrupted hole surfaces for better dimensional accuracy						
23. Drilling and tapping: Chip removal can be a significant problem, especially in drilling and tapping, and lead to tool breakage; a proper cutting fluid and periodic reversal and removal of the tap from the hole are effective means of chip removal and improve the quality of the hole						
24. Face turning, taper turning, and slot cutting: The important turning process parameters, such as tool geometry and material removal rate, relief angles, cutting edge angles, forces in turning, and tool materials, are important in ensuring quality of the components in the braking system						
25. Face turning, taper turning, and slot cutting: Reduce vibration and chatter Minimize tool overhang Support workpiece rigidly Use machine tools with high stiffness and damping capacity						

(*Continued*)

Table 9.29 **(Continued)**

	Opinion disagree–agree					Comments
	1	2	3	4	5	
When tools begin to vibrate and chatter, modify process parameters, such as tool geometry, cutting speed, feed rate, and depth of cut 26. Face turning, taper turning, and slot cutting: Burrs may interfere with the assembly of parts and cause jamming and misalignment; burrs also may reduce the fatigue life of components; when the wheel cylinder is assembled, cup may heat, leading to a reduction in its life; pay special attention to deburring What is the overall functionality of the brake system in terms of its design for quality in this section?						

Table 9.30 **Environmental friendliness guidelines to enhance the functionality of an automotive braking system**

	Opinion disagree–agree					Comments
	1	2	3	4	5	
1. Avoid using asbestos for brake lining materials 2. Avoid the use of toxic materials in the product and its manufacturing processes 3. Use biodegradable lubricants 4. Ship the braking systems components in disassembled form to reduce packaging materials 5. Provide for part numbering during manufacture instead of using adhesive labels 6. If adhesive labels are used, use adhesive material that is compatible with materials being recycled 7. Standardize product components, such as piston and spring, so they can be salvaged and reused 8. Use a rubber cup that can be recycled 9. For brake system, use metallic materials, since they can be recycled easily 10. Reduce the number of components of different materials to make sorting easier for eventual disposal 11. Minimize using liquids such as acids, alkalis, and solvents during manufacturing						

(*Continued*)

Table 9.30 (Continued)

	Opinion disagree–agree					Comments
	1	2	3	4	5	
12. Reduce the amount of energy consumed during operation of the product 13. Design for processes that minimize material scrap and achieve the benefits of designing smaller, lighter parts 14. Avoid using ozone-depleting substances in the product 15. Design the processes to reduce manufacturing residue, such as mold scrap and cutting scrap 16. Minimize the amount and variety of packaging materials used 17. Reduce the use of radioactive materials in the product 18. Design the brake system and its components to be reusable, refurbishable, or recyclable, in that order 19. Design components so fasteners are easily visible and accessible to aid in disassembly 20. Large quantities of chips and grinding sludge are produced and the process must be conducted to make recycling feasible and economical What is the overall functionality of the brake system in terms of its design for environmental friendliness in this section?						

Table 9.31 Performance guidelines to enhance the functionality of an automotive braking system

	Opinion disagree–agree					Comments
	1	2	3	4	5	
1. Select appropriate material that meets the functional requirements of the users with various product features 2. Select appropriate brake lining materials that meet functional requirements at different temperature ranges 3. Determine whether certain maintenance practices or lack thereof require redesign to ensure adequate component performance life 4. Decide braking system component size based on fatigue loading and overload 5. In the brake system, design the wear measure in the form of a product of lining friction coefficient and mechanical pressure						

(*Continued*)

Table 9.31 **(Continued)**

	Opinion disagree–agree					Comments
	1	2	3	4	5	
6. Determine whether wear or use affects brake force distribution, hence braking stability, due to premature rear brake lockup						
7. Design should consider brake fluid volume: Master cylinder bore and piston travel for each brake circuit Wheel cylinder piston travel						
8. To ensure sufficient wear life, thermal performance, and low noise, maximum allowable brake diameter is limited by rim size and determined by vehicle weight						
9. Factors such as boiling point, freezing point, temperature, and lube characteristics should be considered for brake fluid						
10. Brake fluid needs high boiling point and low vaporization pressure						
11. Design should consider compatibility of brake fluid and rubber						
12. Design should consider lobe properties of brake fluid for the braking system components						
13. Design should consider stability properties of the brake fluid						
14. Design should consider the mixability properties of the brake fluid; no commercial product should cause malfunction on mixing with the brake fluid						
15. Design should consider the water-acceptable properties of the brake fluid						
16. Design should consider the metal-compatibility properties of the brake fluid						
17. Brake design should consider braking effectiveness: Minimum stopping distance without wheel lockup Minimum stopping distance without loss of directional control with wheel lockup for dry and cold brakes and for cold and heated brakes Minimum stopping distance without wheel lockup while turning						
18. Design should consider partial failure; when this happens, partial braking effectiveness should be maintained: Braking effectiveness with service system circuit failure						

(*Continued*)

Table 9.31 (Continued)

	Opinion disagree–agree					Comments
	1	2	3	4	5	
Braking effectiveness with partial or complete loss of power assist Braking effectiveness with braking in thermal fade conditions Directional stability with diagonal split failure Increased pedal travel with service system circuit failure Increased pedal force with service system circuit failure						
19. Design should consider thermal analysis based on: Heat transfer coefficient for drum or rotor Brake temperature during continued and repeated braking and maximum effectiveness during a stop Reduced braking effectiveness during thermal faded conditions Thermal stresses to avoid rotor cracking and heat cracking Brake fluid temperatures in wheel cylinders to avoid brake fluid vaporization						
20. Design should consider the effectiveness of the parking brake: Maximum deceleration by application of emergency brake lever on level and sloped roadways Maximum grade-holding capacity Determination under what conditions an automatic emergency application should occur						
21. The mass of an individual part should be no more than the function or strength required of it and is achieved by minimizing the mass/strength ratio						
22. Design of braking system should consider heat flux into drum or rotor surface						
23. Brake design should consider maximum straight line wheels' unlocked deceleration						
24. Select parts with verified reliability						
25. Design a braking system to maintain the independence of functional requirements and adopt a modular design approach to achieve this						

(*Continued*)

Table 9.31 **(Continued)**

	Opinion disagree–agree					Comments
	1	2	3	4	5	
26. Adopt a robust design procedure to determine the settings of the product design parameters that make the product's performance insensitive to environmental variables, product deterioration, and manufacturing irregularities						
27. Design consideration: For air brakes, application and release time lags; for hydraulic brakes, pedal force boosts lag						
28. For ease of maintenance, design self-adjusting devices to reduce the labor and frequency of servicing						
29. The basic functions of a braking system have to be performed during normal operation and, to a lesser degree of braking effectiveness, during a braking system failure						
30. Determine whether the operating environment is adverse to any components of the braking system						
31. Design of components should consider limiting heat transfer into the fluid in absence of which brake fluid would boil, leading to vapor lock						
32. If a partial failure should occur, try to minimize its effects on pedal force and pedal travel						
33. The design of a new brake system begins with the selection of brake force distribution. The optimum distribution is a function of basic vehicle dimensions and weight distribution						
34. To ensure sufficient wear life, thermal performance, and low noise, maximum allowable brake diameter is limited by rim size and determined by vehicle weight						
35. Casting: The different cooling rates within the body of a casting cause residual stresses; stress relieving may be necessary to avoid distortions in critical applications such as piston sticking						
36. Casting: Heat treat the cylinder body to relieve the stresses and increase toughness of the cylinder wall						
37. Honing: The machining allowance for honing should consider piston and hole matching requirements						
38. Boring: Fixture of workpieces for boring is critical for dimensional accuracy						

(Continued)

Table 9.31 (Continued)

	Opinion disagree–agree					Comments
	1	2	3	4	5	
39. Ultrasonic cleaning: Cleaning is essential for more effective application of metal working fluids, coating and painting, brazing, and soldering; clean, reliable functioning of parts is critical in machinery and assembly operations for wheel cylinder and master cylinder						
40. Drilling and tapping: Design should allow for holes drilled on flat surfaces, perpendicular to the drill motion, else drill tends to deflect, leading to dislocation of the hole; exit surfaces for the drill should be flat						
41. Drilling and tapping: Avoid or minimize interrupted hole surfaces for better accuracy						
42. Face turning, taper turning, and slot cutting: The primary factors in basic turning operation are speed, feed, and depth of cut; other factors, such as material and tool type, also are important						
43. Face turning, taper turning, and slot cutting: Consider important process parameters, such as tool geometry and material removal rate, relief angles, and cutting edge angles, in braking system design						
44. Face turning, taper turning, and slot cutting: Reduce vibration and chatter Minimize tool overhang Support workpiece rigidly Use machine tools with high stiffness and damping capacity When tools begin to vibrate and chatter, modify process parameters, such as tool geometry, cutting speed, feed rate, depth of cut, and cutting fluid						
What is the overall functionality of the brake system in terms of its design for performance in this section?						

Table 9.32 Usability guidelines to enhance the functionality of an automotive braking system

	Opinion disagree–agree					Comments
	1	2	3	4	5	
1. Ergonomic considerations and driver acceptance limit pedal force and pedal travel exerted by the right foot, for the 5th percentile female population is approximately 22 N						
2. Brake design should consider maximum straight line wheels' unlocked deceleration						
3. Design should consider if wear or use affects brake force distribution, hence braking stability, due to premature rear brake lockup						
4. Design should consider brake fluid and rubber compatibility						
5. Fit the product to the user: Conform operation of the product to users, physically and mentally. Keep the static strength requirements <10% of the maximum volitional strength exertion capability when muscle loading is protracted. Keep dynamic strength requirements <5% of the maximal volitional strength exertion capability when muscle loading is protracted						
6. Provide feedback: The braking system should provide the users a response to any actions taken, informing the users how the product works						
7. Avoid awkward and extreme motions: Design operating controls and other elements to provide the force or power needed rather than rely on human power. Design machines to accommodate the body measurements and capabilities of the potential user population; if critical, provide an adjustment, since no one size will be optimal for all users. If vibration is present, controls should be isolated from the vibration as much as possible, also provide damping and improve dynamic balance. Minimize force required to activate a control by providing more leverage, optimize the pedal shape and surface by providing power assist						
8. Check adjacent parts for function and attempt to integrate them to produce a multifunctional single part						

(*Continued*)

Table 9.32 (Continued)

	Opinion disagree–agree					Comments
	1	2	3	4	5	
9. Design considerations for air brakes, application, and release time lags: For hydraulic brakes, use pedal force boost lag						
10. Casting: The different cooling rates within the body of a casting cause residual stresses; relieving stress may be necessary to avoid distortions in critical applications, such as piston sticking						
11. Boring: Fixture of workpieces for boring is critical for dimensional accuracy						
12. Honing: The machining allowance for honing should consider the piston and hole matching requirements						
What is the overall functionality of the brake system in terms of its design for usability in this section?						

Table 9.33 Test values for the braking system designer/manufacturing checklist

Section	Cronbach alpha
Reliability	0.8715
Safety	0.8902
Quality	0.9275
Manufacturability	0.9551
Environmental friendliness	0.9124
Performance	0.9705
Usability	0.9705

average score. The method used is the Pearson's product moment correlation and the *t*-test. All values are listed in Table 9.34.

All the correlation coefficients are over 0.5. According to the rule of large correlation, a correlation value ranging from 0.40 to 0.69 suggests modest correlation, high correlation is denoted by a value ranging from 0.70 to 0.89, and very high correlation is depicted by a value ranging from 0.90 to 1.00. Similarly, all the *t*-test values about the correlation values of the seven sections of the test, as depicted in Table 9.33, are highly significant. In conclusion, all seven sections are valid. The results are discussed in the following section.

Table 9.34 **Validity test results for the braking system designer/manufacturing checklist**

Section	Correlation coefficient	Significance
Reliability	0.523	0.015
Safety	0.563	0.008
Quality	0.697	0.000
Manufacturability	0.812	0.000
Environmental friendliness	0.794	0.000
Performance	0.832	0.000
Usability	0.536	0.012

9.6.9 Discussion of the results

9.6.9.1 The reliability test

Since the Cronbach alpha values of all sections are over 0.85, all sections can be reliably used in the braking system functionality guidelines. Since the braking system is more complex in nature, the guidelines are designed accordingly, catering to the level of complexity of the system. As for the reliability test, note that, if the number of items in the questionnaire is increased, the alpha value rises correspondingly. In this situation, higher Cronbach alpha values can be chosen as a standard for acceptance or rejection.

9.6.9.2 The validity test

All the questionnaires have an acceptable correlation. The *t*-test signifying the correlation values illustrates that all sections are significant. Hence, we conclude that the checklist has valid guidelines. Designers have ideas on the functionality of the braking system. The guidelines are useful in improving braking system functionality.

9.6.9.3 Conclusions

The recognition that design decisions made early in the product development cycle can have a significant effect on the manufacturability, quality, product cost, and product introduction time ultimately ensures success in the marketplace. Here we tried to evaluate the guidelines for ensuring product functionality only through questionnaires. These are subjective measurements. For instance, one question in the validity tests was, What is the standard acceptance significance level? This level has the greatest effect on the validity test result. There is no such standard, however. To remove this deficiency, more objective measures can be included to evaluate functionality. Similarly, new directions for developing the mechanism of product function review are needed. The function review is a system that involves gathering and evaluating objective knowledge about product design functionality and concrete plans for making it a reality, suggesting improvements at each point, and confirming that the process is ready to proceed to the next phase. Also, manufacturing measures resulting in conflicting outcomes need further studies. Efforts should be focused on identifying design guidelines to provide better solutions for different situations.

References

Akiyama, K., 1991. Function Analysis—Systematic Improvement of Quality and Performance. Productivity Press, Cambridge, MA.

Asiedu, Y., Gu, P., 1998. Product life cycle cost analysis: state of the art review. Int. J. Prod. Res. 36 (4), 883–908.

Bakerjian, R., 1992. Design for manufacturability Tool and Manufacturing Engineering Handbook, vol. 6. Society of Manufacturing Engineers, Dearborn, MI.

Billatos, S.B., Nevrekar, V.V., 1994. Challenges and practical solutions to design for the environment. In: Mason, J. (Ed.), Design for Manufacturability ASME, New York, NY, pp. 49–64.

Boothroyd, G., 1994. Product design for manufacture and assembly. Comput. Aided Des. 26 (7), 505–520.

Boothroyd, G., Dewhurst, P., 1983. Design for Assembly. Boothroyd and Dewhurst, Amherst, MA.

Bracewell, R.H., Sharpe, J.E.E., 1996. Functional descriptions used in computer support for qualitative scheme generation—schemebuilder. Artif. Intell. Eng. Des. Anal. Manuf. 10 (4), 333–346.

Bralla, J.G., 1996. Design for Excellence. McGraw-Hill, New York, NY.

Brauer, R.L., 1990. Safety and Health for Engineers. Van Nostrand Reinhold, New York, NY.

Chakrabarti, A., Blessing, L., 1996. Special issue: representing functionality in design. Artif. Intell. Eng. Des. Anal. Manuf. 10 (4), 251–370.

Chakrabarti, A., Bligh, T.P., 1994. An approach to functional synthesis of solutions in mechanical conceptual design. Part I: Introduction and knowledge representation. Res. Eng. Des. 6, 127–141.

Chakrabarti, A., Bligh, T.P., 1996. An approach to functional synthesis of mechanical design concepts: theory, applications and emerging research issues. Artif. Intell. Eng. Des. Anal. Manuf. 10 (4), 313–331.

Chu, X., Holm, H., 1994. Product manufacturability control for concurrent engineering. Comput. Ind. 24 (1), 29–38.

Cross, N., 1989. Engineering Design Methods. John Wiley & Sons, Inc., New York, NY.

General Electric Company, 1960. Manufacturability Producibility Handbook. Manufacturing Service, General Electric Company, Schenectady, NY.

Goel, A., Stroulia, E., 1996. Functional device models and model based diagnosis in adaptive design. Artif. Intell. Eng. Des. Anal. Manuf. 10 (4), 355–370.

Govindaraju, M., 1999. Development of Generic Guidelines to Manufacture Usable Consumer Products (Ph.D. dissertation). University of Cincinnati, Cincinnati, OH.

Green, C., 1966. Eli Whitney and the Birth of American Technology. Little Brown, Boston, MA.

Gupta, S.K., Nau, D.S., 1995. Systematic approach to analyzing the manufacturability of machined parts. Comput. Aided Des. 27 (5), 323–342.

Gupta, S.K., Regli, W.C., Das, D., Nau, D.S., 1997. Automated manufacturability analysis: a survey. Res. Eng. Des. Theor. Appl. Concurrent Eng. 9 (3), 168–190.

Hammer, W., 1980. Product Safety Management and Engineering. Prentice-Hall, Englewood Cliffs, NJ.
Hermann, F., 1994. Environmental considerations for product design for the German market. In: Mason, J. (Ed.), Design for Manufacturability ASME, New York, NY, pp. 35–48.
Huang, G.Q., 1996. Design for X—Concurrent Engineering Imperatives. Chapman and Hall, New York, NY.
Huang, G.Q., Mak, K.L., 1998. Re-engineering the product development process with Design for X. Proc. Inst. Mech. Eng. B: J. Eng. Manuf. 212 (B4), 259–268.
Hubka, V., Eder, W.E., 1992. Design Science. Springer, New York, NY.
Hundal, M.S., 1994. DFE: current status and challenges for the future. In: Mason, J. (Ed.), Design for Manufacturability ASME, New York, NY, pp. 89–98.
Jansson, D.G., Shankar, S.R., Polisetty, F.S.K., 1990. Generalized measures of manufacturability. In: Rinderle, J.R. (Ed.), Design, Theory and Methodology—DTM'90, pp. 85–96.
Jo, H.H., Parsae, H.R., Sullivan, W.G., 1993. Principles of concurrent engineering. In: Parsaie, H.R., Sullivan, W.G. (Eds.), Concurrent Engineering: Contemporary Issues and Modern Design Tools Chapman and Hall, New York, NY, pp. 3–23.
Kusiak, A., He, D.W., 1997. Design for agile assembly: an operational perspective. Int. J. Prod. Res. 35 (1), 157–178.
Lacey, R., 1986. Ford: The Man and the Machine. Little Brown, Boston, MA.
Miles, L.D., 1961. Techniques of Value Analysis and Engineering. McGraw-Hill, New York, NY.
Mital, A., Anand, S., 1992. Concurrent design of products and ergonomic considerations. J. Des. Manuf. 2, 167–183.
Miyakawa, S., Ohashi, T., Iwata, M., 1990a. The Hitachi assemblability evaluation method (AEM). In: Proceedings of the International Conference on Product Design for Assembly, Newport, RI.
Miyakawa, S., Ohashi, T., Iwata, M., 1990b. The Hitachi new assemblability evaluation method (AEM). Trans. North Am. Manuf. Res. Inst. SME, 23–25.
Nevins, J.L., Whitney, D.E., 1989. Concurrent Design of Products and Processes—A Strategy for the Next Generation in Manufacturing. McGraw-Hill, New York, NY.
Nielsen, J., 1993. Usability Engineering. Academic Press, San Diego, CA.
Nof, S.Y., Wilhelm, W.E., Warnecke, H., 1997. Industrial Assembly. Chapman and Hall, New York, NY.
Pahl, G., Beitz, W., 1988. Engineering Design: A Systematic Approach. Springer-Verlag, Berlin.
Peien, F., Mingjun, Z., 1993. The research on development of catalogs for conceptual design pp. 101–104 Proceedings of the Third National Conference on Engineering Design. Industrial Press, Beijing, China.
Peien, F., Guorong, X., Mingjun, Z., 1996. Feature modeling based on design catalogs for principle conceptual design. Artif. Intell. Eng. Des. Anal. Manuf. 10 (4), 347–354.
Priest, J.W., 1990. State of the art review of measurement procedures in product design for manufacturing. In: Proceedings of Manufacturing International '90, Atlanta, GA, pp. 25–28.
Qian, L., Gero, J.S., 1996. Function–behavior–structure paths and their role in analogy based design. Artif. Intell. Eng. Des. Anal. Manuf. 10 (4), 289–312.
Rodenacker, W., 1971. Methodisches Konstruieren. Springer-Verlag, Berlin.
Runciman, C., Swift, K., 1985. Assembly Autom. 5 (3), 17–50.
Sanchez, J.M., Priest, J.W., Soto, R., 1997. Intelligent reasoning assistant for incorporating manufacturability issues into the design process. Exp. Syst. Appl. 12 (1), 81–88.

Sturges, R.H., O'Shaughnessy, K., Kilani, M., 1990. Representation of aircraft design data for supportability, operability and producibility evaluations. EDRC Report No. 14513, Carnegie Mellon University Engineering Design Research Center, Pittsburgh, PA.

Sturges, R.H., O'Shaughnessy, K., Kilani, M., 1996. Computational model for conceptual design based on extended function logic. Artif. Intell. Eng. Des. Anal. Manuf. 10 (4), 255–274.

Taguchi, G., Elsayed, E.A., Hsiang, T., 1989. Quality Engineering in Production Systems. McGraw-Hill, New York, NY.

Taylor, G.D., 1997. Design for global manufacturing and assembly. IIE Trans. 29 (7), 585–597.

Ullman, D.G., 1997. The Mechanical Design Process. McGraw-Hill, New York, NY.

Umeda, Y., Tomiyama, T., 1997. Functional reasoning in design. IEEE Exp., 42–48.

Umeda, Y., Taketa, H., Tomiyama, T., Yoshikawa, T., 1990. Function, behavior and structure. In: Gero, J.S. (Ed.), Application of Artificial Intelligence in Engineering V (Design), Proceedings of the Fifth International Conference, vol. 1, Boston.

Van Hemel, C.G., Keldmann, K., 1996. Applying DFX experience in design for environment. In: Huang, G.Q. (Ed.), Design for X—Concurrent Engineering Imperatives Chapman and Hall, New York, NY.

Wall Street Journal, April 26 and 27, 1999.

Welch, R.V., Dixon, J.R., 1992. Representing function, behavior and structure during conceptual design Design Theory and Methodology, vol. 42. ASME, New York, NY.

Welch, R.V., Dixon, J.R., 1994. Guiding conceptual design through behavior reasoning. Res. Eng. Des. 6, 169–188.

Ziemke, M.C., Spann, M.S., 1993. Concurrent engineering's roots in the World War II era. In: Parsaei, H.R., Sullivan, W.G. (Eds.), Concurrent Engineering: Contemporary Issues and Modern Design Tools Chapman and Hall, New York, NY, pp. 24–41.

Design for Usability

10

10.1 Introduction

Product design is the process of creating new and improved products for people to use. Consumer products are designed to facilitate use by the general public, whereas commercial products are used to produce goods and services. Consumer products are different from commercial products in several respects:

- The user generally is untrained.
- The user often works unsupervised.
- The user is a part of a diverse population (Cushman and Rosenberg, 1991).

The process of designing and manufacturing consumer products is influenced greatly by the needs and demands of the customers. In the early twentieth century, consumer products were designed primarily to provide functionality. Later, product features, such as form and appearance, began to be emphasized. Although this resulted in good-looking products with an array of features, such products often were difficult to use (Ulrich and Eppinger, 1995).

During the 1980s, designers began emphasizing the user friendliness of products. Requirements such as product–user interface and safety were incorporated into design. In recent years, concern for the environment and resource utilization has stimulated new awareness among users to seek products that pose minimal risk of environmental pollution, consume less energy, have very little toxic emissions during use, and are recyclable when disposed. To make products usable by making them environmentally friendly, designers need to emphasize energy efficiency, recyclability, and disposability. This calls for considering all life cycle phases of a product simultaneously—design, production, distribution, usage, maintenance, and disposal/recycling—in determining the usability.

Of late, designers have begun emphasizing the customizability of products to meet demand from users to satisfy individual tastes and preferences. The following may be regarded as the criteria for designing and manufacturing usable consumer products:

- Functionality
- Ease of operation
- Esthetics
- Reliability
- Maintainability/serviceability
- Environmental friendliness
- Recyclability/disposability
- Safety
- Customizability.

These customer needs are linked to product design and manufacture. To fulfill these needs, consumer products need to be designed to incorporate those features that

Product Development. DOI: http://dx.doi.org/10.1016/B978-0-12-799945-6.00010-7

meet user requirements followed by manufacturing (accomplished by an appropriate selection of materials, processes, and tools).

10.2 Criteria for designing and manufacturing usable consumer products

10.2.1 Functionality

Most design activities are preceded by obtaining information regarding the needs of users through market research (McClelland, 1990). Figure 10.1 depicts a structured approach to obtaining information pertaining to user needs in the development of consumer products.

Conceptual design deals with the activities that happen early during product development (Dika and Begley, 1988). It involves the creation of synthesized solutions in the form of products that satisfy users' perceived needs through the mapping of

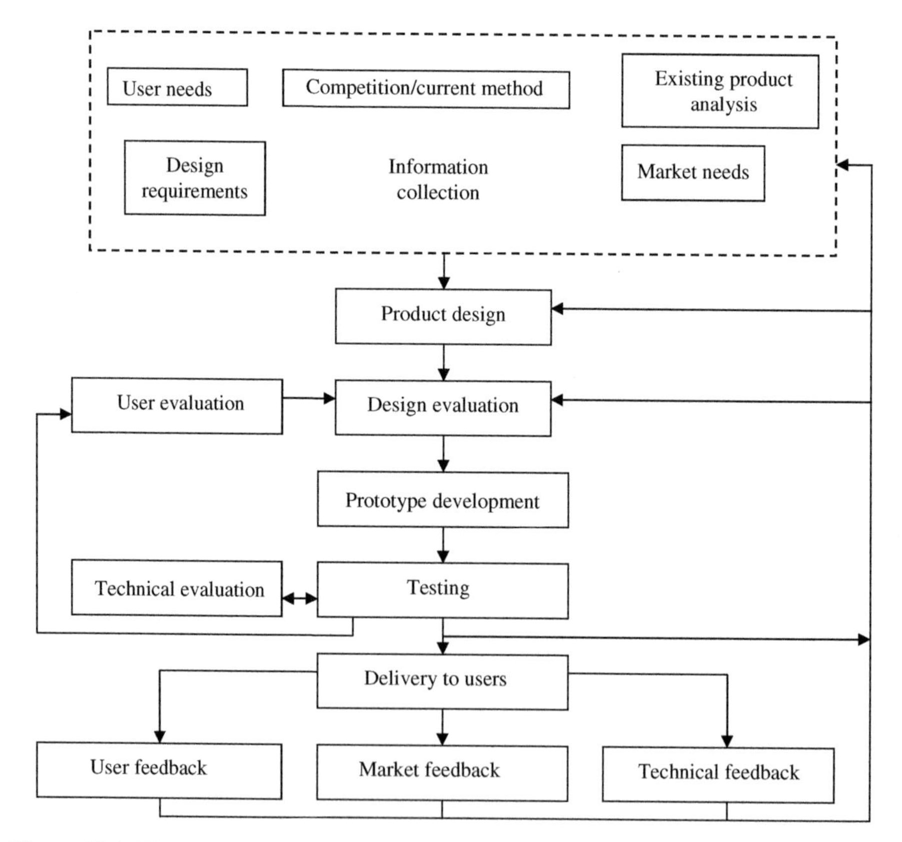

Figure 10.1 The process of design for usability.
Source: Adapted from Mital and Anand, 1992.

functional requirements in the functional domain, the design parameters in the physical domain, and the proper selection of design parameters that satisfy the functional requirements. This mapping is not unique, and the outcome depends on the creative process of the individual designer. Many techniques have been advanced to enhance the creative process, including the trigger word technique, checklist technique, morphological technique, attribute seeking technique, Gordon technique, and brainstorming technique (Suh, 1990).

In the trigger word technique, the verb in the problem definition statement is analyzed recursively to create different sets of connotations and ideas to solve the problem. The checklist method consists of a standard set of questions; each question can have more, related questions. The checklist focuses on various ways of addressing the problem, to stimulate the imagination to explore less obvious concepts surrounding the problem. The morphological chart technique involves analyzing the problem to determine the independent parameters involved. Each parameter is considered separately for possible alternative methods. All methods are tabulated in a matrix, which can be cross correlated to produce possible solutions to the problem. In the attribute seeking technique, all essential characteristics are singled out and analyzed individually, using either the trigger word or the checklist approach. The Gordon technique deals with the basic underlying concepts involved in the situation, instead of considering the obvious aspects of the given problem. This approach compels the designer to take a much broader view, analyzing the reasons why the problems exist in the first place. For instance, when designing a home disposal appliance, one may seek to eliminate the cause of trash rather than dealing with its disposal. Brainstorming is a group-ideation technique usually consisting of six to eight individuals who are well conversant with the field. A moderator defines the situation and provides an interpretation of the problem. The success of this technique depends on the compounding effect of each person in the group responding to the ideas expressed by others.

10.2.2 Ease of operation

A product is considered user friendly if the functions allocated to humans are within the limitations of their abilities and constraints, and the product–user interface is physically comfortable and not mentally stressful (Haubner, 1990; Nielsen 1993a,b). The system should be easy to learn, easy to remember, and relatively error free (Nielsen, 1992, 1993b).

As the user–product interaction continues becoming less physical and more cognitive, it is essential to understand the cognition of product semantics, that is, the symbolic interaction between users and products. Lin et al. (1996) used multidimensional scaling (MDS) to present an approach that can be used to study product semantics in product design. MDS is a process whereby a matrix of distance, either psychological or physical, among a set of objects can be translated into a representation of those objects in space. The results from MDS analysis provide designers with an idea of how to concentrate their efforts in using product semantics in consumer product design.

In recent years, consumer electronics products have intensified their graphical user interface (Schneiderman, 1998). This is made possible by incorporating embedded software of increasing size (Tervonen, 1996). Product quality in such products is greatly influenced by software quality (Kitchenham and Pfleeger, 1996). Zero-defect software can be obtained only by emphasizing quality during all phases of the software development cycle, including requirement analysis, prototype software development, realization, and testing (Rooijmans and Aerts, 1996). Finally, the consumer product can be tested for its ease of use by usability testing procedures, such as the thinking aloud method, where users work on a prototype (Jorgensen, 1990).

10.2.3 Esthetics

A customer's perception of a product's value is based in part on its esthetic appeal (Logan et al., 1994). An attractive product may create a sense of high fashion, image, and pride of ownership (Akita, 1991). The design of products should induce a positive sensual feeling (Hofmeester et al., 1996).

Kansei engineering (KE) is a technology that translates consumers' feelings and image of a product into design elements (Nagamachi, 1995). KE technology is classified into three types: KE Types I, II, and III.

KE Type I deals with design elements of new products. Customers' feelings about a product are expressed in a tree structure to get the details about the design of the product.

KE Type II utilizes current computer technologies, such as expert systems, neural network models (Ishihara et al., 1995), and genetic algorithms (Tsuchiya et al., 1996), and is called the computer-assisted *kansei* engineering system (KES). The KES architecture basically has four databases: *kansei* database, image database, knowledge database, and design and color database. The consumer enters his or her image words concerning the desired product in KES. The KES receives these words and tries to recognize them. The inference engine in this stage works by matching the rule base and the image databases. Later, the inference engine determines the design details, and the KES controller displays the part and color details of the product on the screen.

KE Type III is the mathematical logic model (Nagamachi, 1995).

Functionality and user friendliness designed into a product implies that the product can perform the desired functions without posing excessive demands on the user; the ability of the product to function satisfactorily over a period of time is indicated by its reliability.

10.2.4 Reliability

The reliability of a product is the probability that it will perform satisfactorily for a specified period of time under the stated set of conditions (Anderson, 1991). Mean time to failure, the average of mean lifetime for a population of products, is used as the measure of reliability.

Improved reliability usually is achieved through continuous improvement in materials, product design, manufacturing processes, and use environment (Alonso, 1992). Reliability growth test management is a critical component of the product assurance function (Bieda, 1992). Computer applications, such as a knowledge decision support system, often are used to assist in quantification and monitoring reliability growth during the product development process (Nasser and Souder, 1989). The following choices are available to a product designer to optimize reliability:

- Simplify the design as much as possible. The design with the least complexity and fewest parts generally exhibits higher reliability during operation. The reliability of the individual components of the product should be improved. Standard parts and materials with verified reliability ratings should be used (Priest, 1988).
- Design products with redundant, duplicate, or backup systems to enable continued operation should a primary device fail. The operation of a part at less severe stresses than those for which it is designed is known as derating (Alexander, 1992). Component derating should be used to improve the ratio of load to capacity of the components.
- Give higher priority to improving weak components than other parts. Design to avoid fatigue failures, such as corrosion fatigue (Rao, 1992). Stress concentration points are most prone to fatigue failure. Designers should eliminate sharp internal corners, since they concentrate stress. The prime reason for reduced service life of electronic products is overheating. Adequate means, such as ventilation or heat sinks, must be provided to prevent overheating.
- Adverse effects from the environment need to be reduced by providing insulation from heat sources; providing seals against moisture; using shock absorbing mounts, ribs, and stiffeners to make the product rugged against shock; and by providing shielding against electromagnetic and electrostatic radiation (Bralla, 1996).

It is either technically difficult or prohibitively expensive to produce fail-proof products. Every consumer is aware that, during the lifespan of a product, some repair or maintenance will be needed. However, when a product fails, it should fail safely and the downtime should be as short as possible. A product that can be repaired or serviced easily and quickly has a high degree of maintainability. *Serviceability* and *maintainability* generally are considered to be equivalent terms.

10.2.5 Serviceability and maintainability

Maintainability or serviceability is the element of product design concerned with assuring the ability of the product to perform satisfactorily throughout its intended useful lifespan with minimum expenditure of effort and money. Maintenance can be either preventive or breakdown maintenance. Designing for good serviceability involves providing for ease of both kinds of maintenance (Blanchard et al., 1995). There is a strong overlap between the objective of achieving high product serviceability and other desirable design objectives, such as reliability and ease of assembly and disassembly. Easy serviceability often can compensate for lower reliability. If a component is prone to failure but can be replaced or repaired easily, the consequences of failure are less severe. The availability of the product for use depends on both reliability and serviceability. High availability implies that the product is ready for full use a high percentage of the time, because the failure of components is rare, the replacement of failed parts is quick, or both (Smith, 1993).

The many options available to the designer to facilitate effective and economical service include:

- Use quick-disconnect attachments and snap fits to join high mortality parts or those that may need frequent replacement or removal for service. Funnel openings and tapered end and plug-in or slip fits facilitate easy disassembly. Press fits, adhesive bonding, riveting, welding, brazing, and soldering need to be avoided.
- Consider the use of modules that are easily replaced when necessary and easily tested to verify their operability. Modular design facilitates the identification of faults. If spare modules are available, the defective ones can be removed and repaired while they are replaced with a spare, putting the product back into service much more quickly.
- Products need to be designed for easy testability. Some testability principles include the following:
  1. Design product components such that tests can be done with standard equipment.
  2. Incorporate built-in test capability and, if possible, built-in self-testing devices in the product.
  3. Make the tests easy and standardized, capable of being performed in the field.
  4. Provide accessibility for test probes; for instance, make test points more prominent and provide access parts and tool holes.
  5. Make modules testable while still assembled in the product (Anderson, 1991).

10.2.6 Environmental friendliness

The accelerated flow of waste and emission due to the proliferation of industrial activities spurred by rising demand for consumer products is causing increased pollution of the ecosystem. A design that has minimal or no harmful effects during manufacture, use, and disposal is considered environmentally friendly (Kaila and Hyvarinen, 1996). Life -cycle assessment (LCA) tools analyze and compare the environmental impact of various product designs (Hoffman and Locasio, 1997). LCAs review a product by summing up the influence of all the processes during the life of a product on various environmental impact classes, such as ozone depletion, global warming, smog, acidification, heavy metals, pesticides, and carcinogens. The disadvantage with life cycle analysis is that, to evaluate the environmentally responsible product rating, every LCA tool requires a substantial database for process information of all stages of the life cycle and various impact classes with weighting factors for all materials, emissions, and other reaction products during the product design stage itself (Nissen et al., 1997).

Some of the options available to the designer in enhancing environmental friendliness are as follows:

- Avoid use of toxic materials in the product or production process. All use of substances such as CFCs or HCFCs needs to be eliminated. Minimize equipment cleanouts that generate liquid or solid residues (Billatos and Basaly, 1997).
- Avoid materials that are restricted in supply as well as those that are problematic to dispose of. Use recycled materials instead of virgin materials if possible (Ashley, 1993). Minimize the periodic disposal of solid cartridges, containers, and batteries. Design to utilize recycled consumables from outside suppliers. Design products to minimize liquid replenishment, such as coolants and lubricants.

- Design products to consume less energy. Also, choose the form of energy alternative that has the least harmful effect on the environment. Design should include features such as a sleep mode, which conserves energy during the time that the product is not in use.
- Avoid designs requiring spray paint finishes (Lankey et al., 1997). The need for environmentally damaging solvents can be avoided by using powder coating, roll coating, or dip painting for surface finishing of metals. Plastic parts that are molded in color eliminate the need for painting.

10.2.7 Recyclability and disposability

Every day, thousands of consumer products reach the end of their useful lives and join the waste stream. To deal with the millions of tons of landfill, it is imperative that products be designed with recycling in mind. Product recycling reduces the adverse impact on the environment by reducing the volume of materials deposited in landfills and conserves scarce natural resources (Pnueli and Zusman, 1996; Tipnis, 1994). The steps involved in a recycling program are as follows:

- Collection of worn-out products
- Disassembly of the product and sorting of incompatible materials
- Cleaning, shredding, and grinding of materials as necessary, accompanied by the separation of high value materials, such as steel, for reclaiming
- Conversion into quality-consistent usable material
- Discarding the fluff to the waste stream or landfill.

The guidelines that help reduce cost and increase revenue due to recycling are as follows:

- The product and its components should be designed such that they can be reused. The major components should be designed to be remanufactured or refurbished rather than reclaimed only for its materials.
- Minimize the number of parts a product contains, since fewer parts make material sorting for recycling easier. Avoid the use of separate fasteners if possible. Snap fit connections between parts are preferable, since they do not introduce a dissimilar material and often are easier to disassemble with a simple tool. The number of screw head types and sizes in a product should be minimized to minimize the need for constant tool changing. Use of fewer fasteners reduces the disassembly time. Modular design simplifies disassembly.
- Minimize the amount of material in the product. The less material involved, the simpler is the disposal problem. This also means that the product eventually takes up less landfill space. By designing for near-net shape manufacturing processes that minimize material scrap, designers can achieve benefits comparable to designing smaller, lighter parts.
- Reduce the number of materials in a product. Avoid use of dissimilar materials, which are difficult or impossible to separate from the basic material (Berko-Boateng et al., 1993). Since thermoplastic materials can be recycled by melting, they are preferred to thermosetting materials. As far as labels are concerned, water-soluble adhesives facilitate separation during recycling.

10.2.8 Safety

The increasing number of personal injury lawsuits filed each year in courts involving consumer products indicates that safety may be the most basic consideration in product design from both the human and cost perspectives (Heideklang, 1990; Ryan,

1983). Safety implies the absence of hazards or minimal exposure to them during the entire life cycle of the product.

Standard techniques such as fault tree analysis, failure mode analysis, and sneak circuit analysis can be used to design safety into a product (Hammer, 1980). Safety concerns often overlap reliability and ease of use. The following considerations are intended to aid the designer in creating a safe product:

- Products should be fail safe. Since users occasionally make mistakes in the operations of a product, the design should allow for human error. When such errors happen or the mechanism fails, it should not result in an accident.
- Parts that require service should be freely accessible, easily repairable, and replaceable without causing interference with other assemblies and without posing hazards to the user. Design should replace sharp corners with liberal radii, as sharp external corners present hazards during operation and maintenance of the product.
- The design of the product should be robust enough to withstand any adverse environment in which it might be used and provide safeguards against environmental factors, such as corrosion, vibration, pressure changes, radiation, and fire. Such factors could create safety hazards. The level of noise and vibration needs to be reduced. Adequate ventilation and lighting need to be provided.
- The product needs to be made from high impact or resilient materials so that it does not break into small fragments when dropped accidentally. Minimize the use of flammable materials, including packaging materials. Avoid the use of materials that are a hazard when burned, discarded, or recycled (Bralla, 1996).

10.2.9 Customizability

Customers often are willing to pay more if their individual needs are better satisfied. Design for mass customizability (DFMC) is a new approach catering to an increasing variety of customer requirements without a corresponding increase in cost and lead time (Tseng, 1996). Providing products and services that best serve customers' needs while maintaining mass production efficiency is a new paradigm for industry.

The core of DFMC is to develop a mass-customization-oriented product family architecture (PFA) with meta-level design process integration as a unified product creation and delivery process model. The inherent repetition in product marketing, design, and manufacturing can be recognized through the establishment of patterns. Once patterns are identified and formulated into a PFA, economies of scale can be applied for efficiency. The formulation of PFA enables the optimization of reusability and commonality in both product design and process selection from the product family perspective. It also provides the basis to facilitate the front end configuration to fulfill individual customer requirements.

While product customizability enables the design of products and processes to meet individual customer needs, such needs change frequently, which entails frequent modification in product design. This calls for a dynamic reconfiguration of manufacturing systems to accommodate swift changes in product design. Development of integrated manufacturing systems aimed at multiproduct, small batch production, fast and optimized design, speedy product development, and just-in-time delivery, made possible by strategic information systems, has been a strategy to achieve this (Hitomi, 1991).

In this context, agile manufacturing systems are an emerging concept in industry that seeks to achieve flexibility and responsiveness to changing customer needs.

10.3 Design support tools and methodologies

In addition to the design approaches and guidelines discussed so far, numerous other design methodologies and tools are also widely used. These include design for producibility, design for assembly, robust design, group technology, and quality function deployment (QFD).

10.3.1 Design for producibility

The design of a component has a strong effect on the attributes of the product in which it is used. Design for producibility emphasizes that design of detailed parts cannot be independent of the manufacturing process (Burhanuddin and Randhawa, 1992). Design principles and guidelines for a part that is made with one process may not apply if another process is used. For instance, if a part is to be die cast, the suitable materials, wall thickness, shape, complexity, size, dimensional tolerances, and other characteristics are significantly different from those made using metal stamping or from metal powder. The resultant part attributes, such as strength, temperature resistance, and corrosion resistance, may also be different. The selection of part features and processes should occur simultaneously. The following design principles can be applied when designing parts for producibility:

- Simplify the design of each part as much as possible (Stoll, 1988). Use simple shapes instead of complex contours, undercuts, and elaborate appendages. Parts made of simple shapes have less opportunity to be defective. Use the most liberal tolerances allowable, consistent with the quality and functional requirements of the part and the capabilities of the manufacturing processes involved.
- Select near-net shape processes, which are capable of producing a part to or near final dimensions with a limited number of operations, particularly minimum machining, such as injection molding and powder metallurgy. An injection molded part can have all its final dimensions, identifying nomenclature, finish, and color provided in a single operation.
- Avoid designs that require machining operations. Often another process can be substituted for one that primarily involves machining, with significant savings. For instance, sheet metal processes can be used to provide parts with bearing surfaces, holes, reinforcing ribs, and the like. Extruding, precision casting, cold rolling, or other similar net shape processes may provide the precision needed for elements and surfaces that otherwise would require machining.

10.3.2 Design for assembly

In this approach, the entire assembly is analyzed to determine whether components can be eliminated or combined, leading to simplified product assembly. Service and recycling are facilitated when a product is simplified. A product that is easy to assemble normally is easy to disassemble during maintenance, repair, or disassembly (Eversheim and Baumann, 1991). Simpler assemblies often can be brought to market sooner because of fewer parts to design, procure, inspect, and stock, with less probability that a delay might occur. Products with fewer parts also tend to have higher reliability (Boothroyd, 1994; Boothroyd and Alting, 1992).

Processes such as injection molding and die casting permit the very complex parts that result when separate parts are combined. By selecting flexible material and making wall sections thin, hinges and springs can be incorporated into plastic parts. Integral snap fit elements, tabs, crimped sections or catches, press fits, and rivets can be used to replace threaded fasteners (Joines and Ayoub, 1995). With some manufacturing processes, it is possible to incorporate elements such as guides and bearings in the basic piece by selection of appropriate materials and processes.

Modularity improves reliability and serviceability. Therefore, a design should include modular subassemblies while avoiding too many levels of subassembly (Karmarkar and Kubat, 1987). Adopt a layered, top-down assembly where each successive part in the product can be added to the assembly from above rather than from the side or the bottom. Design parts such that they are self-aligning and cannot be inserted incorrectly. Design very small or highly irregular parts that are manually assembled for easy handling by adding a grasping element to the parts.

10.3.3 Robust design

The goal in robustness of design is a product that will never fail to perform its intended function during its useful life. Robust design methodology, popularly known as the *Taguchi technique*, provides a way to develop specifications for robust design using the design of experiments theory. The procedure attempts to find out the settings of product design parameters that make the product's performance insensitive to environmental variables, product deterioration, and manufacturing irregularities. Controlling the causes of manufacturing variations often is more expensive than making a product or process insensitive to these variations (Juran et al., 1974).

Taguchi separates off-line quality planning and improvement activities into three categories: system design, parameter design, and tolerance design. System design is the application of scientific and engineering knowledge to produce a functional prototype. The prototype model defines the basic product or process design characteristics and their initial settings. Tolerance design is a method for scientifically assigning tolerances so that total product manufacturing and lifetime costs are minimized (Nevins and Whitney, 1989).

10.3.4 Group technology

The group technology procedure classifies the system into subsystems and subdivides them into part families based on design attributes and manufacturing similarities (Chang et al., 1991). Group technology can be used for product design and manufacturing system design. Components having similar shape are grouped together into design families, and a new design can be created by modifying an existing component design from the same family. Using a coding method, each part is given a numeric or alphabetic code, based on its geometrical shape, complexity, dimension, accuracy, and raw material. Using this concept, composite components can be identified. Composite components are parts that embody all the design features of a design family or subfamily (Farris and Knight, 1992).

For manufacturing processes, parts with similar processing requirements constitute a production family. Since similar processes are required for all family members, a machine cell can be built to manufacture the family of parts. Production planning and control, as a result, is made much easier and the cycle time to manufacture a product is greatly reduced, even while maintaining product variability.

10.3.5 Quality function deployment

QFD is a method of translating customer requirements into product and process design (Akao, 1990). The QFD technique uses the concept of the "house of quality." It translates customer views systematically into key engineering characteristics, planning requirements, and finally production operations (Bergquist and Abeysekara, 1996). This is achieved through four key documents: the product planning matrix, product deployment matrix, component deployment matrix, and the operating instruction sheet.

The purpose of the product planning matrix is to translate customer requirements into important design features. Individual customer needs are ranked for importance, and the cumulative effect on each of the design features is obtained. A product deployment matrix is made for each of the product features, all the way down to the subsystem and component level. The product deployment matrix depicts the extent to which the relationship between component and product characteristics is critical and affordable. If a component is critical, it is further deployed and monitored in the design, production planning, and control. The component deployment matrix expands the list of components or the exact parameters required to design a complete component. The operating instruction sheet is the final document that defines operator requirements as determined by the actual process requirements, process checkpoints, and quality control points (Day, 1993). Thus, QFD tries to achieve high quality products by using the philosophy of concurrent engineering (Parsei and Sullivan, 1993), which integrates product design, process design, and process control (Maduri, 1993).

10.4 Design methodology for usability

Here we provide a comprehensive methodology to ensure product usability. Case studies relating to real-life consumer product design also are presented to demonstrate the use of the method.

The design method consists of the following distinct sections: development of a set of checklists for evaluating product usability using all criteria and a set of checklists with guidelines for designing and manufacturing usable products. Each set consists of nine checklists, one for each of the individual usability criteria: functionality, ease of operation, esthetics, reliability, maintainability and serviceability, environmental friendliness, recyclability and disposability, safety, and customizability.

The checklist for evaluating usability can be used to identify deficiencies in a product or prototype. The checklist for design and manufacturing for usability can be used to find product- or process-related design options, either for developing a new product or removing deficiencies from an existing product. Figure 10.2 depicts the entire process of checklist development.

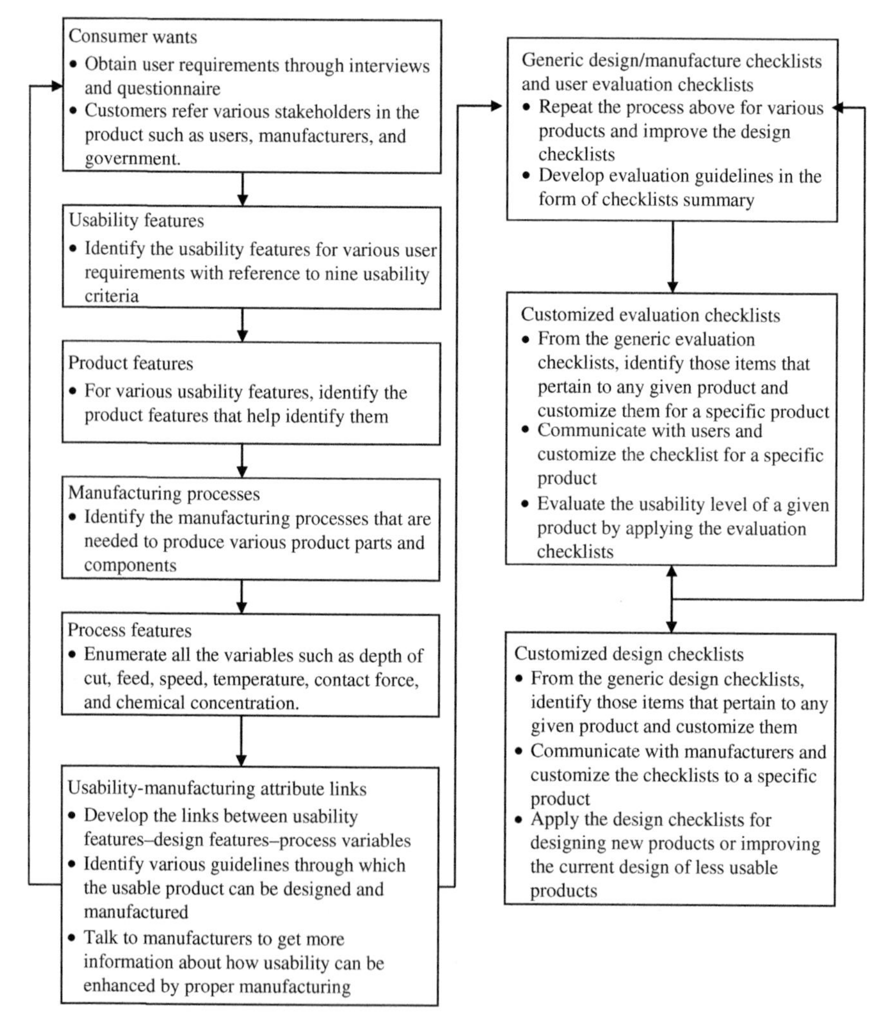

Figure 10.2 Complete process of checklist design.

10.4.1 Development of generic usability evaluation checklists

We conducted an extensive survey of existing guidelines, checklists, and questionnaires on each of the usability criteria. Handbooks, published literature, data manuals, and the like were searched to compile such data. Also, an evaluation of industrial procedures were collected on each of the usability criteria by contacting leading companies involved in product design for usability. Each criterion consists of several related items. Each question is designed to be answered on a 7-point rating scale. At the end of each checklist, a general question is included to obtain the overall evaluation of particular usability criteria.

10.4.2 Development of generic design and manufacturing checklists

Based on the guidelines collected from the preceding sources, we developed a set of checklists to address design and manufacturing guidelines. These guidelines were based on surveys of existing literature, handbooks, data manuals, and information provided by those in the industry. The design guidelines provide a listing of options available to the designer in developing appropriate product characteristics through design. By running through them, the designer can understand the product and process options available to enhance a particular usability criterion.

Checklists developed like these, however, are limited by the level of information available. Incorporating new design data, made available by future studies, can enhance the utility of a design checklist.

10.4.3 Reliability and validity testing

The Cronbach alpha method was used to measure the internal consistency of the questionnaire items. A score of 0.4 or above is considered an acceptable measure of significance correlation. To test the validity of the questionnaires, it is standard practice to compare the scores with another set of questionnaires considered to be standard. However, if no standard is available for the product under consideration, the validity of the questionnaires is tested by comparing the average of all scores in the questionnaire with the overall score for each of the usability criteria.

10.4.4 Testing the effectiveness of the design/manufacturing guidelines

If the design and manufacturing guidelines are useful in developing a more useful product, the outcome of usability evaluation for different makes of the same product should correspond with the outcome of implementing the design and manufacturing guidelines. On an aggregate level, for a given sample of users and designers, the means should not be significantly different if they are related (Figure 10.3). A t-test is performed between the average score of the usability evaluation questionnaires and the average scores of the design and manufacturing questionnaires. Similarly, if the outcome measures are grouped by product models and the usability evaluation scores

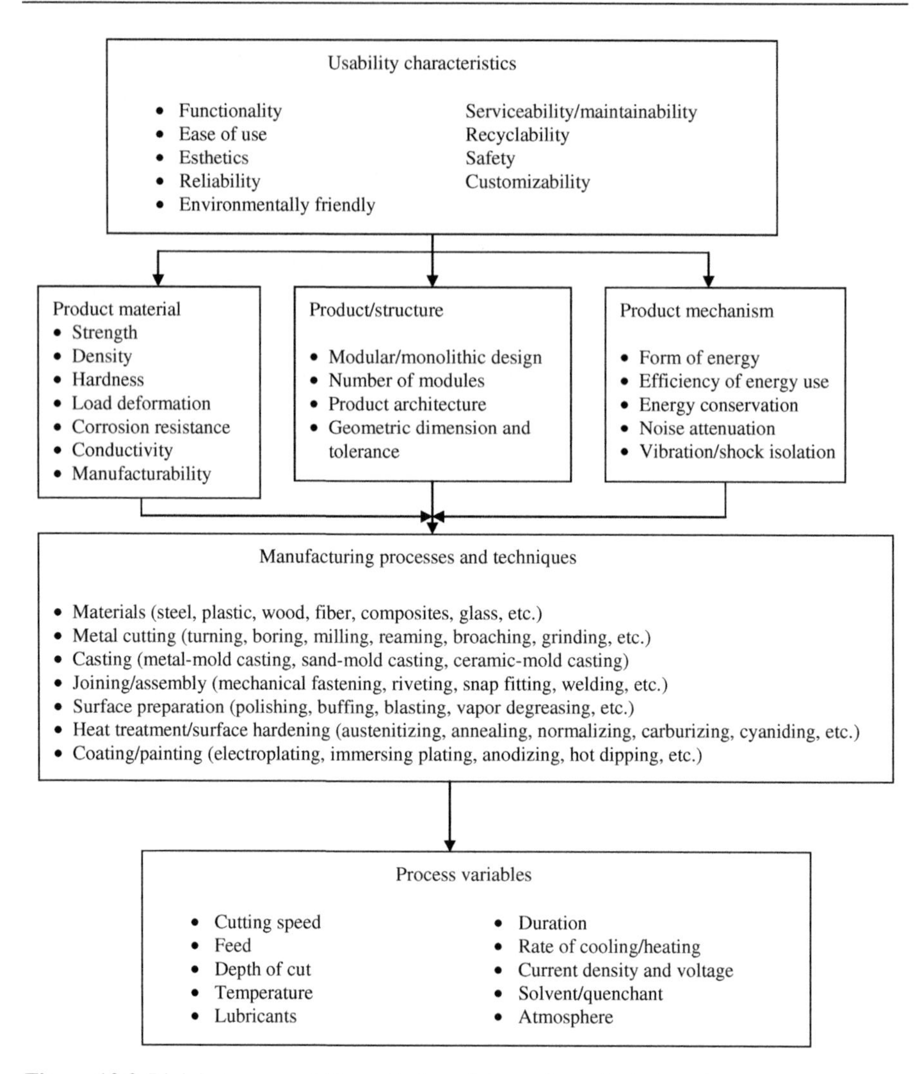

Figure 10.3 Link between usability criteria and design/manufacturing phases.

are correlated with design and manufacture scores, a positive and significant correlation implies a relationship. A correlation analysis was performed between the average usability evaluation scores and the average design/manufacture guideline implementation after the data were grouped by product model.

10.5 Generic checklist design: methods and case studies

The information regarding manufacturing processes that enable the designer to produce usable products is sought from numerous sources, such as users, manufacturers, manufacturing handbooks, and other published literature. Usability-manufacturing links are obtained using usability transformation matrices similar to QFD tables.

A usability transformation matrix is a structured approach to define customer requirements and translate them into specific steps to develop the required products. It takes into account customers' requirements, desires, and preferences throughout all processes, beginning with the concept design activities and continuing throughout the production operations on the factory floor. Transformation matrices use a series of relationship matrices to document and analyze the relationships among various factors. While the details of the matrices vary from stage to stage, the basic underlying concepts remain the same. In the product planning stage, customer requirements are identified and translated into technical design requirements. In the part deployment stage, technical requirements are converted into part specifications. During the process deployment stage, product characteristics are converted into process characteristics. During the manufacturing deployment stage, manufacturing processes are related to specific variables that control the processes under consideration. Through such a stepwise transformation of customer usability requirements into process variables, it is possible to control the usability of a product. Readers can refer to Chapter 9 on product functionality for a detailed explanation of this topic. To demonstrate the relationship between usability attributes and the manufacturing attributes to which they are linked, two case studies have been presented. The first case study involves a manual can opener product design; the second study involves a bread toaster.

10.5.1 Product development for the usability of a can opener

A study was carried out to demonstrate the practical utility of the product design methodology for a manual can opener. Fourteen individuals participated in the study. The preferences of these users were obtained through one-on-one interviews. Each interview lasted approximately 20 min and was conducted in the home of the user. The questions were open-ended. All users owned manual can openers and were very familiar with what would make it usable. The questions pertained primarily to extracting the users' primary needs. The users also were asked to rate the needs based on a scale of 1–10, 10 being most important.

The interviews revealed that users wanted a can opener that would open without skipping, jamming, or rolling off the track. The can opener should be able to sever lids completely and cleanly without leaving behind slivers of metal. A good can opener should not jiggle the lid and cause it to splatter or submerge the lid in the liquid as the cutter progresses around the can.

The users considered a manual can opener to be usable if the following criteria were met:

- It easily pierces the lid.
- It is easy to turn.
- It is comfortable to grip.
- It rolls without slipping.
- It cuts smoothly and neatly.
- It is durable.
- It has a good appearance.

10.5.1.1 Technical requirements deployment

The usability requirements and overall rankings obtained are in the horizontal portion of the first stage of the QFD process (Table 10.1). The usability requirements and rankings are translated into the language a company can use to describe a product for design, processing, and manufacture. The objective of this step is to develop a list of technical requirements that should be worked on to satisfy the customer's voice. These technical requirements, listed in the vertical portion of the QFD, are low crank torque, low handle grip force, tight grip of drive sprocket on can rim, good ripping action of cutting wheel, high corrosion resistance, good surface finish of the frame, and high structural rigidity.

Next, the relationships between the technical requirements and user requirements were established to identify the relative importance of the technical requirements. Every user requirement in the horizontal portion was compared with each technical requirement. The degree of relationship is market at the intersection. Depending on the degree of correlation, a strong relationship is assigned a score of 9 points, a moderate relationship is assigned a score of 3 points, and a weak correlation is assigned a score of 1 point. A strong relationship between a user requirement and a technical requirement indicates that changing one would greatly influence the other. The overall weighting of each of the technical requirements was obtained by multiplying the customer weighting and the numerical weighting of the relationship and summing all values of all relationships. The values of all technical requirements thus obtained are simplified further by dividing and rounding up by a factor of 10. The purpose of calculating the overall weighting is to identify the technical characteristics that influence usability to the greatest extent.

The technical requirements may have some relationship with each other. Attempting to achieve or improve one requirement may affect some other requirement positively or negatively. A positive relationship between different requirements implies that a positive change in one can bring about a similarly positive change in the others. Such a relationship is denoted by a plus sign, and a negative relationship is denoted by a minus sign. A principal benefit of generating such co-relationships is that they warn against negative relationships. In other words, a warning indicates that any action to improve one requirement may have an adverse effect on some other requirement. It is necessary to examine each negative relationship to determine how the design can be changed or desensitized to eliminate or reduce detrimental effects.

The competitive analysis portion of the QFD table evaluates how the product satisfies each of the customer or technical requirements and is rated using five-point scale. Competitive analysis can be used in comparing the usability rating of the product with those of its competitors. Thus, we have a list of customer and technical requirements at the end of the first stage of QFD as well as their relationships, their overall weighting, and target values to be achieved, if any. The other three stages proceed along steps analogous to the first stage. These stages are product deployment, process deployment, and manufacturing deployment.

10.5.1.2 Product deployment

The purpose of product deployment is to translate the previously technical requirements into product specifications (Table 10.2). The technical requirements taken from

Table 10.1 **First stage QFD for a can opener: transformation from user requirements to technical requirements**

	Technical requirements								Competitive analysis				
Customer requirements	**Importance**	**Low crank torque**	**Low handle grip force**	**High cutting force of blade on lid**	**High contact force of sprocket on rim**	**Good rustproofing**	**Smooth surface and edge finish**	**Structural rigidity**	**1**	**2**	**3**	**4**	**5**
Easy to pierce	4		S	S							X		
Easy to turn	8	S			S						X		
Comfortable grip	7		S				W	M		X			
Rolls without slipping	5				S			W			X		
Cuts smooth and neat	6			S	W						X		
Leaves no slivers	3			S							X		
Durable	2					S	M	S				X	
Good appearance	1					M	S					X	
Competitive analysis													
1			X				X						
2		X		X	X			X					
3						X							
4													
5			X				X						
Weighting		7	10	12	12	2	2	4					

Relationships: Strong (S) =5; medium (M) =3; weak (W) =1.

Table 10.2 Second stage QFD for a can opener: transformation from technical requirements to product features

Technical requirements	Product features								Comments
	Importance	Material of handle	Surface and edge regularity of handle	Crank surface and edge regularity	Blade hardness	Drive sprocket hardness	Surface finish	Joints of parts	
Crank torque	7			W	S				Lower manual force preferred
Handle grip force	10	W	M			S			Low grip force preferred
Cutting force of blade on can lid	12				S				Blade should be hard enough to cut
Contact force of sprocket on can rim	12					S			Must be high enough to grip
Rustproofing	2						S		Should last the entire life
Frame finish	2		M	M			S		Should be smooth
Structural rigidity	4	W						S	Should not deform when gripped
Competitive analysis									
1									
2		X					X		
3			X	X		X		X	
4					X				
5									
Weighting		1	4	1	17	20	4	4	

Relationships: Strong (S) =5; medium (M) =3; weak (W) =1.

the vertical column of the previous QFD stage are listed as rows in this stage. Based on previous design experience, the product features necessary to satisfy these technical requirements were identified and listed in the vertical column. The relationships among product feature requirements were identified, and the overall weighting of various product features was established as before.

10.5.1.3 Product architecture

The can opener has five main parts: upper handle, lower handle, blade, crank, and the drive sprocket. The upper handle is joined to the crank and drive sprocket to form a subassembly. The lower handle is joined to the blade to form a second subassembly. The two subassemblies are joined to form the overall assembly. The upper and lower handles are used to hold the opener and provide the gripping force. When mounted properly on the can and gripped with adequate pressure, the blade pierces the can and the sprocket wheel holds on to the top of the can. The crank wheel is used to apply the torque that helps the blade cut the can lid and rotate the can until the lid is completely severed.

10.5.1.4 Process deployment

The purpose of process deployment (Table 10.3) is to determine the manufacturing processes necessary to actually produce the product by relating various product features to specific manufacturing operations. The critical product features identified in the previous stage are listed in the horizontal portion of the matrix. The major process elements necessary to develop the product, extracted from the process flow diagram, are shown at the top of the column section of the matrix. The relationships between the column and row variables are then determined. The overall weight of the process features is determined to find which features affect usability the most.

10.5.1.5 Manufacturing processes

The process description of the can opener was obtained from the manufacturer. The upper handle is cut from an SAE 1008 steel strip and the lower handle is cut from an SAE 1008 steel wire by blanking. The handles are individually subjected to stamping operations using progressive dies. The piercing and bending operations produce two holes and a twist in the upper handle. Two protrusions are formed using die pressing on the lower handle. The blade is cut from an SAE 1050 steel strip, swaged at the top to produce a sharp cutting edge, bent 90° twice, and pierced to produce a hole at the center and two holes at the bottom, using a progressive die in a punch press. The crank is cut from an SAE 1008 steel strip and trimmed to achieve the desired shape. These pieces are tumbled to remove the burrs resulting from the stamping operations on the upper handle, lower handle, blade, and crank as well as surface roughness. The drive sprocket is cut from SAE 1050 steel and stamped to obtain a gear shape. The blade and drive sprocket are heat treated. All five parts are then nickel plated to promote corrosion resistance and appearance.

Table 10.3 Third stage QFD for a can opener: transformation from product features to process features

Product features	Importance	Upper handle (part 1)		Lower handle (part 2)		Blade (part 3)			Crank (part 4)		Drive sprocket (part 5)		Edge and surface preparation	Surface treatment	Subassembly 1 (parts 1, 4, 5)	Subassembly 2 (parts 2 and 3)	Overall assembly	Comments, achieved by
		Blanking	Stamping	Blanking	Stamping	Blanking	Swaging	Heat treatment	Blanking	Stamping	Stamping	Heat treatment	Tumbling	Nickel plating	Swaging	Swaging	Riveting	
Smooth handle surface and edge	4	S	M	S	M								S	M				Deburring and tumbling
Smooth crank surface and edge	1								S	M			S	M				Coating and tumbling
High blade hardness	17					W	M	S										Proper heat treatment
High drive sprocket hardness	20										W	S						Proper heat treatment
Good surface finish	4												W	S				Proper surface coating
Strong joints of parts	4														S	S	S	Riveting and swaging
Cost level																		
Low		X	X	X	X	X	X		X	X	X		X		X	X	X	
Medium								X				X		X				
High																		
Weighting		4	1	4	1	2	5	15	1	0	2	18	5	5	4	4	4	

Relationships: Strong (S) =5; medium (M) =3; weak (W) =1.

The drive sprocket and crank are assembled with the upper handle and swaged to produce a subassembly. The blade is inserted into the two protrusions on the lower handle and swaged to produce another subassembly. The two subassemblies are riveted together to produce the final assembly. The main manufacturing operations involved are blanking, piercing, bending, heat treatment, nickel plating, riveting, swaging, and tumbling. The processes having the most influence on the manufacture of the can opener are the progressive die operations, heat treatment, and inorganic surface coating.

10.5.1.6 Manufacturing deployment

Planning the manufacturing is the culmination of the work done in the previous three stages (Table 10.4). During this stage, the various manufacturing techniques necessary to manufacture the product are related to process attributes that affect them positively or negatively. For instance, the hardness of the blade is affected by the rate of cooling during the heat treatment process. The rate of cooling, in turn, is controlled by the properties of the quenchant. Although the manufacturing process adopted for producing the can opener is affected by numerous process variables, only those variables that affect the usability of the product are considered. The manufacturing techniques are listed in the horizontal portion and the process variables are listed the vertical portion of the QFD matrix. The relationship between them is shown in the matrix, along with any appropriate target values.

The blanking, punching, trimming, and piercing operations are performed using a progressive die on a punch press. The progressive die performs its operations as the stock passes through the press. The following progressive die press operations influence several basic design parameters:

- Dwell is the portion of the press stroke necessary to completely clear the die from the part so that the part may be advanced.
- Pitch is the distance between work stations in the die and must be constant between all stations.
- Press stroke (in inches).
- Drawing speed (fpm) for the material to be fabricated.
- Press speed (strokes per minute).
- Ram speed (fpm).
- Die materials include wrought and cast tool and die steels, wrought and cast carbon and low alloy steels, plain and alloyed cast irons, plain and alloyed ductile irons, sintered carbides, cast aluminum bronzes, zinc alloys, and various nonmetallic materials (Cubberly and Bakerjian, 1989).

Heat treatment is applied to components to impart properties that make a product more usable:

- Increased surface hardness
- Production of necessary microstructure for desired mechanical properties, such as strength, ductility, and toughness
- Releasing residual stresses
- Removal of inclusions, such as gases
- Alteration of electrical and magnetic properties
- Improvement of wear and corrosion resistance.

Table 10.4 **Fourth stage QFD for a can opener: transformation from process features to process variables**

Process features	Process variables											Comments
	Importance	**Pressure**	**Temperature**	**Chemical concentration**	**Duration**	**Quench solution**	**Speed**	**Punch-die clearance**	**Rate of cooling**	**Rivet-hole clearance**	**Current density**	
Blanking	11	S					M	S				Progressive die operation
Stamping	4	M					S	S				Progressive die operation
Swaging	13	S					S					Sharpens the blade
Heat treatment	33		S	S	S	S			S			Increases blade sharpness
Tumbling	5			M	M		M					Removes burrs
Nickel plating	5			S	M		W				S	Improves appearance
Riveting	4	S					W			S		Assembly operation
Range	Units	kPa	°C	ppm	min/s	kg/ft.^{3}	in./min	µin.	°/min	µin.	A	

Relationships: Strong (S) =5; medium (M) =3; weak (W) =1.

Some heat treatment processes are austenitizing, equalizing, quenching, annealing, normalizing, carburizing, nitridring, carbonitriding, chromizing, boronizing, resistance hardening, induction hardening, flame hardening, electron beam hardening, and laser hardening. The factors that affect the result of heat treatment processes include:

- Composition of the metal to be treated
- Critical time–temperature transformation relationship of the metal
- Response of the metal to quenching
- Method of quenchant application
- Temperature control and timing.

Inorganic coating is intended to impart the following properties to the surface of the metal substrate:

- Corrosion protection
- Prepainting treatment
- Abrasion resistance
- Electrical resistance
- Cold forming lubrication
- Antifriction properties
- Decorative final finish
- Easy cleanability.

Some of the methods of surface coating are conversion coating, anodizing, thermal spraying, hard facing, porcelain enameling, chemical vapor deposition, vacuum metalizing, ion vapor deposition, sputtering, flexible overlays, and ion implantation (Dallas, 1976).

Conversion coating forms the coating on a ferrous or nonferrous metal surface by controlled chemical or electrochemical attack. Anodizing is the electrolytic treatment of metals that forms stable films or coating on the metal surface.

Thermal spraying is the process of depositing molten or semimolten materials so that they solidify and bond to the substrate. Porcelain enameling forms highly durable, alkali borosilicate glass coatings that are bonded by fusion to various metal substrates at temperatures above 800°F. Hot dipping is a process by which the surface of a metal product is coated by immersion in a bath of molten metal. Chemical vapor deposition is a heat-activated process relying on the reaction of gaseous chemical compounds with suitably heated and prepared substrates. The ion vapor deposition process takes place in an evacuated chamber in which an inert gas is added to raise the pressure and ionize when a high negative potential is applied to the parts to be coated. In vacuum metalizing, a metal or metal compound is evaporated at high temperature in a closed, evacuated chamber then allowed to condense on a workplace within the chamber. Ion implantation is a process by which atoms of virtually any element can be injected into the near surface region of a solid by a beam of charged ions.

10.5.1.7 Discussion

Through QFD analysis, a clear progression of relationships linking product usability features and process variables affecting manufacturing are established. The QFD matrices indicate that the overall usability of a product can be enhanced by adopting an optimum range of values for the process variables. Thus, to enhance usability, those process variables that affect highly ranked usability features must be controlled.

The critical can opener usability features and manufacturing processes necessary to achieve them are as follows:

- **Ease of cutting**: A hard blade makes cutting easier. The desired hardness can be achieved by proper heat treatment of the blade. Low-carbon steels lack hardenability. For this reason, it is imperative to use medium-carbon steels for the blade and drive sprocket. Both are made of SAE 1050 steel with a carbon content of between 0.48% and 0.55%. The temperature and application of quenchant are the other major process variables. These parameters should be chosen such that problems, such as decarburizing, scaling, quench cracking, residual stresses, and dimensional changes, normally associated with heat treatment are controlled.
- **Smoothness of lid cutting**: The cutting edge should be sharp enough to rip through the can lid. This can be achieved by appropriate swaging of the blade's edge. Swaging improves the tensile strength and surface hardness, as it results in improved flow of metal and a finer grain structure.
- **Slip-free operation**: This characteristic depends on how tightly and smoothly the drive sprocket rolls on the outside of the can rim. Movement of sprocket roll depends on the hardness and wear resistance of the drive sprocket. These properties can be improved by heat treatment and nickel plating.
- **Safety in handling**: The can opener should be free from sharp edges and burrs. Various manufacturing processes, such a blanking, produce burrs and sharp edges. This necessitates secondary operations, such as tumbling, for the removal of burrs from corners, holes, slots, and surfaces of parts and smoothening of edges by means of radiusing.
- **Comfortable grip**: A comfortable grip is ensured by smoothing the surface of the handle and radiusing the edge. Tumbling results in a general reduction of the plane surface roughness from 20 to 5 μin. (rms) and generation of radii to the order of 0.015 in. on the exposed edges.
- **Appearance and durability**: This depends on the surface preparation and coating. Nickel plating is used to impart luster and corrosion resistance. While the former improves the appearance, the latter enhances durability.

Some of the processes produce results that reduce the usability of the product, and care needs to be taken to minimize the detrimental effects of such processes. Some examples follow:

- The blanking operation produces burrs. These are sharp edges along the shearing lines of cut parts. Production of burrs depends on excessive die-punch clearance and dull cutting edges of the die. The die-punch clearance needs to be properly designed and worn die edges eliminated.
- When specifying the rivet for joining the two subassemblies of the can opener, the relationship between the diameters of the rivet body and the work hole must be determined to maximize shear strength. An oversized work-hole diameter prevents the rivet from filling the hole to form a solid assembly and results in improper fastening. On the other hand, an undersized work-hole diameter slows the automatic feeding and clinching of the rivet.

10.5.2 Product development for the usability of a toaster

This section describes the practical utility and implementation of the design for usability method on another widely used consumer product: a bread toaster. Thirteen users participated in this case study.

A toaster works by applying radiant heat directly to the bread slice. When the bread's surface temperature reaches about 310°F, a chemical change known as the Maillard reaction begins ("Three Ways to Brown Bread," 1990). Sugars and starches start to caramelize and turn brown, beginning to take on the intense flavors characterizing toast. If the heating is excessive, the underlying grain fibers start turning to carbon—in essence, burning the toast. A toaster's primary job is to control the amount and placement of the heat it delivers to a slice of bread.

10.5.2.1 User requirements

Based on interviews with users, the following were identified as the primary usability requirements:

1. Functional requirements:
   - A toaster should produce a toast of ideal golden brown color.
   - The toast should be uniformly browned over the entire slice on both sides.
   - Once color control is set, toast of the same color should be turned out every batch.
   - The toaster should accept thin bread slices as well as thicker ones, such as bagels.
   - The toasting action should not take too long.
2. Ease of use:
   - The controls such as push-down lever and sliding knob should be easy to operate.
   - The controls should be conveniently located (in some designs, the slide-down lever is placed along the side panel. This position of the lever disturbs the balance of the gadget while in use and a downward push tends to tip over the toaster).
3. Esthetics:
   - The overall external appearance should be pleasant and stylish.
4. Maintainability:
   - Removal of bread crumbs should be easy.
   - The outside of the toaster should not stain permanently. It should be possible to clean the outside by wiping with a moist towel.
5. Reliability:
   - Since a toaster is intended for daily use by the entire family, it should be sturdy enough to perform reliably. It should be durable.
6. Energy consumption:
   - Electrical energy consumption should be minimized. Although a toaster does not cost too much in terms of energy costs, it does draw a fairly large amount of current. Most two slicers draw 5–9 amperes.
7. Safety:
   - The outside of the toaster should not be so hot as to burn the skin.
   - The toaster should protect the user from being electrocuted. Using a toaster simultaneously with another electric appliance, such as an electric frying pan, could blow a fuse or trip an electric circuit breaker. The toaster should not cause electric hazards of this sort.
8. Recyclability:
   - At the end of its useful life, the toaster should be easy to dismantle and dispose of. The parts of the toaster should be recyclable.

10.5.2.2 Technical requirements deployment

The customer requirements were translated into appropriate technical requirements as follows (Table 10.5):

The temperature attained inside the toaster should be just adequate to brown the slice, not too hot to burn the bread nor too cold to attain the color.

- The application of heat should be uniform enough to brown the slice uniformly.
- The width of the toaster slot should be adequate enough to accept sliced bread as well as thicker bagels.
- The time of exposure for the slice to heat should be adequate enough to produce a fine toast.
- The force required to push the lever should be within acceptable limits for small children and the elderly.
- The location of controls and various knobs should facilitate easy access.
- The color and gloss of external finish and the product shape should be pleasant in appearance.
- The outer casing of the toaster should be stain resistant and cleanable with water.
- The outer casing and inner frames should have high impact resistance and resist breakage on falling.
- The thermal insulation should be high enough so the toaster does not feel hot on the outside.
- The thermal insulation should be such that it avoids heat dissipation, thereby conserving electrical energy.
- The electrical insulation should be good enough and the electrical circuit should never come in contact with the metal frame (responsible for causing electrocution).
- The toaster material should provide adequate structural integrity to the product. At the end of its useful life, the toaster should be recyclable or reusable.

10.5.2.3 Product deployment

During the next stage, the conceptual design of the product is chosen to implement the technical requirements as detailed in the preceding discussion. This involves the functional mechanism, the component subassemblies related to these functions, and the product architecture (Table 10.6).

10.5.2.4 Product architecture

The toaster used in this study is operated mechanically. It has an inner U frame. On one end of this frame, a steel rod is attached. The push-down lever moves on this rod. The pull force exerted by a steel coil attached to the toast lever always tries to keep the lever on top. As the lever is pushed down, the bread holder inside the frame is lowered. A copper rod runs through the bottom of the toaster with one of its ends connected to a contact switch. This contact switch assembly comprises a heating coil and the completed electrical circuit. The other end of the copper rod is activated by the toast lever as the latter is lowered to the bottom. The heating filament is wrapped on two sheets of mica held in place vertically by attachments on the U frame.

As the toast lever is lowered to the bottom, the metal strip moving against the vertical rod locks it in place. The lowering of the lever also closes the circuit and heats the coil. A bimetallic strip attached to the bottom of the U frame acts as a thermostat. This strip expands when heated and pushes a rod that releases a weight. The falling weight, in

Table 10.5 First stage QFD for a toaster: transformation from user requirements to technical requirements

Customer requirements	Technical requirements								Competitive analysis				
	Importance	**Uniformity of heat application**	**Duration of heat application**	**Force on toast lever**	**Color of exterior**	**Thermal/ electrical insulation**	**Accessibility**	**Material properties**	**1**	**2**	**3**	**4**	**5**
Functional requirement	8	S	S			W		W		X			
Ease of use	7			S								X	
Esthetics	4				S			M		X			
Maintainability	3							S			X		
Reliability	5	M	M	M				M				X	
Energy conservation	2					S					X		
Safety	6					S						X	
Recyclability	1							S			X		
Competitive analysis													
1													
2				X	X								
3		X				X	X						
4			X					X					
5													
Weighting		9	9	8	4	8	3	7					

Relationships: Strong (S) =5; medium (M) =3; weak (W) =1.

Table 10.6 **Second stage QFD for a toaster: transformation from technical requirements to product features**

Technical requirements	**Product features**								**Comments**
	Importance	**Thermostat**	**Bread holder**	**Toast lever assembly**	**Color control knob**	**Side and end covers**	**Crumb tray**	**Material of parts**	
Duration of heat application	9	S		W	M				Should be able to caramelize the bread
Uniformity of heat application	9	M	S						Should be able to produce uniform color on the toast
Force on toast lever	8			S	W				Low enough to be operated by children
Exterior finish	4					S		W	Improves appearance
Thermal and electrical insulation	8					S		M	Exterior should not be hot
Accessibility	3					M	S		Should be easy to remove crumbs and for cleaning
Structural rigidity	7							S	Should withstand heat
Competitive analysis									
1									
2		X			X				
3			X	X		X		X	
4							X		
5									
Weighting		11	8	8	4	12	3	9	

Relationships: Strong (S) =5; medium (M) =3; weak (W) =1.

turn, releases the metal strip holding the lever down. On being released, the lever moves up, carrying the weight along with it, and causes the bread holder to pop up. Adjusting the color control knob changes the length the bimetallic strip has to expand before it releases the toast lever. The duration of heating is controlled by the color control knob.

The U frame housing is enclosed by two side covers and two end covers. At the bottom of the frame is a crumb tray to retain falling bread crumbs.

10.5.2.5 Process deployment

The product features developed through the selection of the concept design can be implemented only through appropriate selection of process features such as materials, machines, and tools as depicted in Table 10.7.

10.5.2.6 Manufacturing processes

Since the end bracket imparts structural rigidity, it is made of galvanized steel 40 milli-inches thick. The end brackets and base plate are stamped using a punch press. The end brackets then are mounted vertically on both ends of the base plate using spot welding. The assembly containing the coil and the weight is tabbed onto the left end bracket. The vertical rod is clamped between the coil assembly and the base plate. It holds the toast lever. The bread holder rod is made of steel and coated with copper to prevent rusting and contamination of food. Three mica sheets with heating coil wound on them are mounted onto the base plate, one at the center and the other two on the sides. The heating element is made of Ni-chrome and wound on mica sheets 15 milli-inches thick. Mica is highly heat resistant and withstands the temperature of the heating coil. Since it is a nonconductor of electricity, it can be safely mounted onto the end bracket. The electrical switch assembly is tabbed to the right side of the end bracket and connected to the electric cord. The thermostat assembly is tabbed with the base plate. The crumb tray is made of galvanized steel sheet 18 milli-inches thick and pivoted with the base plate using a thin steel rod.

The U frame attached to the end bracket serves as the intermediate enclosure, covering the top and sides of the toaster. It is made of galvanized steel 25 milli-inches thick and chrome plated on top of a nickel coating. The chrome and nickel coating make the cover rustproof, and the polish helps retain heat inside the toaster due to high reflectivity. The U frame and side covers are made by cold rolling. The end covers are made by plastic molding using thermoset polyester. The entire assembly is held in place by screws. The steel covers are painted on the outside to enhance appearance. High reflectivity due to aluminized paint on the inside of the side covers keeps the external body temperature down. The thermosetting material is a poor heat conductor, enabling the cover to remain cool. The color of the end covers is chosen to meet customer preferences.

10.5.2.7 Manufacturing deployment

The processes and machine tools used to manufacture the toaster components need to be closely controlled in order to achieve the desired quality. The desired goal of

Table 10.7 Third stage QFD for a toaster: transformation from product features to process features

Product features	Process features												Comments
	Importance	Progressive die operation			Compression molding	Assembly operation			Painting and inorganic coating			Shredding/ melting	
		Blanking	Stamping	Tabbing		Spot welding	Screw fastening	Riveting	Painting	Chromium coating	Nickel coating		
Thermostat assembly	11	M	M	M									Assembly purchased from vendor and checked for quality
Bread holder	8	W	W	W		W				M			Should be rust proof and not contaminate toast
Toast lever assembly	8	W	W	W									Pressing toast lever should not tip the toaster
Color control knob	4	W		W		W							Color of toast should be consistent for a batch of six
Side and end covers	12	W		W	S		W		M	M		M	Covers should be easy to assemble/ disassemble
Crumb tray	3	W	W		W		M	W			W		Should be adequately accessible for easy removal
Material of parts	9											M	Should have needed thermal and electrical properties
Cost level													
Low		X	X	X			X	X	X				
Medium						X				X	X	X	
High					X								
Weighting		7	6	1	11	2	2	2	4	6	2	6	

Relationships: Strong (S) =5; medium (M) =3; weak (W) =1.

producing usable products will not be realized unless the manufacturing process is made robust by means of controlling the manufacturing variables (Table 10.8).

Some of the important processes relevant to toaster manufacture are progressive die operations, plastic molding, assembly processes, and painting and inorganic coating. The factors controlling progressive die operations and inorganic coating already have been discussed.

The end covers are made of thermoset polyester by means of compression molding. Thermosets are used as insulators in electrical/electronic appliances, and they offer a wide range of capabilities in terms of resistance to heat and other environmental conditions. Typical distinctive properties of thermosets include dimensional stability, low-to-zero creep, low water absorption, good electrical properties, high heat deflection temperatures, minimal values of coefficient of thermal expansion, and low heat transfer.

The kinds of thermosetting materials capable of being molded include phenolic urea, melamine, and melamine-phenolic. The manufacturing processes for making thermosetting plastic components are compression molding, transfer molding, injection molding, rotational molding, thermoforming, casting, and foam molding.

The external surface of the side covers is painted with an organic coating, whereas the inside is coated with aluminized paint through one of the following coating methods:

- Dip coating
- Flow coating
- Curtain coating
- Roll coating
- Spray coating
- Powder coating.

The material components affecting the quality of paint are the binder, pigment, thinner, and additives. The binder usually is a resinous material dispersed in a liquid diluent, holding the pigment on the surface. Binders can be classified as oils and oleoresins, phenolic resins, alkyd resins, and polyurethanes. The pigment is a solid material that is insoluble in the binder and its thinner. It adds color to the finish and increases the opacity of the coating. The thinner initially dissolves the resin binder but, when added in varying amounts and types, may control viscosity and evaporation rate of the coating film. Broad classes of thinners are aromatic, terpenes, and acetates. Additives are compounds added in small quantities to impart special properties to the coating. Some important types are paint dryers, metallic soaps, and plasticizers. The quality of painting depends on the surface preparation, ambient temperature, flow rate of paint, conveyor speed, and so forth.

Assembly in manufacturing often involves some type of mechanical fastening of a part to itself or two or more parts or subassemblies together to form a functional product of a higher level subassembly. Selection of a specific fastening method depends on the materials to be joined, the function of the joint, strength and reliability requirements, weight limitations, component dimensions, and environmental factors. Other important factors to consider include available installation equipment, appearance, and whether the assembly needs to be dismantled for repair, reuse, or recycling.

Table 10.8 Fourth stage QFD for a toaster: transformation from process features to process variables

Process features	Process variables											Comments
	Importance	Pressure	Temperature	Chemical concentration	Duration	Pigment	Speed	Punch-die clearance	Rate of cooling	Rivet-hole clearance	Current density	
Blanking	7	S					M	M				Progressive die operation
Stamping	6	S					M	M				Progressive die operation
Spot welding	2										S	Joins frame components
Tabbing	2	S						M				Joining components
Screw fastening	2											Manual operation
Riveting	2	S					W			S		Assembly operation
Compression molding	11	S	S		M							Purchased from vendor
Painting	4			M	M	S	M		M			Organic coating
Chromium coating	6			M	M		M				M	Improves appearance and corrosion resistance
Nickel coating	2			M	M		M				M	Improves rustproofing
Shredding/ melting	6		S									Material recovery process
Range	Units	kPa	°C	ppm	min/s	kg/ft.^{3}	in./min	μin.	°/min	μin.	A	

Relationships: Strong (S) =5; medium (M) =3; weak (W) =1.

Some of the more commonly used fastening methods include

- **Integral fasteners**: These constitute formed areas of the component part or parts that function by interfering or interlocking with other areas of assembly. This type of fastening method is commonly applied to formed sheet metal products and generally is performed by lanced or shear forming tabs, extruded hole flanges, embossed protrusions, edge seams, and crimps.
- **Threaded fasteners**: Threaded fasteners are separate components having internal or external threads for mechanically joining the parts. Commonly used threaded fasteners include bolts, studs, nuts, and screws. Such fasteners are used for joining or holding parts together for load carrying, especially when disassembly and reassembly may be required.
- **Rivets**: Rivets are used for fastening two or more pieces together by passing the body through a hole in each piece, then clinching or forming a second head on the other end of the rivet body. Once set in place, a rivet cannot be removed except by chipping off the head or clinched end.
- **Industrial stitching and stapling**: Stitching and stapling are fastening methods in which U-shaped stitches are formed from a coil of steel wire by a machine that applies the stitch. This low-cost joining method is not applicable when repetitive, fast, and easy removal of fasteners is required.
- **Shrink and expansion fits**: A shrink or expansion fit usually is composed of two normally interfering parts in which the interference has been eliminated during assembly by means of a dimensional change in one or both parts through heating one part or heating one part and cooling the other.
- **Snap fit and slide fit**: Snap fit or slide fit assembly is commonly used in plastic parts when frequent disassembly is needed for parts replacement.

10.5.2.8 Discussion

By carefully choosing manufacturing parameters and their levels, a high degree of product usability can be achieved. The following discussion expands on this claim:

- **Functional requirement**: The bread holder keeps the slice in a vertical position, enabling it to receive heat uniformly. The width of the slot is at least 1.375 in. to accept even thick bagels. By controlling the composition of the bimetallic strip, consistent temperature control can be obtained.
- **Ease of use**: The tension of the toast lever coil is determined by the width, thickness, and material. By controlling these factors, the force of application to lower the lever can be controlled. The temperature control knob can be located along the left end of the toaster and below the lever to facilitate access and clear visibility.
- **Esthetics**: The color of the thermosetting plastic material for the end covers and the paint on the side covers can be chosen to match user preferences.
- **Maintainability**: The crumb tray design enables the removal of bread crumbs. A self-cleaning inner lining facilitates removal of bread crumbs. Use of moisture-proof paint for the side cover and the thermosetting plastic adds ease to exterior cleaning. A removable bread crumb holder tray also enhances ease of appliance cleaning.
- **Energy conservation**: Heat loss is reduced by the reflectivity of the U frame and the aluminized coating on the inside of the side cover, helping conserve energy.
- **Safety**: Use of an aluminized coating on the inside of the frame reduces heat transfer and prevents the exterior from getting very hot. Use of a three-pin cord improves electrical safety.

- **Recyclability**: The most important material constituents are steel, thermosetting polyester, and mica. Use of screws makes separation of steel components easy. Use of thermoplastics, in place of polyester, improves recyclability. Insulating sheets made of mica can easily be separated and reused.

The case studies presented here indicate how the concerns of users can be addressed by manufacturing attributes. The usability transformation method was employed to develop the transformations linking usability requirements to manufacturing attributes. The transformation was gradual. Such transformations and relationships between process variables and usability features can help designers and manufacturing engineers develop products that a customer wants and appreciates.

The information obtained through design manuals and manufacturing handbooks is converted into design guidelines and generic usability evaluation guidelines. Bear in mind that the generic design guidelines are very broad in nature and need to be modified to enhance their utility as a design tool for a specific product.

The following section presents a series of these guidelines. This is followed by a case study that involves customized product design and the corresponding modifications in usability guidelines.

10.5.3 Checklists for evaluating the usability of a consumer product

Tables 10.9–10.26 show generic checklists for product usability, accompanied by QFDs wherever appropriate, to illustrate the design process. The following section examines the development and use of customized checklists to facilitate usable product design and presents a case study to illustrate the use of the method.

10.6 Case study for development of customized checklists

Achieving the twin goals of increasing usability while reducing cost is neither solely a marketing problem nor solely a design/manufacturing problem. It is a product development problem involving all activities simultaneously. The simultaneous engineering design philosophy is based on concurrent integration of the appraisal of consumer needs, development of product concept design, and its manufacture (Akao, 1990).

A four-step transformation matrix method similar to the QFD approach is used in defining customer requirements and translating them into specific steps to develop the needed products. It allows customers' requirements, desires, and preferences to be taken into account throughout all processes, beginning with concept design activities and continuing throughout the production operations on the factory floor (Govindaraju and Mital, 1998).

For the sake of explaining the conceptual methodology in tandem with a real-life case study, a hybrid bicycle is the chosen product under consideration. The basic structure and elements of the QFD matrices are as depicted. During the product

Table 10.9 Checklist for the evaluation of the functionality of a consumer product

	Scale: Bad–good							Comments
	1	2	3	4	5	6	7	
1. List the major user requirements and rate how well the product features satisfy each of them * * *								
2. Do you feel some features are not necessary to meet user needs? How well have such necessary features been reduced to a minimum?								
3. Are there too many models of this product? How well have such unnecessary variations in product function and style been minimized?								
4. Do you feel malfunction in any one feature does not affect proper functioning of the rest of the product? How well has functional independence been built in?								
5. Do you want to expand the scope and power of some features in the future? How good is the scalability of the product (particularly for consumer electronics)?								
6. Do you want to add features to the product in the future? How good is the upgradability of the product?								
7. Does the product function satisfactorily in anticipated but unusual field conditions, such as high heat, cold, humidity, and vibration? How robust is the product?								
8. What is the overall usability of the product in terms of functionality?								

Table 10.10 Checklist for the evaluation of the ease of use of a consumer product

	Scale: Bad–good							Comments
	1	2	3	4	5	6	7	
1. How comfortably are the controls and displays located? 2. How obvious are the operations of the controls?								

(*Continued*)

Table 10.10 **(Continued)**

	Scale: Bad–good							Comments
	1	2	3	4	5	6	7	
3. How easy is it for a novice user to intuitively understand the displays and controls of the product? 4. How well have the control tasks been simplified? 5. While making the product controls and displays very simple, does the product still have enough complexity to motivate the user? 6. How well is the possibility of moving the controls in the wrong direction been eliminated? 7. How well are knobs and handles differentiated? 8. How adequately is the feedback provided during product use? 9. How well is information displayed? 10. How well does the product design anticipate human errors in operation? 11. What is the overall usability rating of the product in terms of ease of use?								

Table 10.11 **Checklists for the evaluation of the esthetics of a consumer product**

	Scale: Bad–good							Comments
	1	2	3	4	5	6	7	
1. How good is the color or finish of the product exterior? 2. How good is the texture of the exterior and interior of the product? 3. How soft and flexible is the contact surface of the product? 4. How inviting or attractive is the product shape to the user? 5. How good is the product in imparting a feeling of cleanliness? 6. How good is the product design in incorporating intricate and appealing shapes? 7. What is the overall usability of the product in terms of esthetics?								

Table 10.12 Checklist for the evaluation of the reliability of a consumer product

	Scale: Bad–good							Comments
	1	2	3	4	5	6	7	
1. How well has the design been simplified?								
2. How well are proven, standard, or existing components and design approaches being used?								
3. How well is the product protected against environmental factors, such as heat and moisture?								
4. How fail safe is the design of the product?								
5. How good is the design in permitting continued operation when a critical component fails?								
6. How high is the reliability of the parts or components of the product?								
7. How good is the design in protecting against corrosion and thermal extremes?								
8. How well are the sensitive and weak components protected from damage during use or maintenance?								
9. What is the overall usability of the product in terms of reliability?								

Table 10.13 Checklist for the evaluation of the maintainability and serviceability of a consumer product

	Scale: Bad–good							Comments
	1	2	3	4	5	6	7	
1. How visible or accessible are the maintenance-prone components or assemblies?								
2. How easy is it to identify faulty components and replace them?								
3. How well is the product designed for testability?								
4. How good is the design in structuring the product such that the high mortality parts are more accessible than the rest?								
5. How well is the product designed to require a minimum of cleaning and maintenance?								
6. How simple and easy is it for the customer to remove and replace parts?								
7. How good is the design in avoiding toxic substances and safety hazards during the maintenance procedure?								
8. How well does the product design enable ready maintainability rather than disposability after a malfunction?								
9. What is the overall usability rating of the product in terms of maintainability and serviceability?								

Table 10.14 Checklist for the evaluation of the environmental friendliness of a consumer product

	Scale: Bad–good							Comments
	1	2	3	4	5	6	7	
1. How good is the design in removing toxic materials and incorporating those that are environmentally preferable for the desired function?								
2. How well does the design avoid the use of materials that are restricted in supply?								
3. How well is the product designed to utilize recycled materials wherever possible?								
4. How well does the product incorporate measures to eliminate the use or release of greenhouse gases?								
5. How does the product design avoid producing liquid or solid residues whose recycling is difficult or energy intensive?								
6. How well does the design minimize the use of energy-intensive process steps, such as high heating differentials, heavy motors, and extensive cooling?								
7. How well is the product designed to minimize the use of materials whose extraction is energy intensive?								
8. How good is the design in minimizing the number and volume of different packaging materials?								
9. How well does the product design avoid using materials whose transport to the facility require significant energy use?								
10. What is the overall usability of the product in terms of environmental friendliness?								

Table 10.15 Checklist for the evaluation of the recyclability and disposability of a consumer product

	Scale: Bad–good							Comments
	1	2	3	4	5	6	7	
1. How well does the design minimize the number of materials used in manufacturing?								
2. How well does this product minimize the use of toxic materials?								

(*Continued*)

Table 10.15 (Continued)

	Scale: Bad–good							Comments
	1	2	3	4	5	6	7	
3. How easy is it to identify and separate toxic materials, if used?								
4. How good is the design in avoiding joining dissimilar materials in ways that are difficult to reverse?								
5. How well does the design incorporate thermoplastics in use of thermosetting materials?								
6. How good is the design in avoiding painting and the use of plated metals?								
7. How good is the design in using fasteners, such as clips or hook and loop, in the assembly in place of bonds and welds?								
8. How good is the identification by ISO markings of various plastics or other materials as to their content?								
9. How good is the design in minimizing or eliminating fillers?								
10. How good is the design in avoiding assemblies that are difficult to separate?								
11. How good is the design in eliminating materials that are difficult to recycle?								
12. What is the overall usability of the product in terms of recyclability and disposability?								

Table 10.16 Checklist for the evaluation of the safety of a consumer product

	Scale: Bad–good							Comments
	1	2	3	4	5	6	7	
1. How fail safe is the product design?								
2. How well have sharp cutting edges been removed?								
3. How good is the protection against rotating or moving elements?								
4. How good is the protection from crushing or shearing the fingers or hands of the users?								
5. How good is the protection against electrical hazards?								
6. How effective and visible are the warning devices?								

(Continued)

Table 10.16 (Continued)

	Scale: Bad–good							Comments
	1	2	3	4	5	6	7	
7. How good is the protection against flammable materials?								
8. How safe is the product from heavy elements, such as lead and arsenic?								
9. How well is the product designed to prevent injuries due to breakage on impact?								
10. How good is the product layout in preventing stress due to awkward posture?								
11. How good is the design in avoiding hazards such as cumulative trauma disorders?								
12. What is the overall usability rating of the product in terms of safety?								

Table 10.17 Checklist for the evaluation of the customizability of a consumer product

	Scale: Bad–good							Comments
	1	2	3	4	5	6	7	
1. How well has modular design been adopted?								
2. How well is the design based on PFA?								
3. How well does the product incorporate standard components?								
4. How well does the design incorporate standardized design features?								
5. How good is the adoption of technologies such as CAD/CAM and group technology in the design?								
6. How effectively has the number of product varieties been reduced?								
7. How good is the adoption of near-net shape processes?								
8. How well have the secondary machining processes been eliminated?								
9. How good is the adoption of manufacturing technologies such as cellular manufacturing?								
10. What is the overall usability of the product in terms of its customizability?								

Table 10.18 Checklist for the design and manufacture of consumer parts for functionality

	Implemented? Bad–good					Comments
	1	2	3	4	5	
1. Design the product so that it maps the functional requirements of the users with various product features						
2. Design a product to maintain independence of functional requirement; adopt a modular design approach to achieve this						
3. Minimize the information content in a product by minimizing the number of parts in a product to achieve this; parts can be minimized as follows: Check whether the part or subassembly moves relative to its mating parts during normal functioning Check whether it is essential that the mating part or subassembly be of a different material than its mate to fulfill its function Check whether a combination of certain parts would not affect the assembly of other parts Check whether field service does not require their disassembly						
4. Check all parts for function and eliminate redundant parts wherever possible						
5. Check adjacent parts for function and attempt to integrate them to produce a functional single part						
6. Avoid variations in the product function and styles as far as possible						
7. If product variations are unavoidable, incorporate features of all the product variants in each core component						
8. Introduce part variants of different designs into the product as late as possible within the assembly process						
9. Design a solid base for the product that can provide integral part location, transport, orientation, and inherent strength sufficient to withstand stress during operation						
10. Structure the core assembly into as many subassemblies as possible and make the individual subassemblies self-sufficient						
11. Identify and label all functional surfaces; a functional surface provides support, transmits forces, locates components in the assembly, or transmits motion						
12. Minimize the number of functional surfaces						
13. The mass of an individual part should be no more than the function or strength required of it and is achieved by minimizing the mass/strength ratio						

(*Continued*)

Table 10.18 (Continued)

	Implemented? Bad–good					Comments
	1	2	3	4	5	
14. Try to avoid ceramic and glass components whenever possible						
15. Avoid using materials that have moisture sensitivity, static electricity, or are magnetic						
16. Do not overtolerance nonfunctional or functional dimensions; perform a reversed chain tolerance analysis to establish sensible tolerances of components at progressive steps in the chain						
17. Use materials that give the greatest possibility of part integration						
18. Design the software component of a product so that the software is upgradable						
19. Adopt a robust design procedure to find out the settings of the product design parameters that make the product's performance insensitive to environmental variables, product deterioration, and manufacturing irregularities						
20. What is the overall usability of the product in terms of its design for functionality?						

Table 10.19 Checklist for the design and manufacture of consumer parts for ease of use

	Implemented? Bad–good					Comments
	1	2	3	4	5	
1. Fit the product to the user: The operation of the product should conform to the users both physically and mentally Keep the static strength requirements <10% of the maximum volitional strength exertion capacity when muscle loading is protracted Keep dynamic strength requirement <5% of the maximum volitional strength exertion capability when muscle loading is protracted Avoid lifting, holding, or carrying loads over 50 pounds for men and 44 pounds for women						

(*Continued*)

Table 10.19 **(Continued)**

	Implemented? Bad–good					Comments
	1	2	3	4	5	
2. Simplify the task: Control operations should have a minimum number of steps and be straightforward They should minimize the amount of planning, problem solving, and decision making required The designer should use technology to simplify tasks, particularly if the task involves processing information 3. Make things obvious: Make the controls simulate the arrangement of actual mechanism Place controls for a function adjacent to the device it controls 4. Use mapping: Have the control reflect or map the operation of the mechanism 5. Utilize constraints: Design controls so that an incorrect movement or sequence is not possible 6. Provide feedback: The product must provide users with a response to any actions taken, informing the user how the product works 7. Provide good displays: Displays should be clear, visible, interpretable, and consistent in direction Data displays should be large enough for easy readability Analog displays are preferred for quick reading and to show changing conditions; avoid multiple and nonlinear scales Use digital displays for more precise information Locate displays where viewing would be expected 8. Design controls carefully: For precision control knobs, the diameter should be 8–13 mm and length at least 100 mm Displays and controls should be matched and move in the same direction Displays and controls should be differentiated so that the wrong one is not used; shape knobs and handles differently so they are distinguishable by look and touch Have controls fit the shape of the hand						

(Continued)

Table 10.19 **(Continued)**

	Implemented? Bad–good					**Comments**
	1	**2**	**3**	**4**	**5**	
Organize and group controls to minimize complexity						
Do not require much force for controls unless they are used only in emergencies or occasionally						
Controls should be easy to reach and protected against accidental movement or activation						
Put the controls in the same sequence as they are used, for instance, from left to right in a reading direction						
Key controls are best located close to the user's normal hand position						
Place controls in accordance with their frequency of use, the most commonly used controls being the closest						
9. Anticipate human errors:						
Understand the cause of potential errors and design to minimize them						
Make it possible to undo an error quickly						
If the error cannot be reversed easily, design the equipment so that it is harder to commit such an error						
Provide warnings to the user before the erroneous control is actuated; use an alarm or flashing light if the wrong control is activated						
10. Avoid awkward and extreme motions:						
Group product elements that may involve reaching by the user so that forward reaches are short in length						
Design operating controls and other elements to provide the force or power needed rather than relying on human power						
Design handles with smooth edges and provide high friction so that gripping is easy; handles should be large enough and shaped so that forces are distributed over a large area; the surface should be nonconductive						
Design controls and tools so that the wrist of the operator does not have to bend; the wrist should be in a neutral position throughout its range of use when movement or force is required; change the elevation of the operating elements						
Closing tools, such as scissors, should have a spring-loaded mechanism to lessen muscle force and provide better tool control						
Design tools to be used by either hand						

(Continued)

Table 10.19 (Continued)

	Implemented? Bad–good					Comments
	1	2	3	4	5	
Design equipment to accommodate body measurements and capabilities of the potential user population; if critical, provide an adjustment, since no one size is optimum for all users If vibration is present, controls should be isolated as much as possible from the vibration; also provide damping, improve dynamic balance, and change the machine speed Forces required to activate a control should be minimized by providing more leverage, optimizing handle shape, and surface by providing power assist and decreasing the weight of the item moved 11. What is the overall usability of the product in terms of its design for ease of use?						

Table 10.20 Checklist for the design and manufacture of consumer parts for esthetics

	Implemented? Bad–good					Comments
	1	2	3	4	5	
1. Design the color of the product so that it is pleasant to look at. The surface color can be changed by selecting appropriate inorganic coating, painting, and plating techniques: Select phosphate conversion coatings for blue to red color, chromate conversion coatings for bright clear, yellow, bronze, and olive drab and oxide coatings for gray to black color Choose the color by appropriate combinations of pigment elements, solvents, binders, and additives						

(Continued)

Table 10.20 **(Continued)**

	Implemented? Bad–good					Comments
	1	2	3	4	5	
Various machining techniques, such as grinding, polishing, and lapping, impart luster, depending on level of surface finish						
2. Surface of the product that comes in contact with the body should be smooth and have some texture; appropriate choice of material, such as leather, fabric, and synthetic covers, provides required smoothness and texture						
3. Surface of the product should be soft and not too flexible when touched: Materials such as flexible rubber, foam, and velvet provide required softness Provide a soft surface like the skin of a peach by selecting an appropriate leather and finish Surfaces that are smooth and warm or matte are also perceived as soft						
4. Design of the product shape should be such that it invites one to touch: The shape should be round with smooth transitions The external contour of the product can be designed to arouse male/female feelings by appropriate choice of shapes and protrusions, such as convexity, concavity, bulge, and transitions of such shapes						
5. External metallic parts should be rustproof to provide a feeling of cleanliness, healthiness, and elegance: Select stainless steel parts and apply a rust-resistant coating over plain steel or other metallic parts Select wear-resistant coatings: phosphate coatings promote a continuous oil film that are not subject to rupture to maintain the new look and gloss on the product even after continued use						
6. Use plastic parts, since they can be molded in any color to avoid the need for painting and make products with intricate, appealing shapes						
7. What is the overall usability of the product in terms of its design for esthetics?						

Table 10.21 Checklist for the design and manufacture of consumer parts for reliability

	Implemented? Bad–good					Comments
	1	2	3	4	5	
1. Simplify the design and minimize the number of parts 2. Provide insulation from sources of heat; minimize thermal contact and incorporate fins; lower the surrounding temperature and provide conduction paths 3. Provide seals against moisture; use silica gel to reduce moisture content inside the product 4. Make the product rugged against shock 5. Provide shields against electromagnetic and electrostatic radiation 6. Provide rustproofing to prevent corrosion in a saline atmosphere 7. Use standard parts and materials 8. Select parts with verified reliability 9. Design to avoid fatigue and corrosion failures; avoid stress concentration points and sharp corners 10. If threaded fasteners are used, use lockable types or a locked washer trapped on a fastener 11. Use redundancy: Provide duplicate components, assemblies, and systems that are critical to operation Arrange redundant elements in parallel 12. Provide standby redundancy 13. Use derating (the operation of a part at less severe stresses than those for which it is rated) 14. Provide load sharing so failure of one unit does not place a greater stress on the remaining units 15. Use a burn-in process (the process of operating items at elevated stress levels) to reduce the population of defective components 16. Use a screening process (an enhancement of the quality control process whereby additional detailed visual and electrical/mechanical tests seek defective features) to identify weak items 17. Provide a generous margin of error for a large factor of safety 18. Protect sensitive components and adjustments from accidental change during shipping, service, repair, and operation						

(Continued)

Table 10.21 (Continued)

	Implemented? Bad–good					Comments
	1	2	3	4	5	
19. Protect products with fuses, shear pins, circuit breakers, and the like						
20. Design products to accommodate thermal expansion						
21. Provide sealing against dust and reactive gases: Dust leads to long-term degradation of insulation and increases contact resistance in electronic assemblies; reactive gases lead to corrosion of electrical contacts						
22. Design to reduce component overheating: Provide ventilation or heat sinks to prevent damage to components due to overheating Locate sensitive parts, such as semiconductors and capacitors, remote from high temperature parts Insulate sensitive parts from heat sources Design with larger conductors in printed circuit boards where feasible Provide cooling fins and heat sinks where possible and position heat sinks with fins positioned in the direction of air or coolant flow Locate resistors, transformers, and other heat-producing parts favorably for convection cooling Provide mechanical clamping and other good heat paths for transfer of heat from these devices to heat sinks Use short lead resistors Minimize thermal contact resistance between semiconductor devices and their mountings by using a large area and smooth contact surfaces						
23. Identify the weakest components and give priority to improving them rather than other parts						
24. Design the product and its components for easy testability						
25. Review and analyze data regarding field failures and redesign components that fail often						
26. Perform FMEA, fault tree analysis, and sneak circuit analysis to identify and eliminate reliability problems						
27. Minimize the use of sockets in electronic assemblies; mechanical connections have less reliability than soldered joints						
28. What is the overall usability of the product in terms of its design for reliability?						

Table 10.22 **Checklist for the design and manufacture of consumer parts for maintainability and serviceability**

	Implemented? Bad–good					Comments
	1	2	3	4	5	
1. Design the product so components that require periodic maintenance or are prone to failure are easily visible and accessible 2. Make the components handy for inspection, testing, and easy replacement when necessary 3. Design the covers, panels, and housings so they are easy to remove and replace 4. Locate maintenance-prone components for easy access 5. Assemble highly reliable products first and in a lower, less accessible position, with high mortality parts in an exposed, accessible position when the cover is removed 6. Design high mortality parts and those that may need replacement or removal for service to other parts for easy detachment and replacement: Use quick disconnect attachments and snap fits of the types designed for disassembly; the orientation of the hooking element should be visible and easily retractable Avoid press fits, adhesive bonding, and riveting of parts Funnel openings and tapered ends and plug-in or slip fits are advisable 7. Design high mortality parts so that they can be replaced without removing other parts or disturbing their adjustment 8. Design with field replacement in mind: When tools are required, they should be standard, commonly available types Designs requiring the fewest variety of such tools are advisable 9. Locate maintenance-prone components on the same side of the product 10. Simplify the product so that a user rather than specialist repairs it 11. Adopt modular design to facilitate ease of replacement 12. Adopt modular design to facilitate ease of testing to verify operability and enable isolation of faults						

(*Continued*)

Table 10.22 (Continued)

	Implemented? Bad–good					Comments
	1	2	3	4	5	
13. Design the product for easy testability: Design the product and its components so that these tests can be performed with standard instruments Incorporate built-in test capability and, if possible, self-testing devices Make the tests easy and standardized, capable of being performed in the field Provide for accessibility of test probes by making the test points prominent and providing access ports or tool holes Make modules testable while still assembled Test points and their associated labels and controls should face the technician for best visibility; use color-coded test points for each location Combine test points into clusters for multipronged connectors, where similar clusters occur frequently Locate routine test points so they can be used without removal of cabinet cover or chassis						
14. To facilitate interchangeability of parts, use standard commercially available parts; if these are not available, use parts common to all the company's products						
15. Reduce the number of part sizes and varieties, increasing their availability for field repair						
16. Provide for malfunction annunciation by designing indicators that inform the operator of a malfunction and indicate which component is malfunctioning						
17. Design the product to signal a warning whenever a component requires periodic replacement or servicing before it can fail						
18. The parts requiring replacement during service should be clearly identified with part numbers or other essential designations						
19. Design replacement parts to prevent their incorrect insertion during maintenance						
20. Design for fault isolation and traceability						
21. Provide anticipated spare parts with the product, such as fuses, shear pins, and lightbulbs						
22. When access covers are not removable, they should be self-supporting when open						
23. Remove safety hazards during repair, service, or maintenance						

(Continued)

Table 10.22 **(Continued)**

	Implemented? Bad–good					Comments
	1	2	3	4	5	
24. Eliminate sharp corners and burrs inside the product 25. Protect the user against hazardous fumes, electric shocks, and mechanisms that can pinch or catch fingers or clothing 26. Incorporate automatic timing or counting devices in the product to signal the need for replacement of high wear or depletable parts 27. Provide room for drainage of fluids changed periodically; drainage plugs must be accessible 28. Components apt to be replaced or adjacent to those that are should not be fragile 29. Use self-lubricating components where practicable 30. Use sealed and lubricated components and assemblies where feasible 31. Use gear-driven accessories to eliminate belts and pulleys 32. What is the overall usability of the product in terms of its design for maintainability and serviceability?						

Table 10.23 **Checklist for the design and manufacture of consumer parts for environmental friendliness**

	Implemented? Bad–good					Comments
	1	2	3	4	5	
1. Avoid the use of toxic materials in the product and manufacturing processes; do not use Freon as a refrigerant 2. Minimize the amount of material in the product: Design for processes that minimize material scrap and produce smaller, lighter parts Reduce the amount of packaging materials 3. Design to minimize the use of energy-intensive process steps, such as high heating differentials, heavy motors, and extensive cooling 4. Design the process to reduce manufacturing residue, such as mold scrap and cutting scrap 5. Minimize the amount and variety of packaging material used; design package to be recycled rather than landfilled or incinerated 6. Avoid using materials restricted in supply or likely to become scarce during product manufacture						

(*Continued*)

Table 10.23 (Continued)

	Implemented? Bad–good					Comments
	1	2	3	4	5	
7. Reduce the use of toxic or radioactive materials 8. Avoid using ozone-depleting, global warming substances 9. Design product packaging for shipping in bulk as opposed to or in addition to shipping individually 10. Design refillable or reusable containers where appropriate 11. Reduce the use of solid material components, such as cartridges, containers, and batteries, which require periodic disposal 12. If the product or a part of it is to be dissipated during use, design it to have minimal environmental impact 13. Reduce the amount of liquid materials, such as coolants and lubricants, that require periodic replenishment 14. Develop processes to reduce the gaseous emissions, such as carbon dioxide, during product use 15. Reduce the amount of energy consumed during product operation; provide enhanced insulation or other energy-conserving design features in the product 16. Minimize using liquids, such as acids, alkalis, and solvents, during manufacturing processes 17. In electronic products, use sleep mode to conserve energy when the product is not in use 18. Avoid using heavy metals in the product 19. What is the overall usability of the product in terms of its design for environmental friendliness?						

Table 10.24 Checklist for the design and manufacture of consumer parts for recyclability and disposability

	Implemented? Bad–good					Comments
	1	2	3	4	5	
1. Design the products and its components to be reusable, refurbishable, or recyclable, in that order 2. Minimize the number of parts and adopt a near-net shape approach: Fewer parts make sorting materials during recycling easier When a number of parts are combined into one complex part, both factory assembly and disassembly are aided						

(*Continued*)

Table 10.24 **(Continued)**

	Implemented? Bad–good					Comments
	1	2	3	4	5	
3. Avoid the use of separate fasteners; some portions of these fasteners may be retained in basic parts and contaminate them during recycling 4. Use of snap fit connectors between parts is preferable, as these connectors do not introduce a dissimilar material and are easy to disassemble 5. Utilize the minimum number of screw head types and sizes used in fasteners in one product or portion of the product; the recycler need not change tools to loosen and remove fasteners 6. Use the fewest number of fasteners to reduce disassembly time 7. Design parts to be easily visible and accessible to aid in disassembly 8. Design the product to be easily disassembled even if some parts are corroded 9. Minimize the number of materials in the product to reduce the sorting of parts for recycling: Standardize materials as much as possible Avoid the use of multiple colors in a part Avoid the use of dissimilar materials that cannot be separated or are difficult to separate from basic materials Use of thermoplastic material is preferable to thermosetting plastic materials Solvent, friction, or ultrasonic welding of plastic is preferable to adhesive welding If adhesive bonding is used, find an adhesive material that is compatible when the components are recycled Water-soluble adhesives for labels and other items are preferred Welded joints are preferred to brazed or soldered joints 10. If the number of different materials cannot be reduced, choose materials that are compatible and can be recycled together 11. Avoid the use of composite materials such as glass or metal reinforced plastics 12. Avoid metal-plated plastics 13. Standardize the product components to aid in eventual refurbishing of the products; if major elements are standardized, they can be salvaged and reused more easily 14. Use molded-in nomenclature rather than labels or separate name plates for product identification						

(Continued)

Table 10.24 **(Continued)**

	Implemented? Bad–good					Comments
	1	2	3	4	5	
15. If a separate label must be used on a plastic part, choose a label material and adhesive that are compatible with the material of the base part						
16. Use modular design to simplify assembly and disassembly						
17. Design the modules so that they are upgradable to a new technology or application						
18. In molded and cast parts, identify the material form which the part is made on the part itself Bar coding of material designation and its incorporation in the mold or die from which the part is made facilitates classification and separation of the material during reclamation Color coding of parts, especially plastic parts where color can be incorporated into the material, can be a useful means of material identification						
19. Make separation points between parts as clearly visible as possible						
20. Avoid designs that require spray-painted finishes Use powder coating, roll coating, or dip painting to avoid the need for environmentally damaging solvents If the parts are made of plastic, use molded-in color that is solvent free and more compatible with the base material						
21. Where fasteners or other parts cannot be easily removed, provide predetermined break areas so that the contaminated fastener can be separated from the material to be recycled						
22. Use a woven-metal mesh instead of metal-filled material for welding thermoplastics						
23. Design the product to use recycled materials rather than virgin materials						
24. Avoid threaded metal inserts in plastics						
25. Avoid the use of plated materials						
26. Eliminate or minimize filler material						
27. Use recyclable metals, thermoplastics, and thermosetting plastic materials						
28. Avoid the use of materials that are uneconomical to recycle, such as laminate materials, steel, thermosetting materials, and ceramic materials						
29. What is the overall usability of the product in terms of its design for recyclability and disposability?						

Table 10.25 Checklist for the design and manufacture of consumer parts for safety

	Implemented? Bad–good					Comments
	1	2	3	4	5	
1. Design the product to be fail safe; design mechanisms and features so that a failure will cause an accident						
2. Generous radii should be used wherever possible						
3. Parting lines of molds should be located away from corners and edges						
4. Provide guards or covers over sharp blades and similar elements						
5. Provide guards over power transmission mechanisms and other moving parts, including rotating and reciprocating motions						
6. Guards must: Prevent contact between persons and moving parts Be firmly attached to the product Prevent the insertion of foreign objects Provide protection during maintenance and operation						
7. Parts that may require servicing should be freely accessible and easily repairable or replaceable without interfering with other components or posing hazards to repair personnel						
8. Provide clearances between moving parts and other parts to avoid shearing and crushing points; the space should be too small to admit a child's fingers or have enough clearance not to pinch an adult's finger or hand						
9. Arrange controls so the operator need not stand or reach them in an unnatural awkward position; provide ample clearance from hand levers to other machine elements that the operator could scrape or strike						
10. Anticipate the environment in which the product will be used and provide safeguards against those environmental factors						
11. Electrical products operating on household current should have a grounding or double insulation; utilize electrical properties of plastic to reduce shock hazards						
12. Use electrical interlocks in circuits with potentially injurious voltage to prevent accidental flow of current						
13. Make small components that can be separated from the product bulky enough that they cannot be accidentally swallowed by children						

(*Continued*)

Table 10.25 (Continued)

	Implemented? Bad–good					Comments
	1	2	3	4	5	
14. Make products from high impact or resilient materials to make them child safe						
15. Allow a reasonable factor of safety for stressed or otherwise critical components						
16. Do not use paints or other finishing materials with more than 6% content of heavy metals						
17. Use warning devices that are actuated if hazardous materials in the product are released						
18. Point-of-operation guards should be convenient and not interfere with the user's movement or affect the output of the product						
19. Plastic bags used in packaging should not be too thin; minimum wall thickness for bags large enough to cause suffocation is 0.0015 in.						
20. Minimize the use of flammable materials, including packaging materials; many nonflammable materials will burn if sections are thin enough; avoid paper thin sections of plastics or other potentially flammable materials						
21. Cuts from paper edges can be eliminated by serrating the edges						
22. Markings, especially safety warnings, should be clear, concise, and long lasting						
23. Avoid the use of hazardous materials, including those that are a hazard when burned, recycled, or discarded						
24. Products that require heavy or prolonged user operations should be redesigned to avoid cumulative trauma disorders Avoid awkward positions of hand, wrist, arm, or other body members Avoid the need to apply heavy force Reduce the frequency of repetitive motions Reduce the vibration levels of handled objects						
25. Do not design parts with unguarded projections that can catch body members or clothing						
26. Minimize cables, wiring, and tangled parts						
27. What is the overall usability of the product in terms of its design for safety?						

Table 10.26 Checklist for the design and manufacture of consumer parts for customizability

	Implemented? Bad–good					Comments
	1	2	3	4	5	
Product structure						
1. Adopt modular design: Incorporate all new features in the same module if possible, leaving other modules of the product unchanged Use modular design by grouping parts intended to provide a necessary function, so when an altered function is needed, the entire product need not be modified 2. Develop a product platform design for connecting the modules with an underlying architecture, called *PFA*: The product platform should contain a network structure that describes how different modules are connected The product platform should be configured such that the modules are compatible						
Short time to market						
1. Use standardized components rather than specially designed ones 2. Use standard existing systems, procedures, and materials 3. Do not redesign more than necessary 4. Design conservatively 5. Design it right the first time; apply a concurrent engineering philosophy 6. Design for processes require no tooling lead times or are made with standard tooling 7. Use CAD/CAM with a broad database with quick and easy access to the data						
Low quantity production						
1. Use standard hole sizes, slot widths, filet radii, chamfer dimensions, groove dimensions, bend radii, surface finishes, snap fit tabs, and reinforcing ribs 2. Design like parts to be as identical as possible so only those portions that have to be different can be designed accordingly						

(*Continued*)

Table 10.26 **(Continued)**

	Implemented? Bad–good					Comments
	1	2	3	4	5	
Parts						
1. New part designs should never be made if an existing part meets all user needs 2. When new parts are needed, design them with features similar to existing parts 3. Minimize the total number of part varieties 4. Never design a part that can be obtained from a catalog; use of catalog parts provides more standardization than companywide standardization 5. Use standard materials for common applications, minimizing the number of materials						
Processes						
1. Adopt similar processes to manufacture similar parts 2. Adopt the group technology approach in manufacturing postcomponent standardization 3. Adopt flexible manufacturing cells to operate more effectively when parts and processes are standardized 4. If it is possible to make a special part by modifying a standard part, do so 5. For low production levels, design for inexpensive manufacturing processes with low-cost tooling 6. For machined parts, use computer-controlled machine tools; design parts so that they can be processed on such equipment rather than require special cutting tools and fixtures 7. For machined parts with short runs, use materials with good machinability, such as free machining alloys and steels 8. For other processes, use easily processed materials 9. Use stock material shapes as much as possible to avoid machining 10. Avoid designs that require secondary machining, which is expensive and time consuming; use liberal tolerances consistent with functional requirements and capabilities of manufacturing processes 11. What is the overall usability of the product in terms of its design for customizability?						

planning stage, customer requirements are identified and translated into technical design requirements. During the part deployment stage, technical requirements are converted into part specifications. During the process deployment stage, product characteristics are converted into process characteristics. During the manufacturing deployment stage, manufacturing processes are related to the specific process variables that control them. Through such a stepwise transformation of customer usability requirements into process variables, it is possible to control the usability of a product.

The case study of a hybrid bicycle demonstrates the relationship between product usability attributes and the manufacturing attributes to which they are linked. Bicycles can be roughly categorized into three types: road bikes, mountain bikes, and hybrid bikes. While road bikes are used primarily on paved roads and mountain bikes are used in off-road terrains, such as mountain trails, hybrid bicycles can be used under all sets of conditions. Figure 10.4 illustrates the different components of a typical bicycle.

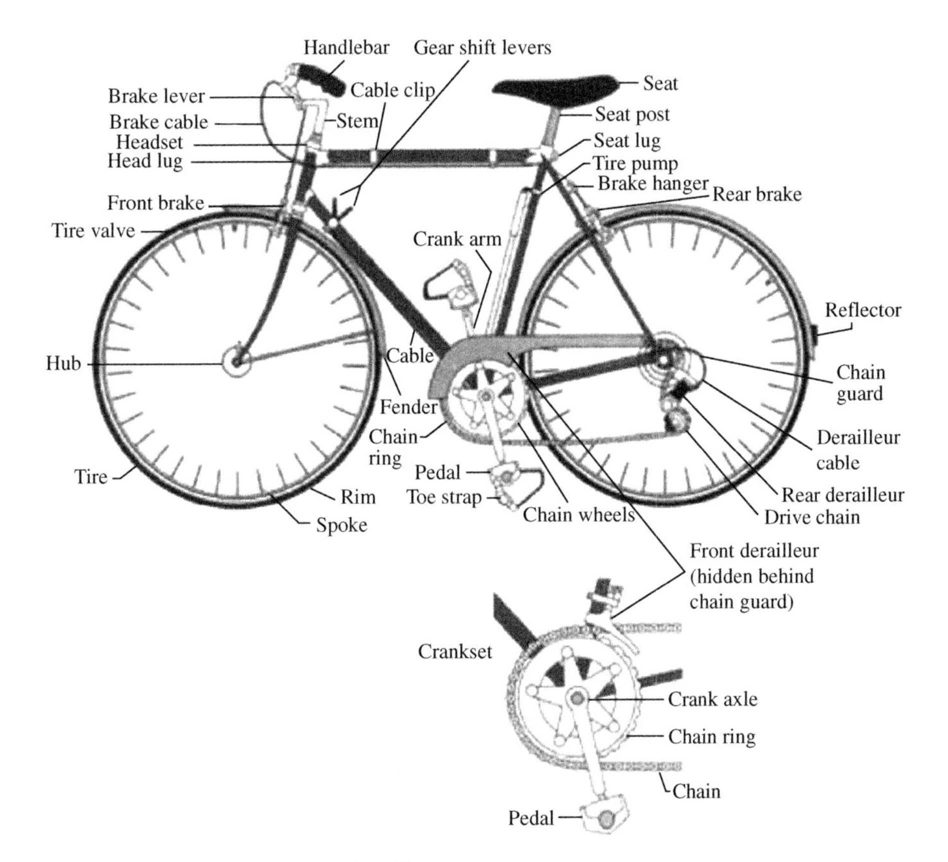

Figure 10.4 Components of a typical bicycle.

10.6.1 Gauging user requirements

A study was conducted to gauge user wants and needs. Eleven subjects participated in the study. The preferences of these 11 users were obtained through one-on-one interviews, each of which was approximately 75 min long. Each question was designed to be open-ended. Each subject owned a hybrid bicycle and was familiar with what would make it usable. The questions were aimed chiefly to extract the primary needs of the users. Users also were asked to rate their needs on a scale of 1–10, 10 being most important. The following outcomes were observed:

- **Stiffness**: The bicycle should have adequate stiffness to withstand the weight of the user without flexing too much. It should be able to withstand the force exerted by the rider while pedaling.
- **Smooth shifting**: Shifting is considered good if it requires less force and dexterity. It should crisply fall into gear. The rider should be able to shift in both directions and in multiple gears.
- **Steady handling**: A bicycle handles well if it is nimble, quick, and steady. A bike with good handling characteristics should respond to a flick of the handlebars and yet feel stable at most speeds.

Ease of use requirements include:

- **Comfort**: The physical contact points between the rider and the bike must be smooth and comfortable. It should not require the user to assume an awkward posture.
- **Adjustability**: Users of the bike differ in gender and age and hence in body dimensions. The bicycle saddle and handlebar should be adjustable to fit any rider. The bike for a small adult should be adjustable in such a way that the brake levers accommodate the hands.
- **Quick adjustment of seat height**: Climbing a hill requires a higher seat position so the rider can extend his or her legs fully and pedal more efficiently. Going down mountain trails requires a low seat for better control. The bike should have a quick release cam lock at the seat post so that the adjustment can be made without getting off the bike.
- **Good shock absorption**: Good bicycles insulate the rider from off-road bumps and drops. The bike should have adequate shock absorption on bumpy rocks and dirt trails and good traction over sand, loose rock, and hard-packed dirt. The bicycle should be both sure footed and shock absorbing.

Esthetics requirements include:

- **Appearance**: The bike should have pleasant external shape and color.
- **Shape**: The structure of the bicycle should have a streamlined shape.
- **Cleanliness**: The exterior of the bicycle should look clean. It should be free from rust and corrosion.
- **Smoothness**: The physical contact points between the user and the bicycle, such as the handlebar grip, pedal, and seat, should have a smooth surface and be soft to the touch.

Reliability requirements include:

- The bicycle should be able to function under varied road surface conditions. Critical components, such as front and rear derailleurs, should not have a tendency to go out of alignment when the frame is impacted on rough surfaces.

- The bicycle shifting should be crisp and smooth. The shifter should have good control over the derailleur. The derailleur should not shift gear more or less than what the rider attempts to shift using the shifter.
- The brake cable should be sturdy. It should be highly dependable whenever braking is needed.

Maintainability and serviceability requirements include:

- **Low maintenance**: The bicycle should require low levels of maintenance, such as cleaning and lubrication. The design should not necessitate constant care.
- **Easy removal**: The bicycle should have quick release levers for wheels and the seat post to reduce the time necessary for making repairs.
- **Accessibility**: Parts of the bicycle requiring regular lubrication, such as the chain, freewheel, and hubs, should be easily accessible.

Environmental friendliness requirements include:

- Material used in the construction of the bicycle should not be a scarce resource.
- The production processes adopted in the manufacture of the bicycle should not harm the environment. Often, paints used for coating the bicycle framing contain solvents that produce greenhouse gases, such as CFCs.
- The lubricant used for bicycle maintenance should be biodegradable.

Recyclability and disposability requirements include:

- Composite materials cannot be melted and reused at the end of their useful lives. Frame materials, such as carbon composites, should not be resorted to unless a very high strength/weight ratio is desired by the user.
- The rubber used in the construction of tires and tubes should be recyclable.
- Avoid using thermosetting plastic materials in bicycle components, as far as possible.
- Critical components, such as derailleur, crank sets, and hubs, should be mounted onto the frame such that they can be removed and reused even if the bicycle is to be discarded.

Safety requirements include:

- **Braking**: Braking brings the bike to a stop smoothly and predictably. If braking is too hard it can throw a rider to the front and lift the rear wheel off the ground. Slow braking is equally hazardous. Braking needs to be consistent under both dry and wet conditions. Ideally, the brakes should be able to stop a bike going 15 mph within 15 ft. Good brakes should stop the bike quickly but smoothly, allowing the rider to control the pressure of the brake pads on the rim. The rear wheel should not lift during panic stops, and the rider should be able to maintain control.

Customizability requirements include:

- The manufacturer should be able to design a bicycle with dimensions that would be appropriate for any given individual, whether male or female and young or old.
- Components, such as chain wheels, and rear sprocket, should be changeable, so that different gear ratios can be obtained based on the needs of the user.
- The bicycle design should permit other accessories, such as a water bottle and chain wheel guards, to be attached at the request of the user.

10.6.2 Technical requirements

Table 10.27 depicts the first stage of the transformation matrix related to the usability requirements coupled with technical requirements. The major technical requirements of the materials for the bicycle are as follows. The material and structural property requirements are:

- The material must be strong enough not to yield under a load. The tubes in a frame usually are loaded under a combination of bending, shear, tension, and compression and hence require adequate strength to withstand the stress. Most failures occur not on account of the material, but because of fatigue. Fatigue-limit stress rather than the ultimate tensile stress should be used as the criterion of acceptable stress.
- The density of the material should be such that the resulting structure is light.
- The resulting structure should not be unduly flexible. The property defining flexibility is elasticity of Young's modulus.
- Failure should be gradual rather than sudden. The property that gives some indication of the failure mode is the elongation at failure.
- Joining one piece to another should be possible without loss of strength in the parent material or the joint.
- The material should be intrinsically resistant or easily protected from corrosion.

For transmission efficiency, friction should be minimized. Under normal riding conditions, the pedals are operated at an average of 60–90 revolutions per minute. This is equivalent to 3600–5400 revolutions per hour. Even a small amount of friction can make a significant difference in the amount of energy exerted to ride the bike every hour. All bearings have some friction and therefore waste some amount of energy. The selection and design of bearings should aim to reduce friction between moving parts.

The structure of the bicycle should use modular architecture instead of monolithic architecture. Modular design improves maintainability. Since the subassemblies are easily removed without affecting other components, they can be replaced very easily. A modular approach also aids in customization. If the rider prefers to change from a three-chain wheel to a two-chain wheel and back to suit a specific need, it can be achieved at minimal expense.

10.6.3 Product and process characteristics

The product features designed into the bicycle are listed in Table 10.28. Table 10.29 transforms these product features into product features. The important parts of a bicycle are the frame, fork, wheels, drive train, and brakes. The frame provides the structural base for the entire bike. The fork constitutes the front end of the bike and is integrated with the frame to provide maneuverability. Wheels are the components that enable motion. One wheel is connected to the fork and the other is connected to the rear end of the frame. The drive train transfers the energy from the rider to the bicycle wheel. The brake mechanism transforms kinetic energy into heat to stop the bike. The most important components in the manufacture of a bicycle are the frame, fork, wheels, and tires.

Table 10.27 **Transformation from customer requirements to technical requirements for a bicycle**

Customer requirements	Technical requirements								Competitive analysis				
	Importance	**Light material**	**Tough material**	**Less torque in cranking**	**Less force to shift gears**	**Less force to apply brakes**	**Long material life**	**Should be adjustable**	**1**	**2**	**3**	**4**	**5**
Impact resistance	1	M	S							X			
Vibration absorption	6		S									X	
Lightness	4	S		S						X			
Ease of pedaling	3	S		S							X		
Ease of shifting gears	5				S		W					X	
Ease of applying brakes	2					S					X		
Maneuverability	7	M		M		S					X		
Long life	8		M				M				X		
Should fit the user	9							S					X
Competitive analysis													
1													
2			X	X									
3	X				X	X		X					
4		X					X						
5													
Weighting	9	9	8	4	8	3	8	9					

Relationships: Strong (S) =5; medium (M) =3; weak (W) =1.

Table 10.28 Transformation from technical requirements to product features for a bicycle

Technical requirements	Product features									Comments
	Importance	Frame	Tube	Handlebar	Wheel	Brake	Derailleurs	Gears	Bearings	
Frame should be light	9	S	S	M						High strength/weight ratio
Frame should be tough	9	M	W	W						Should bend but not yield
Less torque in cranking	8				W				S	Bearings should be smooth
Less force to shift gears	4						S	S		Cable sealing should be friction free
Less force to apply brakes	8				M	S		M		Cable sealing should be friction free
Long material life	3				S					Should be rustproof
Should be adjustable	8			W						Interface should be adjustable
Competitive analysis										
1										
2	X			X				X	X	
3		X	X		X		X			
4						X				
5										
Weighting	11	9	4	4	7	4	6	7	11	

Relationships: Strong (S) =5; medium (M) =3; weak (W) =1.

Table 10.29 **Transformation from product features to process features for a bicycle**

Product features	**Process features**									**Comments**
	Importance	**Welding of tubes**	**Heat treatment of tubes**	**Machining of hubs**	**Surface finish**	**Metal forming**	**Organic painting**	**Inorganic coating**	**Injection Molding**	
Frame	11	S	S		W	M	M			Made of titanium, chrome-moly steel, and aluminum alloy
Tube	9	S	W		W	M	W			Seamless tubes are cold drawn and heat treated
Handlebar	4	W	W			W	M			Cold drawn and attached to the stem
Wheel	4					W		M		Made of aluminum and double walled
Brake	7						M			Brakes usually are cantilever type
Derailleurs	4			S					S	Bought from Shimano
Gears	6							M		Made of chrome-moly steel
Bearings	7			S						Fine finished with borozon
Cost level										
Low		X	X			X	X	X	X	
Medium					X					
High				X						
Weighting		18	12	11	2	7	7	3	4	

Relationships: Strong (S) =5; medium (M) =3; weak (W) =1.

Frames are made by welding or brazing together tubes (which constitute the frame members). Tubes are manufactured by hot extrusion or roll forming. Billets are heated to forming temperatures in a furnace, then extruded into a seamless tube. In roll forming, hot slabs are rolled into circular tubes and electrically welded at the seams. Tubes made in this fashion are not as strong, due to the seam, which is an area of weakness. This process manufactures members of the frame, such as all the tubes forming the frame, chain stays, seat stays, fork blade, and steering column. The final frame assembly consists of three subassemblies. The first subassembly comprises the chain stay and seat stay welded together to form the rear dropout to form the rear angle. The rear angle is the first subassembly. The builder inserts the rear dropout into a slot cut in the back of a chain stay that has been precut to length. The end with the dropout is filled with brass to join the dropout to the chain stay.

When a required number of subassemblies for a given frame size have been completed, the builder assembles them on a jig. As the builder slides the tubes into the lugs and assembles the frame on a jig, each tube is coated with a white paste (flux). When heated during brazing, the flux removes the last of the oxide residues on the tubes and lugs, ensuring a good joint. The fit of the tubes into the lugs is quite snug. The assembly plant often resorts to using a rubber mallet to get everything to come together and stay together. At this point, each lug is heated to a temperature high enough to melt the brass used to braze the frame. Each lug is coated with a small amount of brass, tacking the joint in place. The built-up frame generally is not brazed on a jig. The jig is used as a fixture to assemble the frame.

The tacked frame is removed from the jig, put on an alignment table, and checked for straightness. It then is mounted on a holder on which it can be turned, raised, and lowered to gain complete and easy access to all parts of the frame. The lugs are brazed. The joint is heated with a torch. A rod of low temperature brass is dipped in a bucket of flux and touched to the end of the lug. The brass melts and flows inside the lugs through capillary action, between the lug and the seat tube. The brass follows the heat of the torch as it is moved around the lug. A good assembler usually introduces brass at one point and guides it around the lug.

Once brazing is accomplished, the bicycle is cleaned through secondary finishing activities, such as filing and polishing (with an emery cloth). Once this step is accomplished, the frame is sent for chroming or painting. Before all parts can be assembled on the frame, several different manufacturing activities need to take place, which are detailed as follows.

The fork assembly consists of the steering tube, fork crown, fork blades, and fork tip. The steering tube is brazed to the fork crown on the fork jig. The flat fork tips are inserted into fork blades, which are cut and slotted beforehand. The ends are filled with brass into a fork blade assembly. The fork blade subassemblies are assembled into the crown subassembly and brazed. The fork receives a cleanup similar to the frame. The fork then is raked. It is placed on a mandrel and bent cold to obtain the final shape.

Wheel rims are produced by extruding aluminum through a carefully shaped die. The extruded rims then are roll formed and cut into wheel rings. The ends of the rims are closed by resistance welding. The rims are subjected to a T3 hardening process, which hardens the surface of the rim to a depth of 40–50 microns, giving it high strength and a dark gray color. The rims can be anodized to improve their appearance. Once the frame is aligned, it is ready for assembly with the fork and the wheel.

10.6.4 Manufacturing process attributes

From the foregoing discussion, it is clear that the most important processes involved in the manufacture of the bicycle are welding, machining, heat treatment, painting, inorganic coating, and tube drawing. Effective control of the process variables in Table 10.30 is a prerequisite to obtaining a usable product. A discussion of the process variables associated with these processes follows.

Heat treatment is applied to parts to impart properties that make the product more usable: increased surface hardness; production of necessary microstructure for desired mechanical properties such as strength, ductility, and toughness; releasing residual stresses; removal of inclusions such as gases; alteration of electrical and magnetic properties; and improvement in wear and corrosion resistance.

Some of the heat treatment processes are austenizing, equalizing, quenching, annealing, normalizing, carburizing, nitriding, carbonitriding, chromizing, boronizing, resistance hardening, induction hardening, flame hardening, electron beam hardening; and laser hardening. The factors that affect the result of heat treatment processes are

- Composition of metal to be treated
- Critical time–temperature transformation relationship of the metal
- Response of the metal to quenching
- Method of quenchant application
- Temperature control and timing.

The two most important joining methods adopted in bicycle manufacture are arc welding and brazing. Welding is a metal joining process where localized coalescence is produced either by heating the metal to suitable temperatures, with or without the application of pressure, or by the application of pressure alone, with or without the use of a filler metal. Welding filler metal (flux) has a melting point either approximately the same as or below that of the base metal but above 800°F. Brazing employs nonferrous filler metals with melting points below that of the base metal but above 800°F. The filler is distributed in the closely spaced joint by capillary action. The factors that affect weld quality involve the preparation of the weld areas, the filler metal, and the flux.

Many of the bicycle components, such as hubs, cranks, fork bearing, and pedals, have bearings that need to be machined properly to minimize energy losses. Losses due to mechanical friction are reduced by using bearings in these components. Bearing design affects not only mechanical energy losses but also the smoothness of operation at joints such as forks, where losses otherwise can be minimal. Crude bearings produce rough riding conditions at the pedals and fork. The two most important elements of a bearing are its inner and outer races. The bearing races are produced by operations such as cold stamping and fine machining. Low surface roughness and tight tolerance

Table 10.30 Transformation from process features to process variables for a bicycle

Product features	Process features									Comments
	Importance	Temperature	Pressure/ torque	Duration	Speed/ feed/ depth of cut	Grain size	Die dimensions	Current density	Pigment	
Welding of tubes	18	S						S		TIG or arc welded
Heat treatment of tubes	12	S		S		W				Heat treated for stress relieving
Machining of hubs	11		S							Forged and turned for better dimensional accuracy
Surface finish	2								S	Bearings super finished for smooth operation
Metal forming	7		S		M	M	S			Tubes cold drawn using die-mandrel
Organic painting	7								M	Enhances appearance and corrosion control
Inorganic coating	3								M	Reduces the abrasive nature of aluminum frames
Injection molding	4				M					Used to make composites
Range	Units	°	t	min	m/s	mm	in.	A	g/cc	

Relationships: Strong (S) =5; medium (M) =3; weak (W) =1.

make for good bearings. Machining operations such as grinding and finishing operations such as honing and lapping produce bearings superior to stamped bearings.

The tubes used in bicycles are manufactured by hot working, the plastic deformation of a metal under the influence of an applied external force to change the shape by working the metal above its recrystallization temperature. The factors that affect hot working are the temperature of deformation, rate of deformation or strain rate, and amount of deformation. Success also depends on the method and tooling of hot working. Roll and die quality affect the tube quality. The two major hot working processes involved in tube manufacture are hot rolling and hot extrusion. Hot rolling needs a secondary operation of electric welding to close the seams. Tubes made by hot rolling are weak at the seams, whereas extruded tubes are free of this limitation.

Inorganic coating is intended to impart the following properties to the surface of the metal substrate:

- Corrosion protection
- Prepainting treatment
- Abrasion resistance
- Electrical resistance
- Cold forming lubrication
- Antifriction properties
- Decorative final finish
- Easy cleanability.

Some of the methods of surface coating include conversion coating, anodizing, thermal spraying, hard facing, porcelain enameling, and ion vapor deposition.

The outside of the frame and fork is painted with an organic coating to improve its appearance and prevent corrosion. The material components that affect the quality of painting are binder, pigment, thinner, and additives. The binder usually is a resinous material dispersed in a liquid diluent, and it holds the pigment to the surface. The pigment is a solid material, insoluble in the binder, and its diluent. The diluent initially dissolves the resin binder, but when added in varying amounts and types, it acetates. Additives are compounds added in small quantities to impart special properties to the coating. Some important types are antiskinning agents, preservatives, paint dryers, metallic soaps, wetting agents, viscosity and suspension control agents, fungicides, and moldicides. The quality of painting depends on the surface preparation, ambient temperature, flow rate of paint, and speed of conveyor holding the frame.

10.6.5 Development of usability and design checklists

Input obtained from user interaction, as outlined in the preceding section, was used to determine the wants and needs of customers and to obtain design/manufacturing links, determined by analysis based on usability transformation matrices. Usability and design/manufacturing information through this interaction was translated into appropriate questionnaires.

A study was conducted to test the usability evaluation and design/manufacturing guidelines. Twenty-two users participated in the study involving 10 models of hybrid bicycles made by 10 manufacturers. Eight of these bicycles were tested by two users each, and the remaining two bicycles were tested by three users each. The users were named U_1 to U_{22} and the bicycle models were named C_1 to C_{10}. The assignment of code names for users and bicycle models was random.

10.6.5.1 Data collection

The users who participated in the study had used bicycles for at least a couple of years. All were familiar with the use, repair, and maintenance of bikes. The subjects belonged to both genders. They evaluated the usability of the bicycles that they used regularly for commuting or trekking. All the technical terms and definitions pertaining to various usability criteria were explained to them in an easily understandable manner. In general, the awareness of the users of the nine usability criteria, excepting the environmental friendliness and recyclability and disposability, was very good. Whenever the awareness of the users on these two criteria was found to be deficient, they were instructed as to what to expect in a bicycle that was supposed to be environmentally friendly and recyclable as well as disposable. The subjects took an average of around 90 min to fill out the questionnaires.

Reliability was tested by measuring the internal consistency of the score for each questionnaire. The Cronbach coefficient alpha method was used to test for reliability. Validity of any questionnaire usually is obtained by comparing the scores obtained from the users with those obtained using a questionnaire considered to be a standard. No such standard is available to evaluate and design bicycles. Hence, for the sake of this study, validity was tested by comparing the average scores of the questionnaire items with the overall score of that questionnaire.

10.6.5.2 Results

The Cronbach coefficient alpha values for the evaluation questionnaires for usability attributes (functionality, ease of use, esthetics, reliability, maintainability and serviceability, environmental friendliness, recyclability and disposability, safety, and customizability) are shown in the tables that follow. Since a value of 0.4 is considered to be significant, all usability criteria were found to be reliable. The correlation coefficient values testing the validity of the questionnaires for these attributes were found to be statistically significant at the 0.05 level of significance.

The correlation between the average score and the overall score for each of the nine usability evaluation questionnaires was found to be significant. For functionality, ease of use, reliability, maintainability and serviceability, environmental friendliness, and safety, the statistical significance level was 0.01, whereas the level was 0.05 for reliability, recyclability and disposability, and customizability.

The checklists used to evaluate each of the usability criteria under question follow in Tables 10.31–10.51.

Table 10.31 **Usability evaluation of a hybrid bike for functionality**

	Opinion: Bad–good							Comments
	1	2	3	4	5	6	7	
1. How easy is the bike to pedal when riding at a normal speed of 15 mph on a flat road?								
2. How comfortable is the pedaling rate when riding at a normal speed of 15 mph on a flat road?								
3. How easy is it to shift the gear to adjust to a comfortable pedaling thrust?								
4. How effective is the braking under wet conditions?								
5. How much vibration (due to rough road conditions) is transferred to you?								
6. How easy, quick, and stable is the bike when changing direction?								
7. Does the bike have adequate stiffness to withstand shocks without flexing too much?								
8. What is the overall usability of the bike in terms of functionality?								

Table 10.32 **Usability evaluation of a hybrid bike for ease of use**

	Opinion: Bad–good							Comments
	1	2	3	4	5	6	7	
1. How comfortable is the seat height?								
2. How easy is it to adjust the seat?								
3. How comfortable is the height of the handlebar?								
4. How soft and firm is the grip of the handlebar?								
5. How easy is it to rotate the handlebar?								
6. How much force do you need to apply to shift gears?								
7. How comfortable is the seat cushioning?								
8. How slip free is the pedal?								
9. How much force do you need to engage the brakes?								
10. How portable is the bike?								
11. What is the overall usability of the bike in terms of ease of use?								

Table 10.33 Usability evaluation of a hybrid bike for esthetics

	Opinion: Bad–good							Comments
	1	2	3	4	5	6	7	
1. How good is the color or finish of the bike? 2. How good is the texture of the bike exterior? 3. How soft and flexible is the contact surface of the bike? 4. How attractive and inviting is the bike shape to the user? 5. How "clean" does the bike "feel?" 6. How well does the bike design incorporate intricate and appealing shapes? 7. What is the overall usability of the bike in terms of esthetics?								

Table 10.34 Usability evaluation of a hybrid bike for reliability

	Opinion: Bad–good							Comments
	1	2	3	4	5	6	7	
1. Does the front derailleur often fail in shifting the chain to the chain wheel? 2. Does the rear derailleur often fail in shifting the chain onto the rear cassette? 3. Does the rear derailleur go out of alignment whenever the bike is impacted? 4. Does the brake cable or shifter cable often break? 5. Does the braking distance vary in dry and wet conditions? 6. Do the crank and hub perform differently in clean conditions as opposed to on a dirty or muddy track? 7. What is the overall usability of the bike in terms of reliability?								

Table 10.35 Usability evaluation of a hybrid bike for maintainability and serviceability

	Opinion: Bad–good							Comments
	1	2	3	4	5	6	7	
1. Does it take less time to remove the front or rear wheel?								
2. Does it take a small amount of time to remove the front and rear derailleur for maintenance?								
3. Is it easy to apply lubricant to the crank, front wheel hub, rear wheel hub, and the sprockets of the derailleur?								
4. Does the bike need special tools to repair critical components, such as the wheels and derailleur?								
5. Is it easy to assemble and disassemble the bike without special training?								
6. How often is lubrication of critical components necessary?								
7. What is the overall usability of the bike in terms of maintainability and serviceability?								

Table 10.36 Usability evaluation of a hybrid bike for environmental friendliness

	Opinion: Bad–good							Comments
	1	2	3	4	5	6	7	
1. Does the bike frame use materials that are restricted in supply?								
2. Does the bike use a high percentage of components that have been reused or refurbished?								
3. Does the manufacturing process of the bike avoid the use of greenhouse gases, such as CFCs or HCFCs?								
4. Does the lubricant recommended for the bike have no harmful effect on the environment when it is ultimately washed out?								
5. Does the manufacture of the bike avoid the use of energy-intensive processes?								
6. Does the manufacture of the bike avoid the use or release of liquid or solid materials that are not environmentally friendly?								
7. Has the design minimized the number and volume of different packaging materials?								
8. What is the overall usability of the bike in terms of environmental friendliness?								

Table 10.37 Usability evaluation of a hybrid bike for recyclability and disposability

	Opinion: Bad–good							Comments
	1	2	3	4	5	6	7	
1. Are the bike parts and its coatings of different materials?								
2. Do the components, such as the derailleur, use parts of different materials joined together in ways that are easy to separate?								
3. When synthetic plastic materials are used for parts, such as shifter, brake level, and derailleur brackets, are they recyclable?								
4. Are important parts of the bike identified by ISO markings regarding their material content?								
5. Are the rubber tubes and tires recyclable?								
6. What is the overall usability of the bike in terms of recyclability and disposability?								

Table 10.38 Usability evaluation of a hybrid bike for safety

	Opinion: Bad–good							Comments
	1	2	3	4	5	6	7	
1. Does the bike stop within at least one car distance when moving at 15 mph?								
2. Is the bike provided with reflectors behind the pedals and seat, and are those reflectors visible at night?								
3. Is the bike provided with a protective guard to protect from sharp chain wheels?								
4. Are the brake levers designed to avoid pinching the user's fingers when applying brakes?								
5. Is the lubricant recommended for the bike safe for handling?								
6. Is the bike design free of sharp edges, points, and corners that can injure a user?								
7. Over an extended time horizon, does the bike expose the user to health risks?								
8. What is the overall usability of the bike in terms of safety?								

Table 10.39 Usability evaluation of a hybrid bike for customizability

	Opinion: Bad–good							Comments
	1	2	3	4	5	6	7	
1. Can a user get a customized bike to meet his or her specific wants and needs?								
2. Can a user request a retailer to attach add-on accessories, each of which serves a different purpose?								
3. Can a user change a chain wheel and cassette to obtain a different gear ratio, if required?								
4. Can a user change critical components, such as crank, gear wheel, or derailleur, when improved components are available in the future?								
5. What is the overall usability of the bike in terms of customizability?								

Table 10.40 Checklist for design and manufacture of hybrid bicycle for functionality

	Implemented? Bad–good					Comments
	1	2	3	4	5	
1. Select appropriate materials that meet the requirements of strength/weight ratio						
2. Choose appropriate frame structures that are rigid						
3. Use double-butted tubes instead of single-butted tubes						
4. Use seamless tubes instead of seamed tubes						
5. Use superior welding techniques, such as tungsten inert gas welding instead of arc welding, for superior toughness of joints						
6. Adopt brazing techniques so that temperature does not exceed 800 °F; rate the brazing with respect to the temperature attained						
7. Select design dimensions such as fork angle, tread, tire width, and wheel diameter so that bike handles easily						
8. Heat treat the frame to relieve the stresses and increase the toughness of the frame joints						
9. Use aluminum rims to prevent wet faze; avoid steel rims to avoid deterioration in brake performance under wet conditions						
10. Provide gears with a range of speeds						
11. Use a center pull design instead of a side pull design when designing brakes						
12. Use cantilever brakes instead of caliper brakes						
13. Use suspension brakes to dampen the vibration						

(*Continued*)

Table 10.40 (Continued)

	Implemented? Bad–good					Comments
	1	2	3	4	5	
14. Provide toe caps to make it easier to pedal 15. Make the wheel spokes double butted 16. Provide indexed shifters for shifting gears 17. What is the overall rating of the product in terms of its design for functionality?						

Table 10.41 Checklist for design and manufacture of hybrid bicycle for ease of use

	Implemented? Bad–good					Comments
	1	2	3	4	5	
1. Use a closed frame design for males and a mixed design for females 2. Use indexed shifters so that shifting occurs in discrete steps 3. Provide visible numbering for the gears so that user knows in which gear he or she is 4. For persons with less dexterity, use a rapid-fire shifting mechanism requiring less force and easy to use 5. Keep the distance between the handlebar and the seat such that the user need not assume an awkward posture 6. Keep the height of the stem such that the user stays erect and need not lean too much 7. Use frictionless cable sealing so that the braking force requirement is minimal 8. Use silicone lubricant that adheres to the chain better for enhanced lubrication and reduced mechanical friction 9. Use bearings packed and sealed with grease so that minimal rotation torque is required by the user 10. Do not weld the seat stem to the frame; make is adjustable 11. Provide adequate padding to the seat for cushioning 12. Provide adequate spring support for the saddle to absorb vibration and shocks of the road 13. Anodize the aluminum frame tubes so there is no physical injury while riding 14. Design pedals for slip-free performance 15. Make tread width around 1.15 in. for higher traction 16. When portability is preferred, use lightweight materials such as aluminum alloy or composites instead of steel alloy; typical weight of a hybrid bike is 28–29 pounds 17. What is the overall usability of the product in terms of its design for ease of use?						

Table 10.42 Checklist for design and manufacture of hybrid bicycle for esthetics

	Implemented? Bad–good					Comments
	1	2	3	4	5	
1. Design the bike frame to impart a flowing feel and aerodynamic appearance						
2. Provide the frame a clear coat to impart a glossy appearance						
3. Provide a wide spectrum of colors to meet user preference						
4. Polish interior parts such as rear and front wheel hubs						
5. Make shifters of rubber, which is soft to touch and easy to use						
6. Make shift levers, brake levers, etc., of synthetic materials with coloring agents that match individual taste						
7. Design the seat out of leather or any material that provides softness of touch						
7. What is the overall usability of the product in terms of its design for esthetics?						

Table 10.43 Checklist for design and manufacture of hybrid bicycle for reliability

	Implemented? Bad–good					Comments
	1	2	3	4	5	
1. Provide a mechanism to the rear derailleur to allow for adjustment during any misalignment						
2. Use a strong material, such as stainless steel, to manufacture brake and shifter cables						
3. Make the rim out of aluminum alloy to enable reliable braking performance						
4. Provide the brake cables with seals to protect them from dirt						
5. Simplify the design and minimize the number of components, such as derailleur						
6. Use a burn-in process to reduce the chance of defective components						
7. Use X-ray machines to screen tubes with structural defects						
8. Provide generous margins of safety for frames, crankshaft, stem, hubs, and derailleur						
9. What is the overall usability of the product in terms of its design for reliability?						

Table 10.44 Checklist for design and manufacture of hybrid bicycle for maintainability and serviceability

	Implemented? Bad–good					Comments
	1	2	3	4	5	
1. Provide easy clamp mechanisms for easy removal and repair of front and rear wheel assemblies						
2. Provide derailleurs with easy clamp mechanisms to enable repair when needed; use brackets instead of brazing them on the frame						
3. Design components so that they require fewer types and numbers of tools to repair						
4. Adopt modular designs for assemblies, so the entire unit can be removed, tested, and replaced						
5. Provide easy access to areas such as hub and crank bearings for lubrication						
6. Use standard commercial parts to facilitate interchangeability of components						
7. Eliminate sharp corners and burrs, which pose a threat during repair						
8. Use sealed bearings, if possible, to avoid the need for regular lubrication and prevent collection of debris						
9. What is the overall usability of the product in terms of maintainability and serviceability?						

Table 10.45 Checklist for design and manufacture of hybrid bicycle for environmental friendliness

	Implemented? Bad–good					Comments
	1	2	3	4	5	
1. Avoid using paints with ozone-depleting liquids, such as CFCs; instead, use paints with catalytic converters						
2. Use molded-in color systems for plastic components						
3. Prefer the use of biodegradable lubricants						
4. Prefer using materials such as aluminum alloy and steel for frames instead of rare materials such as titanium						
5. Use refurbished components, such as crankshafts and derailleurs, as much as possible						
6. Design packaging for shipping the bike in disassembled form to reduce the need for packaging materials						
7. What is the overall usability of the product in terms of environmental friendliness?						

Table 10.46 **Checklist for design and manufacture of hybrid bicycle for disposability and recyclability**

	Implemented? Bad–good					Comments
	1	2	3	4	5	
1. Avoid the use of coating materials different from the base metal; use processes such as anodizing for aluminum frames						
2. Reduce the number of components of different materials, such as derailleurs, to make sorting easier during disposal						
3. Avoid using multiple colors when painting the frame						
4. Use metallic materials for frames due to their ease of recycling						
5. For frames, avoid using composite materials, such as glass or carbon reinforced plastics						
6. For composite frames, select thermoplastic materials instead of using thermosetting plastic materials						
7. Add numbers to parts during processing (such as casting and forging) instead of using adhesive labels						
8. If adhesive labels are used, use adhesive material that is compatible with components when they are recycled						
9. Standardize product components, such as derailleurs, hubs, and crankshafts, so that they can be salvaged and reused						
10. Use rubber tubes and tires that can be recycled						
11. What is the overall usability of the product in terms of its design for disposability and recyclability?						

Table 10.47 **Checklist for design and manufacture of hybrid bicycle for safety**

	Implemented? Bad–good					Comments
	1	2	3	4	5	
1. Make the rims out of aluminum alloy, not steel alloy, which show wet faze						
2. Provide the bike with reflectors behind pedals and seats, so they are visible especially at night						
3. Provide the bike with protective guards to protect the user from sharp chain wheels						
4. Use lubricant that is harmless to the hands and eyes						
5. Provide ample clearance to the brake lever, so it does not pinch when the brakes are applied						

(Continued)

Table 10.47 (Continued)

	Implemented? Bad–good					Comments
	1	2	3	4	5	
6. Make the saddle comfortable; hard, narrow seats reduce blood circulation and produce excessive pressure at the point of contact, which may cause impotence over a prolonged period of time						
7. Avoid using paints that contain heavy metals, such as lead, arsenic, and cadmium						
8. Minimize excessive cable length for brakes and shifters, so they do not tangle and interfere with user movements						
9. What is the overall usability of the product in terms of its design for safety?						

Table 10.48 Checklist for design and manufacture of hybrid bicycle for customizability

	Implemented? Bad–good						Comments
	1	2	3	3	4	5	
1. Adopt a modular design for important assemblies, such as frame, front and rear gear, and derailleurs, so they can be modified to suit individual preferences							
2. Use standard components using standard processes and procedures							
3. Select processes that do not require long tooling times or could be made with standard tooling							
4. Use standard hole sizes, slot width, filet radii, and other design attributes when designing components							
5. Provide locations for add-ons, such as a dynamo, water bottle, or chain wheel guard, should the user want them							
6. Use parts from a catalog instead of newly designed to provide a high degree of standardization							
7. Use near-net shape processes, such as powder metallurgy for making pulleys used in derailleurs, instead of machining							
8. For machined or formed parts, such as hubs, use steel with high machinability or formability							
9. What is the overall usability of the product in terms of its design for customizability?							

Table 10.49 Assignment of bicycle models to specific users, by code

Bicycle models	Users
C_1	U_1, U_{11}
C_2	U_2, U_{12}
C_3	U_3, U_{13}, U_{20}
C_4	U_4, U_{14}
C_5	U_5, U_{15}
C_6	U_6, U_{16}
C_7	U_7, U_{19}
C_8	U_8, U_{17}, U_{22}
C_9	U_9, U_{21}
C_{10}	U_{10},U_{18}

Table 10.50 Reliability test values for usability evaluation questionnaires

Usability criteria	Cronbach coefficient alpha
Functionality	0.629
Ease of use	0.723
Esthetics	0.670
Reliability	0.635
Maintainability/serviceability	0.413
Environmental friendliness	0.628
Recyclability/disposability	0.708
Safety	0.742
Customizability	0.574

Table 10.51 Validity test values for usability evaluation questionnaires

Usability criteria	Correlation coefficient	*f* Test	
		f Value	*p* Value
Functionality	0.722	21.723	0.0001
Ease of use	0.541	8.283	0.0093
Esthetics	0.570	9.620	0.0056
Reliability	0.438	4.767	0.0411
Maintainability/serviceability	0.651	14.672	0.0010
Environmental friendliness	0.576	9.927	0.0050
Recyclability/disposability	0.452	5.512	0.0349
Safety	0.635	13.516	0.0015
Customizability	0.488	6.262	0.0211

10.7 Concluding remarks

In an increasingly competitive world, design lead time is on a continuous decline in an ever-increasing push to get new products to market quicker. This means that the onus is now on the manufacturer to develop a highly usable product right, the first time. The checklist-based approach provides a heuristic method to tell the designer what to expect in a product to adopt the most appropriate design and relate manufacturing processes to user requirements. The cases presented in this chapter developed and tested questionnaires for each usability dimension. The practical implication of the overall usability score is that it can be used as a criterion for product selection.

The study presented in the preceding pages, however, concentrated on suggesting measures aimed at product usability. It did not study the influence such measures have on product cost. For instance, using a graphite matrix frame in a bicycle results in a superior product, but such a product often is prohibitively expensive. This can be an area future product designers can focus on to increase a product's overall appeal.

References

Akao, Y., 1990. Quality Function Deployment: Integrating Customer Requirements into Product Design. Productivity Press, Cambridge, MA.

Akita, M., 1991. Design and ergonomics. Ergonomics 34 (6), 815–824.

Alexander, S.M., 1992. Reliability theory. In: Hodson, Industrial Engineering Handbook McGraw-Hill, New York, NY.

Alonso, R., 1992. Continuous Improvement in Reliability, pp. 106–114.

Anderson, D.M., 1991. Design for Manufacturability. CIM Press, Lafayette, CA.

Ashley, S., 1993. Designing for the environment. Mech. Eng.

Bergquist, K., Abeysekara, J., 1996. Quality function deployment (QFD)—a means for developing usable products. Int. J. Ind. Ergon. 18, 269–275.

Berko-Boateng, V., Azar, J., de Jong, D., Yander, G.A., 1993. Asset recycle management—a total approach to product design for the environment. In: IEEE International Symposium on Electronics and the Environment, Piscataway, NJ.

Bieda, J., 1992. Reliability Growth Test Management in the Automotive Component Industry, pp. 387–393.

Billatos, S.M., Basaly, N.A., 1997. Green Technology and Design for Environment. Taylor and Francis, Washington, DC.

Blanchard, B.B., Verma, D., Peterson, E.L., 1995. Maintainability: A Key to Effective Serviceability and Maintenance Management. John Wiley & Sons, Inc., New York, NY.

Boothroyd, G., 1994. Product design for manufacture and assembly. Comput. Aided Des. 26 (7), 505–520.

Boothroyd, G., Alting, L., 1992. Design for assembly and disassembly. Ann. CIRP 41 (2), 625–636.

Bralla, J.G., 1996. Design for Excellence. McGraw-Hill, New York, NY.

Burhanuddin, S., Randhawa, S., 1992. A framework for integrating manufacturing process design and analysis. Comput. Ind. Eng. 23 (1–4), 27–30.

Chang, T., Wysk, R.A., Wang, H., 1991. Computer Aided Manufacturing. Prentice-Hall, Englewood Cliffs, NJ.

Consumer Reports, 1990. Three Ways to Brown Bread, June, pp. 380–383.
Cubberly, W.H., Bakerjian, R., 1989. *Tool and Manufacturing Engineer's Handbook*, Desk Edition. Society for Manufacturing Engineers, Dearborn, MI.
Cushman, W.H., Rosenberg, D.J., 1991. Human Factors in Product Design. Elsevier Science, Amsterdam.
Dallas, D.B., 1976. Tool and Manufacturing Engineer's Handbook. McGraw-Hill, New York, NY.
Day, R.G., 1993. Quality Function Deployment: Linking a Company with its Customers. ASQC Quality Press, Milwaukee.
Dika, R.J., Begley, R.L., 1988. Concept Development through Teamwork—Working for Quality, Cost, Weight and Investment, pp. 277–288.
Eversheim, W., Baumann, M., 1991. Assembly oriented design process. Comput. Ind. 17, 287–300.
Farris, J., Knight, W.A., 1992. Design for Manufacture: Expert Processing Sequence Selection for Early Product Design.
Govindaraju, M., Mital, A., 1988. Design and manufacture of usable consumer products. Part II: Developing the usability-manufacturing linkages. Int. J. Ind. Eng.
Hammer, W., 1980. Product Safety Management and Engineering. Prentice-Hall, Englewood Cliffs, NJ.
Haubner, P.J., 1990. Ergonomics in industrial product design. Ergonomics 33 (4), 477–485.
Heideklang, H.R., 1990. Safe Product Design in Law, Management and Engineering. Marcel Dekker, New York, NY.
Hitomi, K., 1991. Strategic integrated manufacturing systems: the concept and structures. Int. J. Prod. Econ. 25, 5–12.
Hoffman III, W.F., Locasio, A., 1997. Design for environment development at Motorola. IEEE, 210–214.
Hofmeester, G.H., Kemp, J.A.M., Blankendaal, A.C.M., 1996. Sensuality in product design: a structured approach. CHI, 13–18.
Ishihara, S., Ishihara, K., Nagamachi, M., Matsubara, Y., 1995. An automatic builder for a kansei engineering expert system using self organizing neural networks. Int. J. Ind. Eng. 15, 13–24.
Joines, S., Ayoub, M.A., 1995. Design for assembly: an ergonomic approach. Ind. Eng., 42–46.
Jorgensen, A.H., 1990. Thinking aloud in user interface design: a method promoting cognitive ergonomics. Ergonomics 33 (4), 501–507.
Juran, J.M., Gryna, F.M., Bingham Jr., R.S., 1974. Quality Control Handbook. McGraw-Hill, New York, NY.
Kaila, S., Hyvarinen, E., 1996. Integrating design for environment into the product design of switching platforms. IEEE, 213–217.
Karmarkar, U.S., Kubat, P., 1987. Modular product design and product support. Eur. J. Oper. Res. 29, 74–82.
Kitchenham, B., Pfleeger, S.L., 1996. Software quality: the elusive target. IEEE Trans. Softw. Eng., 12–21.
Lankey, R., McLean, H., Sterdis, A., 1997. A case study in environmentally conscious design: wearable computers. IEEE, 204–209.
Lin, R., Lin, C.Y., Wong, J., 1996. An application of multidimensional scaling in product semantics. Int. J. Ind. Eng. 18, 193–204.
Logan, R.J., Augaitis, S., Renk, T., 1994. Design of simplified television remote controls: a case for behavioral and emotional usability. In: Proceedings of the Human Factors and Ergonomics Society 38th Annual Meeting, pp. 365–369.

Maduri, O., 1993. Design planning of an off highway truck—A QFD approach. In: Kuo, W., Pierson, M.M. (Eds.), Quality Through Engineering Design Elsevier Science Publishers, Amsterdam.
McClelland, I., 1990. Marketing ergonomics to industrial designers. Ergonomics 33 (4), 391–398.
Mital, A., Anand, S., 1992. Concurrent design of products and ergonomic considerations. J. Des. Manuf. 2 pp. 197–183.
Nagamachi, M., 1995. Kansai engineering: a new ergonomic consumer oriented technology for product development. Int. J. Ind. Ergon. 15, 3–11.
Nasser, S.M., Souder, W.E., 1989. An interactive knowledge based system for forecasting new product reliability. Comput. Eng. 17 (1–4), 323–326.
Nevins, J.L., Whitney, D.W., 1989. Concurrent Design of Products and Processes. McGraw-Hill, New York, NY.
Nielsen, J., 1992. The usability engineering life cycle. Computer, 12–22.
Nielsen, J., 1993a. Usability Engineering. Academic Press, San Diego, CA.
Nielsen, J., 1993b. Iterative user interface design. Computer, 32–41.
Nissen, N.F., Griese, H., Middendorf, A., Pottor, M.H., Reichl, H., 1997. Environmental assessments of electronics: a new model to bridge the gap between full life cycle evaluations and product design. IEEE, 182–187.
Parsei, H.R., Sullivan, W.G., 1993. Concurrent Engineering: Contemporary Issues and Modern Design Tools. Chapman and Hall, London.
Pnueli, Y., Zusman, E., 1996. Evaluating end of life value of a product and improving it by redesign. Int. J. Prod. Res. 35 (4), 921–942.
Priest, J.W., 1988. Engineering Design for Producibility and Reliability. Marcel Dekker, New York, NY.
Rao, S.S., 1992. Reliability Based Design. McGraw-Hill, New York, NY.
Rooijmans, J., Aerts, H., 1996. Software quality in consumer electronics products. IEEE Trans. Softw., 55–64.
Ryan, J.P., 1983. Human factors design criteria for safe use of consumer products. In: Proceedings of the Human Factors Society, 27th Annual Meeting, pp. 811–815.
Schneiderman, B., 1998. Designing the User Interface: Strategies for Effective Human Computer Interaction. Addison-Wesley, Reading, MA.
Smith, D.J., 1993. Reliability, maintainability and risk Practical Methods for Engineers. Butterworth-Heinemann, Oxford, UK.
Stoll, H.W., 1988. Design for manufacture. Manuf. Eng., 67–73.
Suh, N.P., 1990. The Principles of Design. Oxford University Press, New York, NY.
Tervonen, I., 1996. Support for quality based design and inspection. IEEE Trans. Softw., 44–54.
Tipnis, V.A., 1994. Challenges in product strategy, product planning and technology development for product life cycle design. Ann. CIRP 43 (1), 157–162.
Tseng, M.M., 1996. Design for mass customization. Ann. CIRP 45 (1), 153–156.
Tsuchiya, T., Maeda, T., Matsubara, Y., Nagamachi, M., 1996. A fuzzy rule induction method using genetic algorithm. Int. J. Ind. Ergon. 18, 135–145.
Ulrich, K.T., Eppinger, S.D., 1995. Product Design and Development. McGraw-Hill, New York, NY.

Concurrent Consideration of Product Usability and Functionality

11.1 Introduction

Consumers have always been meant to be at the center of all product design activities. Their needs, wants, and requirements have led companies to consistently work at improving their products. Product design today has moved beyond the symbolic "shape" and "appearance" considerations. The Industrial Designers Society of America defines *design* as "the professional service of creating and developing concepts and specifications that optimize the function, value appearance of products and systems for the mutual benefit of the user and the manufacturer."

This definition is broad enough to encompass a wide range of activities of the product development team. The consumer product market is highly dynamic, and the variety of design goals have been influencing and shifting the product design paradigm. The concepts of assembly lines, interchangeable parts, and standardization brought to the fore the concept of mass production. The rapid increase in green marketing and research in the 1990s coincided with the drastic and inevitable shift of consumers toward green products. The global consumer boycott of CFC-driven aerosol products is an excellent example of this trend.

The present day concurrent engineering design process gained importance with the concept of design for "X." The process focuses on a number of design goals where X could stand for assembly, manufacturability, quality, life cycle, and so on. This process is intended to ensure the consideration of all phases of the product development cycle, i.e., design and other downstream processes such as manufacturing, distribution, maintenance, and disposability, among others, thereby ensuring a better match of structural to functional requirements.

The design perspectives have been ever-changing, but the provision of both usability and functionality has been a cornerstone of the product design and development process; these represent two of the most important product attributes from the consumer's perspective. The development of products that remain usable and functional throughout the product life cycle recognizes these as essential design requirements (Chiang, 2000; Han and Kim, 2003).

Product usability consists of several underlying design dimensions, and there exists a functional relationship between them (Kim and Han, 2008). Several design goals

Product Development. DOI: http://dx.doi.org/10.1016/B978-0-12-799945-6.00011-9

such as reliability, safety, ecological design, and customizability have been considered in the concurrent design and manufacturing of consumer products. Manufacturing variables have been successfully linked to a product's usability features, and design guidelines that ensure that a usable product can be manufactured have been developed (Govindaraju, 1999). These design guidelines were presented in the form of design checklists. Babbar et al. (2002) used affinity diagrams to sift through customer experiences of different consumer products.

Despite considerable progress in developing design methodologies, it is still an uphill task for a design team to develop a multifunctional usable product, since existing design processes lack the means to integrate multiple design criteria; instead, design criteria need to be considered one at a time. Integration of multiple design criteria remains desirable because it would reduce the occurrence of conflicting design features or those that favor only one design objective. This simultaneous consideration does, however, mean that design cannot be optimized from the viewpoint of each and every criterion; the final design is simply the best possible compromise. The work presented in this chapter has sought to simultaneously consider multiple design criteria defining both the overall functionality and overall usability of consumer products in order to produce highly functional and usable products. Such considerations require a clear and concise understanding of consumer requirements and a comprehensive elaboration and identification of critical issues pertaining to those two design goals and how they impact product design.

11.2 Design methodology

The methodology presented in this chapter integrates separately existing design guidelines for usability and functionality already presented in this book. The intent is to generalize these integrated design guidelines and then demonstrate their application through examples. The design guidelines for usability and functionality were developed using linkages with products' manufacturing attributes. Figure 11.1 presents a schematic of the procedure used to achieve the aforementioned integration.

The integration of design guidelines can be classified into three different stages:

1. Development of generalized guidelines for various usability and functionality criteria
2. Deriving customized checklists for a specific product or a family of products from generic checklists
3. Testing the integrated design criteria checklists.

Phase 1:

As demonstrated in previous chapters on designing for usability and functionality, multiple design criteria are valid across all consumer product families, and customized design checklists can be developed for a specific product. This makes it all the more important to develop integrated generic design guidelines which would provide the designer with suitable product and process design options. The information necessary to develop these guidelines was accumulated from sources such as consumer

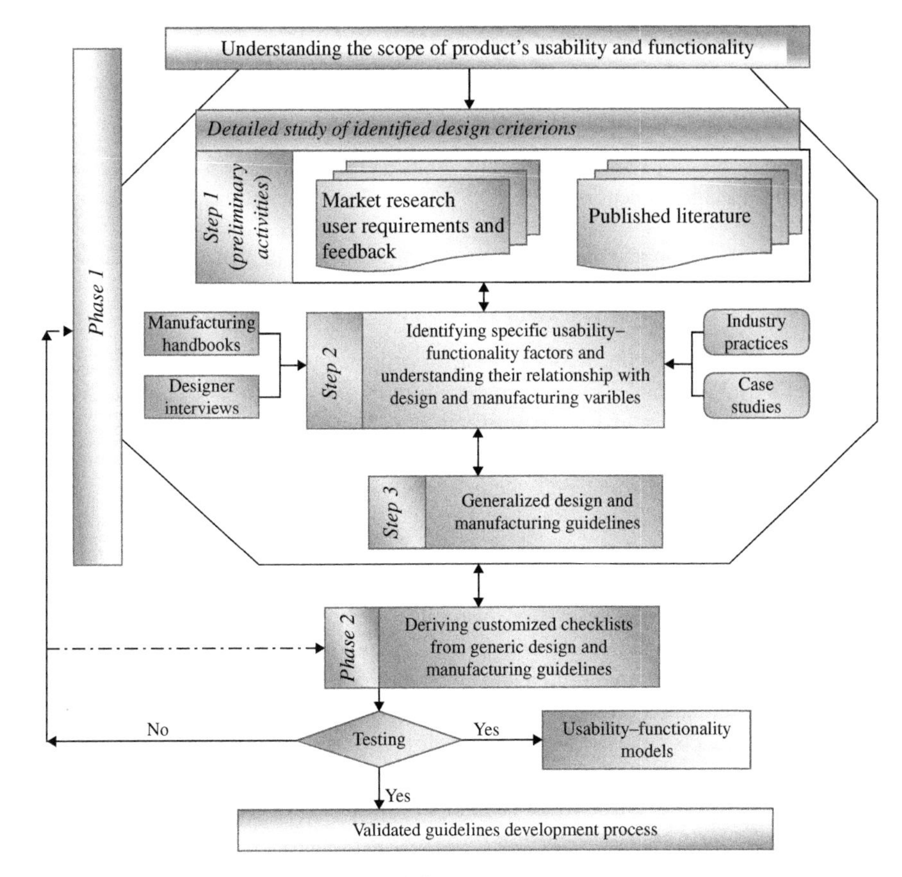

Figure 11.1 Development of design guidelines.

surveys, questionnaires, published literature, design and manufacturing handbooks, and user complaints databases.

Phase 2:

The generic design guidelines must be customized for a specific product before being used by a designer. The guidelines developed in phase 1 were translated for the specific product family. Additional information was collected from users and manufacturers of the product, given the fact that the usability and functionality requirements are different for specific product families. The inputs obtained included customer requirements, manufacturing processes, etc.

Phase 3:

The customized design checklists prepared in phase 2 were statistically tested. Data were gathered from users of the product. The checklists were tested for their reliability and validity. The reliability test measured the consistency of the instrument, whereas the validity test determined whether the instrument fulfilled its purpose. Various quantitative tools such as principal component regression (PCR) were also created for prioritizing the design dimensions.

11.2.1 Developing generic integrated design guidelines

The activities carried out for developing integrated design guidelines consisted of four iterative steps, as detailed here:

1. Understanding the scope of the product's usability and functionality
2. Developing specific dimensions for each of the design goals, namely usability and functionality
3. Identification of design and manufacturing variables and linking them to the usability and functionality requirements
4. Development of guidelines in the form of checklists.

Step 1: To start out with, it is essential to properly understand the full scope of the product's usability and functionality. In order to deliver a user-friendly and functionally superior product, it is imperative to concurrently design for several design criteria. After thoroughly reviewing existing customer surveys, guidelines, etc., the different design criteria identified were usability, performance, reliability, safety, maintainability, serviceability, esthetics, etc.

Step 2: It was assumed that the design criteria identified in step 1 could be split into latent factors that are also known as design categories. Table 11.1 depicts the categories for each criterion. These latent variables are not directly measurable but can be explained by various indicators or specific dimensions. The necessary information regarding these categories and dimensions was gathered from a variety of sources such as case studies, published literature, and databases of user experiences and complaints.

Table 11.2 depicts specific dimensions for each criterion. Their classification under various design categories was deliberately avoided due to their high correlation and overlapping nature. These dimensions are controlled through different variables discussed in the next step. The efforts of any design team should be fully directed toward optimizing these dimensions, and thus improving the product's usability and functionality.

Step 3: Various design and manufacturing variables are essential in controlling the factors considered under various design criteria. Manufacturing variables consist of

Table 11.1 Grouping dimensional categories based on design criteria

Design criteria	Design categories
Ease of use	Interaction methods, task simplification, learning, instinctive design
Performance	Material, part features, product integration
Reliability	Protection, diagnostics, redundancy, components
Safety	Failure rules, packaging, protective devices, clearances
Esthetics	Image description, subjective perception, basic shape/sense
Environmental affinity	After life, raw materials, disposal procedures, energy and water conservation
Maintainability/serviceability	Standardization, product order and layout, location and joints

Table 11.2 **Constitution of various usability–functionality design criteria**

Performance	Ease of use	Safety	Reliability	Environmental affinity	Maintainability	Esthetics
Appropriate material	User friendliness	No sharp edges	Number of parts	Recyclable	Modular design	Good displays
Product ordering	Memorability	Provision of guards	Redundancy	No toxic material	Diagnosis system	Satisfy
Environment	Simplicity	Modular design	Material strength	Disposability	Standardization	Attractive
Solid base	Modelessness	Abuse	Diagnosis	Disassembly consideration	Identification of weak components	Preference
Minimizing mass/strength ratio	Ergonomics	Interlock provision	Loads and capacity	Degradable scrap	Accessible parts	Enticing shape
Geometric sizes and dimensions	Utilizing constraints	Redundancy	Testability	Reusable	Joints	Rust proof
Functional variability	Robust	Poka-yoke	Failure rules	Heavy metals	Toxic gases/liquids	Smooth surface
Subassemblies	Controls	Interaction	Geometric variability	Manufacturing processes used	Location of failure prone parts	Brightness
Multifunction parts	Error prevention	Identification	Wear of hazardous components and other sources	Energy consumed	Sharp parts	Gentle touch experience
Modular design	Predictability	Moving parts	Serviceable	Water usage	Testability	Translucency
Effectiveness of function	Feedback provision	Effective diagnosis system	Identification of weak components	No of materials used	Provision of inspection	Metaphoric image

(*Continued*)

Table 11.2 **(Continued)**

Performance	Ease of use	Safety	Reliability	Environmental affinity	Maintainability	Esthetics
Tolerances	Speedy performance	Clearances	Abuse	Material used	Interchangeable	Salience
Redundancy	Use of mapping	Detailed safety guidelines procedures	Controlling environmental conditions	Joint of parts	Part labeling	Granular
Product structure analysis	Balance	Flying objects	Failure analysis	Paints	Traceability of defects	Clean
Profile of functional surfaces	Interactive displays	Toxic metals	Product integration		Repair guidelines	Harmonious design
Surface finish	Flexibility	Warning devices	Load sharing		Disposability	
Stiffness	Leverage	Fuse	Standard parts			
Steadiness	Functional variability	Safe disposal	Scope of thermal expansion			
Consistency	Compatibility adaptability		Protection			

Source: Adapted and modified from Chiang et al. (2009) and Han et al. (2001).

Table 11.3 **Manufacturing and design variables**

Material variables	Process variables	Design variables
Strength	Temperature	Tolerance
Hardness	Pressure	Height
Density	Time	Tools
Fatigue resistance	Depth of cut	Product structure

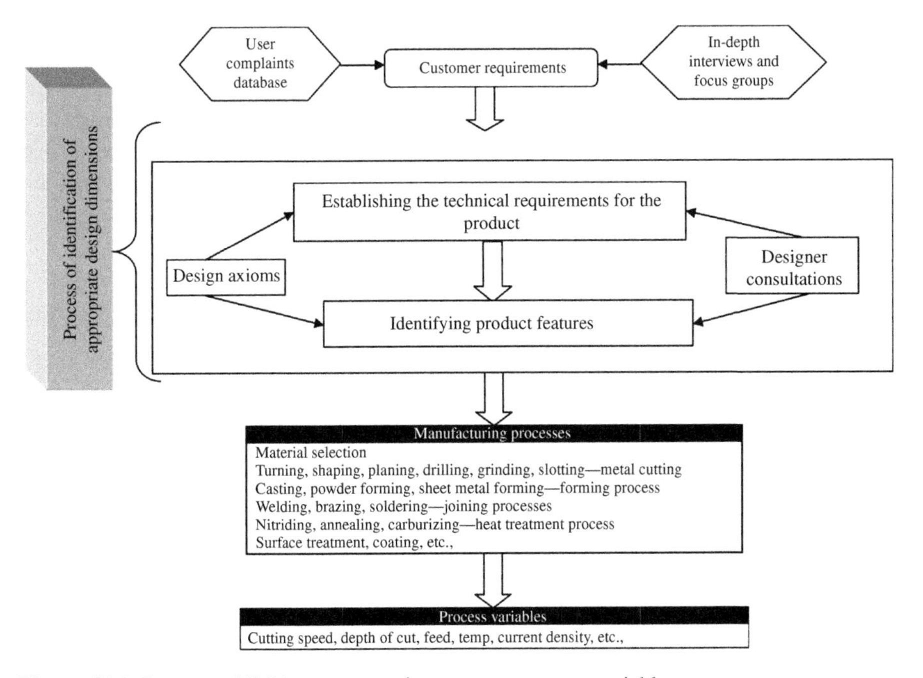

Figure 11.2 Process of linking user requirements to process variables.

material and process variables. Customers' usability/functionality requirements dictate the materials to be used. Material properties in turn determine appropriate manufacturing processes. Design variables are under the direct control of designers. They are highly subjective in nature and are influenced by factors such as the designer's experience and type of design (creative or adaptive). Design variables are treated as constraints. Table 11.3 presents some material, process, and design variables.

It is difficult to quantify the optimum values of these variables. However, their impact in terms of specific dimensions could be determined. In order to increase the utility of design guidelines, it is important to add to this current level of knowledge. A case study was performed for this reason. It also illustrates the procedure which will facilitate future researchers and users in developing more information and data. Figure 11.2 depicts the entire procedure for linking usability/functionality features to manufacturing attributes of a product.

Various techniques could be used for effectively transferring usability/functionality requirements into the corresponding process variables.

11.2.2 Case study: can opener

The example of a can opener is used here with the twin purpose of adding to the existing design data and for better illustration of relationships. The development of generalized checklists was not the purpose of this case study. Rather, the guidelines for usability and functionality were integrated with the aid of this case; the generalization of the guidelines was performed subsequently. Transformation diagrams and matrices were used for establishing the relationships.

The usability/functionality requirements were determined after one-on-one interviews with users of this product. The can opener, according to users, is a product that pierces the lid of the can in an efficacious manner without causing discomfort to the user in any way. The principal functionality and usability requirements are listed below:

1. Neatly pierces the lid
2. Rolls without slipping
3. Effective on thick surfaces
4. Silent operation
5. Easy to grip
6. Avoids awkward motion
7. Durable.

11.2.2.1 Identifying linkages

Transformation diagrams and matrices were used to correlate the specific dimensions to manufacturing variables. The various stages involved in the process are described in the following sections.

11.2.2.2 Establishing technical requirements and generating product features

This process entailed translating specific criteria into technical requirements and then into product features (Figure 11.3). User requirements were listed in the upper row and the technical requirements in the row below it. The next row contains product features that embody the design requirements. The design dimensions (or specific usability and functionality factors) identified after breaking down the customer requirements into technical requirements and corresponding product features are listed in Table 11.4.

Similarly, the requirements could be mapped to specific usability/functionality factors in other design criteria.

11.2.3 Manufacturing process

The process description was obtained from the manufacturer and involved blanking, piercing, heat treatment, bending, nickel plating, riveting, tumbling, and swaging.

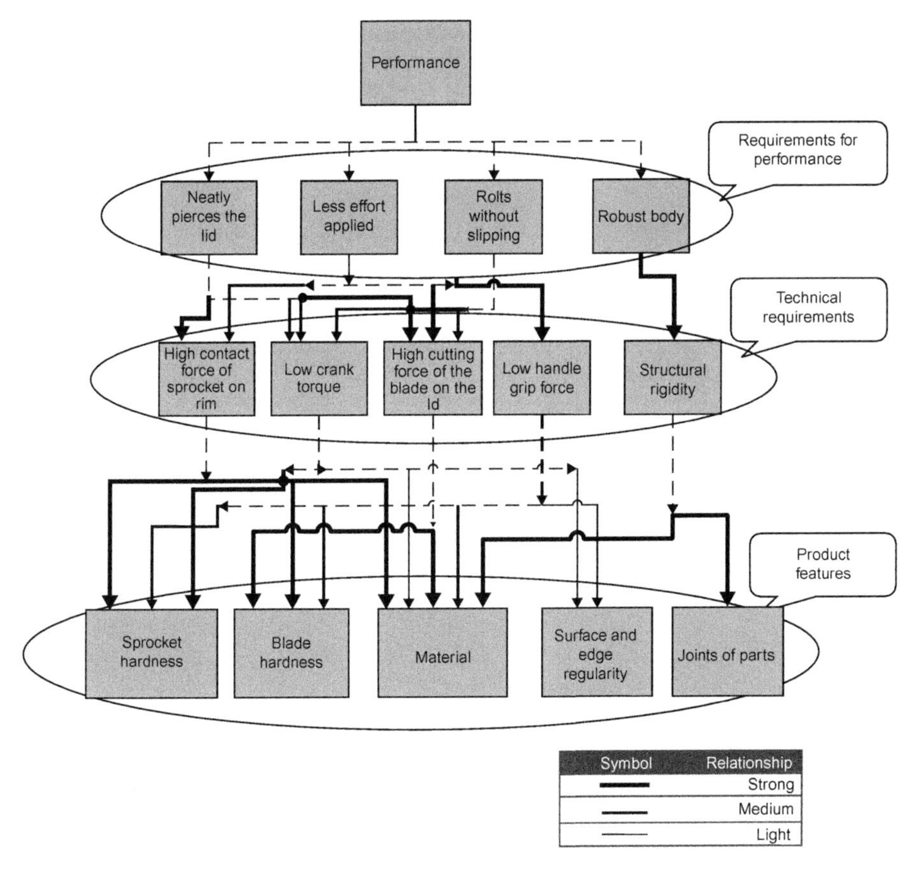

Figure 11.3 Transformation diagram for can opener.

The upper handle is cut from SAE 1008 steel by blanking. Both handles are individually subjected to stamping operations using progressive dies. The piercing and bending operations produce two holes and a twist in the upper handle. Two protrusions are formed using a die operation in the lower handle. The blade is cut from an SAE 1050 steel strip, swaged at the top to produce a sharp cutting edge. The crank is cut from an SAE 1008 steel strip and trimmed to achieve the desired shape. These pieces are tumbled in order to remove the burrs resulting from the stamping operation. The drive sprocket and blade are heat treated. All parts are nickel plated to promote corrosion resistance and enhance appearance. The drive sprocket and crank are assembled together with the upper handle and swaged to manufacture another subassembly. The two subassemblies are riveted together to produce the final assembly.

11.2.3.1 Process deployment

In this step (Figure 11.4), we bring into use the concept of transformation matrices, similar to the QFD technique. This phase depicts the manufacturing processes

Table 11.4 Usability–functionality factors

Performance	Ease of use
Appropriate material	Learnable
Effectiveness of function	Task simplicity
Tolerances	User friendliness
Subassemblies	Avoids awkward posture
Material strength	

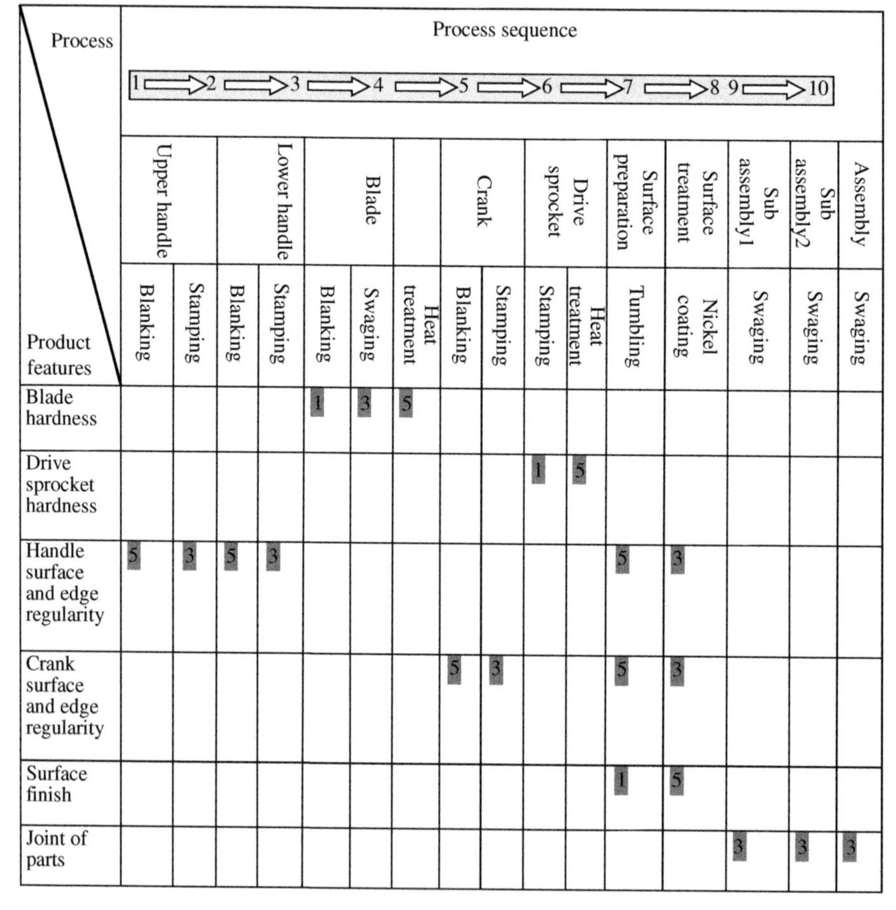

Figure 11.4 Manufacturing deployment.

required to produce the product. The product features identified in the previous step are listed in the horizontal section and the manufacturing processes required to achieve them are listed in the vertical section. The strength of relationships is indicated using a numeric value—the greater the value, the stronger the correlation.

For instance, the blade needs to be swaged to improve its sharpness. Swaging increases the surface hardness as it results in a finer grain structure. Similarly, nickel plating is used to enhance corrosion resistance and improve luster (Table 11.5).

11.2.4 Can opener assembly

This was obtained from the manufacturer. The critical components of this assembly are: upper handle, lower handle, blade, crank, and drive sprocket. The upper handle is joined with the crank and drive sprocket. The blade is joined to the lower handle. The aforementioned subassemblies are joined to form the overall assembly. The sprocket wheel holds onto the outer rim of the can when the blade pierces the lid once the opener is mounted onto the can. Torque is applied through the crank wheel, thus helping the blade cut the lid until it is completely severed.

11.2.4.1 Manufacturing deployment

The method used is similar to the one used in the previous step. This is shown in Figure 11.5. The successful implementation of this step requires detailed knowledge of the manufacturing processes and underlying dimensions that control the processes.

The manufacturing processes generally have numerous controlling variables, but only those which affect the can opener are considered for this case study. For instance, the die dimensions and die clearance are the variables that must be controlled during the stamping operation as they will determine the burrs produced and uniformity of thickness.

11.2.4.2 Development of generic guidelines

Using the knowledge gained, manufacturing and design guidelines are now developed. The procedure is recursive, and new information can be continuously added to make these guidelines more efficacious. Tables 11.6–11.12 depict detailed guidelines in the form of checklists. The items in the questionnaire could be evaluated by the designers on a 1–5 scale, with 1 being the least important and 5 being the most important.

11.2.4.3 Inferences

The work presented in this chapter provides a systematic procedure for integrating and generalizing design guidelines for product usability and functionality. The procedure may be extended by including more design criteria and considering new factors evolving over time using this approach as a premise.

Table 11.5 Process deployment

Process sequence	Process variables	Punch die clearance	Revert hole clearance	Duration	Cooling rate	Operation rate	Chemical	Pressure	Current density	Quenching solution	Comments
Blanking		5				3		5			Progressive die operation
Stamping		5				5		3			Progressive die operation
Tumbling				3		3	3				Secondary operation: burr removal
Heat treatment				5	5		5			5	Strength improvement
Swaging						5		5			Blade sharpening
Riveting			5			1		5			Tertiary operation: assembly operation
Nickel plating				3		1	5		5		Corrosion resistance: image enhancement

Source: Adapted from Chiang et al. (2001).

Figure 11.5 Mountain touring bicycle.

Whenever multiple design criteria are used, it is not possible to achieve the same level of product performance for each criterion as one would when using a single design criterion. This is due to the compromises necessitated by a methodology based on multiple design criteria.

One may also ask whether the integrated design guidelines developed in this chapter provide a better design than that resulting from considering guidelines for functionality and usability separately. In other words, would the result be the same or different if the checklists were considered separately or together? An answer to such a question is impossible because that would require product design by considering the guidelines separately as well as together and then ascertaining how such a design would fare on the market. Given that consideration of multiple criteria is a practical necessity, integrated checklists make more sense. This provides the designer a better vision in identifying conflicts in the requirements, thereby allowing one to give more or less weight to a specific design criterion when integrating them and thus control the design process. It is worth noting that the generalized guidelines presented here can be customized for specific product families as needed. The next section presents a case study to elucidate this fact.

11.2.5 Case study: mountain touring bike

The objective of this case study is to manifest how the methodology and the generic guidelines presented in the previous part of this chapter could be applied to a specific consumer product. An MTB (mountain touring bike) was selected for this study. The generic guidelines developed earlier were modified for the bike after relevant inputs were obtained from designers, users, and the published literature. Not all dimensions present in the generic guidelines were applicable to the MTB. Interaction with bike users was important in understanding and incorporating their requirements in the bike design. These requirements were analyzed using usability–functionality

Table 11.6 Design guidelines for usability

Questions	Additional comments	Rate at item
Simplify the tasks	**Choose an item for each numerical category**	
1. The control process should simulate the actual mechanism 2. The control operations should be straightforward and easy to understand 3. Mapping: The control should map the actual operation of the mechanism 4. Reduce the complexity in the design of the product; usability increases with simplicity in design 5. Should eliminate or minimize the decision making or the prior planning process involved in controlling an operation 6. Control of an operation should be provided adjacent to the part which performs it		
Learnable		
1. The user interactive methods should be intuitive 2. Design should adopt familiar controls, user interface, and interactive methods 3. Design for modelessness, i.e., each control should have only one designated behavior or output i. Should conform to the users choose an item 4. Limit the static strength requirements to 5% and dynamic to 10% of the maximal volitional strength exerted when muscle loading is protracted 5. Try to minimize the lifting load and the parts choose an item i. Avoiding awkward postures 6. Provide means to leverage the force applied by a human 7. The controls and the parts requiring a frequent physical user touch should be placed at accessible location and should be grouped 8. The handles and grips should be free of burrs or sharp corners 9. Should allow for the uniform distribution of force over the entire area		
Controlling errors		
1. Reverse or erroneous movement of the controlling parts should be restricted 2. Allow for reversal of error choose an item 3. Anticipate the most common human error and design to minimize them		

(*Continued*)

Table 11.6 **(Continued)**

Questions	Additional comments	Rate at item
Simplify the tasks	**Choose an item for each numerical category**	
Provide clear, consistent, and interpretable displays		
1. Use digital displays wherever precise information is required, as it enhances readability 2. Minimize the number of controls used or activated to operate the part 3. Use analog, if monitoring the changing conditions is important 4. Use displays to provide the status of the product and response to any user action taken 5. The controlling parts like knob should be sufficient wide and long, so that they fit into hands comfortably 6. Different controls should be designed differently, e.g., knobs, push buttons should be different in shape 7. Match the knob with its corresponding display		
Provide clear, consistent, and interpretable displays		
1. The operating force should not be high except for the emergency parts 2. Group the controls operating the same part 3. The display of the control should be matched to its direction of movement 4. Provide controls in the accessible location 5. The most often used control should be provided nearest to the operator's location and so on 6. Identify the potential errors and design to minimize them, phase out all the hazardous ones 7. Provision for reversing the error 8. Minimize the variability in functions of the product; the functional variance undermines the usability of the product 9. The product should be internally and externally compatible; using standard sizes could help reduce this		
Additional points importance		
1. Add your point here 2. Add your point here Aggregative rating for design in terms of ease of use rating		

Table 11.7 Design guidelines for performance

Questions	Additional comments	Rate an item
Material properties	**Rate each numerically listed item**	
i. Select the material operable in all kind of environment 1. Wide range of temperature, i.e., provision for thermal expansion 2. Insensitive to the moisture content in the air 3. Are not prone to static electricity and are nonmagnetic ii. User requirements to features correspondence 1. Minimize the number of different materials used 2. Use a material which successfully maps the maximum no. of user requirements into a corresponding product features 3. The material used should permit the concept of maximum no. of part integration; the processes like die casting and powder metallurgy allows numerous shapes		
Material properties		
1. Maximize the number of subassemblies in the design; these should be structurally self-sufficient 2. Reduce the product's complexity and redundancy 3. Try to integrate the parts into multifunctional single parts 4. The part or subassembly should not interfere or undermine other part or subassembly's performance 5. Reduce the number of functional surfaces 6. Mating parts and subassemblies should be of similar material as far as possible 7. Design to minimize geometric and functional variability in style, shape, and function of product 8. Carefully monitor the different manufacturing variables affecting the performance, during a production stage 9. Adopt the modular design approach; this approach promotes functional requirement independence; in this the parts and the subassemblies performing specific functions are combined in a self-contained single unit 10. Minimize the mass–strength ratio; the mass of the part should not be more than the strength required of it 11. Carefully narrow down on the sizes and geometric dimensions which optimize the performance of the product 12. All the surfaces transmitting forces and motion should have a smooth profile 13. Avoid over or under tolerance of components; provide sensible tolerances using tolerance analysis		

(*Continued*)

Table 11.7 **(Continued)**

Questions	Additional comments	Rate an item
Material properties	**Rate each numerically listed item**	
14. During the design stage pay attention to controlling both dimensional and functional tolerances 15. Provide a solid base in order to improve the durability and also making the product easier to be oriented or transported 16. Identify all the manufacturing processes which also generate some aberrations on the surface of the product; provide suitable finishing process to remove the abnormalities		
Additional points importance		
1. Add your point here choose an item 2. Add your point here choose an item Aggregative rating for design in terms of performance rating		

Table 11.8 **Design guidelines for reliability**

Questions	Additional comments	Rate an item
Protection	**Rate each numerically listed item**	
i. Overheating 1. Incorporate fins and minimize the direct contact with the heat sources; position them in the direction of air flow 2. Provide heat sinks for the heat disposal Choose an item 3. Use ventilation to prevent overheating 4. Provide heat conducting paths and place them so that the heat transfers through them to the heat sink 5. Components highly sensitive to heat like semiconductors should be provided extra protections and be placed away from high temperature points 6. Use coolants locate the functional surfaces close to the coolant flow or in the air field 7. Use resistors and larger area conductors		

(Continued)

Table 11.8 **(Continued)**

Questions	**Additional comments**	**Rate an item**
Protection	**Rate each numerically listed item**	
8. Keep important electrical components protected by providing fuse, shear pins, etc. 9. Provide seals to protect against moisture and other foreign impurities ii. Protection from environment 1. Keep the product or components protected from radiation, using shields 2. Design to accommodate for other environmental influences like fire, vibrations, electrical hazards etc.; anticipate the possible environmental impacts and provide measure to protect against those factors 3. Use galvanization or rust-free coating for protection against corrosion 4. Simplify the part design and minimize the number of parts		
Strength of material		
1. Phase out brittle materials like glass and ceramic etc. 2. Use materials with greater shock resistance and impact strength; also provide some cushioning or other shock absorbing materials The product should be properly packaged. Provide two different layers. Primary packaging should envelope the product by being in direct contact and the secondary packing is outside the primary. This protects it from compression, shock vibration, and unfavorable temperature conditions		
Redundancy		
1. Critical components and subassemblies should be duplicated 2. Provide standby components in case of failure of the primary one		
Failure analysis and rules		
1. Use FMEA, fault tree analysis, etc. to diagnose the problems associated with reliability 2. Identify the weakest component in the assembly or subassembly and work toward improving their reliability 3. Label all the functional surfaces 4. Collect and analyze the data related to different kind of failures; record the part no. and estimated time of failure; this would simplify the detection of erroneous mechanisms and parts 5. Design so as to fail safely in case of a sudden breakdown		

(*Continued*)

Table 11.8 **(Continued)**

Questions	Additional comments	Rate an item
Protection	**Rate each numerically listed item**	
Testability and inspection		
1. Identify the defective parts by testing them at the escalated stress levels 2. Use the product function monitoring system to test the overall reliability		
Field replaceable units		
1. The components involved in performing a specific function should be combined together in a single self-contained unit; these field replaceable units (FRU) are easy to be replaced and are installed as a single subassembly 2. Make the identification of components, modules and interface easier. Use a suitable numbering, color coding scheme, etc. 3. Reduce the complexity of design; simpler designs tend to have more reliability 4. Applications like coating, soldering, welding, etc. are more efficient if the surfaces are clean 5. Increase customer awareness through the use of service manual and instruction manuals 6. Design component for higher stress level than it actually experiences 7. Use engineering techniques like reliability-centered maintenance; this makes the failure diagnosis easier and warns about impending system or part failure; e.g., fuel gauge used in cars 8. Design the product for allowing load sharing; in case of failure of a component the other parts keep the product in a functional state 9. The rate of wear out is to a greater extent controlled by degree of surface finish 10. Avoid sharp corners and stress concentration points to avoid fatigue failures		
Additional points importance		
1. Add your point here 2. Add your point here Aggregative rating for design in terms of reliability rating		

Table 11.9 Design guidelines for safety

Questions	Additional comments	Rate an item
	Rate each numerically listed item	
1. Design the product to fail safe; for both mechanical and electrical equipment a feature or device should be provided that eliminates or minimizes the harm to users and other parts in case of a failure; e.g., a fuse, circuit breaker, a hand crank, etc. 2. Provide both active safety devices (like seat belt) and passive safety devices (like air bags), etc., which work toward minimizing the chances of human injury in case of a failure		
Moving parts and clearances		
1. The surfaces, parts, or subassemblies moving relative to each other should be placed away from the human reach provide sufficient clearances between the moving parts; this would decrease the chances of crushing or shearing of fingers 2. Use modular design approach; this allows use of field replaceable units, which could be removed or installed quickly. This design approach makes the unit more accessible, easily repairable, more mobile, and relatively risk free		
Material		
1. Minimize the use of flammable material 2. The product should not contain toxic substances; keep it away from direct contact, if their use is unavoidable 3. None of the material used should emit any harmful fumes, vapors, radiation 4. Use resilient materials like toughened glass; avoid the use of brittle materials as they would break down into fragments on breaking 5. The materials should be chemically stable 6. The outer body should be made from a nonconducting material 7. The product should be structurally stable; it is not tilted, has an upright stance, and should not wobble 8. The product requiring frequent relocation should not be heavy; for bigger products provide wheels at its base to increase mobility 9. There should be no projection or exposed electrical cables. Both of them have a potential threat of harming a user		

(*Continued*)

Table 11.9 **(Continued)**

Questions	Additional comments	Rate an item
	Rate each numerically listed item	
Packaging		
1. Mention all the handling instructions on the packing of the product; provide pictorial markings for handling; provide special symbol for hazardous materials 2. Try to minimize sharp corners; replace them with the contours of liberal external radii 3. Mention all handling instructions on the packing of the product; provide pictorial markings for handling; provide special symbol for hazardous materials 4. Use child-resistant packing; this prevents children from ingesting dangerous items 5. The plastic bags used for packing should not be either too thin or too thick; only bags with thickness more than 0.0015 in. can cause asphyxiation 6. Provide a separate safety instructions manual with the product; provide adequate warnings for the dangerous to use parts		
Consideration for ergonomics of design		
1. Design the product so as to operate within human capabilities in terms of force 2. User should not have to assume an awkward posture to operate the product 3. Eliminate redundant, repetitious motion required 4. Phase out all the lengthy user control operations; they increase the chances of cumulative trauma disorders 5. The disposal of the product after its use should be safe; e.g., it should not explode on accidentally catching fire		
Provide guards		
1. To cover sharp blades or other sharp edges 2. To keep the foreign particles out; e.g., a windshield keeps unwanted flying objects out of the car 3. Should not scrape the user's hand or cause any other danger during the time of operation 4. Provide sufficient factor of safety to all the under critical stress components; this reduces the risk of injury to user at a sudden failure of the product		

(*Continued*)

Table 11.9 **(Continued)**

Questions	Additional comments	Rate an item
	Rate each numerically listed item	
5. Use interlocks, a device which prevents the working of machine in undesired state; e.g., interlock switches provided in microwaves disables the magnetron in case the door of microwave is opened 6. If the product is dissipated during the use, then it should not emit any harmful/toxic gases during the period it is in use 7. Minimize the use of heavy metals like mercury, lead, or asbestos products in the products; use alternatives instead 8. The body of the product should not become scathing hot; provide adequate heat sinks and sources for ventilation 9. Provide audiovisual warning indicators which flashes when user has not completed all the safety measures; e.g., door lock indicators, seat harness buzzer, etc. 10. The body of the product should not become scathing hot; provide adequate heat sinks and sources for ventilation		
Additional points importance		
1. Add your point here choose an item 2. Add your point here choose an item Aggregative rating for design in terms of safety rating		

Table 11.10 **Design guidelines for maintainability and serviceability**

Questions	Additional comments	Rate an item
	Rate each numerically listed item	
1. Adopt modular design approach for expediting the repair of faulty systems; as replacing the faulty modules would repair the unit		
Easy diagnostics, testing, and inspection of the system		
1. Provide the warning indicators like brake fuel gauge which signals the imminent maintenance requirement 2. Design the product components for them to be tested using standard instruments		

(*Continued*)

Table 11.10 **(Continued)**

Questions	Additional comments	Rate an item
	Rate each numerically listed item	
3. Provide built in diagnostic capabilities 4. Ease the accessibility of test probes to various components and test points by providing ports, tool holes, and giving roper markings 5. Design for malfunction annunciation; provide a means for identifying the malfunction in the product; e.g., fuel gauge or oil pressure gauge 6. For easy fault isolation provide multipronged connectors with test points so that they could be connected to external equipment and checked for operability 7. Design so that the field replacement units do not need to be disassembled for testing 8. Provide fault detection system in the product 9. Incorporate the modular design approach for easy tests of operability and easy fault isolation 10. Use modular design approach for the products which requires fast replacements and consists of relatively inexpensive parts		
Standardization		
1. Standardize the tests and make them easy so that they can be performed in the field 2. Use standard parts to promote interchangeability 3. Design for compatibility among the mating parts and minimize number of different designs; this reduces spare part requirements, reduces maintenance time required, and makes it more cost-effective 4. Standardize the assembly and disassembly procedures with standard sizes, shapes, and interface locations for modules 5. Tolerances, both dimensional and functional, should be controlled to make the replacement easier in the field 6. Use predictive maintenance techniques like vibration analysis, sound level measurements, oil analysis, etc; they help in determining the state of in-service equipment; maintenance therefore can be performed when it is most cost-effective		
Accessibility		
1. Components or modules requiring periodic maintenance should be placed at accessible locations		

(Continued)

Table 11.10 **(Continued)**

Questions	Additional comments	Rate an item
	Rate each numerically listed item	
2. Fastening devices should be at visible and accessible locations 3. For modular design the higher reliability parts should be assembled before the lower reliability or high mortality rate components 4. Spatial arrangements should be controlled for providing better accessibility to the items 5. Assembly should be simple and should not make workers assume awkward postures while servicing		
Joint of parts		
1. Use slips fits, funnel openings, snap fits, and tapered ends; this makes the disassembly easier 2. Design the product so that the operator does not come in contact with toxic elements present, if any 3. Provide explicit warning over the dangerous components or critical components which should only be serviced or repaired by a specialist 4. Minimize the sharp points or edges in the inside of product 5. Provide protection to all the critical or fragile components likely to be damaged while performing routine service tasks 6. Provide sufficient outlets for the drainage of fluids; provide necessary drainage plugs		
Additional points importance		
1. Add your point here 2. Add your point here Aggregative rating for design in terms of maintainability/serviceability rating		

Table 11.11 **Design guidelines for environmental friendliness**

Questions	Additional comments	Rate an item
	Rate each numerically listed item	
Recyclability and disassembly consideration		
1. Minimize the number of parts; makes it easier to sort products for recycling 2. Avoid using dissimilar materials; they either cannot be or are difficult to separate 3. Avoid the use of fasteners; connections like snap fits make disassembly easier and also reduce dissimilar materials 4. Minimize screw head types and different sizes of fasteners, as it will take more time and more tools to disassemble them 5. Use ultrasonic friction welding over adhesive bonding, or use the material compatible to adhesive bonding 6. Give identification to different materials and parts; color coding, bar coding, or labeling could be used for easy recognition 7. Use welding over other joining processes like soldering or brazing 8. Use materials which are easier to recycle; thoroughly review all the metals, plastics, and other materials which fulfill the user requirements and are easy to recycle 9. Use water-based solvents paints (latex), as they are easy to recycle 10. Minimize the number of fasteners used and locate them at visible and accessible places; this reduces disassembly time after its end of life 11. Design components so that they can be remanufactured and not only reclaimed for materials 12. Use standard sizes and parts; this makes the refurbishing process easier		
Paints and their disposal		
1. Avoid paints with environmentally harmful solvents; use water-based solvent as it has less volatile organic content compared to the oil-based solvents 2. Oil-based paints and paint thinners should be disposed of as hazardous materials; they should never be discharged in storm drains 3. Close all the inlets to drains; this stops material from reaching the water system in case of accidental spills		

(Continued)

Table 11.11 **(Continued)**

Questions	Additional comments	Rate an item
	Rate each numerically listed item	
4. Avoid using aerosol paint cans, as the propellants used in them pose huge environmental risks; classify the used-up cans as hazardous waste, as they possess flammable contents 5. Avoid using paints consisting of heavy metals and toxic ingredients; refer to green seal standard for paints before finalizing 6. Use reblended paints as they are cost-effective and also energy-efficient 7. Clean the floor of any spillage after use of paints; otherwise it may flow onto streets, drains, catch basins, etc. 8. Avoid using heavy metals like mercury or lead, as their emission into the environment even in small concentration poses serious health risks 9. Use highest purity raw materials 10. Avoid using greenhouse or ozone-dissipating gases		
Disposal		
1. Materials like batteries, other heavy metal-containing products which could emit heavy metals into environment on disposal should be treated as hazardous waste and special measures should be taken before their disposal in landfills; pack them in nonflammable containers and wrap the ends to prevent sparking 2. Have proper hazardous waste cleanup equipment available; e.g., mercury spill kits should be used for mercury spills 3. Follow specific state guidelines for the disposal of universal paints; educate workers about rules and methods 4. Design the products to near net shape; this will reduce the amount of material used; less material used corresponds to less landfill space consumed at the end of the life of the product 5. Reduce the amount of packaging material, or use refillable, consumable packaging		
Manufacturing process		
1. The process used should produce minimum of the scrap 2. Plan the process so that it significantly reduces water consumption		

(*Continued*)

Table 11.11 (Continued)

Questions	Additional comments	Rate an item
	Rate each numerically listed item	
3. If possible, eliminate or minimize emission of harmful gases (like carbon monoxide, unburned hydrocarbons, etc.), chemicals, heavy metals (like lead), or other solid pollutants (like charcoal) during the manufacturing process 4. Separate residues into different classes before discharging them; hazardous materials should be separated from recyclables 5. Minimize the use of all those materials which are limited in supply in nature 6. Minimize the use of materials like paper, coal, etc., which directly consume ecological wealth; numerous alternatives are available for each of them 7. Use materials that are not toxic or radioactive 8. Make the process energy-efficient; develop the process such that it minimizes dependence on different energy sources 9. All materials requiring periodic replenishment or disposal should be avoided; materials like batteries, coolants come under this category 10. Buy bulk quantities of raw material in large containers, as this would reduce the quantity to be disposed 11. Lessen the use of alkalis, acids, and solvents during the manufacturing process		
Additional points importance		
1. Add your point here 2. Add your point here Aggregative rating for design in terms of environmental friendliness rating		

Table 11.12 **Design guidelines for esthetic appeal**

Questions	Additional comments	Rate an item
	Rate each numerically listed item	
1. The colors and shape of the product should be decided prudently; using appropriate market research techniques, select the apposite color–shape combinations for the segment targeted 2. The contour should be round and should fit the profile of human hand 3. Avoid sharp corners 4. Use suitable inorganic coating techniques to color the product; for example, different conversion coatings can be used to impart variety of colors 5. The suited color can also be given by painting; select right mixture of pigments, binder, additives, etc. 6. Use techniques like Kensei engineering to translate the consumer's perception about the product into its design 7. The product should have a smooth surface finish; use techniques like grinding, polishing, honing, etc. to give the highest level of surface finish		
Gentle touch experience		
1. Use materials like rubber, velvet, etc.; they will give a soft touch experience 2. Provide leather finish, wooden finish at the exterior 3. The style should be appealing; the product should appear to be compact with a mix of style and simplicity 4. Use coatings like zinc phosphate, which besides slowing down the wear process also gives a glossy appearance		
Cleanliness		
1. The surface should be free of nooks or small pockets; these places gather dust and are harder to clean 2. The product should give a feeling of cleanliness; this increases elegance and is perceived as more hygienic 3. Provide coating or other corrosion resistant finish to give its surface a long-lasting life 4. The components should appear to be well matched with each other; the exterior should be worked out with great care and should exhibit fine details 5. Judiciously decide on the translucency (opaque, translucent, or transparent) and the level of brightness of the product		
Additional points importance		
1. Add your point here 2. Add your point here Aggregative rating for design in terms of aesthetic appeal rating		

transformation matrices. The information extracted was integrated in the customized checklists. It is expected that the generic guidelines provided earlier in the chapter can be similarly customized to enhance the usability and functionality of other consumer products (Figure 11.5 and Tables 11.13–11.15).

There are different types of bikes available on the market. These include road bikes, recumbent bikes, hybrid bikes, and mountain bikes. Road bikes are designed for use on paved roads, while recumbent bikes provide an ergonomic design by having the rider in a laid-back position rather than the more common upright position. Hybrid bikes are a cross between road and mountain bikes. MTBs are specifically designed for off-road conditions including bumpy terrain. Both recumbent bikes and hybrid bikes are also available in a mountain bike variation. For purposes of this chapter, we will focus on the MTB. For the enthusiast, riding an MTB is purely an emotional experience. Mountain bikes are currently available in rigid frame and suspension frame designs. The most important components of the bike are the frame, fork, brake, wheels and derailleur, and crank set. The brake, derailleur, and crank set are bought from the vendor and assembled.

A brief description of the manufacturing processes for each component is outlined in the following paragraph.

Frame: There are as many different frame materials as variations in frame design. The most commonly used materials include steel, aluminum, carbon composites, metal matrix, titanium, etc. The frame is made by welding together seamless tubes. These tubes are manufactured through hot extrusion and are joined together using TIG welding, brazing, or lugs. The joining procedure of choice depends on frame material and design variable selection. For steel, aluminum, and titanium, TIG welding is the most cost-effective and efficient method. For carbon fiber frames, lugs are most commonly used. Lugs are manufactured through investment casting. After the joining process, various components such as front derailleur, hanger, brake noses, and water bottle bosses are added. The joints for these are produced using silver solder. The frame is then sandblasted to remove any remaining flux. Fluxes from the tough parts are removed using taps. This includes bottom brackets, derailleur hanger, mounted bosses, etc. The frame is then ready for coating and painting.

The frame is fixed to a jig and inspected for alignment, which is done at room temperature, referred to as cold setting.

Fork: The fork assembly consists of a steering tube, fork crown, fork legs, brake bosses, and fork ends. Fork legs and the steering column are manufactured using hot extrusion. The fork crown is brazed to the steering tube on a jig. The fork ends are inserted in the fork legs. The two sun assemblies are assembled using the same joining techniques as are used for the frames. The fork is thereafter bent cold using a mandrel. The same finishing process is then applied to the fork.

Wheel: The rim is produced through the extrusion process by squeezing aluminum through a die. It is then roll formed and cut into wheel rings. The end is closed using resistance welding. The rim is hardened using the T3 hardening process. Anodization is used to improve appearance. The wheel is then tensioned, trued, and dished.

Table 11.13 **Technical requirements deployment for a bicycle**

	Force applied to brake	Force applied to shift	Material strength	Material weight	Material protection	Adjustability	Crank torque	Gripping	Seat shape	Stiffness	Traction	Tubing	Appearance	Dimensions and distances	Subassemblies
Efficient braking	5			1	3		3				5			3	
Smooth gear shifting		5												1	
Robust body			5	5	5							5			
Comfortable seating									5					3	
Shock resistance			5	3						5					
Pedaling						5									
Stability	5						3				3				
Handling							5							3	
Light weight				5								3			
Life			5		5					3					
Stylish					5								5		
Fit the user						5		3	1					3	
Easy removal														3	5
Small down time															5

Table 11.14 **Product features deployment for a bicycle**

	Shift lever	Brakes and brake lever	Handlebar	Crank sets	Frame	Chain rings	Derailleurs	Saddle	Seat post	Suspension	Wheels (tires)	Wheels (rims and spokes)	Headset and stem	Fork
Force applied to brake		5									3	1		
Force applied to shift	5						3							
Material strength			3		5	3			3				3	5
Material weight					5									5
Protection			3		5				1				5	
Adjustability			3						5					
Crank torque				5										
Gripping	3	3	5											
Comfortable seat								5	5					
Stiffness					5					3				5
Traction											5			
Tubing			3		5								3	5
Appearance			1		5			3						3
Dimensions and distances	3	5							5					
Subassemblies		5	5				5							5

Table 11.15 Manufacturing process deployment

	Welding	Heat treatment	Organic painting	Inorganic coating	Extrusion	Surface finish	Roll forming	Anodization
Handlebar	5	5	5		5			
Frame	5	5			5	5		
Seat post	3		3		5			
Fork	5	5	5		3			
Wheel				5		3	5	3
Handset and stem	5	3			5			

11.2.5.1 Customized design and manufacturing guidelines

11.2.5.1.1 Development procedure

The generic guidelines formulated in the first part of this chapter were customized for the bicycle. For example, the ease of use guideline "Provide means to leverage the force applied by a human" was customized as "The lever on the left handlebar must operate the front brake and the right handlebar lever must operate the rear brake." It is essential to understand that not all the generic guidelines could be transformed into corresponding customized guidelines.

Consultation with designers and users also played a crucial role in the development of the customized guidelines. Recommendations by designers and users were incorporated into the study during the process of developing design/manufacturing linkages. The usability–functionality requirements gathered from customers were analyzed using usability–functionality matrices; this facilitated the development of corresponding linkages.

11.2.5.2 User requirements

User requirements were obtained from potential users of the product. One-on-one interviews were conducted to gather information. All users had been riding mountain bikes for at least 5 years and had knowledge of the different parts. All users were experienced in servicing and repairing their bikes to a large extent. The following is an elaboration of the aforementioned requirements:

- **Adequate stiffness**: The bike should be able to withstand the weight of the rider.
- The bike should remain steady when pedaling force is applied.
- The saddle should be designed to provide maximum comfort to the rider.
- The handlebar should have smooth contact points and should not require the rider to assume an awkward position.
- The bike should provide good traction over sandy and slippery surfaces.
- The bike should protect the rider from shocks when riding on bumpy roads and dirt trails.
- The seat height and handle height should be easily adjustable. Different types of riders within a specific age range should be able to effectively operate the brake lever.
- The gears should shift smoothly in both directions at multiple speeds.
- Handling should be steady and responsive. It should stay steady at different speeds and tracks but should respond to the slightest of flicks.
- Braking should be smooth, and the stopping distance should be optimum. It should be effective in both wet and dry conditions.
- The manufacturing process and afterlife of the bicycle should be environmentally friendly.
- The frame should be robust and sturdy yet light.
- Critical components such as chain, crank set, and cassette cogs should be easily accessible.
- The bike should require low maintenance.
- The seat and handlebars should be easily removable.
- The bike should provide good protection from mud and other substances that may splash onto the rider while riding.
- The bike should have a trendy and stylish look in terms of color and shape.
- The bike should have a streamlined appearance.
- The bike should have good vibration and impact resistance.
- There should be a provision for safety measures such as reflectors and chain guards.

11.2.5.3 Mapping design dimensions

As discussed in the first part of this chapter, the usability–functionality requirements can be mapped onto the following design dimensions (Table 11.16). Each of these dimensions is dependent on one or another of the design, manufacturing, and material variables and must be tightly controlled for optimal product design.

Table 11.16 A description of a typical mountain bike

Component	Description
Frames	Common types are rigid frame and suspension frame, which are available in numerous variations. The members of the frame are down tube, head tube, top tube, seat tube, chain stay, seat stay, bottom bracket shell, shock (for suspension frames), main and dropout pivot, and braze-ons like shock mount, etc. There are two important suspensions in the full suspension frame design, viz. the front (fork) and the rear suspension. The frame consists of the front triangle, which includes fork, head tube, top tube, down, seat tube, and bottom bracket shell. The rear assembly includes the swing arm, which consists of chain stays and the seat stays
Forks	Connect the handlebar to the front wheel. The fork subassembly consists of steering tube, fork crown, outer and inner fork legs, brake bosses, fork boots, and the suspension, comprised of spring and dampening system
Wheels	Comprises of rims, spokes, tires, tubes, cassette, and hubs. Rim, spoke, nipple, and hub are required to assemble the wheel. The spoke is connected to the rim with the help of nipples, which are connected to the rim through eyelets. The head of the spoke is connected to the hub shells. Hub is connected to the rim through spokes. The wheel is mounted to the frame and the fork via dropouts
Crank set and pedals	It consists of crank arms, chain rings, bottom bracket, chain ring bolts, and crank bolts. The crank arm has spider arms protruding out, to which the chain rings are bolted using the chain ring bolts. Pedals are screwed to the crank arm at the end
Brakes	Disc and rim brakes are the most popular types of brakes. The brake assembly consists of brake caliper, brake lever, and cables. The levers are mounted on the handlebar using a bolt. Disk brakes have the pads squeezed against the hub-mounted disc. Disc brakes have the rotor bolted to the hub of the wheel; the caliper is then mounted onto the fork (Zinn, 2005).
Steering assembly	Consists of the handlebar, stem, shifters, grips, and bar ends. The bar ends are connected to the handlebar using the bar end bolts. Grips are twisted onto the handlebar. The handlebar is connected to the frame through the stem to which it is bolted.
Derailleurs and shifters	The rear derailleur is bolted to the hanger on the rear dropout. The front derailleur is mounted either on the face of the bottom bracket shell or to braze-on boss. Shifters are mounted on the handlebar. The shifter cables connect the shifters to derailleurs. The chain is routed through chain rings, cog, and the jockey wheel of the derailleur.

11.2.5.4 Linkage identification

This section uses transformation matrices to establish linkages between usability–functionality requirements and product–process requirements. It was accomplished by using a series of transformation matrices.

11.2.5.5 Technical requirement deployment

The various usability–functionality requirements were correlated to the technical requirements in this step. The strength of the relationship is shown using numerals 1 (weak), 3 (medium), and 5 (strong). The material properties and structural rigidity are vital as well. For instance, the material should be strong enough so as to not yield under load and should be adequately stiff. Also, to facilitate easy removal and reduce downtime, the design is of the modular type. Other issues such as transmission efficiency and friction characteristics were also considered.

11.2.5.6 Product feature generation

Features including functional mechanisms, subassemblies that are required to fulfill technical requirements, were designed into the bicycle. The technical requirements were then linked to product features.

11.2.5.7 Process characteristics

A product of the highest quality can only be made if tight controls are established for processes and machines. The manufacturing processes will yield the best results only if the manufacturing variables are tightly controlled. This step involved process deployment and manufacturing deployment.

11.2.5.8 Checklist development

Based on the information garnered using the steps above, design and manufacturing guidelines were developed. These guidelines were prepared in the form of checklists. Tables 11.17 and 11.18 depict guidelines for ease of use and performance only. The items in the questionnaire were evaluated by designers on a 1–5 scale, with 5 being the most important, and vice versa.

Table 11.17 Design dimensions for a mountain bike

Performance	Ease of use
Appropriate material	User conformity
Effectiveness of function	Learnability
Consideration of operating environment	Modelessness
Minimizing mass/strength ratio	Simplicity
Profile of functional surfaces	Ergonomics
Responsiveness	Balance
Mating parts	Leverage
	Adaptability

Table 11.18 Design guidelines for performance for a mountain bike

Questions	Additional comments	Rate an item
	Rate each numerically listed item	
1. Select a suitable material that optimizes mass/strength ratio 2. Appropriate materials for frame are composites made from carbon, boron, and Kevlar fibers, as well as titanium magnesium alloys, aluminum alloy, chromium molybdenum steel alloy, etc.; composites are most preferred because of their high strength/density ratio 3. Appropriate material for crank sets, chain rings, derailleurs are usually made of aluminum or stainless steel 4. Use titanium, aluminum, or chrome-moly steel for stems 5. Use superior joining techniques such as tungsten inert gas welding over traditional arc welding. TIG produces stronger joints; it produces lighter joints compared to fillet brazing joint, and is cheaper 6. Brazing can also be used for joining purposes, as it could be done at lower temperatures (at around 800°) and produces more ductile joints; can be used to join tubes of various diameters 7. Ensure compatibility between wheel and bike by selecting correct rim width, wheel diameter, hub axle width, and tire 8. Prefer ball bearing over plain bearings, as they require less starting torque and also have less coefficient of friction 9. Use tubeless tires as they provide better traction and also better suspension 10. Use appropriate finishing processes to give a smooth profile to all mating surfaces and components transmitting forces such as chains, cassette cogs, ball bearings, bottom bracket axle, brake pivot, suspension fork legs, threads, and wheel axles 11. In case metal is used for the frame material, heat treat the frame (anneal) to relieve it of stresses and increase toughness 12. Use a one-piece seat post as they are more efficient than two-piece seat posts 13. Use double-butted tubes instead of single-butted tubes 14. Use cup and cone model headsets; they are simple in design, efficient, and are not expensive 15. For front derailleurs use the cage, which not only pushes the chain sideways but also lifts it; this increases the speed of shifting from small to large chain rings		

(*Continued*)

Table 11.18 **(Continued)**

Questions	Additional comments	Rate an item
	Rate each numerically listed item	
16. For mountain bikes use threadless type stem 17. Use maximum number of subassemblies possible; use subassemblies for headsets, crank sets, suspension, brakes, seat posts, derailleurs, etc. 18. Prefer disc brakes over rim brakes; disc brakes perform consistently in different weather conditions; they also have more mechanical advantage than rim brakes 19. Pay close attention to the height of cage plates, the distance between cage plates, and rigidity when designing a derailleur 20. Use double pivot rear derailleurs, as they can handle a much wider range of cassette cogs than one with the single pivot 21. Use cassette cogs for holding gears on the rear wheels, as they are more efficient than freewheels 22. Use wider gear ranges in order to minimize the customization of drive train; speeds are usually 19, 22, or 25; for mountain bikes use 3 chain rings and use 7, 8, or 9 speed cassettes 23. Try to minimize "pogoing" and "bio pacing" by using an appropriate rear suspension design; the multiple pivot design and saddle suspension systems can minimize them (Langley, 1999) 24. Use seamless tubes for frames; if required, increase the strength or decrease their weight by altering the thickness of tube walls; techniques like butting should be used for this purpose as this also increases the resilience of the frame 25. Use hubs with medium-sized flanges, as this gives the right mix of stiffness and comfort 26. The stem of a handlebar must have a permanent mark representing the minimum depth the handle stem must be inserted in bicycle fork; the circle should be made at 2.5 times the diameter of the stem from the bottom of stem 27. While designing a frame, pay close attention to various measurements such as seat angle, head angle, fork rake, chain stay length, and drop; these factors collectively determine the bike's wheelbase 28. Distance between wheel and a frame should be no more than 3 mm		

(Continued)

Table 11.18 **(Continued)**

Questions	Additional comments	Rate an item
	Rate each numerically listed item	
29. Use suspension forks to minimize vibrations; be careful about factors such as preload and spring stiffness 30. Use indexed shifters for shifting gears; either the trigger or twist shifter could be used 31. To optimize performance of the derailleurs, use shift levers and front and rear derailleurs manufactured by the same company 32. For best results, use similar materials in the mating components 33. Use narrow width chains with higher speed drive trains 34. Use and design the parts so that there is minimum friction between mating surfaces; this requires superior surface finish and proper machining processes 35. Any subassembly or part used should be compatible with all other parts and subassemblies, and in no way should undermine their performance 36. Use die-drawn brake cables as they operate with less friction 37. Use diamond, double diamond, full, or rear suspension frame with larger tubing diameter compared to road bikes 38. Provide a higher bottom bracket in mountain bike frames; this gives more ground clearance and thus makes it more suitable for the bumpy rides 39. Rim brakes are cheap, so if using rim brakes prefer side pull cantilever/V-brake design		
Additional points importance		
1. Add your point here 2. Add your point here Aggregative rating for design in terms of ease of use rating		

11.2.5.9 Survey deployment and testing

11.2.5.9.1 Data collection and analysis

In order to test the effectiveness of the guidelines, a survey was undertaken and subsequent analysis was performed. The objective was to evaluate whether or not the guidelines ensured the fulfillment of the users' usability–functionality requirements.

The test was administered to 18 people. The population included product users, mechanics, and a few designers. Each participant was given an overview of the study and was informed of the design and features of the bike under consideration.

Reliability and validity tests were conducted on the data collected through surveys. The reliability test was conducted to verify whether the test yielded similar results irrespective of the person in question. The internal consistency of the test was determined through the reliability coefficient, which in this case was the value of Cronbach alpha. Cronbach alpha values of 0.4 or above meant the measure was reliable.

Validity establishes how well an instrument measures what it is intended to measure. In this case, it is essential to compare the average score of items with the overall score of the questionnaire. To determine the degree of relationship between these two variables, correlation analysis was performed. Pearson's r is the most common measure to determine the association between two variables in terms of direction and strength of relationship.

Once the reliability and validity tests were performed, it was necessary to determine which of the identified design dimensions were most influential or important for customers. The objective was to quantify the importance of these dimensions. This knowledge should help designers focus their efforts more effectively on more important design dimensions. This will reduce the subjectivity involved in the selection of design variables. These dimensions are highly correlated to each other and render normal statistical techniques ineffective. Thus, PCR was used for analysis.

11.2.5.10 Test results

11.2.5.10.1 Reliability and validity

The checklists were analyzed using SAS and Microsoft Office applications. The values are presented in Table 11.19.

A value of 0.4 or above for Cronbach alpha was considered acceptable. All the sections were considered reliable according to this rule. As discussed earlier, the validity of the questionnaires was measured by comparing the average score of the questionnaire items with the overall score of the questionnaire. Upon analyzing the scores, it was concluded that all the design criteria were valid. The correlation coefficient for "performance," "safety," "ecological affinity," "reliability," and "ease of use" was modestly valid, as their values lie between 0.4 and 0.69. All sections were found to be significant at the 5% level of significance.

11.2.5.11 Variable screening

The next step was to screen the dimensions for their importance using PCR. Based on the MINEIGEN (which was set to be equal to unity) criteria, only three and four principal components (PCs) were selected for the two criteria. The cumulative variance explained by them was 66.14% and 88.46%, respectively. The questionnaire items for ease of use were named from V1 to V8, and for performance P1 to P7. Tables 11.20 and 11.21 depict the variance explained by each PC and the loading of each variable (design dimension) on each PC (see also Tables 11.22 and 11.23).

Almost all the dimensions represent an equal degree of importance,though for "ease of use" it was observed that users found ergonomics to be more important than other factors.

Table 11.19 Design guidelines for ease of use of a mountain bike

Questions	Additional comments	Rate an item
1. Adults of normal intelligence must be able to understand the assembly of the bicycle easily 2. The bicycle should not have any sharp protrusions like a sheared metal edge that may cut the user's hands or legs; all the sharp edges must be freed of feathering or burrs through rolling or processes like iodization of aluminum frame 3. Unless specified by the customer, the distance between handle lever and the handlebar should not be more than 3.5 in. 4. The lever on the left handlebar must operate the front brake and similarly the right handlebar lever must operate the rear brake 5. The brake pads must contact the braking surface on the wheel if the force of 10 pounds or less is applied 1 in. from the end of the lever; use frictionless cable sealing, as it will reduce the force required 6. To ensure slip-free performance, pedals must have treads on both sides; make the tread width around 1.15"; this will increase traction 7. Ends of the handlebar must be capped or covered 8. No part of the seat, seat support, or accessories attached to it must be 5 in. above the surface of the seat 9. Quick-release devices with the lever must be adjustable, allowing the lever to be set for the tightness 10. Use rubber hood covers that fit around the lever body; they increase comfort as the hands can rest on the lever hoods 11. Use set of screws running through the clamp to set the saddle tilt instead of serrations; this kind of design increases seat adjustability 12. Add bar ends to the handlebar; it provides rider with more traction when shifting body positions and also eliminates the discomfort on the jarring rides 13. For lighter weight use tires with tubes as tubeless are heavier because of sealant in them 14. Use clipless pedals as they are easier to use and are more comfortable; these could be used with any of the following types of pedals: nylon block, rattrap, quill, and platform		

(*Continued*)

Table 11.19 **(Continued)**

Questions	Additional comments	Rate an item
15. For providing cushioning effect to the rider the seat should be adequately padded; saddles should generally be padded with vinyl bases; vinyl or leather 16. Provide sufficient spring support to the saddle to absorb vibrations and shocks from the road 17. Do not weld seat stem to the frame; use clamps to make the seat adjustable; once the seat is tightened at particular position it should not move from there during normal use 18. Bearings should be packed and sealed with grease; this minimizes the rotation torque required by the user 19. For gears, use indexed shifters; this will help in discrete shifting 20. Rapid-fire shifting mechanism can be used for people with less dexterity 21. For mountain bikes, use 1 ×1/8 stem diameter; this reduces the flex required for steering 22. To avoid awkward posture, handlebar should be made symmetrical; the vertical distance between the seat in its lowest position and the handlebar in its highest should be no more than 16 in. 23. Keep the seat tubes shorter for more stand-over clearance 24. No part of the seat, seat support, or accessories attached to it can be 5 in. above the surface of the seat 25. Use light weight but resilient materials; the weight of the bike should be less so that the biker does not experience any discomfort in pulling the bike 26. Use either a standard flat or rise handlebars, as they allow more upright position while riding 27. Use stem made of titanium, aluminum, or chrome-moly steel; they permit greater adjustment of the handlebar		
Additional points importance		
1. Add your point here 2. Add your point here Aggregative rating for design in terms of ease of use rating		

Table 11.20 **Reliability and validity test values**

Design criteria	Reliability test	Validity test	
	Cronbach alpha	Pearson correlation coefficient	Significance
Ease of use	0.578198	0.68462	0.001
Performance	0.660777	0.46691	0.044
Safety	0.719717	0.50381	0.029
Esthetics	0.511939	0.83429	0.000
Ecological affinity	0.579859	0.49301	0.032
Reliability	0.833865	0.53742	0.018
Maintenance	0.575608	0.75181	0.000

Table 11.21 **PC variances and loading for each dimension (ease of use)**

Principal components	Loading								
	EV	V1	V2	V3	V4	V5	V6	V7	V8
PC1	2.85	0.29	0.38	0.43	0.43	0.32	0.43	0.31	0.11
PC2	1.36	0.40	−0.34	0.06	0.07	−0.47	−0.18	0.37	0.58
PC3	1.08	0.41	0.01	−0.34	−0.16	0.46	−0.10	0.48	0.49

Table 11.22 **PC variances and loading for each dimension (performance)**

Principal components	Loading							
	EV	P1	P2	P3	P4	P5	P6	P7
PC1	4.89	0.18	0.17	0.05	0.08	0.34	0.33	0.21
PC2	2.94	−0.11	−0.26	0.54	0.19	0.27	0.06	−0.14
PC3	2.49	−0.51	0.44	0.19	−0.03	−0.13	−0.27	0.26
PC4	2.13	0.05	−0.15	−0.09	0.60	−0.23	−0.17	0.34

11.2.6 Automatic transmission: case study

This part of the chapter presents another case study to illustrate the effectiveness of the methodology presented in this chapter. Automatic transmissions are gaining wide popularity throughout the world. They are ideal for family cars as the driver is relieved of operating the clutch pedal during start-up, gear shifting, and repeated starting/stopping. It entirely eliminates the need for the driver to coordinate the operations and

Table 11.23 **Degree of influence of identified dimensions**

Design dimensions (ease of use)	DOI	Design dimensions (performance)	DOI
User conformity	0.12	Consideration of operating environment	0.15
Learnability	0.11	Profile of functional surfaces	0.14
Leverage	0.13	Appropriate material	0.15
Modelessness	0.11	Effectiveness of function	0.14
Ergonomics	0.15	Minimizing mass/strength ratio	0.15
Balance	0.11	Mating parts	0.14
Simplicity	0.11	Reactiveness	0.14
Adaptability	0.14		

identify the most suitable moment for gear shifting. For fully automatic gearboxes, all the functions ranging from start-up functions to shifting gears are automated. Automatic transmissions were produced with the express purpose of increasing the comfort of the driver and passengers by making the acceleration changes smooth, thereby minimizing sudden changes in acceleration.

This part of the chapter focuses on developing design guidelines for improving the overall performance and usability of automatic transmissions. The purpose is to test whether the methodology developed earlier can successfully be applied to a complex consumer product with limited user interface. The requirements were obtained from users and from experts in the field. The requirements were analyzed using flow diagrams along the lines of QFD, thus converting customer requirements into engineering characteristics.

11.2.6.1 Components of an automatic transmission

The typical automatic transmission system consists of a transmission gearbox, a torque converter, and an electronic hydraulic control unit. The automatic transmission is expected to improve mechanical efficiency and ride quality. The function of each component is described here:

- **Torque converter**: Responsible for transferring power to the transmission gearbox through the automatic transmission fluid. Besides multiplying torque, it is required to transfer engine power smoothly to the transmission gearbox unit. The torque converter consists of a pump impeller, which is connected to the engine crankshaft; there is a turbine runner connected to the transmission gearbox. The other components are the stator and the slip-controlled lockup clutch. The stator has an important function in improving hydraulic efficiency.
- **Hydraulic gear unit**: Actuates the transmission gearbox and lockup clutch. Hydraulic pressure is regulated electronically, the control parameters being the speed and throttle position. The ATF is the hydraulic well.
- **Transmission gearbox**: The function of this component is to increase or decrease the engine torque through appropriate gear ratio selection and to reverse rotation. It consists of multiple wet disc clutches and planetary gear sets. Variation of speeds depends on the number of planetary gear sets used. One planetary gear set can produce three gear ratios.

11.2.6.2 Performance and usability of automatic transmissions as perceived by users

Users desire that the car transmission convert the energy from the engine into forward and reverse motions with sufficient torque and speed. The transmission should provide these functions without any noise, vibration, or other harshness with an uncompromising overall shift quality. Also, automotive safety requires that the transmission provide the engine's braking action.

The interview process with users to establish product requirements also included soliciting information about the most common problems encountered during a transmission's operation. These included incessant transmission noise, excessive heat generation, incorrect torque, incorrect gear ratio selection, and excess change of ratios in dynamic conditions or ratio busyness and ratio hunting, i.e., the continuous change of ratio under steady conditions.

11.2.6.3 Usability–functionality design criteria

Since the product under consideration has a fairly limited direct contact with the user, user opinion provides only a superficial view in terms of the requirements for the car's transmission. Therefore, the consumers' views were integrated with the experts' opinions to obtain a complete view of the product. Based on these surveys, a list of design dimensions was selected and then applied within the framework of a systematic design process, which includes the stages of task clarification, conceptual design, and detailed design for a specific product.

11.2.6.4 Description of group

User requirements were obtained from 13 different users. This group included five general users (from a university setting) who were moderately familiar with the functioning of a transmission. The remaining eight users included six auto mechanics and two product designers; they were classified as experts. The auto mechanics were the employees of car servicing stations (all of which had nationwide presence).

The issues faced by users were initially mapped to a relevant design dimension and the corresponding generic guidelines. The identified customer issues were then discussed with the product experts. For instance, the issue of "incorrect torque" was associated with generic guidelines such as material properties and design of functional surfaces. After discussion with designers and identification of the relevant features, the specific guidelines generated were "ATF should have antishudder properties" and "use of slip-free lockup clutches." Strictly speaking, due to product complexity (and limited human interface), a substantial amount of information came from the product designers and auto mechanics. Some of the information obtained through experts was used as such due to the difficulty in associating it with generic guidelines. As for the remainder, generic guidelines were modified based on the inputs. Table 11.24 provides a starting point in the design and development of a successful product.

Table 11.24 Usability–functionality factors necessary for the development of a successful product

Performance	Ease of use
Responsiveness	Provision of controls
Appropriate characteristics for the specified operating environment	Accessibility of controls
Minimizing mass/strength ratio	Error prevention (utilizing constraints)
Product structure (modular/integrated)	Noise free
Mating parts	Mapping features for facilitating memorability and learnability
Effectiveness of the function	Feedback provision
Consistency	Interactive displays
Smoothness	

11.2.6.5 Development of linkages using flow diagrams

The development of linkages involved associating customer attributes to measurable engineering metrics and product features. Specific usability–functionality criteria were used as a starting point. Tight control of manufacturing and design parameters such as part dimensions, shape, and materials could optimize consumer attributes. Figures 11.6 and 11.7 depict how the design criteria for performance and reliability were related to product attributes. For instance, the criterion "response" reflects the power and pickup of the entire power train. This is determined by the gear ratio selected for the given condition. If the right ratio is not selected, the vehicle will perform fairly poorly. Also, "smoothness" refers to the quietness of the transmission and smooth acceleration changes. Inconsistent transmission changes will lead to jerks experienced by passengers. After the matrices were determined, the next step involved selecting suitable product attributes. The development of product attributes involved the development of concept design and functional mechanisms. For a complicated product such as a transmission, there can be several dependencies.

A single engineering metric could be dependent on several product attributes. Thus, the relationship between the two is of a highly complex nature. Also, functional performance and usability are influenced not only by controllable parameters but by other noise factors such as customer usage, wear of the product, and environmental factors such as impact, braking, and busy throttle.

11.2.6.6 Development of design guidelines

The next step was the establishing of design guidelines after acquiring the required information on the product. Guidelines for only two design criteria, namely functionality and usability, were established and tested. The objective was to test if specific design guidelines could be derived from the generic guidelines using the framework presented in the first part of this chapter. Table 11.25 presents the design guidelines for "performance."

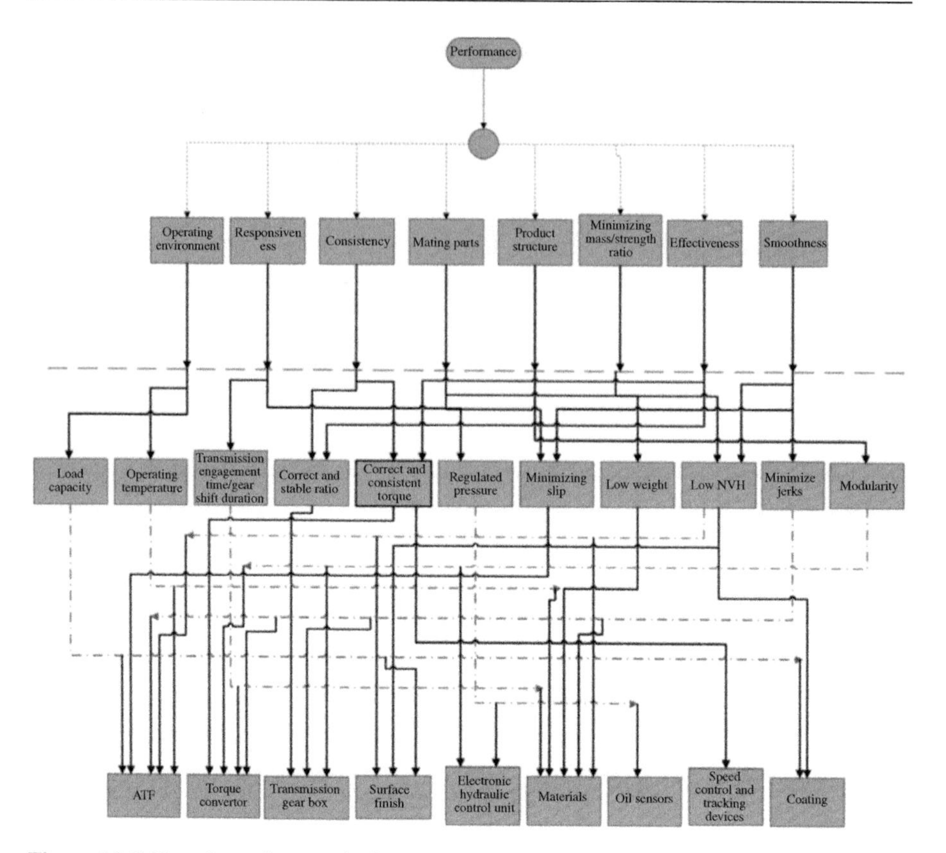

Figure 11.6 Transformation matrix for automatic transmissions (design for functionality).

11.2.6.7 *Survey deployment and analysis*

The questionnaire was circulated among 10 experts, including auto mechanics and students with a matching field of research. Data collection and analysis procedures were similar to the case study for the bicycle presented earlier in this chapter. Reliability and validity of the questionnaire were tested after the scores were obtained. Unlike the mountain bike study, individual design dimensions were not evaluated given the complex nature of the product.

11.2.6.8 *Test results: reliability and validity*

The values of Cronbach alpha and the Pearson correlation coefficient for all the design criteria are given in Table 11.26.

The values of Cronbach alpha for both design criteria lie in the acceptable region. Therefore, the test was considered reliable. The Pearson correlation coefficient for usability was 0.46, which is modest. However, the *p*-value was found to be statistically insignificant. The design criterion "performance" had a high correlation coefficient and was significant at the 0.05 level. Therefore, the checklist for the section "performance" was considered valid.

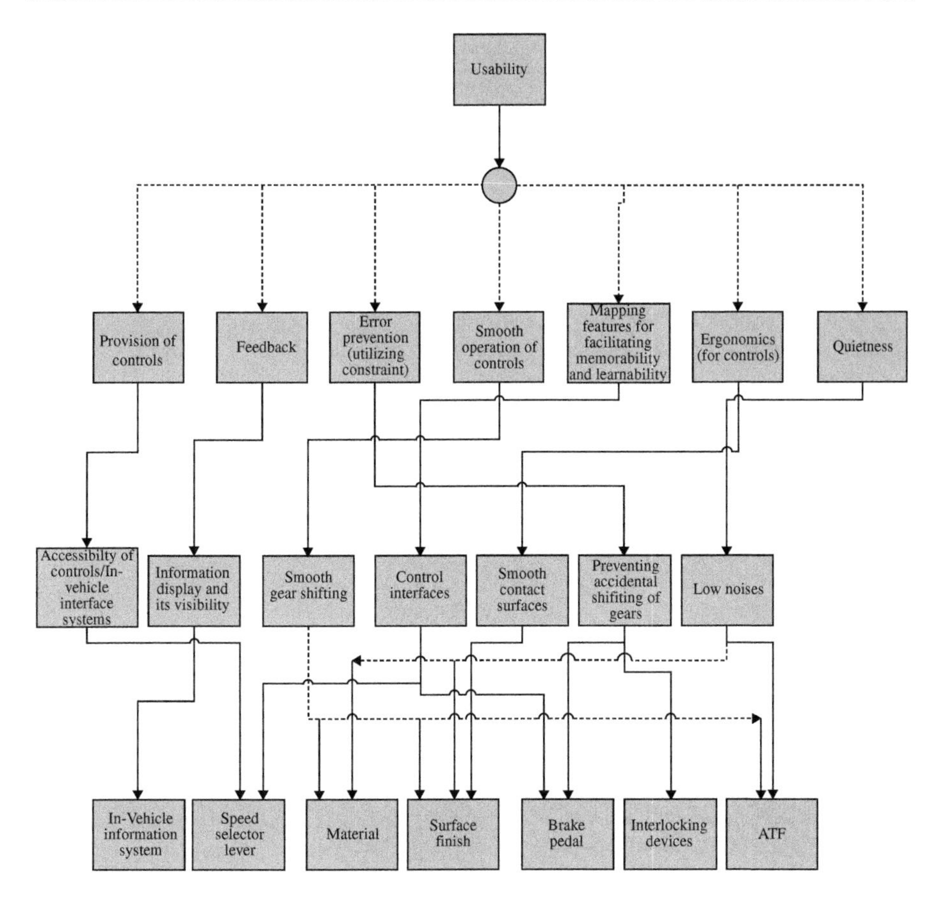

Figure 11.7 Transformation matrix for automatic transmissions (design for usability).

Table 11.25 Design guidelines for performance; product: automatic transmission

Questions	Additional comments	Rate an item
	Rate each numerically listed item	
1. Use materials suitable for fulfilling the functional requirements of the user 2. Design should consider the ATF used should have the antiwear properties which should prevent the wear of steel surface in pressure conditions; phosphate compounds such as zinc dithiophosphates, or phosphorous compounds should be used as antiwear agents (Kugimiya et al., 1998)		

(Continued)

Table 11.25 **(Continued)**

Questions	**Additional comments**	**Rate an item**
	Rate each numerically listed item	
3. The ATF should have the antioxidant agents to minimize its oxidation and also should have additives like metallic sulfonates, metallic phenates, etc., which disperse any sludge formed due to its oxidation; sludge could block the passage of oil in hydraulic control unit (Kugimiya et al., 1998) 4. Design should consider adding suitable additives to ATF for making its friction properties correspond to wet plate clutches; fatty acids, amides, phosphate ester, etc. are suitable friction modifiers 5. Close attention should also be paid to viscosity temperature properties, corrosion inhibition properties of the ATF to be used 6. The shift clutches used should be tested for their friction capacities against the standard test machines; the dynamic torque capacity plays the role in determining shift quality 7. The ATF used should have the antishudder properties; the shudders, which are unstable vibrations, can be produced due to the continued slipping of the lockup clutch; this could create power loss and lower fuel economy 8. The friction properties of wet friction material of clutches play an essential role in improving the smoothness of shift; suitable friction material increases the power transfer and minimizes the relative motion; it also keeps interface temperature low 9. The sintered brass lining applied to the steel disk can be used as friction material; they are compatible with most of the ATF, and cost little more than paper-based friction material 10. The clutch face temperature during engagement is highly correlated with smooth shifts and clutch life; the clutch material and the wet friction materials are the important parameters 11. The carefully selected friction material is important for keeping the normal load low at the clutch interface; load is positively correlated to the temperature; high temperature negatively impacts friction material, thereby adversely affecting clutch engagement 12. Auxillary devices should be provided to prevent moving more than one selector bar at the same time; this could be done by providing the plunger interlocking devices		

(*Continued*)

Table 11.25 **(Continued)**

Questions	**Additional comments**	**Rate an item**
	Rate each numerically listed item	
13. Control hydraulic pressure at hydraulic control unit using electronic controllers; optimum pressure is required for better operation of the lockup clutch and transmission gear unit		
14. Consider using speed sensors to monitoring engine speed, turbine speed, and output shaft speed; the operating torque of torque converter depends on the turbine and impeller speed		
15. The gear geometry should be carefully chosen so as to minimize transmission error; transmission error is cause of excitation for gear noise		
16. Design should consider the factors important for minimizing noise originating from the gears, such as increased face width, decreased lead crowning, and threaded wheel grinding (Joachim et al., 2004)		
17. The electro valves in the hydraulic control system should be minimized by increasing the number of functions to a single valve		
18. Regulate engine torque during shifts using appropriate control mechanisms as they are crucial for clutch's life and also smooth acceleration change		
19. The load-carrying capacity of the gears and its efficiency greatly improves by applying proper finishing process and using right coatings; it also reduces coefficient of friction and provides corrosion protection		
20. Use same number of proportional valves as the number of components to be actuated with controlled pressure; have at least one proportional valve for the lockup clutch on torque convertors, next speed engagement		
21. Use sensors to monitor the gearbox oil temperature; the response of the oil actuator is correlated with the viscosity of oil		
22. The coating of tungsten carbon carbide or amorphous boron carbide is most appropriate for gears; preferably apply it through the process of PVD as it keeps the material temperature below 200°C (Joachim et al., 2004)		
23. Preferably use the slip-free lockup clutch in torque convertors in order to minimize loss of engine power to the transmission choose an item		
24. Use parts with established reliability rating		

(Continued)

Table 11.25 **(Continued)**

Questions	Additional comments	Rate an item
	Rate each numerically listed item	
25. Minimize the mass–strength ratio; the mass of the part should not be in correspondence to the strength required of it 26. Adopt the modular design approach as it will maintain independence of functions 27. Design should be sufficiently robust to be insensitive to small manufacturing irregularities and operating environment variations 28. Robust integrated controllers should be developed for estimating the turbine torque for closed-loop control of power train components, thereby improving shift quality 29. Design should consider adverse situations arising due to customer usage such as vehicle overload and sudden braking		
Additional points importance		
1. Add your point here 2. Add your point here Aggregative rating for design in terms of ease of use rating		

Table 11.26 **Reliability and validity test values**

Design criteria	Reliability test	Validity test	
	Cronbach alpha	Pearson correlation coefficient	Significance
Usability	0.628224	0.45502	0.1864
Performance	0.728694	0.75478	0.0116

11.3 Conclusion

It can be stated that the methodology presented in this chapter can be successfully applied to ensure the functionality of a complex consumer product with relatively few interface elements. As far as usability is concerned, some users did not concur with some of the proposed design guidelines, while some considered the section incomplete, resulting in a poor correlation between average and overall values. Thus, the usability guidelines could be improved further with a more thorough analysis.

References

Arora, A., Mital, A., 2011a. Concurrent consideration of product usability and functionality. Part 1: Development of integrated design guidelines. Int. J. Prod. Dev. 15 (4), 177–204.

Arora, A., Mital, A., 2011b. Concurrent consideration of product usability and functionality. Part 2: Customizing and validating design guidelines for a consumer product. Int. J. Prod. Dev. 15 (4), 205–221.

Arora, A., Mital, A., 2011c. Concurrent consideration of product usability and functionality. Part 3: Customizing and validating design guidelines for a the subassembly of a complex consumer product. Int. J. Prod. Dev. 15 (4), 222–231.

Babbar, S., Behara, R., White, E., 2002. Mapping product usability. Int. J. Oper. Prod. Manage. 22 (10), 1071–1089.

Bergquist, K., Abeysekera, J., 1996. Quality function deployment (QFD)—a means for developing usable products. Int. J. Ind. Ergon. 18, 269–275.

Bralla, J.G., 1996. Handbook of Product Design for Manufacturing: A Practical Guide to Low-Cost Production. McGraw-Hill Book Company, New York, NY.

Chiang, W.-C., 2000. Designing and Manufacturing Consumer Products for Functionality (Doctoral dissertation). University of Cincinnati, OH.

Chiang, W.-C., Pennathur, A., Mital, A., 2001. Designing and manufacturing consumer products for functionality: a literature review of current function definitions and design support tools. Integr. Manuf. Syst. 12 (6), 430–448.

Chiang, W.-C., Mital, A., Desai, A., 2009. A generic methodology based on six sigma techniques for designing and manufacturing consumer products for functionality. Int. J. Prod. Dev. 7 (3–4), 349–371.

Cubberly, W.H., Bakerjian, R., 1989. Tools and Manufacturing Engineers Handbook,Desk Edition. Society for Manufacturing Engineers, Dearbron, MI.

Dunteman, G.H., 1989. Prinicipal Component Analysis. Sage Publications, Inc., Newbury Park, CA.

Genta, G., Morello, L., 2009. Automatic gearboxes. Automotive Chassis 1, 543–592.

Govindaraju, M., 1999. Development of Generic Design Guidelines to Manufacture Usable Consumer Products (Doctoral dissertation). University of Cincinnati, OH.

Han, S.H., Yun, M.W., Kwahk, J., Hong, S.W., 2001. Usability of consumer electronic products. Int. J. Ind. Ergonomics 28 (3–4), 143–151.

Han, S.H., Kim, J., 2003. A comparison of screening methods: selecting important design variables for modeling product usability. Int. J. Ind. Ergon. 32 (3), 189–198.

Joachim, F., Kurz, N., Glatthaar, B., 2004. Influence of coatings and surface improvement on lifetime of the gears. Gear Technol. 21 (4), 50–56.

Kim, J., Han, S.H., 2008. A methodology for developing a usability index of consumer electronic products. Int. J. Ind. Ergon. 38 (3–4), 333–345.

Kugimiya, T., Yoshimura, N., Mitsui, J., 1998. Tribology of automatic transmission fluid. Tribol. Lett. 5 (1), 49–56.

Langley, J., 1999. Bicycling Magazine's Complete Guide to Bicycle Maintenance and Repair for Road and Mountain Bikes: Over 1,000 Tips, Tricks, and Techniques to Maximize Performance, Minimize Repairs, and Save Money. Rodale Press, Emmaus, PA.

Plotkin, W.C., Moon, S.K., 2006. Using expanded QFD matrix analysis to establish and link test instrumentation to customer satisfaction attributes. J. IEST 49 (1), 90–91.

Thompson, R., 2007. Manufacturing Processes for Design Professionals. Thames & Hudson, London, UK.

Zinn, L., 2005. Zinn & the Art of Mountain Bike Maintainance, fourth ed. Velo Press, Boulder, CO.

Part Three

Establishing the Product Selling Price

12

12.1 Why estimate costs?

To survive, businesses must be profitable. In other words, the revenue generated by business activities must exceed the costs incurred in undertaking those activities. And herein lies the motivation for estimating costs. Cost estimation, and thereby profitability, also is necessary for determining the economic advantage of the business, which determines the ability of a company to be competitive. A squeeze on profits, from inaccurate cost estimation, escalating costs, increased market competition, or excessive supply, sets the stage for financial failure. It certainly results in withdrawing new products from the market shortly after they are introduced or abandoning projects regardless of inspiration.

Cost estimation is an established fact and a routine activity. It is critical that the profitability of a product be determined for its success. Note that, in addition to recovering the cost of business activities, profits must be sufficient to pay taxes (local, state, and national), dividends to stockholders, interest on borrowed capital, research and development funding, reinvestment in upkeep and modernization, and investment in exploring other options. The ramifications of poor or inaccurate cost estimation can be very serious, ranging from product withdrawal to business bankruptcy.

The estimation of costs is essential for design improvements and optimization, as shown in Figure 12.1. The goal is to achieve maximum design efficiency at the least cost. The importance of cost estimation in this process is evident.

The designer and the cost estimator also need to realize that our planet has limited nonrenewable resources, such as oil. As the demand for these resources increases and the supply decreases, the cost increases. As the technology improves, the skills and materials required to produce high-technology products also become more costly. In such an environment, it is necessary to accurately estimate the costs of processes, products, services, and projects. With the inflationary nature of money demand and supply, changing interest rates also necessitate frequent and accurate cost estimates.

One way to stretch limited resources is to increase productivity by improving the efficiency of the work output. The planning, scheduling, estimating, and carrying out of an activity, therefore, have a significant impact on the effective use of resources. The development of realistic estimates of the costs of an activity result in adequate allocation of resources and improved chances of project completion; fewer partially completed activities will be canceled for lack of funds, and activities that are started will be more likely to be completed successfully.

Profitability and the most efficient work output, therefore, are two major requirements of accurate cost estimation. A business that produces high cost estimates and

Product Development. DOI: http://dx.doi.org/10.1016/B978-0-12-799945-6.00012-0

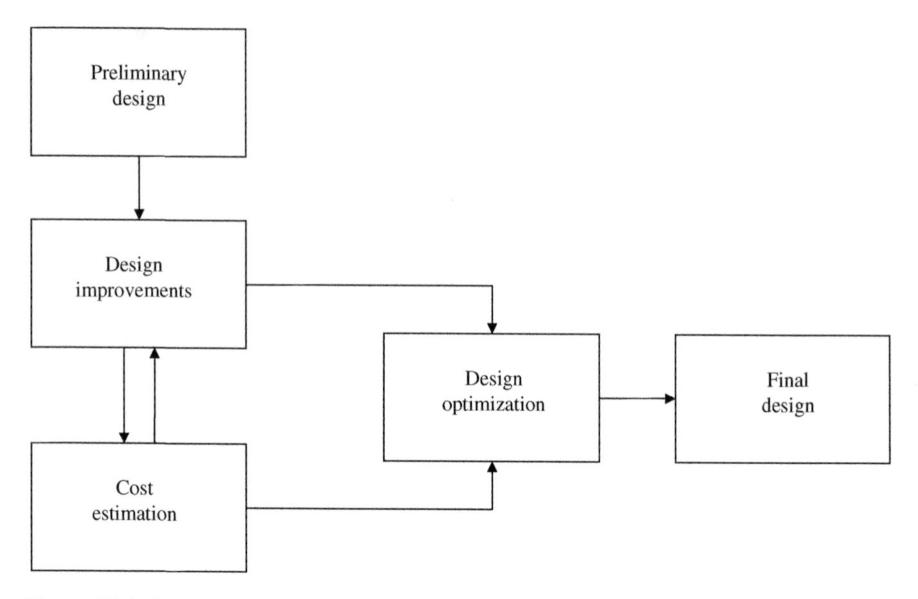

Figure 12.1 Role of cost estimation in design optimization.

bases its price on those estimates is more likely to fail, as it becomes less competitive in the marketplace. By the same token, a business that produces low cost estimates will become unprofitable and fail.

12.2 Cost and price structure

Figure 12.2 shows the basic cost and price structure. Direct labor costs and materials costs add up to make the prime cost, sometimes also known as the *operational cost*. Direct labor cost is the cost of the actual labor used to produce the product. The direct materials cost comprises the cost of raw or semifinished materials that can be directly attributed to the product. When factory overhead is added to the prime cost, it is called the *cost of goods manufactured* or the *manufacturing cost*. Overhead cost includes indirect materials cost (factory supplies and lubricants), indirect labor costs (cost of supervision and inspection and the salaries of factory clerks), and fixed and miscellaneous costs such as rent, insurance, taxes, depreciation, maintenance and repair, utilities, and small tools. When the costs of distribution, administration, and sales are added to manufacturing costs, we get the cost of goods sold (COGS). Addition of the profit to COGS determines the selling price.

Distribution costs typically include the following costs:

- Advertising
- Samples
- Entertainment and travel
- Rent and insurance

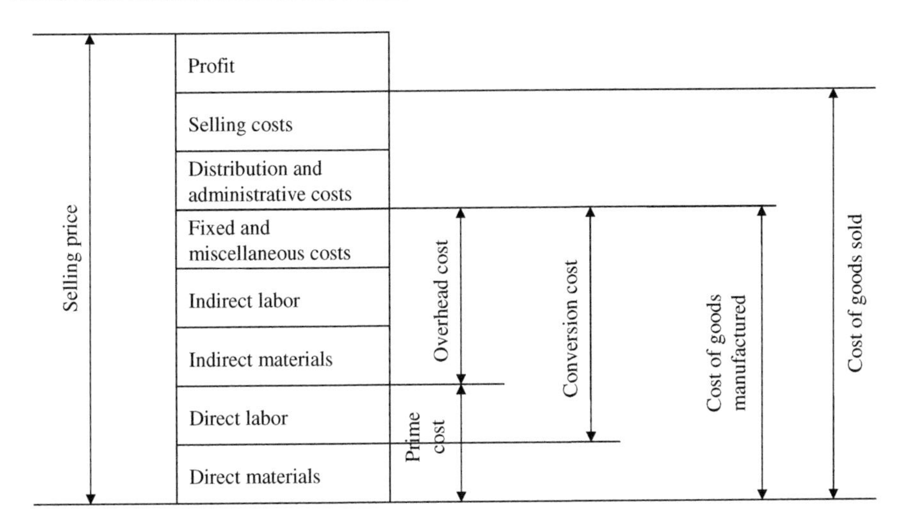

Figure 12.2 A typical cost and price structure.

- Communication
- Office expenses (stationary, postage, etc.)
- Freight
- Miscellaneous selling expenses.

Administrative expenses include:

- Administrative and office salaries
- Rent and insurance
- Accounting and legal expenses
- Communication expenses
- Office expenses
- Engineering
- Miscellaneous administrative expenses.

Selling expenses typically include the cost of sales staff salaries, commissions to dealers, and warranty costs.

Table 12.1 defines the terms used in discussing various kinds of costs. The following are examples of different kinds of fixed costs:

- Depreciation on buildings, machinery, and equipment
- Insurance premiums (fire, theft, flood, and occupational hazards)
- Property taxes (sometimes states and communities give tax breaks to industrial corporations to attract them)
- Interest on investment capital (cost of borrowing money)
- Factory-indirect labor cost (wages of security personnel, secretarial, clerical, janitorial, accounting staff, etc.)
- Engineering cost (design and other engineering personnel, and R&D)
- Cost of rentals or lease (building, equipment, and the like)
- Cost of general supplies (supplies used by indirect labor, e.g., copying, stationery, and forms)

Table 12.1 **Cost definitions (Ostwald, 1974)**

Capital cost	The cost of obtaining capital expressed as an interest rate
Depreciated cost	A noncash tax expense deduction for recovery of fixed capital from investments whose economic value is gradually consumed in the business operation
Detailed cost	The value of the detailed estimate obtained with almost complete disclosure of engineering design data using various methods of estimating
Direct cost	Cost traceable to a unit of output, such as direct labor costs or direct materials costs
Direct labor cost	The labor cost of actually producing goods or services
Direct materials cost	The cost of raw or semifinished materials that can be traced directly to an operation, product, project, or system design
Engineering cost	The total of all costs incurred in a design to produce complete drawings and specifications or reports; included are the costs, salaries, and overhead for engineering administration, drafting, reproductions, cost engineering, purchasing and construction, costs of prototype, and design costs
Estimated costs	Predetermined value of cost using rational methods
Fixed cost	Costs that are independent of output
Future cost	Costs to be incurred at a future date
Historical cost	A tabulated cost of actual cash payments consistently recorded
Indirect cost	That cost not clearly traceable to a unit of output or segment of a business operation, such as indirect labor costs and indirect materials costs
Joint cost	Exists whenever, from a single source, material, or process, there are produced units of goods having varying unit values
Manufacturing overhead cost	This includes all production costs, except direct labor and direct materials
Marginal/incremental cost	The added cost of making one additional unit for an operation or product without additional fixed cost
Measured cost	A cost based on time relationships to dollars using mathematical rules
Operating cost	This comprises two distinct cost elements, direct labor and direct materials
Operation cost	This includes labor, materials, asset value consumed, and appropriate overhead cost pursuant to the operation design
Opportunity cost	The estimated dollar advantage forgone by undertaking one alternative instead of another
Optimum cost	That operation, product, project, or system economic value for which a minimum (or maximum as appropriate) is uncovered for specified design variables using variational methods
Period cost	Cost associated with a time period
Policy cost	A cost based on the action of others; considered fixed for the purpose of estimating

(*Continued*)

Table 12.1 (Continued)

Preliminary cost	The value of a preliminary operation, product, project, or system design estimate; usually obtained quickly with a shortage of information
Prime cost	The total of labor and materials directly traceable to a unit of output
Product cost	Includes operation costs, purchase materials, overhead, general and administrative expenses, and appropriate design and selling costs
Project cost	The investment or capital cost proposed for approval in a single evaluation of an engineered project
Replacement cost	A present cost of the design equipment or facility intended to take the place of an existing design of equipment or facility
Standard cost	Normal predetermined cost computed on the basis of past performance, estimates, or work measurement
Sunk cost	The past or continuing cost related to past decisions that are unrecoverable by current or future decisions
System cost	Usually a hypothetical cost for the evaluation of complex alternatives. It may include elements of operation, product, or project costs
Unit cost	This implies, in manufacturing, the sum of total material, labor, and manufacturing overhead divided by the quantity produced; for an investment, it is the installed cost of the producing unit in convenient units of production
Variable cost	The cost that varies in proportion to the rate of output

- Management and administrative expenses (salaries and wages paid to legal and corporate staff)
- Marketing and sales expenses (salaries and wages paid to marketing and salespeople, transportation and delivery expenses, warehouse rentals, and telephone).

The various variable costs of interest are:

- Materials cost
- Labor costs (including production supervision)
- Energy cost (power, gas, and oil)
- Utilities cost (water and sewer)
- Cost of maintaining production equipment.

12.3 Information needs and sources

Accurate estimation of costs requires reliable information. The flow of information should be uninterrupted, timely, consistent, thorough, and simple. Usually, three kinds of information are important: historical, measured, and policy. The information from internal reports is historical and is typically included in accounting reports. Money expended on materials, labor, and expenses such as utilities is recorded in ledgers, through accounting records, and is available through company accountants. Measured information generally is in dollar or time dimensions, such as material quantities calculated from drawings. Policy information is fixed in nature and includes information such as union management wage settlements, Social Security tax, and liability insurance. Other forms of information include drawings, specifications, production schedules, manufacturing plans, production work hour and cost records, handbooks, other published references, personal knowledge of the operations, and market or industrial surveys.

The sources of information are many and can be classified as internal and external. Among the internal sources are

- Accounting department (cost-accounting ledgers)
- Personnel department (union contracts and wage agreements)
- Operating departments (monitoring, scrap, repairs, efficiency, downtime, and equipment costs)
- Purchasing department (regulations, make or buy, suppliers, material costs, order size, vendors, and shipping policy)
- Sales and marketing (product pricing, market demand, sales, consumer analysis, advertising, brand loyalty, and market testing).

Among the external sources of information are government agencies, international agencies, business firms, trade associations, and publications. Government sources provide a wide variety of basic economic facts and trends. The Bureau of Labor Statistics, for example, provides elements of cost on the prices of materials and labor. The bureau surveys labor markets and provides information on wage movements for as many as 60 occupational groups, the variety of benefits, and labor practices. The Office of Business Economics in the U.S. Department of Commerce provides information on the gross domestic product and a simple input–output ratio, which is a measure of productivity. The output measure is in constant dollars and the input measure is hours worked.

Many international agencies provide similar information about other countries. Among the agencies that provide information on trade, flow of funds, labor rate, and international finance are the International Labor Organization, the United Nations, and many international banks.

Cost and financial data also are available from many business firms through their annual income and expense reports. Economic data are available through the monthly bulletins of the district Federal Reserve banks. Other sources of information are the Chamber of Commerce, Federal Home Loan banks, savings and loan institutions, business analysts and market research organizations, universities, and research and development entities.

Trade associations and business publications are other sources of economic information. Among the well-known ones are the National Machine Tool Builders Association, the American Institute of Steel Construction, Marshall and Stevens Equipment Cost Index, *Engineering News-Record*, and the *Thomas Register*.

12.4 Estimating direct and indirect costs

Regardless of the source for the cost data (historical, measured, or policy), there must be a uniform method of distribution of costs before an engineering design can be estimated. Structuring cost information is a process of grouping like facts about a common reference on the basis of similarities, attributes, or relationships. Once the cost information is classified, it is summarized. Sometimes, a master list of cost codes is used. Materials, supplies, equipment costs, and the like are assigned costs from the original documents and given the appropriate code number.

In the following sections, we discuss how direct labor, direct material, and overhead (indirect) costs are estimated.

12.4.1 Direct labor costs

To determine the direct labor cost, we need to know two things: the time it takes to complete the task and the wage rate. Determination of the time it takes to complete a task falls under the purview of time study and has received considerable attention. Any text on motion and time study can be used to learn the details of time study equipment and the procedure used in determining time it takes to complete the task (Niebel and Freivalds, 2002). The procedure is described briefly and in simple terms below.

The first step in determining the time required is to observe the task being performed and record the time it takes to complete the entire cycle of the task. The time for a number of task cycles is recorded and averaged (in general, the number of task cycles recorded should be sufficient to include most variations in the method and pace of the operator). This average cycle time is called the *observed time*. This observed time is modified by multiplying it by a factor called the *rating*. The rating is the pace at which the operator being observed works. A normal worker (someone who is "suitably motivated" and has "adequate experience") is considered to work at a normal pace or 100% rating. A worker working faster than a normal worker has a higher rating, while someone working at a slower pace has a lower rating. This allows for a longer normal time if the observed worker is working faster, and a shorter normal time if the observed worker is working slower than a normal worker. When the observed time is multiplied by the rating, the resulting time is called the *normal time*. This normal time is modified by adding allowances for personal needs, unavoidable delays, and fatigue (PDF) to determine the *standard time*. The standard time is also known as the *job standard*. The following relationships clarify how the job standard is developed:

$$\text{Average observed time} = (\text{Total time for } X \text{ complete cycles of the task} / X)$$

$$\text{Normal time} = \text{Average observed time} \times \text{Rating}$$

$$\text{Standard time} = (\text{Normal time})/(1 - \text{PDF allowances})$$

The PDF allowances are the fraction of the normal time devoted to personal needs, unavoidable delays, and recovery from fatigue. The standard time may be expressed in seconds, minutes, or hours.

The standard time may be used to determine output per hour in terms of number of pieces as follows:

$$\text{Pieces per hour} = 60 / \text{Standard time in minutes}$$
$$\text{Standard output / Day} = [(\text{Pieces / Hour}) \times 8\,\text{hours}]\text{pieces}$$

From the preceding, the direct labor cost can be calculated as follows:

$$\text{Direct labor cost / Piece} = \text{Wage rate (\$ / hr)} \times \text{Standard time (hr) / Piece}$$
$$\text{Direct labor cost / Day} = (\#\ \text{Pieces / Day})[\text{Standard time (Hour / Piece)}] \times \text{Wage rate (\$ / Hour)}$$

Note that the wage rate may or may not include the cost of fringe benefits. If such cost is not included, the wage rate must be modified to accommodate it. The actual wage rate in that case is

$$\text{Actual wage rate} = \text{Wage rate} \times (1 + F + U + W + H),$$

where

F =FICA fraction,
U =unemployment compensation fraction,
W =workers compensation fraction,
H =health and other insurance compensation fraction.

Sometimes it is necessary to modify the time per unit to account for the effect of learning. This may be done as follows:

$$\text{Time}, T, \text{per piece after a cumulative production}: P = T_0 \times P^n$$

where

P =cumulative production,
T_0 =time to make the first unit,
n =the learning rate.

It may not be possible to observe the job being performed to establish job standards, for instance, when the production facility is being planned and the job does not exist. In such situations, it is recommended that a predetermined motion time system (PMTS) be used. Several kinds of PMTSs are widely available. Among the most popular are MTM PMTS, and the work factor PMTS; MTM is the most popular in the United States.

MTM analysis requires that a job, or operation, be broken down into fundamental elements. The time for these fundamental elements is predetermined and available in widely published tables. The time for the job is determined by adding times for all the elements. The PDF allowances should be added as shown for establishing job standards.

Another technique useful in determining time taken to perform a job when several people work on the job or the work is nonrepetitive in nature is work sampling. This method requires observing the activity at random times and noting if the person(s) is engaged in performing the activity or not. The number of observations can be determined statistically from the binomial distribution. Based on the person being busy or not, a fraction p, indicating the proportion of times the person is busy, is determined as follows:

$$p = (\#\text{ of times person engaged in the activity} / \text{Total}\ \#\text{ of observations})$$

Once the value of p is established, it can be used to determine the actual time for the job. For instance, one may be interested in knowing the time it takes to just paint a house, not cleaning or preparing to paint it or after-painting cleanup. Let us say that the observer found that, over a period of 4 weeks, this person was found actually painting the house 70% of the time. In other words, p was observed to be 0.70. The person worked 8 h for 5 days every week and painted four houses. The time for actually painting the house is

$$[(0.7 \times 8\,\text{hours} \times 5\,\text{days} \times 4\,\text{weeks})/4\,\text{houses}]$$
$$28\,\text{hours}/\text{house}$$

If several people were engaged in the task, the time could be determined similarly.

The MTM Association for Standards and Research, located in Des Plains, IL, developed aggregate standards for handling times for various manufacturing activities. These time standards can be used to estimate labor hours required for different manufacturing activities. Tables 12.2–12.5 provide MTM-based examples of standard times for some manufacturing activities (examples taken from Stewart, 1991).

12.4.2 Direct material costs

The bill of materials is essential for determining the quantities of materials required, as it generally contains the pounds, cubic or square yards, board feet, square feet, gallons, or linear feet of the required materials. The next step is to apply the appropriate material unit price or cost to this quantity to develop the material cost as follows:

$$\text{Material cost for a unit } (\$/\text{unit}) = W(1 + L_1 + L_2 + L_3)P - R$$

where

W = weight in pounds for a unit, or in dimensions compatible to price P,
P = price per pound of material or per unit length or volume,
R = unit price of salvaged material per unit ($),
L_1 = losses due to scrap, in fractions,
L_2 = losses due to waste, in fractions,
L_3 = losses due to shrinkage, in fractions.

Table 12.2 **Examples of cutting time standards for raw materials**

Setup on band saw or power hacksaw: Pickup material, position to saw, take out of saw (part size = 1 in. to 2 ft.) Wet blade for internal cut (band saw) Open saw guard, break blade, remove slide, put blade through, grind ends of saw blade, clamp blade in weld fixture, weld, anneal, unclamp smooth weld, put saw on pulleys and guides, adjust saw, close guard	0.1 h 0.05–0.50 min 3.5 min	
	Min./in.	
	Soft material	**Hard material**
Time required to cut 1 in. of metal (band saw)		
⅛ in. thick stock	0.02	0.50
Time required to cut 1 in. of metal (power hacksaw)		
1 in. thick stock	0.30	1.15
3 in. thick stock	2.55	10.50
6 in. thick stock	10.40	42.50

Table 12.3 **Examples of time standards for cutting/turning on a lathe (Warner Swasey, type 3)**

	Min./job
Setup times	
Fill in time slip, check in, analyze job from blueprint	2.00
Trip to tool crib and return	5.00
Set up measuring instruments, average three (0.70 min. each)	2.10
Install collet or chuck	2.00
Install and square-off stock	3.00
Deliver first part to inspection	0.70
Teardown times	
Remove collet or chuck	1.50
Clean and store measuring tools	1.00
Total	**17.3**
Add per cutting tool	
Install hex turret tools, average six (3 min. each)	18.00
Install cross slide tools, average two (5 min. each)	10.00
Tear down, clean, store, average eight (2 min. each)	16.00
Total	**44.00**

(*Continued*)

Table 12.3 **(Continued)**

	Min./job	
Run time handling time per part (1 in. diameter stock)		
Release collet, chuck advance bar to stop	0.105	
Tighten collet chuck	0.02	
Start machine	0.05	
Position coolant	0.05	
Change spindle speed	0.10	
Cut off and remove part, set aside	0.0325	
Check part	0.04	
Total	**0.35**	
	Min./in.	
	Soft material	**Hard material**
Turn on bore (1 in. diameter stock)[a] back hex turret from work, index to next station, advance tool to work	0.110	0.110
Turn, bore, etc., 0.0075 in. feed × 0.125 in. depth[b]	0.096	0.700
Total	**0.206**	**0.810**
	Min./job	
Tap handling times		
Change to slower spindle speed	0.066	
Reverse spindle direction back out	0.031	
Change spindle direction to tap	0.026	
Brush oil on tap	0.070	
Blow tap clean	0.120	
Total	**0.379**	
Machine time		
Noncollapsing taps (includes back out at 2 × tap): ⅛ in. diameter × NS40 threads per inch	0.240	

[a]Time to bore or turn 1 linear inch of 1-in.-diameter stock may be used as a basic time unit in estimating small machined parts. Used with discretion, it serves as an average time per cut to turn, bore, drill, ream, knurl, form, and cut off.
[b]Feeds for aluminum vary from 0.002 to 0.030 in. and for steel from 0.003 to 0.010 in. A light rough cut is represented by 0.0075 in. feed. Double the times shown for rough and finish cuts.

In case the raw material is in the form of a sheet, the material cost would be as follows:

$$\text{Material cost for a unit (\$/unit)} = \text{Stock width} \times \text{Stock thickness} \times \text{Blank length} \times P$$

where P =price of material in \$/unit volume. Similarly, other measuring units can be used to calculate material cost.

Table 12.4 **Examples of milling time standards**

	Min./job		
	Soft material	**OR**	**Hard material**
Setup times			
Charge time on card and check in			1.00
Analyze drawing			1.00
To tool crib for tools and return tools for previous tasks			5.00
Clean T-slots and table			3.00
Assemble and align vise or holding fixture			5.00
Install cutter to collet			8.00
Adjust table to locate initial cut			2.00
Use measuring tools; deliver first piece to inspection			3.00
Total			**28.00**
Operations per cut			
Start machine and advance work to cutter			0.10
Back work from cutter and stop			0.10
Set table at proper position			0.20
Index dividing head			0.15
Total			**0.55**
Profile or end mill			
Rough profile, ½ in. deep ×¾ in. wide cutter	0.067		0.260
Finish profile, ½ in. deep ×¾ in. wide cutter	0.033		0.130
Total	**0.100**		**0.390**

Sometimes, contractual materials are used. These materials are purchased at different times with different costs specifically to manufacture a certain product. To determine the material cost, one can use several rules: first in/first out cost; last in/first out cost; current cost; or actual cost. The first three of these methods do not provide a true representation of actual cost; even in the first method, if the time in storage is long, the cost is not necessarily the same as the current market price.

The actual price method requires calculating equivalent cost; it works as follows:

$$\text{Cost}_{\text{Equivalent}} = (\sum C_i A_i) / \sum A_i$$

where i (the lot number) $=1$ to n and A is the unit in dimensions compatible to cost C.

Sometimes, in the absence of a specific design, mathematics may have to be used to determine the cost. The following example, taken from Ostwald (1974), demonstrates this point. The shell weight of a pressure vessel is given as

$$t = (PD / 2SE) + C_a$$

Table 12.5 **Examples of drilling time standards**

	Min./job
Setup times	
Fill in job card, check in, analyze drawing	2.00
Go to tool crib for tools and return	5.00
Handle jigs, fixtures, and vises	1.50
Adjust machine, change speeds and feeds	0.80
Adjust feed stop	0.50
Insert drill bit in spindle	1.75
Deliver first piece to inspection	0.70
Total	**13.25**

	Min./in.		
	Constant time[a]	**Soft material**	**Hard material**
Operation			
General purpose (500–2000 rpm)			
Drill ⅛ in. diameter hole	0.05	0.140	0.556
Tap ⅛ in. × NS40	0.05	0.119	0.240
Countersink $\frac{1}{8}$ in. × $\frac{1}{16}$ in. deep	0.05	0.009	0.009
Heavy duty press (1–1000 rpm)			
Drill 2 in. diameter hole	0.05	1.170	1.170
Tap 2 in. × 4½ threads/in.	0.05	0.035	0.555

[a]Constant time is the value for moving the part to align for the next hole plus lowering the drill to the surface.

where

t = vessel wall thickness,
D = average inside/outside vessel diameter in inches,
E = joint efficiency,
P = pressure in pounds/in.^{2},
S = maximum working stress (lb/in.^{2}),
C_a = corrosion allowance.

A second relationship gives: cost per vessel (steel) = $K\Delta CtL$

Δ = material density (lbs/in.^{3}),
K = cost per pound of steel,
C = mean circumference (in.),
L = length (in.).

Given a corrosion allowance for carbon steel to $\frac{1}{16}$ inch, $S = 10{,}000\,\text{lbs/in.}^2$, $D = 22$, $E = 1$ for welded structure, $P = 25\,\text{lbs/in.}^2$, $\Delta = 0.28$, $K = 0.245$, and $L = 18$ ft.:

$$t = (25 \times 22/2 \times 10{,}000) + 0.062 = 0.090\,\text{inch}$$

This gives the material cost (without waste) = 0.245 × 0.28 × 0.090 × 216 = \$92.02.

If a waste allowance of 40% is used, the total material cost = 1.40 ×92.02 = \$128.83.

Sometimes, a past material cost is known but not the current cost. In such situations cost indexing may be used as follows:

$$\text{Current cost} = \text{Cost at previous time} \times (\text{Current index} / \text{Index at the time of previous cost})$$

The Consumer Price Index may be used as an index. The Marshall and Swift Index should be used for industrial equipment.

To determine the cost of two similar but varying size (capacity) equipment, the "six-tenths rule" may be used:

$$C_2 = C_1(S_2 / S_1)$$

where the C values are the costs and the S values are the size (or capacity) of the equipment.

Often, regression equations (see Chapter 13) may be used to estimate material cost. The cost would be determined as follows:

$$\text{Cost} = \textit{function}(\text{horsepower, for example})$$

12.4.3 Indirect or overhead costs

Overhead or indirect cost is that portion of the cost which cannot be clearly associated with a particular operation, product, project, or system and must be prorated among all the cost units on some arbitrary basis. As stated earlier, indirect costs include indirect labor cost and indirect materials cost, and some fixed and miscellaneous expenses. Some of the methods used to distribute factory overhead costs to job costs are

1. Overhead cost applied on the basis of direct labor hours
2. Overhead cost applied on the basis of direct labor cost
3. Overhead cost applied on the basis of prime cost
4. Overhead cost applied on the basis of machine hours.

Using these methods, the overhead rate is calculated as follows:

$$\text{Overhead rate} = (\text{Actual factory overhead} / \text{Actual direct labor hours}) \times 100\%$$

$$\text{Overhead rate} = (\text{Actual factory overhead} / \text{Actual direct labor cost}) \times 100\%$$

$$\text{Overhead rate} = (\text{Actual factory overhead} / \text{Prime cost}) \times 100\%$$

$$\text{Overhead rate per machine hour} = (\text{Actual factory overhead} / \text{Machine hours}) \times 100\%$$

The method of applying overhead on the basis of direct labor cost is the oldest and most popular method and is recommended.

12.4.4 An example

A number of stock bars, each 3.25 in. (81 mm) in diameter and 12 ft. (3.6 m) in length, are to be used to produce 2000 bars, each 2.75 in. in diameter and 12 in. (300 mm) in length. The material cost is \$3.00 per pound (\$6.61/kg), and the density is 0.282 lb/in.^{3} (789 kg/m^3). The total overhead and other expenses are \$195,000. The total direct labor expense for the plant is \$90,000. Estimate the production cost for a piece.

The following assumptions are made

- The facing dimension necessary for a smooth end finish is $\frac{1}{16}$ in. (1.6 mm).
- The width of the cutoff tool is $\frac{3}{16}$ in. (4.76 mm).
- The collet requires 4 in. (100 mm) of length for last part gripping.
- Heavy cuts are followed by a light finishing cut: for two rough cuts, cutting speed is 200 fpm (60 m/min) and feed is 0.005 in. (0.125 mm).
- The time taken to return the tool to the beginning of cut is 15 s.
- Load (setup) and unload time is 1 min.

12.4.4.1 Machining time

Position tool to perform cutoff = 15 s
Cutoff time $=[(D+a)/2] \times (1/\text{radial feed rate})$
$=[(3.25+0.75)/2] \times [(\pi \times 3.25)/(200 \times 12 \times 0.01)] \times 60 = 51\text{ s}$
Position tool to carry out facing operation = 15 s
Facing time $=[(D+a)/2] \times (1/\text{radial feed rate})$
$=[(2 \times \pi \times 3.25)/(300 \times 12 \times 0.005)] \times 60 = 68\text{ s}$
Positioning tool to perform first rough cut = 15 s
First rough cut $= [(12 + \frac{1}{16} + \frac{3}{16} + \frac{4}{16})/\text{Feed rate}] = 320\text{s}$
Position tool to perform second rough cut = 15 s
Second rough cut = 320 s
Position tool for finishing = 15 s
Finishing: (12.5/feed rate) = 360 s
Cutoff: counted for the next piece
Load/unload time per piece = 60 s
Total machining time per piece = 1254 s.

12.4.4.2 Cost of labor/piece

Number of pieces produced from a single bar $= 12 \times 12/12.25 = 11$
Number of stock bars $= 2000/11 = 181.8 = 182$ bars
Total loading time $= 2 \times 182 \times 60$ (2 min/bar)
Share of each piece $= (2 \times 182 \times 60)/2000 = 10.9 = 11$ s
Total average production time/piece $= 1254 + 11 = 1265$ s (direct labor)
Labor/piece $=(1265/3600) \times \$15/\text{h} = \$5.27$ (assuming CNC machine used).

12.4.4.3 Material cost/piece

Material cost/piece (assuming no scrap)
$= [182 \times (\pi/4)(3.25)^2 \times 12 \times 12 \times 0.282 \times 3.00]/2000 = \$91.92.$

12.4.4.4 Overhead/piece

Overhead rate = (195,000 × 100)/90,000 = 216.67%
Overhead cost = 5.27 × (216.67/100) = \$11.42.

12.4.4.5 Total cost/piece

Total cost/piece
= direct labor + direct material + overhead = 5.27 + 91.92 + 11.42 = \$108.61.

12.5 Product pricing methods

Setting the price is quite complex, as it invites reaction from consumers as well as the competition. Often, pricing has to conform to some legal requirement. While a lack of competition encourages hiking prices, competition means pressure to lower prices. Pricing also is a function of well-established practices, such as sales, coupons, introductory offers, and holiday sales. Dean (1951) suggested some basic ways to determine prices:

- Prices proportional to the full cost, that is, produce the same percentage net profit for all products.
- Prices proportional to the incremental costs, that is, produce the same percentage contribution margin over incremental costs for all products.
- Prices with profit margins proportional to the conversion cost, that is, they do not consider purchase material costs. Conversion cost corresponds to the value-added concept.
- Prices that produce contribution margins that depend on the elasticity of demand.
- Prices that are systematically related to the stage of market and competitive development of individual members of the product line.

The first three rules stress costs; the fourth rule is rarely used.

In general, the following four methods of product pricing are used: conference and comparison method, investment method, full cost method, and direct costing or contribution method.

12.5.1 Conference and comparison method

As the name implies, this method involves people knowledgeable about the product and the market. They meet and discuss competitors' prices, the effect of volume, past pricing history, and price changes. They also evaluate the response of salespeople and consumers and the share of the market secured or expected to be secured. Product costs and profitability requirements also are considered. All in all, it is a collective decision.

12.5.2 Investment method

The basic model is as follows:

$$\text{Sales dollars}, S = [\text{Rate of return } i, \text{in percentages} \times \text{Investment}, I, \text{in dollars}]/ (1 - \text{Tax rate}, t) + C, \text{the total cost excluding investment}$$

Dividing the total sales dollars by the expected number of units to be sold determines the selling price.

12.5.3 Full cost method

The method is also known as the *cost-plus* or *markup* method. It is given as follows:

$$\text{Selling price} = (\text{Total cost less investment})/(1 - \text{Markup } \%) \times (\text{Number of units sold})$$

12.5.4 Direct costing or contribution method

This method is used when costs vary closely with volume. All expenses are divided into fixed and variable components, and only the variable costs move with volume changes. As the fixed costs get prorated over the larger volumes, profits increase (increase on increments). With increased profits, one can manipulate the markup in the full cost method (keep it the same or reduce it with increasing volume) to determine pricing.

12.6 Summary

This chapter summarizes methods for product cost estimation and pricing. The purpose of this chapter is not to provide detailed methods of cost estimation, as many excellent books are available on this topic, but to provide the reader with some idea as to what is involved in the cost estimation process. We also emphasize that cost estimation requires a diverse background on the part of the cost estimator; at the very least, the person should have a background in business, economics, finance, and engineering. It is very unlikely that one person would have all the necessary tools for this purpose, and as a result, estimating cost is often done by a team.

Note that the techniques used in cost estimation come from different engineering disciplines as well. For instance, time study is the purview of industrial engineers, while estimating machining times from theoretical models may require expertise in manufacturing engineering. Estimating costs from theoretical models or design/cost optimization may fall in the realm of mechanical engineers and operations researchers. The reader, therefore, should consider this chapter to be a brief overview of the topic and consult references, such as those that follow, for detailed procedures.

References

Dean, J., 1951. Managerial Economics. Prentice-Hall, Englewood Cliffs, NJ.
Niebel, B.W., Freivalds, A., 2002. Methods, Standards, and Work Design, eleventh ed. McGraw-Hill Professional, New York, NY.
Ostwald, P.F., 1974. Cost Estimating for Engineering and Management. Prentice-Hall, Englewood Cliffs, NJ.
Stewart, R.D., 1991. Cost Estimating, second ed. John Wiley, New York, NY.

Assessing the Market Demand for the Product

13

13.1 Why assess the market demand?

A business is concerned primarily with the success of the products it introduces to the market. This concern manifests itself in terms of the company's marketing plans for specific products. These marketing plans, in turn, involve promotional efforts, projected product changes, channel placement, and pricing. All these efforts require projection of sales and, therefore, the sales forecast.

In Chapter 1, we talked about the changes in sales of a product over its life cycle (Figure 1.9). These changes were divided into four periods: the product introduction period (the product is new and sales build up slowly), the sales growth period (sales start building up rapidly as consumers get to know about the product from sources such as advertising), the period in which sales have reached maturity (the product is well established), and the period of product sales decline. As Figure 1.9 shows, the sales trend over the product's life cycle changes considerably. Since the sales volume changes considerably over time, we need to forecast sales for

1. Managing the company's sales function
2. Determining the needs of finance and accounting (projection of costs, capital needs, profits, and the like over various time intervals)
3. Determining the needs of purchasing and production activities (material requirements, labor planning, production planning)
4. Determining the needs of logistics (storage, distribution, transportation equipment requirements).

In summary, we can say that sales forecasts, or assessment of market demands, are to determine the needs of the following departments: marketing, accounting, finance, human resources, production engineering, sales, and distribution and transport.

As the needs of different departments differ due to planning requirements for differing time horizons, the need to update forecasts also differs. Further, as shown in Figure 1.9, there is considerable variation in sales patterns over time. Both factors necessitate the use of different techniques for sales forecast at different points in a product's life cycle. The available forecasting techniques can be grouped under the following three categories:

1. Qualitative techniques, also known as intuitive methods (all information and judgment relating to an item are used to forecast the demand; such techniques are used in the absence of any demand history; e.g., *Time*'s panel of economic experts).
2. Casual techniques (cause–effect relationships are sought; a relationship between the product's demand and other factors, such as national indices, is sought).

Product Development. DOI: http://dx.doi.org/10.1016/B978-0-12-799945-6.00013-2

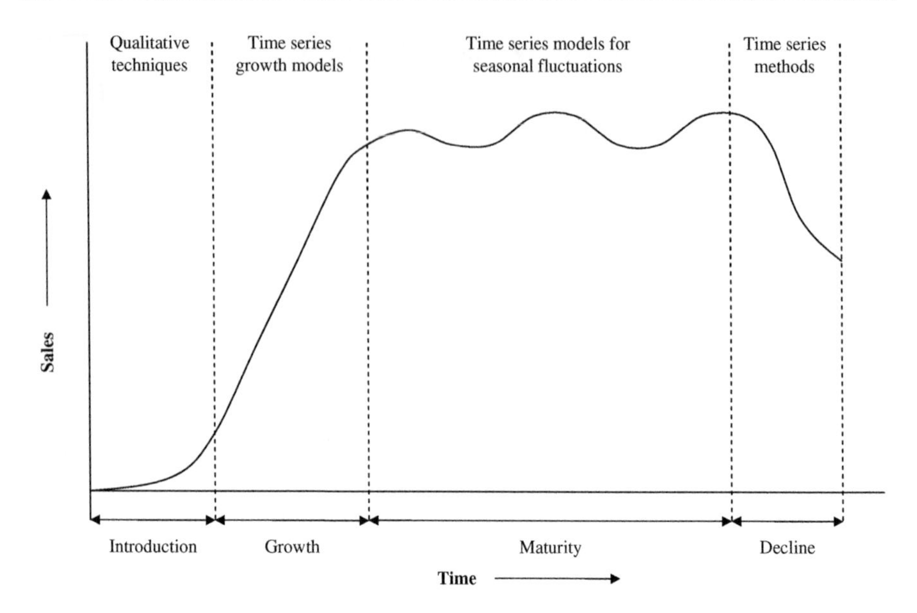

Figure 13.1 A product's life cycle and recommended sales forecasting method.

3. Time series analysis (past demand data are used to develop forecasts, assuming the past trends will continue in the future).

Figure 13.1 shows the most appropriate forecasting technique for different periods in a product's life cycle. Within each period, several techniques are available. Some of the most commonly used techniques are discussed in this chapter. A word of caution: necessary as sales forecasts are, forecasting, at best, is an art and not a science; therefore, forecasts of an event in the future have some degree of uncertainty. However, forecasting is a part of everyday life and a necessity for businesses; as Confucius said, "If a man takes no thought about what is distant, he will find sorrow near at hand." For a good manager, it is necessary not to minimize the effects of past mistakes but to successfully manage the future. He or she should worry about:

- What will next month's sales be?
- How much should be produced this month?
- How big a stock should be kept?
- How much material should be bought?
- When should the material be bought?
- Should the labor force be increased?
- What price should be charged?
- When should the advertising campaign begin?
- How many workers are required?
- Should a second or third shift be added?
- How much will the profit be?
- Do I need to have added transportation capacity?
- Are my storage facilities adequate?

To answer all these questions, we must forecast. We may be wrong, but we must forecast. With time and careful attention, forecasts get better and we improve.

13.2 Methods for assessing the initial demand

When a product is introduced, its life cycle begins. At this stage, there is very limited consumer awareness and no demand database. There also may be difficulties associated with supply lines. As a result, one must rely on qualitative techniques to forecast market demand (sales).

The qualitative techniques have certain advantages as well as disadvantages. The major advantages are their potential for predicting changes that can occur in sales patterns, based on the knowledge and experience of people within and outside the company. This is something quantitative techniques simply cannot do. Another major advantage is the use of the rich experience and judgment of marketing people, salespeople, experienced executives, and other experts.

The major problems with qualitative techniques, however, are the bias introduced by the forecasters, their overconfidence in their forecasts, politics within the company, pressure to make forecasts agree with the business plan, and confusing forecasts with sales quotas. Unfortunately, in the real world, there is no perfect way to deal with these issues; therefore, qualitative sales forecasts should be considered carefully.

13.2.1 Expert evaluation technique

Expert evaluations use the combined experience and judgment of those, such as salespeople, marketing people, distributors, dealers, and experts, who are familiar with the product line, product family, or similar types of products to generate forecasts.

13.2.2 Jury of executive opinion

This method requires executives from different corporate functions to gather to generate the forecast. The meeting is termed a *jury of executive opinion*. This is one of the most frequently used forecasting techniques. Care should be taken to ensure that all basic functions that need the forecast (see the preceding section) are represented and that representatives are experienced individuals. The result is a forecast that represents not only experience but consensus as well. To reduce the bias that builds in such a technique, participants should be provided background information on the jury and each should understand the relationship between the business plan and sales forecast.

13.2.3 Delphi method

In the Delphi method, the input of experts (internal or external to the company) is sought, as follows:

1. Each panel member writes a forecast for the product.
2. The forecasts are summarized and returned to the members of the panel, without identifying the forecast with the person who provided it.

3. After reading the forecast summary, each panel member reviews his or her forecast, revises it if necessary, and resubmits it in writing.

The forecasts are summarized again and returned to the panel. This process continues until the range of the forecast is sufficiently narrowed. Prediction of company's mid-term to long-term sales levels is an appropriate use of this method. The problem with this method is that it depends too much on the panel composition.

13.2.4 Sales force composite

The sales force composite uses the knowledge and experience of a company's salespeople to produce a sales forecast, as is done in a large number of companies. The problem is that the salespeople are not trained in forecasting. The major advantage of the method is that salespeople are the ones closest to the consumers and know them the best.

13.2.5 Supply chain partner forecasting

The supply chain partner forecasting method requires participation of salespeople through their interaction with the company's supply chain partners. Since many manufacturers never directly experience end-user demand and need to provide forecasting information about anticipated end-user demand, they rely on supply chain partners. Supply chain partners, who understand the requirements of effective supply chain management, also understand that accurate forecast of demand is crucial. However, it should be realized that the overall effectiveness of the supply chain is affected if the distributors view this as an opportunity to manage their inventory, increasing the distributor's inventory in response to trade promotions, increasing the manufacturer's inventory as safety stock, or avoiding inventory in the supply pipeline. For this method to work properly, both manufacturers and supply chain members must work to enhance the effectiveness of the supply chain.

13.2.6 Market research

As mentioned in Chapter 2, market research is an effective tool to generate information on anticipated demand. Market surveys can be used to assist in monthly, quarterly, or long-term (1- to 5-year) sales forecasts. These can also be used for short-term sales adjustments. The data for anticipated sales can come from face-to-face interviews, telephone surveys, or mail surveys of a company's business or institutional customers. Independent distributors or a sampling of households or customers also can be used for this purpose. Focus groups are another means for collecting such data.

Common leading economic indicators can also be used to forecast sales. In this case, the key is to determine which economic indicator(s) should be used. The sales forecast data thus developed are called *secondary data*. Among the most commonly used economic indicators are

- Length of the average work week for manufacturing or production workers
- New manufacturing orders for durable goods

- New manufacturing orders in general
- Construction contracts (e.g., construction of new housing)
- Plant equipment purchases
- Capital appropriations
- The magnitude of business
- After-tax business profits
- Stock price indices
- The level and changes in business inventories
- Consumer spending
- Growth in durable goods industries
- Growth in capital equipment industries
- The level and changes in money supply
- Bond prices
- Energy costs.

The statistics that correspond to aggregate changes in economic trends and confirm that a change actually is occurring (also known as *simultaneous* or *coincident indicators*) are

- The unemployment trend (rate)
- An index of help-wanted ads in the newspapers
- An index of industrial production
- Gross domestic product
- Personal income
- Retail sales
- An index of wholesale prices.

For sales data to be useful, a company needs to understand which simultaneous indicators are important for its industry.

13.2.7 Decision tree diagram

The decision tree diagram is a tool rather than a technique. It allows participants in a qualitative forecast process to visualize the context of a complex decision, thereby reducing bias in the forecast. The decision makers consider all alternatives and assign a probability to each alternative. The probabilities can be revised based on experience, judgment, and data. Figure 13.2 shows a decision tree diagram and illustrates the concept.

The probability of success when sales are high is (35/45%) 77.78%; the probability of success when sales are low is (10/45%) 22.23%. The probability of failure when sales are high is (15/55%) 27.27%; the probability of failure when sales are low is (40/55%) 72.27%. Therefore, the ability to forecast success or failure is increased through the decision tree diagram.

13.2.8 Market potential–sales requirement method

The market potential–sales requirement method is depicted in Figure 13.3. Top-down (market potential) and bottom-up (sales requirements) approaches are conducted

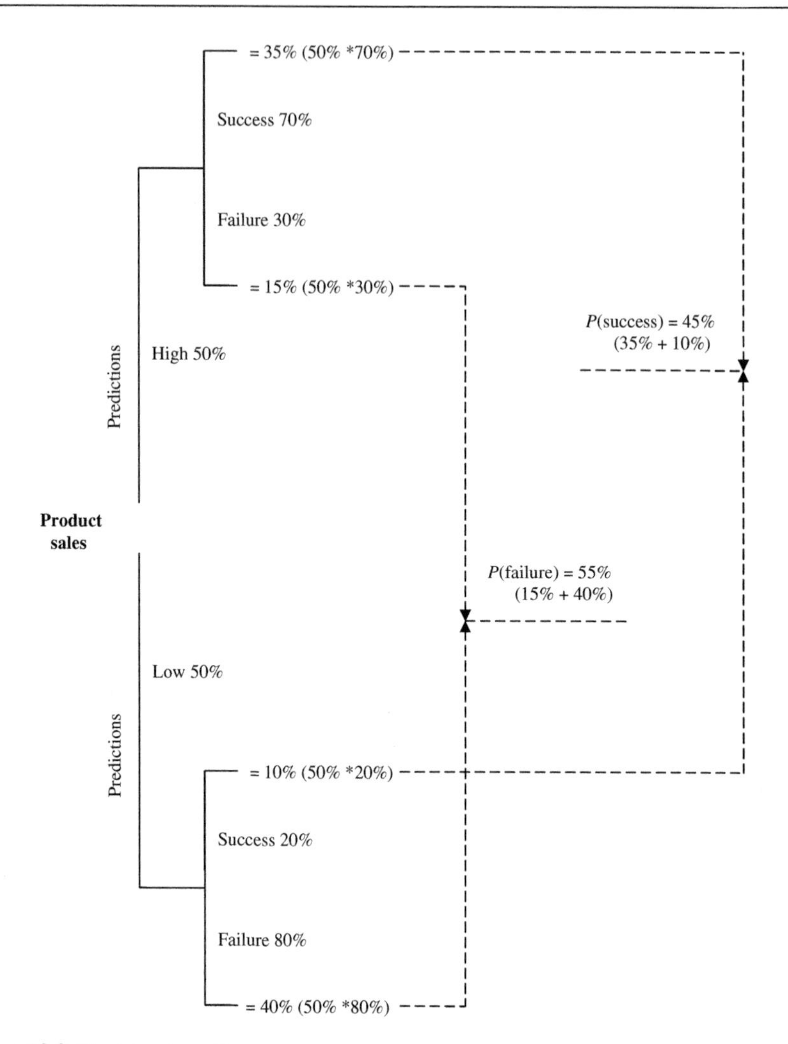

Figure 13.2 Decision tree diagram.

simultaneously and, potentially, by different individuals. Combining the two approaches (step 2) requires judgment and knowledge of competitors and their likely behavior. Also, information on market growth and sources of competitive advantages must be considered during this step. Step 3 is iterative and requires revisions on the sales requirements side of the procedure.

13.3 Methods for determining the annual growth

In this section, we describe procedures that are useful for sales forecasting during the growth phase of a product's life cycle. We look at developing different kinds of time

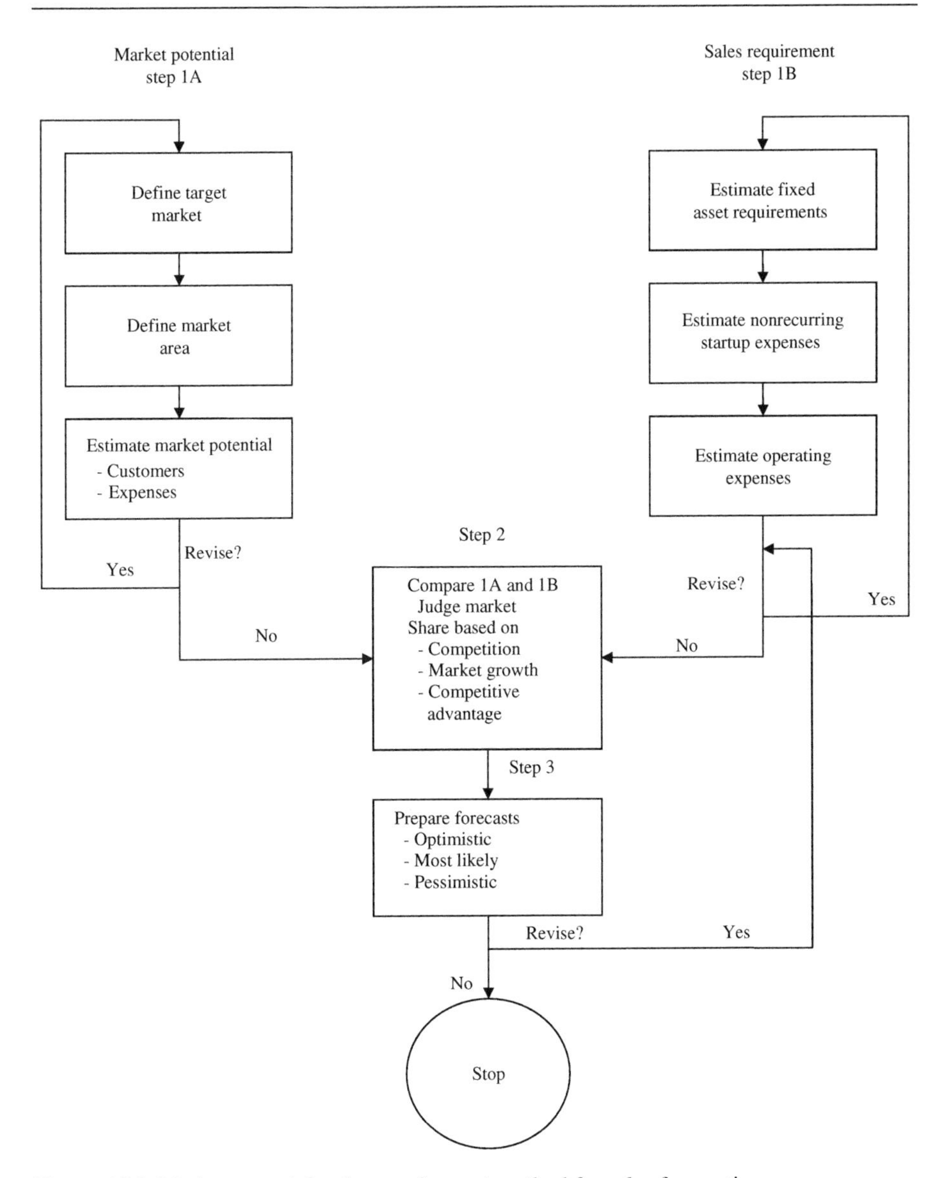

Figure 13.3 Market potential–sales requirement method for sales forecasting.

series growth models that may be applicable during the various phases of the growth, from simple linear models to simple nonlinear models. For more detailed methods, the reader is referred to any text on time series analysis.

The methods described here require the existence of a stable data structure. The first step in the process is collecting the sales data. These data may come from internal sources, such as past sales records, production, and stock inventories, or external sources, such as government or industry sources (e.g., Ministry of International Trade and Industry, Japan). The data are time dependent, or a function of time, as follows:

$$\text{Sales} = \textit{function}\ (\text{Time})$$

Once the data are collected, data reduction, the second step, must be undertaken. The criteria for data reduction generally include: "Are data relevant?" "Are they reliable?" and "Are they recent?"

The third step, to be discussed in some detail in the following sections, is the development of relevant model:

$$\text{Forecast}, Y = f(\text{Independent variables, e.g., Time})$$

The fourth step is model extrapolation, for example,

$$Y = A + Bt, \text{for some future time}, t$$

13.3.1 Graphical displays of data

In any effort to determine the annual growth pattern, plotting data is very important. Generally, such descriptive analysis applies when data are a function of a single variable. The primary reason for displaying data in a graphical form is that it makes interpretation easier than if the data were in a tabular form; the human eye and brain extract information more easily from a graph than a table (the data in Table 13.1 are displayed in Figure 13.4). Graphical displays also reveal relationships among variables. Figure 13.5, for instance, shows several types of frequency curves that can be obtained from the analysis of single-variable data.

However, when several variables are involved, descriptive methods are inadequate. In such circumstances, multiple regression analysis, using the method of least squares, is used. Our discussion here is limited to single-variable regression analysis for developing typical time series, as models shown in Figure 13.6.

13.3.2 Constant mean model

The constant mean model is given as

$$X_t = \mu + \in_t \ (t = 1,2,3,\ldots)$$

where μ is the constant mean and $\in_t$ is error with zero expectation and constant variance Σ^2.

Forecasts for X_1, X_2, ..., X_t have been obtained and a forecast of a future observation X_{t+k} is required.

$$X_{t+k} = \mu + \in_{t+k}$$

Table 13.1 **Unemployment rates for men, by age**

Age	Year 1 (%)	Year 2 (%)
14–19	16.7	13.8
20–24	9.4	7.8
25–34	4.3	3.7
35–44	4.0	3.4
45–54	4.2	3.4
55–64	4.9	4.0
64+	5.5	4.7

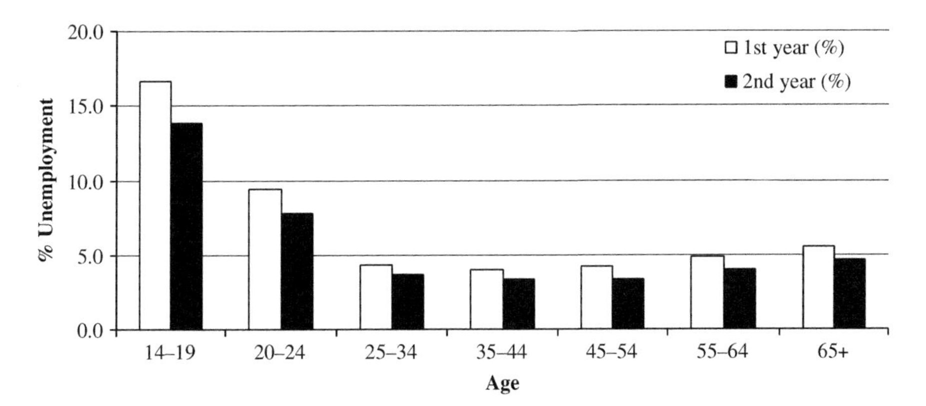

Figure 13.4 Display of unemployment data for men.

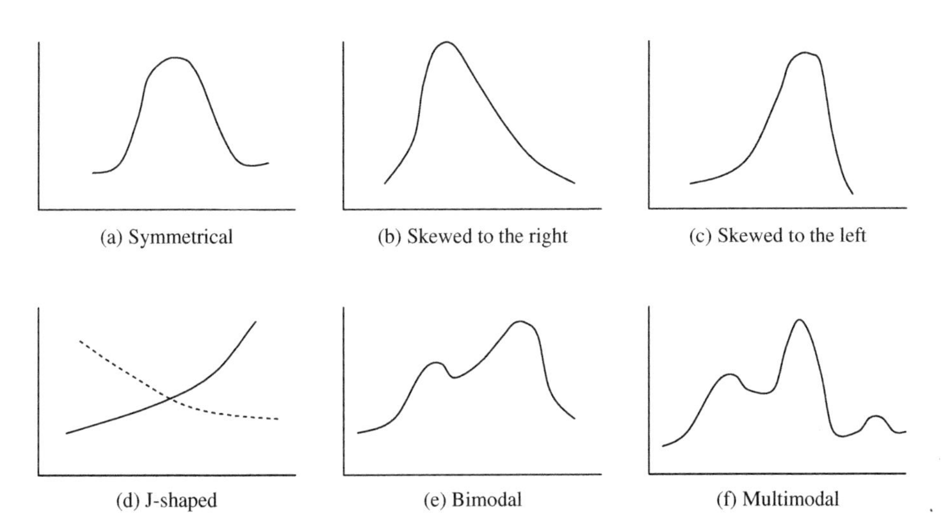

Figure 13.5 Underlying distribution of single-variable data.

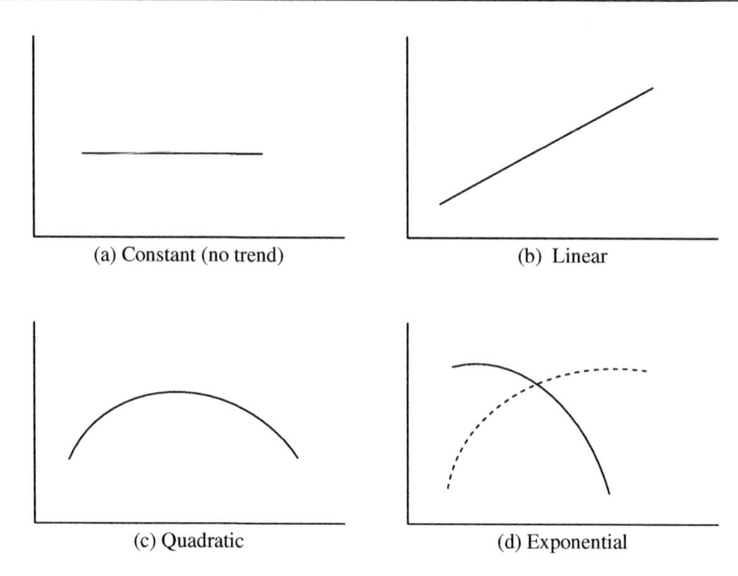

Figure 13.6 Typical time series models.

Table 13.2 Constant mean model example

t	X_t	X_t/t	$\tilde{X}_{t-1}$	$\tilde{X}_{t-1}(t-1/t)$	$\tilde{X}_t$
1	7	7.000	0	0	7.000
2	3	1.500	7.000	3.500	5.000
3	9	3.000	5.000	3.333	6.333
4	6	1.500	6.333	4.750	6.250
5	8	1.600	6.250	5.000	6.600
6	7	1.167	6.600	5.500	6.667

The most natural estimate for μ is the sample mean based on all available data and for $\in_{t+k}$ is zero. This gives

$$\tilde{X}_{t+k} = (X_1 + X_2 + \cdots + X_t) / t$$
$$\tilde{X}_{t+1} = [(X_1 + X_2 + \cdots + X_t) + X_{t+1}] / (t+1)$$

or

$$\tilde{X}_{t+1} = [t\tilde{X}_t + X_{t+1}]/(t+1)$$
$$\tilde{X}_t = (t - 1/t) \times \tilde{X}_{t-1} + (1/t)X_t \times \tilde{X} = 0$$

Table 13.2 shows the use.

13.3.3 Linear model

The linear regression model perhaps is the oldest method of forecasting the future demand for items with a trend demand pattern. Here, the n most recent demand entries are used with equal weight to seek estimates of constants a and b in the following model:

$$Y = a + b \times X$$

where Y is the forecast as a function of some X, usually time, and b and a are constants estimated using the method of least squares by minimizing the square of the residual error. The residual error is given by

$$\epsilon_i^2 = [Y_i - a - bX_i]^2$$

Minimizing this error and setting the partial derivatives (in the above equation) with respect to constants a and b equal to zero, we can generate normal equations. These normal equations, when manipulated algebraically, yield the constants as follows:

$$b = [n\sum X_i Y_i - \sum X_i \sum Y_i]/[n\sum X_i^2 - (\sum X_i)^2]$$
$$a = (\sum Y_i / n) - (b\sum X_i / n)$$

where n = number of observations.

The correlation between Y, the response, and X, the independent variable, could be negative linear, nonexistent, or strong positive, as shown in Figure 13.7. Table 13.3 and the following equations show the use of the model:

$$b = [(7)(1,575) - (315)(28)]/[(7)(140) - (28)(28)] \text{ or } b = 11.25$$
$$a = (315/7) - (11.25 \times 28/7) \text{ or } a = 0.0$$

So $Y = 0.0 + 11.25(t)$, and forecast for $t = 8$ is $Y = 0 + 11.25 \times 8$ or $Y = 90$.

13.3.4 Quadratic model

The quadratic time series model is another common trend polynomial of the form

$$Y(t) = a + b \times t + c \times t^2$$

where

$$t = \text{time}$$
$$a = [\sum Y(t_i)/n] - b[\sum t_i / n] - c[\sum t_i^2 / n];$$
$$b = (\gamma\delta - \theta\alpha)/(\gamma\beta - \alpha^2);\ c = [\{\theta - b\alpha\}/\gamma];$$
$$\gamma = (\sum t_i^2)^2 - n(\sum t_i^4);\quad \delta = [\sum t_i \sum Y(t_i)] - n[\sum t_i Y(t_i)];$$
$$\theta = [\sum t_i^2 \sum Y(t_i)] - n[\sum t_i^2 \sum Y(t_i)];$$
$$\alpha = [\sum t_i \sum t_i^2] - n[\sum t_i^3];\quad \beta = [(\sum t_i)^2] - n[\sum t_i^2]$$

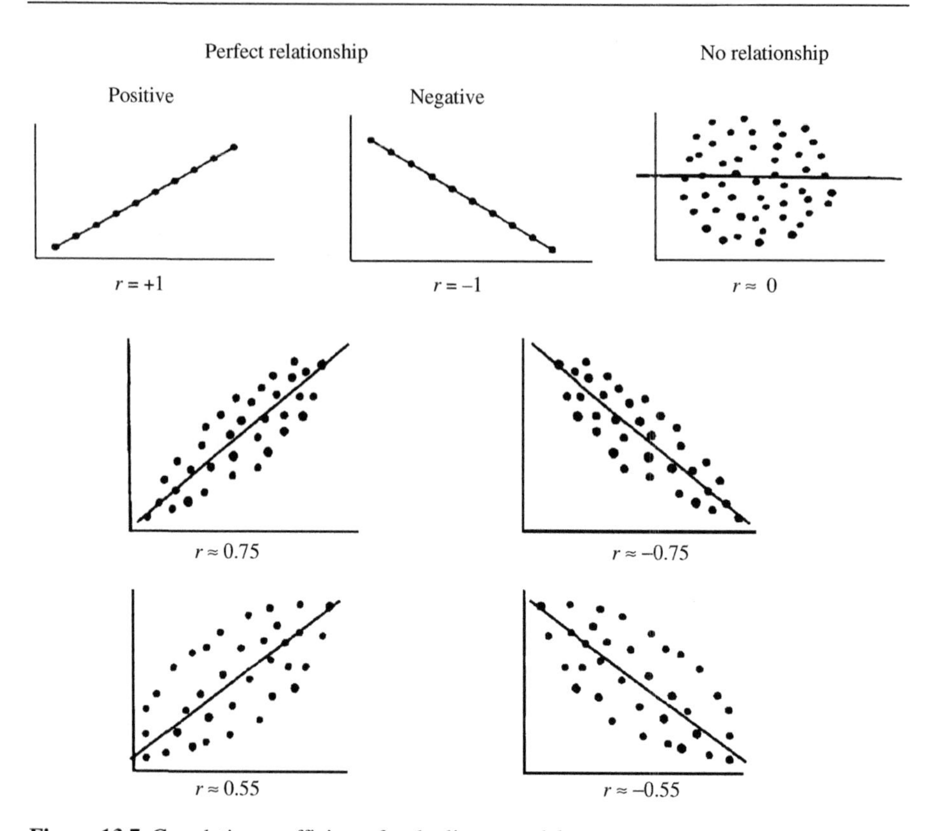

Figure 13.7 Correlation coefficients for the linear model.

Table 13.3 Linear model example

t	$Y(t)$	$t \times Y(t)$	t^2	
1	15	15	1	
2	20	40	4	
3	35	105	9	
4	40	160	16	
5	55	275	25	
6	70	420	36	
7	80	560	49	
28	315	1575	140	Column totals

Table 13.4 Quadratic model example

	t	t^2	t^3	t^4	$Y(t)$	$t \times Y(t)$	$t^2 \times Y(t)$
	1	1	1	1	16	16	16
	2	4	8	16	24	48	96
	3	9	27	81	34	102	306
	4	16	64	256	46	184	736
	5	25	125	625	60	300	1500
Total	15	55	225	979	180	650	2654

Table 13.4 shows the use of the model.

$$\alpha = (15) \times (55) - (5)(225) = -300$$

$$\beta = (15)(15) - (5)(55) = -50$$

$$\gamma = (55)(55) - (5)(979) = -1{,}870$$

$$\delta = (15)(180) - (5)(650) = -550$$

$$\theta = (55)(180) - (5)(2{,}654) = -3{,}370$$

$$b = [(-1{,}870)(-550) - (-3{,}370)(-300)]/[(-1{,}870)(-50) - (-300)(-300)] = 5$$

$$c = (-1{,}870)/(-1{,}870) = 1$$

$$a = (180/5) - (5 \times 15/5) - (55/5) = 10$$

$$Y(t) = 10 + 5t + t^2$$

13.3.5 Exponential model

In many situations, sales increase or decrease at an increasing or decreasing rate. Industries that expand rapidly experience exponential trends. In such situations, a more likely forecast model is the exponential model, which is given as

$$Y(t) = ae^{bt}$$

In this case, a and b cannot be determined directly. However,

$$\ln[Y(t)] = \ln(a) + \ln(e^{bt}) = \ln(a) + bt$$

becomes a linear model and can be solved. Table 13.5 and the following equations show the use of the model:

$$b = [(5)(24.20) - (9.60)(10)]/[(5)(30) - 100] = 0.5$$

$$\ln(a) = [9.60/5] - [(0.5)(10)/5] = 0.92$$

$$\text{or}, a = e^{0.92} = 2.50$$

This gives $Y(t) = 2.5e^{0.5t}$

Table 13.5 **Exponential model example**

	t	$Y(t)$	$\ln[Y(t)]$	$t \ln[Y(t)]$	t^2
	0	2.50	0.92	0.00	0
	1	4.12	1.42	1.42	1
	2	6.80	1.92	3.84	4
	3	11.20	2.42	7.26	9
	4	18.47	2.92	11.68	16
Total	10		9.60	24.20	30

13.4 Adjusting for seasonal fluctuations

The techniques described in this section are applicable when the product growth has matured and remains, more or less, stable (horizontal). The individual demand entries for a product during this stage seldom remain constant but fluctuate around the average in a somewhat random fashion. The role of the forecasting model is to estimate the average demand from the entries of the past and use this average as the forecast of demands for future time periods.

In this section, we describe three of the most commonly used horizontal forecasting techniques: the naive, moving average, and single smoothing technique. The single smoothing techniques, also known as the *exponential smoothing technique*, perhaps is the most widely used horizontal sales forecasting model in industry today.

13.4.1 Naive model

The naive forecasting model takes all future demands to be the same as the most current sales entry. While the method is easy to apply, the forecasts generated are poor and have little practical value. The only benefit of this model is that it serves as a comparison to horizontal models that are more refined.

Let us say that the sales for the year 2006 were 157 units. The sales forecast for each of the future years would then be 157 units as well. If the actual sales for the year 2007 turned out to be 173 units, all future forecasts, from years 2008 onward, are revised to 173 units. In this method, the further in the future a forecast is, the greater the error is. At its very best, this model forecasts sales for the immediate next period and assumes that the sales during this period will be the same as for the immediately preceding period.

13.4.2 Moving average model

In the moving average model, the average demand from the N most recent time periods (days, weeks, or months) is used to forecast demands for the future time periods. To get started, the forecaster must select a parameter N from which the forecasts will be determined. Once this is done, the corresponding sales entries $(X_T, X_{T-1}, X_{T-2}, \ldots, X_{T-N+1})$ are collected and their average (M_T) is calculated as

$$M_T = (X_T + X_{T-1} + X_{T-2} + \cdots + X_{T-N} + X_{T-N+1})/N$$

The variable M_T is called the *moving average* as of time T and gives the forecast for the next period. The method is described with the help of sales data shown in Table 13.6.

In Table 13.6, the first column shows the time (month), followed by actual sales data in the second column. The 12-month moving average is shown in column 3, calculated by averaging sales data for the first 12 months. This yields an average of 216.75, which physically lies between the months of June 2002 and July 2002. The next average in this column is calculated by dropping the sales data for January 2002 and including the sales data for January of 2003. This yields an average of 231.75, which lies between the months of July 2002 and August 2002. Continuing in a similar manner, dropping the oldest sales data and adding the next most recent data, the remaining 12-month averages in column 3 are calculated.

Column 4 reflects the centered averages and is calculated by averaging two consecutive averages in column 3. For instance, when the first two averages in column 3 (216.75 and 231.75) are averaged, the result is 224.25, which physically lies at the center of July 2002. The average of the next two averages, 231.75 and 244.33, is 238.04 and lies at the center of August 2002. In a similar manner, all centered 12-month averages in column 4 are calculated. Column 5 is simply the ratio of values shown in column 4 and column 2.

Once all possible ratios in column 5 are calculated, they are tabulated. For the sample sales data in Table 13.6 (which actually continues until June 2006), all such ratios are shown in Table 13.7. Since there are multiple ratios for each month, these are tabulated by month in Table 13.7.

The sum of column averages is 1188.4. From this, seasonal (monthly) index factors for the following year can be determined. For instance, if the sales for 2007 are 1200 units, the monthly indices may be determined as follows:

$$\text{Adjustment factor} = 1,200 \ / \ 1,188.4 = 1.0098$$

Each column average in Table 13.7 needs to be multiplied by a factor of 1.0098 to determine the monthly sales for 2007. The revised indices are January = 142.2, February = 109.7, March = 92.8, April = 103.7, May = 92.2, June = 90.0, July = 101.9, August = 83.0, September = 79.0, October = 83.1, November = 97.8, December = 124.5; total of all these indices = 1200.

13.4.3 Exponential smoothing

The exponential smoothing model is the most frequently used model in industry today. At each time period, the forecasts are updated in a recursive manner using the most current demand data. The model assigns more weight to more recent data and, in this way, forecasts react more quickly to shifts in the level of demand. Because of the limited need to store data, the method is well suited to forecasting demands for a large number of items. The model also allows the forecaster to change the rate of response, which is difficult to achieve with the moving average model.

The model is given as

$$S_t(X) = \alpha X_{t-1} + (1 - \alpha) S_{t-1}(X)$$

Table 13.6 Sales data for a sample product; calculation of 12-month moving average, centered moving average, and ratio of actual to centered moving average

Month	Sales	12-Month moving average	Centered 12-month moving average	Ratio of actual to centered moving average
January (2002)	236			
February	184			
March	179			
April	190			
May	186			
June	182	216.75		
July	213	231.75	224.25	95.0
August	189	244.33	238.04	79.4
September	186	254.67	249.50	74.5
October	209	266.17	260.42	80.2
November	296	275.00	270.58	109.4
December	351	283.92	279.46	125.6
January (2003)	416	291.42	287.67	144.6
February	335	299.25	295.33	113.4
March	303	305.00	302.12	100.3
April	328	309.42	307.21	106.8
May	292	307.33	308.37	94.7
June	289	304.33	305.83	94.5
July	303	299.33	301.83	100.4
August	283	292.83	296.08	95.6
September	255	285.58	289.21	88.2
October	262	278.08	281.83	93.0
November	271	270.83	274.46	98.7
December	315	263.58	267.21	117.9
January (2004)	356	257.33	260.46	136.7
February	257	249.50	253.42	101.4
March	216	241.83	245.68	87.9
April	238	235.58	238.71	99.7
May	205	230.58	233.08	87.9
June	202	226.83	228.71	88.3
July	228	222.75	224.79	101.4
August	189	222.71	222.46	84.9
September	163	221.92	222.04	73.4
October	187	221.58	221.75	84.3
November	211	222.50	222.04	95.0
December	270		222.62	121.3

Table 13.7 **Ratios of actual to centered moving averages for the sales data shown in Table 13.6**

Year	Jan	Feb	March	April	May	June	July	Aug	Sept	Oct	Nov	Dec
2002							95	79.4	74.5	80.2	109.4	125.6
2003	144.6	113.4	100.3	106.8	94.7	94.5	100.4	95.6	88.2	93.0	98.7	117.9
2004	136.7	101.4	87.9	99.7	87.9	88.3	101.4	84.9	73.4	84.3	95.0	121.2
2005	137.1	111.3	94.5	103.4	95.4	90.0	110.3	78.6	82.0	77.9	94.8	125.3
2006	144.4	106.0	89.4	102.0	86.4	83.7						
Column average[a]	140.8	108.7	91.9	102.7	91.3	89.1	100.9	82.2	78.3	82.3	96.9	123.3

[a]The column average excluding the highest and lowest values.

Table 13.8 **Variation in α**

	Variation in α values		
Drift in actual data	**Small $\alpha \approx 0$**	**Little $\alpha \approx 0.5$**	**Large $\alpha \approx 1$**
None	None	None	None
Moderate	Very small	Small	Moderate
Large	Small	Moderate	Large

where α is the smoothing constant, $0 < \alpha < 1$ and $S_t(X)$ is the smoothed (forecasted) value of the function. One of the requirements of using this method is that $S_{t-1}(X)$ (forecast for period $[t - 1]$) must be known; X_{t-1} is the sales for period $(t - 1)$. If t is the first period, $S_{t-1}(X)$ would be the most recent sales observation.

The choice of α depends upon the drift in the data. Table 13.8 shows variation in α values.

When α is small, $S(X)$ behaves as if the function is providing the average of past data. If α is large, $S(X)$ responds rapidly to changes in trend. The actual value of α is determined on the basis of past experience.

As an example, the sales for November were 300 units and the forecast sales for November were 309 units. If $\alpha = 0.5$, the forecast for the month of December is

$$F_{\text{December}} = \alpha S_{\text{November}} + (1 - \alpha).F_{\text{November}}$$

$$\text{or, } F_{\text{December}} = 0.5 \times 300 + 0.5 \times 309$$

$$\text{or, } F_{\text{December}} = 150 + 154.5 = 304.5 = 305$$

13.5 Summary

This chapter briefly described procedural details of some of the most commonly used qualitative and quantitative forecasting methods. Since the subject matter is vast, it is not possible to cover the details of all time series techniques or multiple linear or nonlinear regression analysis. Some background in statistical methods is necessary to understand those methods. Should the reader wish to learn about regression analysis or time series analysis, many books are available on these topics.

Planning the Product Manufacturing Facility

14

14.1 Introduction

Now that we know what product to manufacture, how to manufacture it, and in what quantity, we need to focus on planning the facility to manufacture it. Facility planning can be divided into two distinct components: facility location and facility design. The first component deals with the determination of the physical location of the manufacturing facility; the second component deals with the actual design of the manufacturing facility. The design of the facility includes the physical layout of the manufacturing equipment and the design of the material handling system (both between and at the equipment).

Primarily, there are two kinds of manufacturing facilities. The first kind is a job shop production facility, suitable for a company that produces many kinds of products but in small quantities. Such a facility requires a collection of general-purpose equipment, highly skilled labor, and general tooling and fixtures. Similar machines typically are grouped together within the plant to form a department, and products move from department to department as the manufacturing sequence requires.

The second kind of manufacturing facility is dedicated to mass (high-volume) production. The entire facility and its equipment are dedicated to the production of a single product. The equipment is specialized and operates at great speed. Automobile production is an example of this kind of production.

There are some variations to these two basic kinds of manufacturing facilities. For instance, some use batch mode production, in which there are multiple but fewer varieties than in a job shop facility and with known and stable demand. The production facility is larger than the demand for a single item and therefore the products are manufactured in batches. Manufactured items are stored at a preplanned level to meet current and future demand. After a product is manufactured, the facility switches to another product. The equipment is somewhat specialized and provides high speed. Furniture manufacturing is an example of this kind of production. In general, most facilities make multiple models of a product and, often, multiple products, in larger quantities than in a job shop.

The facilities design problem deals with either improving an existing facility to manufacture the product or developing a new facility for that purpose. Our discussion is limited to designing new facilities. Since the topic is vast, just like product cost estimation and sales forecasting, it is not possible to present the detailed coverage the subject matter requires. Therefore, our discussion is limited in scope and intended

Product Development. DOI: http://dx.doi.org/10.1016/B978-0-12-799945-6.00014-4

only to give the reader an overview of the facilities planning process. As stated previously, the facilities planning process includes determination of the location of the manufacturing facility in relationship to the customer and the actual design of the manufacturing facility. In the following sections, we cover both topics briefly.

14.2 Determining the location of the manufacturing facility

The basic problem is to determine where to locate the manufacturing facility with respect to customers. This problem has received extensive attention from researchers over the years, for instance, by Sule (2001). Although a manufacturing facility location decision is required infrequently, it is important, as once a facility is located it is very difficult and expensive to relocate.

There are a number of qualitative and quantitative approaches to solving facilities location problems. Most quantitative techniques, however, address location problems within a facility—for instance, where to locate another machine within a department. Many of the quantitative techniques are quite theoretical and, therefore, usually ignored. Often, these techniques are considered of little practical value, due to the difficulty involved in understanding them as well as the various assumptions made in solving them. For all these reasons, we discuss a simple facility location technique based on consideration of a number of factors that might influence the location decision.

The facility location problem must be analyzed at several different levels. According to Konz (1985), one must look at the location within a geographic area, location within a region, and location within a site. For instance, location within a geographic region may involve selecting the country in which to locate the facility. This, in turn, may depend on factors such as political stability, distance between the facility and customers, climate, commitment of the government, adaptability of technology, tariffs, and security. As an example, a multinational corporation might decide to locate the facility in the United States within the "Sunbelt" region. Once the geographic region has been decided on, the next level of decision involves selecting a territory within this region. The decision may be to locate the plant in a rural area as opposed to an urban setting. Factors that should be considered in selecting the territory are:

- Market
- Raw materials
- Transportation
- Power
- Climate and fuel
- Labor and wages
- Laws and taxation
- Community services and attitude
- Water and waste.

The third level of decision involves choosing a location from several sites within the selected territory. Moore (1962) provided a checklist of 36 factors that can be used to compare sites of interest:

- Labor history
- Labor availability
- Influence of local industry on labor
- Maturity of citizens
- Management potential
- Electric power
- Fuel oil
- Natural gas
- Coal
- Water supply
- Water pollution
- Rail transportation
- Truck transportation
- Air transportation
- Water transportation
- Miscellaneous transportation
- Raw material supply
- Residential housing
- Education
- Health and welfare
- Culture and recreation
- General community aspects
- Commercial services
- Specific-site considerations
- Police aspects
- Fire aspects
- Roads and highways
- Trash and garbage
- Sewage
- Planning and zoning
- State taxes
- Community financial picture
- Community business climate
- State business climate
- Community employer evaluation
- Physical climate.

As is evident from this list, a large amount of information needs to be collected to aid in the decision-making process. In the United States, such information can be obtained from a variety of sources, including agencies such as the Department of Labor, the Federal Power Commission, chambers of commerce, state industrial commissions, the Federal Trade Commission, and publications such as *Industrial Development*, *Site Selection Handbook*, and *Plant Location*.

Once information about the various sites under consideration has been gathered, the sites must be compared. A list of critical factors (from the list just given) is

Table 14.1 Comparison of site locations under consideration

Critical factor	Max weight	Site A	Site B	Site C
Raw material	75	25	50	75
Land	100	70	80	90
Labor availability	100	80	75	90
Transportation	100	90	90	80
Power	100	80	80	90
Taxes	60	20	40	60
Education	75	50	50	75
Total	**610**	**415**	**465**	**560**

developed and each factor is assigned a maximum weight. Each site is evaluated for each of the critical factors using a linear rating and scaling method. This method may use a three-, four-, five-, or six-point scale with equal unit value for evaluation purposes (any other linear scale may also be used; different scales may be used for different factors as well). For instance, the *raw material availability* may be a critical factor, with a maximum assigned weight of 75 points. All sites under consideration may be evaluated for this factor using, for instance, a three-point scale with the following values:

Poor (0 scale value) = 0 points
Fair (1 scale value) = 25 points
Good (2 scale value) = 50 points
Excellent (3 scale value) = 75 points.

Similarly, all sites are evaluated for all factors and scores for each site are totaled. The site with the highest score is selected. Table 14.1 shows an example. In this example, site C has the highest score and, therefore, is selected.

The method described involves quite a bit of subjectivity. One way to add objectivity is to consider associated costs. For example, it would be appropriate to consider the costs of available capital, transportation to customers, transporting raw materials, and power in addition to availability and the like. In such cases, sites are compared using the standard engineering economy method *comparison and decision-making among alternatives*. Cost comparison alone may not be sufficient to make the final site selection. For instance, it may be cheaper for procuring raw materials but educational facilities in the area may be lacking for the families of the workers, or taxes in the area may be relatively high. The facility planner needs to look at costs as well as all the critical factors.

Other qualitative facility location techniques are available. For details, consult the text by Sule (2001). As stated earlier, quantitative techniques are less popular and used mostly for location problems within a plant. Should one want to refer to those, the following references, in addition to those already mentioned, may be useful: Heragu (1997), Sule (1988), Tompkins and White (1984), and Tompkins et al. (2003).

14.3 Developing the preliminary design for the manufacturing facility

To design a manufacturing facility, we must know the product design, the manufacturing quantity, the routing the product will take during the course of manufacturing, the support services it will require, the sequence of tasks required to complete manufacturing of the product, and the standard time for various tasks.

Chapters 4–11 discuss the details of product design and Chapter 6 briefly discusses the preparation of the routing sheet, also called *production routing*. It provides details on how the product is to be manufactured, what kinds of machines are needed, what tools are needed, setup times for the machines, the sequence of operations (tasks), and production output in terms of number of units per hour from each machine. Determining the manufacturing quantity is discussed in Chapter 13. To determine the standard time, we must rely on time study, discussed in Section 12.4.1, with a reference provided for more details. There are other methods for determining standard times as well. For instance, predetermined motion time systems such as methods time measurement and work factor, and standard data (such as those tabulated in Chapter 12) may be used for establishing task times. In fact, for new products, this is the only way, as the task does not exist to allow performing a conventional time study. The reference text also provides information on other charts, such as the operation process chart, flow process charts, left-hand, right-hand charts, gang chart, and Gantt chart, which are useful in developing, planning, and analyzing layouts of machines and the plant. The operation process chart and the flow process chart are particularly helpful in visualizing the operation sequence to manufacture and assemble the product.

Once we have the sequence of tasks and standard times for each task, we must determine the space requirements for various departments. These departments can be broken down broadly into production departments, which have production machines, and nonproduction departments, such as offices and cafeterias. In the following sections, we describe how space requirements for production and office areas are established, how the flow of a production line is balanced, and how the preliminary design of the facility is developed.

14.3.1 Determining space requirements

To determine space requirements for a production area, we must know the number and type of machines. We will demonstrate calculations for determining the number of machines for mass production layout (also known as *product layout*) and job shop layout (also known as *process layout*) with the help of an example.

In a product layout, we must know the quantity to be produced, standard times for all tasks, and an estimate of scrap (waste) rate. Let us say that we wish to produce 80,000 units of a product and the management has established that work will be performed during only one standard shift (8 hours per day, 5 days per week, and 50 weeks per year). This gives us a total number of operating hours as follows:

$$\begin{aligned}\text{Number of operating hours} &= 8\,\text{hours/day} \times 5\,\text{days/week} \times 50\,\text{weeks/year}\\ &= 2{,}000\,\text{hours/year}\end{aligned}$$

Table 14.2 Calculation of machines needed

Operation	Machine	Standard time	Production capacity	No. of workers[a]	No. of machines[b]
1	Type A	0.019 h	1/0.019 = 52.6 units/h	0.8 = 1	1
2	Type B	0.064 h	1/0.064 = 15.6 units/h	2.7 = 3	3
3	Type C	0.042 h	1/0.042 = 23.9 units/h	1.7 = 2	2

Notes: Production capacity = 1/standard time.
[a]Number of workers = production rate/production capacity; rounded upward.
[b]Number of machines = number of workers, as we only operate one shift per day.

From the number of hours of operation and annual production, we can establish the demand rate:

$$\text{Demand rate} = 80{,}000\,\text{units per year} / 2{,}000\,\text{hours per year} = 40\,\text{units} / \text{hour}$$

If our scrap rate is, say, 5%, the production rate that would be needed to meet the annual demand is

$$\text{Production rate} = 40/(1 - 0.05) = 42.1\,\text{units} / \text{hour}$$

Let us say that three operations, each carried out on a different type of machine, are needed in sequence to complete the manufacturing, and the standard times for these operations have been established as discussed earlier and are given as

Operation 1 is facing the top and requires a machine of type A
Operation 2 is turning the top and requires a machine of type B
Operation 3 is boring and requires a machine of type C.

The calculation of the number of machines of each type is shown in Table 14.2.

For a job shop layout, calculations of machines requirements are more complicated and require some knowledge of the kinds of operations that can be performed on different kinds of machines and standard times for each operation. As stated earlier, in a job shop environment, a variety of products are manufactured in limited quantities. Here we can begin by establishing the number of operating hours per month, say 60 hours per week. To determine the quantity of a specific type of machine, we need to know what products will be using that particular kind of machine, for what operation, the standard time of each operation, and the production quantity of each product. We also need to know how much time it would take to set up the machine to begin production of each product. Let us assume that a particular type of machine is used by three products and each product undergoes one operation on this machine. The information in Table 14.3 is given for each product.

From this information we can determine the number of hours per month for setup and the number of hours per month for production of each product. The total of the two times gives us the capacity of the machine each product will require (Table 14.4). Since the job shop operates only 240 hours per month, in order to accommodate 504 hours of capacity it would need three machines (504/240 = 2.1 = 3).

Table 14.3 **Quantity of operations per machine**

Product	Setup time	Standard time	Volume	Setups/month
A	0.3 h	0.01 h/unit	5500	4
B	0.25 h	0.035 h/unit	8000	5
C	0.85 h	0.082 h/unit	2000	3

Table 14.4 **Capacity of each machine**

Product	Setup hours/month	Production hours/month	Capacity (hours) required
A	1.2	55	56.2
B	1.25	280	281.25
C	2.55	164	166.55
Total			**504.00**

Once we know the number of machines of each type, we can estimate the space requirements for the production area. The manufacturer of each machine can provide the dimensions of the machine's footprint (length $\times$width). We add to this footprint area the necessary clearance for operator, area for auxiliary equipment, and area for inventory in process (incoming and outgoing). Adding all these areas gives us the area requirements per machine as follows:

Machine Type A has dimensions of 8 ft. $\times$ 3 ft. = 24 ft.^{2}

Clearance for machine A is = 5 $\times$ length + 4 $\times$ width = 40 + 12 = 52 ft.^{2}

Auxiliary equipment (e.g., inspection table) = 2 ft. $\times$ 6 ft. = 12 ft.^{2}

Incoming inventory bin = 3 ft. $\times$ 2 ft. = 6 ft.^{2}

Outgoing inventory bin = 3 ft. $\times$ 2 ft. = 6 ft.^{2}

Total area for one Type A machine = 100 ft.^{2}

If there are 10 machines of type A, the total area requirement = 10 $\times$ 100 = 1000 ft.^{2}

The area for all types of machines is added in a similar manner. To this area we add areas for aisles and so forth, to determine the total area requirements for the production department. Areas for all other production departments may be determined similarly.

For office spaces, we can use the guidelines provided by Sule (1988), which are shown in Table 14.5. Once all areas are determined, we can develop the preliminary facility layout using the procedure described in Section 14.3.3.

14.3.2 Assembly line balancing

In mass production facility layouts, parts are assembled and made into the final product as the unit progresses from station to station. This method is called the *assembly line method of production*. Producing a perfectly balanced assembly line requires that the

Table 14.5 **Typical office space requirements (Sule, 1988)**

Office/occupants	Area (ft.^{2})
President	250
General manager	200
Sales manager	200
Production manager	200
Accountants (for four)	800
Engineers (for six)	775
Sales reps (for six)	600
Secretaries (for seven)	700
Receptionist	150
Conference room	250
Copy room	100
Coffee room	200
Restrooms	350
Total	**4775**

work advance from station to station in the same amount of time. Since a perfect balance is not possible, we attempt to advance the work in *approximately* the same amount of time. The process that helps us achieve that is called *assembly line balancing*.

Numerous procedures are available to accomplish a line balance. Here, however, we discuss only the most popular assembly line balancing method, which is based on the largest candidate rule. The method requires that we know the following:

1. The number of units to be assembled per hour.
2. The number of tasks (elements) that make up the job and the standard time for each task.
3. The list of immediate predecessor task(s) for each task.

From this, our goal is to determine the minimum number of assembly work stations and what tasks are to be performed at each work station.

Table 14.6 shows the tasks, task times, and immediate predecessors for a sample assembly job. It is required that 40,000 units be assembled per year and the plant operates one standard shift (2000 hours per year).

To assemble 40,000 units per year in 2000 h, we must assemble 20 units per hour (40,000/2000). This means that, every 3 min, one unit must be completed and the work must advance from station to station every 3 min. This is known as the *cycle time*. From this, we can determine the minimum number of work stations as follows:

$$\text{Number of work stations} = \text{Total assembly time per unit / cycle time}$$
$$\text{or, Number of work stations} = 10.9/3 = 3.63\ \text{work stations} = 4\ \text{work stations}$$

To determine which tasks (elements) should be assigned to each station, we must:

1. List the tasks in decreasing order of magnitude of task time, the task requiring the longest time being first. The corresponding immediate predecessor tasks for each are listed as well.

Table 14.6 **Predecessor list for the sample assembly job**

Task	Time/unit (min)	Immediate predecessor
1	2.0	–
2	1.0	1
3	0.7	–
4	1.0	3
5	1.5	2
6	1.0	4, 9
7	1.2	5, 6, 10
8	0.4	–
9	2.0	8
10	0.1	–
Total	**10.9**	

Table 14.7 **Task time in decreasing order**

Task	Time/unit (min)	Immediate predecessor
1	2.0	–
9	2.0	8
5	1.5	2
7	1.2	5, 10
2	1.0	1
4	1.0	3
6	1.0	4, 9
3	0.7	–
8	0.4	–
10	0.1	–

2. Designate the first station in step 1 as station 1 and number the remaining stations consecutively.
3. Beginning at the top of the list, assign a feasible task to the station under consideration. Once a task is assigned, all reference to it is removed from the predecessor list. A task is feasible only if it has no predecessors or all its predecessors have been deleted. It may be assigned only if it does not exceed the cycle time for the station, and this condition can be checked by comparing the cumulative time of all the tasks so far assigned to that station, including the task under consideration, with the cycle time. If the cumulative time is greater than the cycle time, the task under consideration cannot be assigned to the station. If no task is feasible, go to step 5.
4. Delete the task that is assigned from the task list. If the list is empty, go to step 6; otherwise return to step 3.
5. Create a new station by increasing the station count by one. Go to step 3.
6. All tasks are assigned.

In our example, we must assign all tasks to four stations. The tasks are listed in decreasing time order in Table 14.7. Applying the procedure, we open station 1 and

Table 14.8 Station–task assignment for the sample job

Station	Tasks
1	1, 2
2	5, 3, 8, 10
3	9, 4
4	6, 7

assign it task 1, which has no predecessor. We eliminate all reference to task 1 from Table 14.7 and compare the cumulative time to the cycle time. We still have 1 min left, so we return to the top of the list in Table 14.7. The next task that can be assigned to station 1 is task 2, as it has no predecessor left unassigned, it has the highest time of all the tasks up to that point that can be assigned, and the cycle time is not exceeded (tasks 9, 5, and 7 have unassigned predecessors). So we assign task 2 to station 1 and eliminate it from Table 14.7. Note that station 1 now has cumulative time equal to the cycle time, and therefore we must open station 2.

Returning to the top of the list in Table 14.7, we note that the first task that can be assigned to station 2 is task 5. This will be followed by assignment of tasks 3, 8, and 10 to station 2. The cumulative time is 2.7, and since there is no task left that has a task time of 0.3 min or less, we open station number 3. Eliminating tasks 5, 3, 8, and 10 from the list, we go back to the top of Table 14.7. We can assign tasks 9 and 4 to station 3 and eliminate these from the list. Following the procedure, we open station 4, which is assigned tasks 6 and 7. We have now assigned all 10 tasks among the four work stations without violating the predecessor list. Table 14.8 shows the final assignment.

14.3.3 Systematic layout planning

The systematic layout planning procedure described here was developed by Muther (1973) and has six steps:

1. Determine the area required for each department (see Section 14.3.1)
2. Develop a relationship chart
3. Convert the relationship chart into a graphical representation (nodal diagram)
4. Convert the nodal arrangement into a grid representation
5. Develop templates to represent each area
6. Arrange the templates in the same fashion as the step 3 relationship chart. Adjust the shape of the departments to fit within the shape of the building.

The second step requires developing a relationship chart, which really is a qualitative description of the degree of closeness that the facility designer feels should exist among the various departments. The following codes are used to describe the closeness between the departments:

A = Absolutely necessary, value = 4
E = Especially important, value = 3

Table 14.9 Relationship chart for a sample problem

	Department							
Department	**Mfg**	**St**	**Of**	**Tr**	**C**	**Q**	**R**	**S**
Mfg	–	A	E	A	E	A	E	E
St		–	O	O	U	O	U	A
Of			–	U	O	O	U	O
Tr				–	O	A	U	U
C					–	U	U	U
Q						–	U	O
R							–	U
S								–

Note: Mfg, manufacturing; St, storage; Of, office; Tr, tool room; C, cafeteria; Q, quality control; R, receiving; S, shipping.

Table 14.10 Value chart for Table 14.9

	Department								
Department	**Mfg**	**St**	**Of**	**Tr**	**C**	**Q**	**R**	**S**	**Total**[a]
Mfg	–	4	3	4	3	4	3	3	24
St		–	1	1	0	1	0	4	11
Of			–	0	1	1	0	1	7
Tr				–	1	4	0	0	10
C					–	0	0	0	5
Q						–	0	1	11
R							–	0	3
S								–	0

[a]Sum of numbers in column and row for each department.

I = Important, value = 2
O = Ordinary, value = 1
U = Unimportant, value = 0
X = Undesirable, value = −1.

The relationship among departments could be based on movement of materials between them, need for supervision, use of common equipment, service needs, avoiding noise, or flow of paperwork. The relationship chart is converted to a value chart next, using the value of the codes. For instance, Table 14.9 shows a sample relationship chart and Table 14.10 shows its conversion into a value chart.

To prepare the nodal diagram, the department with the highest value is placed at the center and other departments are placed around it, beginning with the department with the next higher value, and so on. Departments are connected by converting relationship codes between departments into lines. Departments represented by

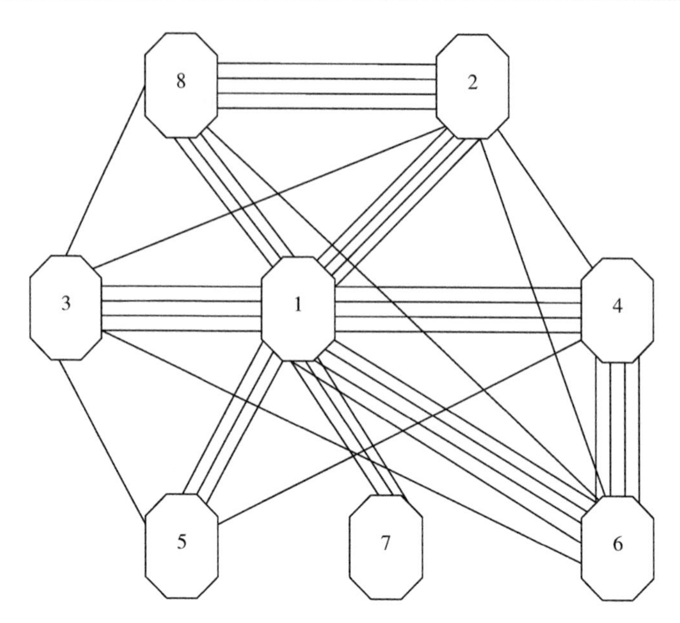

Figure 14.1 Nodal representation.
Source: Adapted from Sule (1988).

relationship code A are connected by four lines, those connected by code letter E are connected by three lines, I relationships are connected by two lines, O by one line, U are left unconnected, and X by a wiggly line. This is the most critical step in the facilities layout procedure and, therefore, the nodal diagram is adjusted, switching departments if necessary, to satisfy all step 2 relationships. Figure 14.1 shows a typical nodal diagram from Sule (1988).

The fourth step is to convert the nodal diagram into a scaled grid arrangement. The area for each department is converted into an approximate number of blocks, each block representing approximately 200 ft.^{2}. A 2000-ft.^{2} department would be equivalent to 10 blocks. Each department is placed in the grid, represented by the blocks for the area and using the arrangement of the nodal diagram. Figure 14.2 shows a grid diagram for the nodal representation from Figure 14.1.

Next, each department is represented by a template on the grid. The resulting layout could be irregular in shape (Figure 14.3) and manipulation of templates is needed until a nice-looking (regular) layout emerges (Figure 14.4). Step 3 of the process is repeated until several layouts emerge. These layouts are evaluated using multiple criteria, such as investment, ease of supervision, ease of expansion, and ease of operation. A weight is assigned to each criterion and all layouts are evaluated for all criteria. The weighted scores for each layout are totaled, and the layout with the highest score is selected. This, however, may not be the final layout; good features from other layouts are incorporated in this layout to come up with the final facilities layout. The evaluation method has been described in Chapter 3.

8	8	2	2	2
8	8	2	2	2
3	8	2	2	4
3	1	1	1	4
3	1	1	1	4
3	1	1	1	6
3	1	1	1	6
3	5	5	7	7

Figure 14.2 Grid representation.
Source: Adapted from Sule (1988).

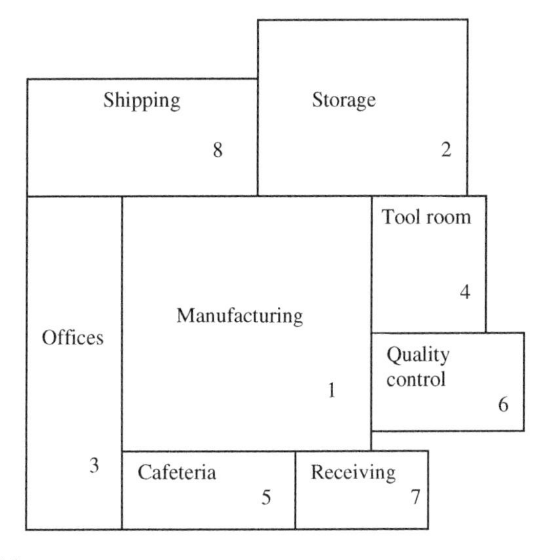

Figure 14.3 Initial layout.

14.4 Summary

This chapter provided a brief overview of the facilities planning procedure and was not intended to replace the details a text on this subject would cover. We covered the very basic details of facilities planning: determining the location of the facility and

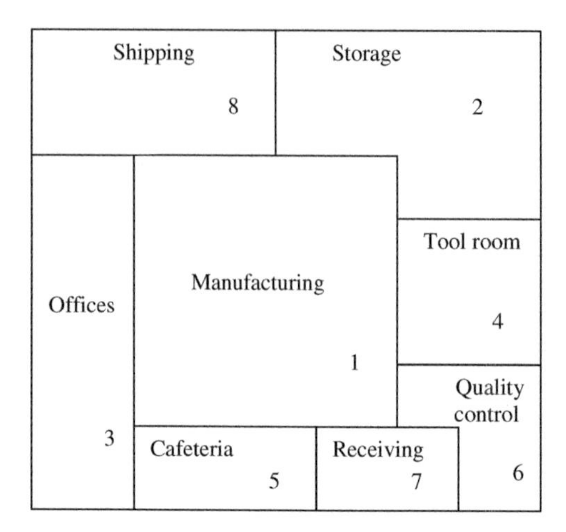

Figure 14.4 Final layout.

developing the layout of the facility. Only those techniques that are widely used were discussed. Other aspects of facilities planning, such as the design of material handling systems, design of warehouses, and trade-offs between facilities layout and material handling systems, have been skipped. These are all critical issues in the overall planning of the facility and must be given due attention. Since our intent was to provide an overview of the entire product design, development, and manufacture spectrum, we provided only a brief coverage of this important topic; providing details of the entire facilities planning process has not been our intention, and we presume that the person or the team involved in this activity would have adequate background in this subject. It is recommended that readers refer to the references that follow for detailed treatment of the subject matter.

References

Heragu, S., 1997. Facilities Design. PWS Publishing Company, Boston, MA.

Konz, S.A., 1985. Facility Design. John Wiley & Sons, Inc., New York, NY.

Moore, J.M., 1962. Plant Layout and Design. Macmillan, New York, NY.

Muther, R., 1973. Systematic Layout Planning. Cahners Books, Boston, MA.

Sule, D.R., 1988. Manufacturing Facilities: Location, Planning, and Design, second ed. PWS Publishing, Boston, MA.

Sule, D.R., 2001. Logistics of Facility Location and Allocation. Marcel Dekker, New York, NY.

Tompkins, J.A., White, J.A., 1984. Facilities Planning. John Wiley & Sons, Inc., New York, NY.

Tompkins, J.A., White, J.A., Bozer, Y.A., Tanchoco., J.M.A., 2003. Facilities Planning, third ed. John Wiley & Sons, Inc., New York, NY.

Printed and bound by CPI Group (UK) Ltd, Croydon, CR0 4YY
08/05/2025
01864790-0001